Periodic Table of the Elements

Group	I	II											III	IV	V	VI	VII	0
Period 1	H 1																	He 2
2	Li 3	Be 4											B 5	C 6	N 7	O 8	F 9	Ne 10
3	Na 11	Mg 12					Transition elements						Al 13	Si 14	P 15	S 16	Cl 17	Ar 18
4	K 19	Ca 20	Sc 21	Ti 22	V 23	Cr 24	Mn 25	Fe 26	Co 27	Ni 28	Cu 29	Zn 30	Ga 31	Ge 32	As 33	Se 34	Br 35	Kr 36
5	Rb 37	Sr 38	Y 39	Zr 40	Nb 41	Mo 42	Te 43	Ru 44	Rh 45	Pd 46	Ag 47	Cd 48	In 49	Sn 50	Sb 51	Te 52	I 53	Xe 54
6	Cs 55	Ba 56	* 57-71	Hf 72	Ta 73	W 74	Re 75	Os 76	Ir 77	Pt 78	Au 79	Hg 80	Tl 81	Pb 82	Bi 83	Po 84	At 85	Rn 86
7	Fr 87	Ra 88	† 89-103	Ku 104	Ha 105													

*	La 57	Ce 58	Pr 59	Nd 60	Pm 61	Sm 62	Eu 63	Gd 64	Tb 65	Dy 66	Ho 67	Er 68	Tm 69	Yb 70	Lu 71
†	Ac 89	Th 90	Pa 91	U 92	Np 93	Pu 94	Am 95	Cm 96	Bk 97	Cf 98	Es 99	Fm 100	Md 101	No 102	Lr 103

CHEMISTRY

CHEMISTRY

Fifth Edition

Michell J. Sienko | Robert A. Plane

Professor of Chemistry
Cornell University

Professor of Chemistry
and President, Clarkson College

McGraw-Hill Book Company

New York   St. Louis   San Francisco   Auckland   Düsseldorf   Johannesburg   Kuala Lumpur   London   Mexico
Montreal   New Delhi   Panama   Paris   São Paulo   Singapore   Sydney   Tokyo   Toronto

THE COVER

The photomicrograph that appears on the cover shows an organic reaction in process—the formation of a polyester. Polymers of ester compounds include many useful synthetic polymers, among the better known being Dacron. The micrograph was made by Dr. Roman Vishniac.

CHEMISTRY

1 2 3 4 5 6 7 8 9 0   VHVH   7 9 8 7 6 5

This book was set in Melior by York Graphic Services, Inc. The editors were Thomas A. P. Adams and Michael LaBarbera; the designer was Merrill Haber; the production supervisor was Dennis J. Conroy.
Von Hoffmann Press, Inc., was printer and binder.

Library of Congress Cataloging in Publication Data

Sienko, Michell J
    Chemistry.

    Includes bibliographical references and index.
    1. Chemistry.   I.   Plane, Robert A., joint author.
II.   Title.
QD31.2.S56   1976   540   75-22142
ISBN 0-07-057335-2

Contents

Preface

Today's students, confronted by more subject matter than faced their predecessors, ask not only "what" but "why." Dedicated teachers welcome this expanded curiosity, but realize that it has two implications. The first is the desire to understand the reasons behind the facts. The second is the student concern that since they can't possibly learn everything, why should they learn this subject. These two implications of "why" form the basis for this extensively revised fifth edition of *Chemistry*.

Our approach is to justify to the student the subject matter of chemistry by reducing it to its most basic fundamentals. By building on these fundamentals, the collection of separate subjects, important for any general chemistry course, hang together naturally. Furthermore, we as teachers are guided as to the depth necessary for the coverage of each topic. Certainly, the most difficult task in presenting the introductory course is the determination of how far to go. At each turn, we have been guided by the principles that the coverage should be intellectually correct in terms that a college freshman can understand, and that it should be sufficient for further development and understanding of the material to be covered in this introductory course.

In preparing this revision, material not required for a chemical understanding of our physical world has been omitted. At the beginning, the students are told the kind of questions which will be discussed and how we will proceed to answer them, and then are lead through the development of the subject. The text thus contains the essential coverage needed for any true introductory chemistry course.

Even so, we recognize fully that not all instructors agree on the chemical applications that should comprise the introductory course. For example, there is little agreement as to what balance between organic and inorganic chemistry should be included. Increasingly, biochemistry is included in general chemistry, a development which parallels the growing biological interests of chemical researchers and of today's students. Thus, it seems perfectly in order for the chemistry teacher to illustrate the subject with examples from biology, a science of great current interest that is directly dependent on the application of principles of chemistry.

In order to provide flexibility for individual instructors to design a course most appropriate for their students' needs, the general outline of the book places the fundamental principles first (Chapters 1–13) with sufficient applications to illustrate the various principles as they are encountered. The chapters contain worked-out examples, with each step of the reasoning explicitly spelled out. There are more questions than previously at the ends of chapters, all of which are different from earlier editions. As in previous editions, a majority of these are qualitative, designed to test the student's thought processes and comprehension. The remainder are quantitative. Of these, numerical answers to about half (the odd-numbered questions) are listed as Appendix 10.

The chemical applications start with Chapter 14, which covers the chemistry of hydrogen, oxygen, and water, fundamental for any of the descriptive chemistry which follows. Inorganic descriptive chemistry, making full use of previously developed principles, comprises Chapters 15–18. Although less frequently encountered elements are but briefly discussed, in shorter courses time may not be sufficient for inclusion of each of these chapters. They are written so that they can be covered in any order or omitted according to individual needs.

Organic chemistry is the topic of Chapters 19 and 20. In recognition of diverse student needs and individual instructor preference, these chapters may be covered in several ways. Both may be used to give a comprehensive introduction to both the classical, systematic approach to organic chemistry (Chapter 19) plus the mechanistic approach (Chapter 20). Conversely, either approach may be used separately because the two chapters each stand alone. For the instructor who prefers a mechanistic approach with minimum emphasis on nomenclature, Chapter 20 begins with a table summarizing the names of functional groups needed for this approach.

Finally, the principles of chemistry as embodied in organic reactions and structures are applied to biological systems and the life processes in the concluding chapters (21–24). Classroom experience has shown that completion of the entire development is a rewarding experience for students with a variety of backgrounds, interests, and futures. However, it was recognized in preparing this text that individual classes differ as to the best selection of topics to be covered. Thus, we have endeavored to provide individual instructors with a maximum degree of flexibility.

We acknowledge the help we have received in preparing this manuscript from our able assistant at Cornell, Florence Bernard, and from Karl Bratin, an able chemistry student at Clarkson.

Michell J. Sienko
Robert A. Plane

CHEMISTRY

Study of Matter

Science strives to make the physical environment understandable to the human mind. It seeks to answer our questions about our world. Of all the sciences, chemistry is perhaps the most abstract, in the sense that it is not generally understood just what questions chemists attempt to answer. The poet Gertrude Stein—herself not a chemist—posed the fundamental problem in a different context. At the very end of her life she asked those gathered around her:

What is the answer?

When no one spoke, she said:

In that case, what is the question?

In any kind of scientific inquiry, the second question is the one that must be answered first. And it is not always clear that the right question is being asked. For chemistry, few have framed the question as well as the great eighteenth-century chemist Antoine Lavoisier. Chemistry may be said to have begun with the publication of his original textbook in 1789. Because this *Traite elementaire de chimie* predates the atomic theory, it is an unequaled source for raising the questions that have led chemists to employ atomic theory and all its subsequent extensions. The very essence of chemistry is contained in Lavoisier's questioning the nature of fermentation of *grape must* (i.e., crushed grapes), the process by which wine is formed. Lavoisier states the question so clearly that it is worth quoting from his text:

This operation [fermentation] is one of the most striking and extraordinary of all those which chemistry presents to us, and we must examine whence comes the disengaged carbonic gas and the inflammable spirit which is formed and how a sweet body, a vegetable oxide, can transform itself thus into two different substances, one combustible and the other highly incombustible. It will be seen that to arrive at a solution to these two questions it is first necessary to know the analysis and the nature of bodies susceptible to fermentation and the products of fermentation, for nothing is created in the operation of art or of nature, and it can be taken as an axiom that in all operations the quantity of matter before is equal to that found after the operation; that the quality and quantity of the principles is the same; and that there are only changes and modifications.

Upon this principle the whole art of making experiments in chemistry is founded. We must always suppose a true equality between the principles of the body which is examined and those which are obtained on analysis. Thus, since must of grapes gives carbonic acid gas and alcohol, I may say that *must of grapes = carbonic acid + alcohol*.† It follows that we may determine by two different ways what takes place in vinous fermentation, first, by determining the nature and principles of fermentable bodies, and second, by observing the products which result from fermentation, and it is evident that the information obtained from one leads to accurate conclusions concerning the nature of the other.

This is a truly remarkable discourse—a cornerstone of the science of chemistry. Several points in it call for special attention. The first is Lavoisier's opening phrase describing fermentation as something "chemistry presents to us." It is chemistry's first priority to pick out natural processes which can be separated for study. In this way, the whole of biological nature can be analyzed into a series of discrete, chemical processes. Each isolated process can be described and understood just as Lavoisier did in the case of fermentation. By viewing chemistry in this light, we see that it is not an artificial science which invents its own problems to solve, but instead it starts from those presented to us by the nature of our physical world. The problems may be biological, geological, or ones of better fitting us to cope with our surroundings, such as finding better materials for covering ourselves. In solving the problems, opportunity is left for chemists to invent artificial (or model) situations which lead to a deeper understanding of the original situation which confronted them. This brings us to the next remarkable feature in the Lavoisier excerpt: ". . . nothing is created in the operation of art or of nature, and it can be taken as an axiom that in all operations the quantity of matter before is equal to that found after the operation . . . ." This statement provides the foundation of the quantitative aspects of chemistry. It has been called the law of conservation of mass, and on it are built other laws of chemical change, chemical formulas, and chemical equations, which in turn lead to calculations of weight relations in chemical reaction. In fact, Lavoisier anticipated future developments in writing the first chemical equation, a concise word description of the natural process of fermenting grape juice.

Lavoisier's second premise caused more difficulty and present day

†H. M. Leicester and H. S. Klickstein, from whose *Source Book in Chemistry, 1400–1900*, this translation comes, note: "This is one of the beginnings of the modern chemical equation."

chemical research still seeks to fully illuminate its intuitive truth. This is the assertion ". . . that the quality [as well as the quantity] of the principles is the same; and that there are only changes and modifications." It is one thing to note that the quantity, or amount, of matter does not change during chemical reaction, but quite another to say that the quality, or *fundamental nature*, does not change. Lavoisier admits that the problem is a profound one when he details the change of properties from "a sweet body" to "two different substances, one combustible and the other highly incombustible." However, he is confident that these different properties arise from more fundamental characteristics which can be understood ". . . by determining the nature and principles of fermentable bodies, and second, by observing the products which result from fermentation, and it is evident that the information obtained from one leads to accurate conclusions concerning the nature of the other."

Since Lavoisier's time, chemists have worked diligently to connect the observed properties—before, after, and during chemical change—to a model of the nature of matter. Such studies have led to the atomic theory, to the discovery of subatomic particles, and to various theories of the changes these particles undergo during a chemical reaction. All such theories must strictly conform to the law of physics. In this sense, chemistry is based on physics and uses physics to explain chemical behavior. Chemistry, in turn, is the basis of other natural sciences: biology, geology, agronomy, etc. Our study of chemistry will recognize this order of things. We will look to physics for our fundamentals in order to understand the phenomena encountered in the other sciences, especially biology.

During our study we will also discuss examples of the application of chemistry to help us better cope with our environment. All too often, examples of applied science are rated inferior to those of pure investigation. One of the true giants of science, Louis Pasteur,† stated: "Nothing gives the scientific investigator greater pleasure than to make new discoveries; but his joy is redoubled when his observations prove to have a direct application in practical life."

Some would argue that current problems with the environment have resulted from applied science, that a moratorium is in order. Such an attitude ignores the fact that the problems are with us and can be solved only through intelligent application of science. A large portion of the world has already outrun its food supply. How can these people be fed? Instead of a science moratorium, the need is for better applied science, which requires a broader base of scientifically educated people. Not only will the problems caused by increasing numbers of people require increased scientific activity, but responsible decisions concerning the application of science will require a society that is scientifically aware.

---

†It was Pasteur, more than anyone to date, who picked up the challenge of Lavoisier to understand the true nature of the fermentation process. Pasteur's contributions to the theory and practice of wine making are as monumental and more extensive than the other contributions more usually associated with him. In fact he "pasturized" wine before milk.

1.1 Chemical Processes in Nature

Of all the observed phenomena in nature, which are the ones that chemists should single out and work to understand? Once it was thought that natural phenomena were of two distinct types: *organic,* arising from living matter, and *inorganic,* arising from inanimate matter. However, this division was destroyed in 1828 by the German chemist Friedrich Wöhler, who wrote that his research ". . . gave the unexpected result that by combination of cyanic acid with ammonia, urea is formed, a fact that is more noteworthy inasmuch as it furnishes an example of the artificial production of an organic, indeed a so-called animal substance, from inorganic materials."

Earlier it could be thought that chemists should confine their attention to the mineral world. However, Wöhler's experiments showing convertibility of inorganic and organic substances proved, in the logic of Lavoisier, that the substances are of the same fundamental nature, subject to the same principles of chemistry. Although inorganic and organic classifications persist, both are specialties within chemistry and neither operates with principles that are not equally applicable to the other. Thus, the mineral world and the biological world both supply appropriate problems for the chemist's consideration.

In looking about us, we might focus first on the components of the earth. The solid rocks and soil, the bodies of water, the air above, each provide countless problems for chemistry. Do all rocks, despite their obvious differences of color, crystallinity, and hardness, illustrate common principles whose understanding will give us knowledge of the origins of the earth? Chemistry confirms and permits further the application of such principles to improve life on earth. Along the way chemistry provides answers asked by the alert observer concerning details such as: Why are there limestone caves containing the spectacular stalactites and stalagmites? Why do some minerals dissolve in water, and hence become badly weathered and washed into the oceans, while others do not? Why are certain minerals brilliantly colored? Why do metals conduct heat and electricity while the minerals from which they are extracted do not? Can we make artificial minerals which conduct electricity in one direction and not in the other? Why do the natural substances graphite and diamond, despite identical chemical composition, differ so greatly that graphite is soft and a conductor, while diamond is the hardest substance in nature and among the poorest conductors?

Similarly, air and water present a multitude of questions: What is the mechanical nature of air that it seems to be so easily compressed, allows ready penetration by other substances, contains dissolved water, escapes to fill a vacuum, causes rusting of iron, becomes liquid at very low temperature, and is essential for life?

Similarly, why does water dissolve many but not all substances? What in its nature accounts for its expansion on freezing so that it forms ice that floats, thereby preventing lakes from freezing solid and destroying the plant and animal life in them? How is it that the cleaning power of water is increased by soap, although so-called "hard water" cannot be so readily increased? How can "hard water" be "softened?" Why does water dissolve some substances to give acid solutions? And how does the body maintain a constant, very low level of acidity for the blood in the presence of such substances?

Above all, our human curiosity demands that we ask of chemistry questions concerning our origin and continued existence. How could life originate? If the sun is the source of our energy, what is the source of the sun's energy? How is sunlight converted to food and fuel? How does the human body convert food back to energy? What is the chemistry involved in the transmission of hereditary information from one generation to the next? How may this hereditary information be utilized in the body to synthesize living matter?

All these questions, and others like them, form the basis for our study of chemistry. It is humanity's inherent curiosity which has contributed most to the development of the science. In the process of seeking answers, chemists often are caught up in worthwhile pursuits of abstract reasoning, on the one hand, and application of knowledge to meet practical ends, on the other. Sometimes, the original goal is lost sight of. Perhaps this is just as well, since, as we shall see, understanding of some of the apparently simplest of questions is an involved process exceeding the capabilities of any single investigator. Eventually, satisfying pathways can be traced from complex problems of nature back to simple, fundamental truths of physics. To follow such a pathway can be a rewarding and fulfilling human experience. It is the task of chemistry to provide the path. Because the path leads ultimately to physical principles, we will discuss such principles early in our treatment of the science of chemistry.

Before starting on our chemical path, one question must be examined. What do we mean by *understanding* an observation? Frequently, we are comfortable with an observation because it matches our experience of similar observations. For example, we are satisfied in our observation of the workings of an ordinary thermometer in noting that the liquid filling expands as the temperature is raised because other substances we know expand at elevated temperature. We note at once that classification of similar phenomena does not explain any of them, and hence "reasoning by analogy" does not provide true understanding. However, summarizing individual observations into generalizations is the first step in the development of science and the first step toward understanding. In science, *generalizations describing the behavior of nature* are called *laws*.

Natural laws, which may be qualitative statements or mathematical formulas, describe observed phenomena. They contrast with legislative laws which require or prohibit, and which may be "broken." There is no room in science for the statement "the exception which proves the rule." A familiar example of a natural law is the law of gravity. Less familiar examples of laws are those which describe the behavior of gases. For example, all gases can be compressed, and Boyle's law states that their volume is inversely proportional to the pressure exerted on them. Boyle's law, like the law of gravity, gives no reason for natural behavior but simply states what the behavior is.

Having observed nature, and summarized the observations as laws, scientists next further their understanding by asking: Why are things that way? Why, for example, are all gases compressible? At this point, scientists depart from observations and begin to make guesses. In accounting for the compressibility of gases, they postulate that all gases consist of sub-microscopic particles (called *molecules*) with relatively large spaces between them. When a gas is compressed, the molecules are pushed closer together. Such a model for a gas is an example of a *theory*. In general, any theory is an *explanation of observed behavior in terms of a simple model which has familiar properties*. The observed facts are thus "explained," but only in the sense that they are made plausible by being related to simpler or more familiar phenomena. Since it is a product of the mind, theory is not infallible. It may have to be modified or even completely discarded in the light of further experiments.

Once a model has been proposed to account for some observations, it may be possible to predict from the model behavior which has not been previously investigated. New experiments can be performed which test the validity of the model and incidentally uncover new facts. Thus, theories serve as a stimulus for the growth of science.

1.2   Matter and Energy

The physical universe (which includes living organisms) is composed entirely of matter and energy, which together are the basis of all objective phenomena. *Matter* is usually defined as anything that *has mass and occupies space*. The term *mass* describes the tendency of an object to remain at rest if it is stationary *or* to continue in motion if it is already moving. (For example, a boulder is harder to move than a pebble, and harder to stop; it has more mass.) The mass of an object can be determined, for example, by measuring its *weight*, the force with which it is attracted to the earth. Because the force of gravity is not the same at every point of the earth's surface, the weight of an object is not constant.† Consequently, the mass of an object can be determined from a direct measurement of its weight

---

†For example, an object weighing 1.00000 lb in Panama weighs 1.00412 lb in Reykjavik, Iceland.

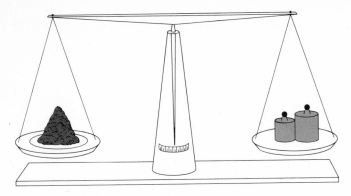

Figure 1.1
Two-pan chemical balance.

only if the earth's gravitational attraction at the point of measurement is also known.

However, the *mass* of an object is constant. It can therefore be determined by comparing its weight with that of a known mass; this may be done, for example, by means of a chemical balance. Figure 1.1 shows a classic type of two-pan balance in which an unknown mass is measured by placing known masses on the opposite pan so as to balance the gravitational force on the unknown mass. A similar but less evident comparison of masses takes place in the modern single-pan balance, such as is illustrated in Figure 1.2. The counterweight exactly balances the known masses plus the empty pan. When the unknown mass is placed on the pan, enough of the known masses are removed to restore balance. The mass of the unknown is then equal to the known masses removed.

Energy is usually defined as the *capacity to do work*, where *work* means the moving of matter against a force. Anything which has the capacity to push matter from one place to another has energy. This energy may be either

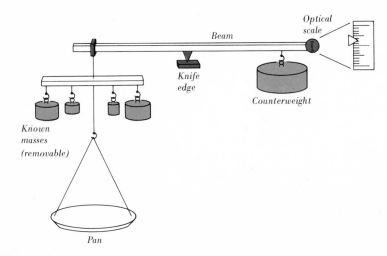

Figure 1.2
One-pan chemical balance.

kinetic or potential. The term *kinetic* is derived from the Greek word *kinein,* to move, and *kinetic energy* is that energy *intrinsic in an object which has motion.* The amount of kinetic energy a body possesses is equal to one-half its mass times the square of its speed, or $\frac{1}{2} ms^2$. *Potential energy* is "stored" energy intrinsic in an object not because of its motion but because of its *position with respect to other objects.*

In any system, the total energy present in that system is partly kinetic energy and partly potential energy. A swinging pendulum, for example, shows continual interconversion of kinetic and potential energy. At the bottom of the swing its speed is greatest and so its kinetic energy is a maximum; at the ends of the swing the speed is zero and all the energy has been converted to potential. Other familiar forms of both kinetic and potential energy are heat, light, sound, electric energy, and chemical energy. These may be transformed one into another, and it is usually these transformations which are observed. The explosions and fireballs that accompany certain chemical reactions represent transformations in which part of the chemical energy is converted into other forms of energy, such as heat, light, and sound. In any energy transformation the total energy remains the same. This is consistent with the *law of conservation of energy,* which states that the *total energy of the universe is constant.* This law, more inclusive than the law of conservation of mass, is a basis of thermodynamics which, as we will see, gives valuable insights into chemical reactions.

In a chemical reaction, the transformation of one substance into another is generally accompanied by a change in the *potential* energy of the matter. This change is observable as a liberation or absorption of kinetic energy in the form of heat. If heat is liberated to the surroundings, the reaction is said to be *exothermic;* if heat is absorbed, *endothermic.* There is a common misconception that any spontaneous chemical reaction necessarily shows a decrease in potential energy and so must be exothermic. Yet we know, for example, that the melting of ice is an endothermic process that absorbs heat from the surroundings and occurs spontaneously above the melting point. How can it be that a substance *spontaneously* goes to a state of higher potential energy? Surely, other factors in addition to energy may be involved in determining the course of spontaneous change. We will become equipped to develop these other factors in subsequent chapters. In the meantime it is important to recognize the role of energy changes in chemical reactions.

1.3   Heat and Temperature

Frequently, there is confusion of the concept of heat with that of temperature. *Heat* is measured as a *quantity of energy,* whereas *temperature* describes the *intensity of heat,* or hotness. The distinction may be made clear by considering that a burning match and a bonfire can be at the same temperature, but there is certainly much more heat in a bonfire. The difference can also be illustrated by noting that a given quantity of heat will raise the temperature of a cup of water more than it will raise the temperature of a tub of water.

Temperature can also be defined as the property which fixes the direction of heat flow, in the sense that heat always flows from a body at high temperature to a body at low temperature. Qualitatively, a hot body has a high temperature, and a colder body has a lower temperature. Quantitatively, temperature can be measured by taking advantage of the fact that most substances expand as they get hot. The mercury thermometer is a temperature-measuring device which works because the volume occupied by mercury increases as its temperature increases. As shown in Figure 1.3, the sensing bulb contains a large volume of mercury which can expand and

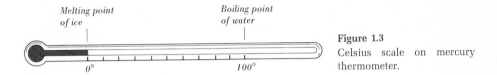

Melting point of ice *Boiling point of water*

0° *100°*

Figure 1.3
Celsius scale on mercury thermometer.

move into a capillary. The volume of the capillary is much less than the volume of the bulb, so that a small expansion of the mercury causes a large movement of the mercury thread.

One temperature scale in common use, the Celsius scale, is obtained by placing the thermometer in melting ice and marking the point at which the mercury thread stops as 0°C. When the thermometer is placed in boiling water, the mercury expands and the thread moves to a new position, which is marked as 100°C. The space between is divided into 100 equal parts. The Celsius scale is often called the *centigrade scale*. The Fahrenheit scale differs in that the ice point is marked 32° and the boiling point 212°, with 180 equal divisions between them. To convert from Fahrenheit temperature to Celsius temperature, we subtract 32 from the Fahrenheit reading and multiply the result by $\frac{5}{9}$.

There is a similar problem with units used to measure the quantity of heat. Traditionally, chemists and biologists have used the *calorie*. This unit is the amount of heat required to increase the temperature of one gram (g) of water one degree on the Celsius scale. The burning of a match liberates approximately 500 cal. It is sometimes more convenient to use the *kilocalorie*, which, as the name implies, is equal to 1000 cal. The kilocalorie is often abbreviated as kcal. The so-called "big calorie," or Calorie, used in nutrition is actually the kilocalorie.

For the future, international agreement has been reached to replace the calorie as the unit of heat. Instead, the joule should be used. This unit emphasizes the fact that heat is a form of energy. One joule, which is defined for mechanical energy, is in fact the amount of energy needed to move a mass of one kilogram from rest through a distance of one meter in one second of time. The joule has units of $kg \cdot m^2/s^2$ and 1 joule is equal to 0.2390 calorie. (Conversely, 1 calorie equals 4.184 joules.) More important is the fact that the joule measures energy or the capacity to do work. The energy may be used in a variety of ways which include the acceleration of matter, lifting of weights, or heating of water. For example, 1 joule can raise the temperature of a drop of water by about 5°C. In the interest of simplicity (and scientific as well as international harmony) we will use the joule as our fundamental unit for heat and energy throughout. (For future reference, the joule and other units are listed in Appendix 4.)

The amount of heat needed to raise the temperature of 1 g of a material by 1°C is a characteristic property of the material. It is commonly called the *heat capacity*. For water, 4.184 joules is required to raise the temperature of 1 g by 1°C. For other substances, the amount of heat required is different. Typical values for some common metals are 0.904 joule/g · deg for aluminum, 0.386 joule/g · deg for copper, 0.128 joule/g · deg for gold, and 0.452 joule/g · deg for iron. Because the heat capacity may vary with temperature, these values refer to the metals at their respective melting points.

The amount of heat liberated or absorbed in a chemical reaction can be measured experimentally by using a *calorimeter,* such as that shown in Figure 1.4. This type of calorimeter consists of an insulated box filled with water in which the reaction vessel is placed, along with a thermometer and a stirrer. The principle on which the calorimeter operates is simply that the temperature change of a mass of water depends on the amount of heat added or subtracted. Suppose that the reaction of A and B is exothermic so that, when A and B are mixed, heat is liberated to the water. (A and B can be mixed by simply inverting the reaction tube.) Measurement of the water

Figure 1.4
A cutaway view of a calorimeter.

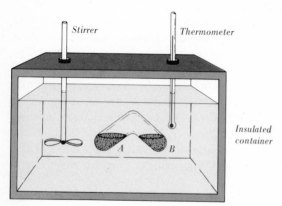

Stirrer

Thermometer

Insulated
container

A   B

temperature before and after reaction gives the temperature rise of the known mass of water. Since 4.184 joules of heat raises the temperature of 1 g of water by 1°C, the total number of joules liberated can be calculated from the mass of water and its temperature rise.

EXAMPLE 1   A sample of metal weighing 400 g is heated to 96.50°C and then dropped into a calorimeter containing 2000 g of water at 20.00°C. If the final temperature becomes 21.50°C, what is the heat capacity of the metal?

The temperature rise of the water is 1.50°C.

Heat absorbed by water = (2000 g of water) (1.50°C) (4.184 joules/g · °C)
= 12,600 joules

The temperature drop of the metal = 96.50 − 21.50°C
= 75.00°C

$$\text{Heat capacity of metal} = \frac{12{,}600 \text{ joules}}{75.0°\text{C} \times 400 \text{ g}} = 0.420 \text{ joule/g} \cdot °\text{C}$$

1.4 The Study of Matter

The concepts of heat and energy are important in chemistry, mainly because they are of invaluable assistance in the primary task of understanding the nature and changes of matter. How do we begin this study of the stuff our world is made of?

In selecting a material sample from nature, it is first necessary to make sure that it is a pure substance in the sense that it has a single set of properties. This is not always the case. For example a sample of granite consists of three *phases, or regions, which exhibit distinct properties.* The three phases are quartz, feldspar, and mica. Substances like granite which *consist of more than one phase* are called *heterogeneous.* For systematic study the sample must be separated so that each phase can be separately studied. For granite, this separation can be achieved by making use of one of the differences of properties of the phases, the *density,* or mass per unit volume. For quartz, feldspar, and mica typical values of the density are, respectively, 2.60, 2.59, and 2.74 g/cm³ (grams of mass per cubic centimeter of volume). On agitation of the ground-up mixture with a liquid of density 2.65 g/cm³, the more-dense mica can be sunk away from the less-dense feldspar and quartz, which will float and can be skimmed off.

Once the homogeneous phase has been separated (note that all the separate pieces of quartz together constitute a single phase since the same set of properties apply to each separate piece of quartz), its properties and composition can be meaningfully measured. Although meaningful, there are an infinite number of such sets of properties and compositions because an infinite number of homogeneous chemical substances are possible. In order to systematize homogeneous substances each must be broken down into its simplest components. The idea that this could be done is a very old one, and it has long been believed that from relatively few elements (once thought to be but four) all more complex substances could be formed. In the case of the three phases of granite, none is an elementary substance. The simplest is the quartz phase which is composed of, and can be decomposed into, the more elementary substances silicon and oxygen. Feldspar may be decomposed into potassium, aluminum, silicon, and oxygen; mica, into potassium, aluminum, silicon, hydrogen, and oxygen. Because these more *elementary substances* cannot be further decomposed into simpler chemical constituents, they are called *elements*. At the present time all the countless materials of our world, and as far as we know, the entire universe, can be resolved into a mere 105 elements.

Chemistry is thus made systematic and manageable through the reduction of all matter, representing countless different materials, to just over a hundred elements. In the shorthand used by chemists, these elements are represented by *symbols,* consisting of letters or combinations of letters derived from the name of the element. The symbol may be just the first letter of the name, as H for hydrogen, O for oxygen, N for nitrogen, and C for carbon. When two or more elements have the same initial letter, a second letter is added to make the distinction. Thus, we have Ni for nickel, Nb for niobium, and Ne for neon, as well as N for nitrogen. Sometimes the symbol is derived from a foreign name for the element. For example, K is the symbol for potassium and comes from the Latin *kalium,* which in turn stems from the Arabic word *qili,* meaning ashes. Thus, the symbol K reflects the fact that ashes of plants are particularly rich in potassium (potash) content. Similarly, the symbol Na for the element sodium, stemming from the Latin *natrium* (from the Greek *nitron*), reflects the occurrence of sodium in the natural mineral called nitre, or saltpeter.

The discovery and isolation of elements still continue. A few of the elements, such as iron (Fe), copper (Cu), silver (Ag), and gold (Au), have been known since antiquity. But most of them were discovered in the nineteenth century. Others are only now being discovered, in the special sense that they are only now being created in particle accelerators. The most recent discovery is the 105th element, hahnium (symbol Ha). Like most of the artificial elements produced since 1943, it has a short lifetime and vanishes almost as fast as it is produced. On the inside back cover of this book is a complete listing of the known elements and their symbols.

In characterizing homogeneous material, we need to know first its *qualitative analysis,* that is, *which of the elements it contains.* But we also need to know its *quantitative analysis, how much, or what percentage, of the elements it contains.* The quantitative analysis of homogeneous substances shows that they all fall into two more or less distinct classes, *compounds* and *solutions.* The distinction between these two classes of substances, either of which may be liquid, solid, or gaseous, is that a compound has a fixed composition (i.e., so much percentage of element A, so much of B, etc.) which is unchanged when part of the sample goes through a change of state, as from liquid to solid in the process of freezing. On the other hand, a solution is continuously variable in composition; i.e., the relative percentage of elements A, B, etc., can be changed at will so that A and B can form an infinite number of solutions. Also, more significantly, when part of a solution is put through a change of state, such as freezing, the solidified part will differ in composition from the part remaining unfrozen. There is a limitation on what the composition of a compound can be and only a limited (but large) number of compounds can be made from the 105 elements. At the present time, it is estimated that several million compounds have been identified, isolated, and characterized. New ones are being synthesized or discovered at an average rate of about 200 per day.

The law of definite composition for compounds stands with the Lavoisier enunciated law of conservation of mass as the basis for chemical study of matter. The *law of definite composition* can be stated: *The composition of the compound formed from reaction of elements is the same, no matter what ratio of starting elements is used.* This principle is illustrated in Figure 1.5 and the examples that follow.

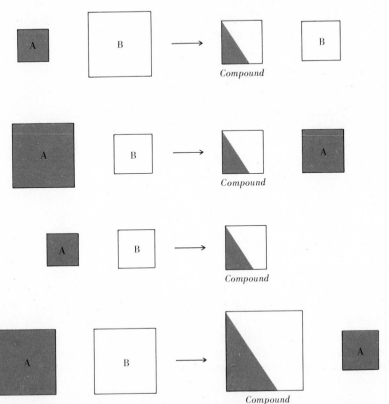

Figure 1.5
Representation of law of definite composition showing that the ratio of elements A and B in final compound is the same no matter what initial amounts of A and B are used.

EXAMPLE 2　A chemical reaction is allowed to occur in a mixture consisting of 4.0 g of Ca and 10.0 g of Cl. After reaction, it is found that there are 11.1 g of calcium chloride and 2.9 g of unreacted chlorine. What is the percentage composition of the compound formed?

$$\text{Calcium used} = 4.0 \text{ g}$$
$$\text{Chlorine used} = 10.0 - 2.9 = 7.1 \text{ g}$$
$$\text{Total weight of compound} = 11.1 \text{ g}$$

$$\% \text{ calcium} = \frac{4.0 \text{ g}}{11.1 \text{ g}} \times 100 = 36\%$$

$$\% \text{ chlorine} = \frac{7.1 \text{ g}}{11.1 \text{ g}} \times 100 = 64\%$$

EXAMPLE 3　How much calcium chloride can be made from a starting mixture containing 10.0 g of Ca and 4.0 g of Cl?

The amount of calcium chloride (36% Ca and 64% Cl) that can be formed is limited by the element not present in excess. To find which element is in excess, calculate the amount of product which could be formed from each.

From 10.0 g of Ca. we could get

$$\frac{10.0 \text{ g Ca}}{36 \text{ g Ca}/100 \text{ g cpd}} = 28 \text{ g cpd}$$

From 4.0 g of Cl, we could get

$$\frac{4.0 \text{ g Cl}}{64 \text{ g Cl}/100 \text{ g cpd}} = 6.3 \text{ g cpd}$$

Clearly, Ca is in excess, the Cl is limiting, and the mass of calcium chloride formed will be 6.3 g.

The laws of chemical change are examples of natural laws in summarizing a large body of individual observations, or data. As laws they explain nothing. But as we shall see, they lead us to ask: Why?, and the answer is a theory of the ultimate nature of matter. This theory, in turn, leads to further systematization and understanding in our study of matter.

QUESTIONS

Here and at the end of succeeding chapters you will find questions and problems designed to draw out important points considered in the foregoing material. The exercises are generally given in order of increasing difficulty, and each is labeled with its principal topic. For the most effective use, answers should be worked out in clear, logical sequence with due attention to the reasoning involved. In most cases, the final numerical answer is not so important as mastery of the steps by which it was reached. An asterisk denotes a problem whose answer appears in Appendix 10 at the end of the book.

1.1 *Lavoisier* In terms of our discussion thus far, in what sense is Lavoisier correct in stating that "must of grapes" and "carbonic acid + alcohol" are of the same nature?

1.2 *Study of Matter* Why isn't a 10-g sample of granite directly suitable for study without first being separated into homogeneous phases?

1.3 *Science Methods* (*a*) In what sense can a law be described as "true" that a theory cannot? (*b*) What quality does a theory possess that a law does not?

1.4 *Lavoisier* State as clearly as you can the basic law of chemical change first advanced by Lavoisier and describe the kind of experiments which could have led to it.

1.5 *Elements* Why is it that millions of compounds can be made from only 105 elements?

1.6 *Matter* Which of the following are compounds: (*a*) ammonia, (*b*) urea, (*c*) water, (*d*) an iron ball, (*e*) wine?

1.7 *Terms* Explain what is meant by each of the following terms: (*a*) matter, (*b*) energy, (*c*) heat, (*d*) temperature.

1.8 *Terms* Many people are confused with the terms mass and weight. (*a*) Give definition of both. (*b*) Which of the two has a constant value at any place on our planet?

1.9 *Phases* Classify each of the following as homogeneous or heterogeneous: (*a*) granite, (*b*) a tree branch, (*c*) water, (*d*) paper, (*e*) quartz, (*f*) saltwater.

1.10 *Heat* 10.2 g of aluminum is heated from 75 to 248°F. (*a*) What is the difference in temperature in °C? (*b*) What is the amount of heat absorbed by the aluminum during the process?

1.11* *Heat* How much heat is necessary to warm: (*a*) 250.0 g of water from 25.1 to 32.3°C; (*b*) 11.1 g of water from 20.3 to 49.1°C?

1.12 *Science and Society* Suppose some chemists suspect that they have learned how to prepare the most deadly substance known. Explain which of the following courses they should take: (*a*) Stop their experiments short of preparing it. (*b*) Prepare it and disclose their results only to trusted people. (*c*) Prepare it and "tell the world."

1.13 *Heat* Describe an experiment which shows clearly the distinction between *heat* and *temperature*.

1.14 *Definite Composition* 20.0 g of element A is heated with 10.0 g of element B. After reaction, half of B remains unchanged, but all of A is gone. (*a*) What is the percentage composition of the compound? (*b*) Show that in arriving at your answer to (*a*) you assumed one law of chemical change and demonstrated another.

1.15 *Definite Composition* 10 g of element Q reacts with 19 g of element R to form a homogeneous product containing no unreacted Q or R. (*a*) Why is this observation not a good demonstration of the law of definite composition? (*b*) Give two possibilities for the nature of the product.

1.16 *Density* A 5000-ml beaker contains: (*a*) one copper rod 6.00 cm long and 2.00 cm in radius with density 7.86 g/cm^3; (*b*) an aluminum ball of radius 4.00 cm and density 2.70 g/cm^3; (*c*) 2.10 liters (1 liter is 1000 cm^3) of water of density 0.998 g/ml. Calculate the average density of matter in the beaker.

1.17* *Density* A bottle having a capacity of 15.0 liters (1 liter is 1000 cm^3) contains 923 g of sand with density 2.67 g/cm^3, a cube of lead 11.0 cm on edge having density of 11.3 g/cm^3, and 4.20 liters of water having density of 0.998 g/ml. The remainder of the interior of the bottle contains air having density of 1.18 × 10^{-3} g/cm^3. Calculate the average density of matter in the bottle.

1.18 *Energy* (*a*) Describe any chemical reaction you have experienced which liberates energy. (*b*) Describe one which requires energy.

1.19* *Heat Capacity* A sample of a solid weighing 500 g is heated to 75.00°C and then dropped into a calorimeter containing 3500 g of water at 21.50°C. If the water temperature becomes 22.75°C, what is the solid's heat capacity?

1.20 *Study of Matter* Describe experiments you might do to decide whether a given sample of matter is (*a*) heterogeneous or homogeneous, (*b*) a compound or a solution, (*c*) a compound or an element.

1.21* *Energy* (a) What kind of energy is obvious in a rolling bowling ball? (b) If the ball is rolled twice as fast, how much does this energy change?

1.22 *Laws of Chemical Change* The carbonic gas mentioned by Lavoisier is a gas composed of 27.3% carbon and 72.7% oxygen. How much of this gas can be made from 5.00 g of C and 10.0 g of O?

1.23 *Units* Which member of each of the following pairs is the larger: (a) 1 deg Celsius or 1 deg Fahrenheit, (b) the number of degrees measuring the temperature of this room in Celsius or in Fahrenheit, (c) 1 cal or 1 joule, (d) the number of joules or the number of calories needed to melt 25 g of ice?

1.24 *Chemical Observation* List three examples of each of the following from your observations of nature outside the laboratory: (a) heterogeneous substances, (b) single phases, (c) solutions, (d) chemical reactions.

1.25 *Mass* Describe qualitatively the relationship between *mass, weight,* and *the law of gravity.*

1.26 *Elements* Suppose that all matter consisted of only four elements. (a) How many compounds could there be which differed in the kinds of elements they contained? (b) How many compounds could there be which differed in the amounts of elements they contained? Explain.

1.27* *Heat Capacity* A 250-g chunk of iron initially at 80.0°C is dropped into a calorimeter whose temperature thereby is raised from 25.00 to 26.25°C. (a) How much water did the calorimeter contain? (b) What weight of copper at 80.0°C would have produced the same temperature rise?

1.28 *Definite Composition* A given sample of a pure compound is found to contain 2.50 g of element A, 3.20 g of element B, and 7.30 g of element C. (a) What is its percentage composition? (b) What weight of compound can be made from 1.50 g of A, 1.60 g of B, and 4.50 g of C? (c) The compound, on heating, forms a different compound containing equal weights of A and B, but no C. After complete conversion of 5.00 g of the original compound to the new one, what weight of which elements remains uncombined?

The Nuclear Atom

2

As far back in time as the early Greek philosophers, thinkers have asked about the ultimate nature of matter. One school of philosophers argued that the process of subdividing matter into progressively smaller units could not be continued indefinitely. Eventually an ultimate, or indivisible, particle would be reached. The ultimate particles were named atoms, from the Greek word *atomos* meaning uncut. However, the Greek philosophers had no evidence for or against the idea that subdivision of matter could continue without limit, and so the atomic nature of matter remained an unsolved philosophic question for many centuries.

Following the stating of the laws of conservation of mass and definite composition, scientists had experimental evidence to guide their speculations concerning the ultimate nature of matter. It was an English schoolteacher, John Dalton, who first presented (ca. 1803) a convincing scientific argument for the atomic theory. He based his case on the repeated observations: (1) Mass is conserved during chemical reaction; (2) chemical compounds have definite composition.

These laws were discussed in Chapter 1. In this chapter we will discuss Dalton's reasoning, the concepts which grew directly from his theory, the further experiments which showed that atoms can be further subdivided.

2.1 Dalton's Atomic Theory

To account for the laws of chemical change, Dalton's atomic theory postulated:

1   All matter is composed of atoms that can be neither created nor destroyed.

2   Atoms of a particular element are identical in size, shape, mass, and all other properties and differ from atoms of other elements in these properties.

3   Chemical reaction is the union or separation of atoms.

A test of any theory lies in how well it accounts for the observed facts. Does the Dalton atomic model account for the conservation of mass in chemical reactions? If chemical change is merely the union or separation of undivided atoms, their mass must be conserved. No new atoms are created, and no old ones are destroyed. For example, in the reaction of hydrogen with oxygen to form water, hydrogen atoms unite with oxygen atoms, but all the atoms initially present, and no others, remain after reaction is complete. The contribution to the total mass by the hydrogen atoms and by the oxygen atoms is the same whether they are present as the original elements or are combined in the compound water.

Does the Dalton atomic theory account for the law of definite composition? As we saw in Example 3 earlier, the mass of product formed from the reaction of two elements is fixed by the mass of one element, which is a statement of the law of definite composition. Suppose we consider the reaction of known amounts of carbon and oxygen to form carbon monoxide. The known amount of carbon corresponds to a definite number of carbon atoms, and the known amount of oxygen to a definite number of oxygen atoms. We assume that, in the formation of carbon monoxide, only one carbon atom unites with each oxygen atom. If, as shown in Figure 2.1, there are 18 carbon atoms and 15 oxygen atoms available, there will be 3 carbon atoms left after reaction. There are not enough oxygen atoms to satisfy all

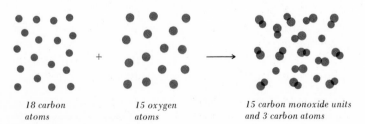

18 carbon
atoms

15 oxygen
atoms

15 carbon monoxide units
and 3 carbon atoms

Figure 2.1
Limitation of product by one reactant.

the carbon atoms; hence, the amount of product is limited only by the oxygen, not by the carbon.

A second way of stating the law of definite composition is to say that compounds have *characteristic* compositions. The explanation can be seen by considering pure carbon monoxide, which is composed of complex units, or molecules,† each containing one carbon atom and one oxygen atom. In each carbon monoxide molecule, the fraction of the mass contributed by carbon is the mass of the carbon atom divided by the mass of the molecule. Since, according to Dalton's theory, all carbon monoxide molecules are alike, any sample containing many such molecules has the same percentage of carbon by mass as the single molecule.

If a scientific theory is any good, it should lead to predictions about behavior of nature that has not yet been recognized. Reasoning from his theory, Dalton was able to predict a regularity in the weight relations for the case of the same two elements forming two different compounds. In such cases, *the mass of one element, A, per unit mass of the other element, B, in one compound is a simple multiple of the ratio g A/g B in the other compound.* This relationship was borne out by repeated experiments and has come to be called the *law of multiple proportions.*

EXAMPLE 1   Under different conditions, carbon and oxygen can combine to form two different compounds. One of these is 42.9% carbon; the other is 27.3% carbon. Show how these data illustrate the law of multiple proportions.

In the first compound, there is 57.1 g of oxygen per 42.9 g of carbon, or 1.33 g of O per g of C.
In the second compound, there is 72.7 g of oxygen per 27.3 g of carbon, or 2.66 g of O per g of C.
The second ratio, 2.66, is just twice the first, 1.33.

In terms of the Dalton theory, the law of multiple proportions can be viewed as follows: Suppose two different reactions are possible between the same two elements, as shown in Figure 2.2. Carbon monoxide is formed when one atom of carbon unites with one atom of oxygen. Under different conditions, one atom of carbon unites with two atoms of oxygen to form carbon dioxide. If atoms are indivisble, either one atom of oxygen or two atoms of oxygen unite with a single carbon atom; there cannot be a fractional number of oxygen atoms combined with one carbon atom. Thus, the number of oxygen atoms per carbon atom in the two compounds is in the ratio 1:2, a ratio expressible by small whole numbers. Likewise, the *mass* of oxygen per unit *mass* of carbon in the two compounds is in the ratio of 1:2.

---

†In his original works, Dalton did not use the term *molecule,* which in its modern sense usually denotes a characteristic aggregate of two or more atoms. Instead, he referred to *compound atoms* such as atoms of water, atoms of sugar, atoms of "sulphuretted hydrogen," etc. An atom of sulphuretted hydrogen, according to Dalton, was composed of one atom of sulfur and three atoms of hydrogen!

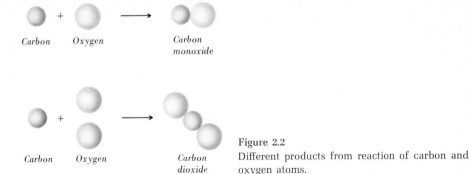

Figure 2.2
Different products from reaction of carbon and oxygen atoms.

Carbon   Oxygen   Carbon
                  monoxide

Carbon   Oxygen   Carbon
                  dioxide

2.2 Atomic Weight†

One of the most important concepts to come from Dalton's work is that of atomic weight. How does the weight of one atom compare with the weight of another atom? Because atoms are so tiny, their individual weights are difficult to measure. However, their *relative* weights can be obtained by measuring the weight of one element combined with another, provided that the relative number of atoms in the compound is known. For example, analysis of water shows that it is 11.19% hydrogen and 88.81% oxygen. This means that in water 88.81/11.19, or 7.937, times as much weight is contributed by the oxygen atoms as by the hydrogen atoms. If there is *one* oxygen atom for *every* hydrogen atom, then the oxygen atom is 7.937 times as heavy as the hydrogen atom. If, however, there is *one* oxygen atom for every *two* hydrogen atoms, the oxygen atom is 7.937 times as heavy as two hydrogen atoms, or 15.87 times as heavy as one hydrogen atom. Since the second formulation turns out to be correct, the oxygen atom is 15.87 times as heavy as the hydrogen atom.‡

There is another complication, which was unknown in Dalton's day, in that not all the atoms of an element are identical: Even though their

† Strictly speaking, this section should be entitled "Atomic Mass" instead of "Atomic Weight." As noted in Section 1.2, *mass* measures the inertial properties of matter whereas *weight* measures the force with which matter is attracted to the earth. What then is the atomic weight of an element when it is in a "weightless" condition, as, for example, in a satellite in orbit around the earth? Surely, the atoms have not lost their mass, even though they may appear to have lost their weight. One might well argue in such a case that the term *atomic weight* should be replaced by *atomic mass*. In practice, there are objections to doing so. For one thing, physicists and chemists generally interpret the term *atomic mass* to mean the actual mass of a single atom. For another thing, the related term *molecular mass* generally means the mass of a single molecule. Thirdly, the terms *atomic weight* and *molecular weight* are internationally accepted for describing the mass behavior for the average of a large number of particles. Fourthly, *weight* has become a common substitute for the more precise term *mass*. Finally, how many of us will be doing chemical experiments in orbiting satellites?

‡ In Dalton's early attempt to set up a scale of atomic weights, he made the mistake of assuming that water contains *one* oxygen atom for *one* hydrogen atom. Consequently, his oxygen atom was underweight by a factor of 2. Furthermore, the percentage composition of water was poorly determined, and suggested 12.5% H and 87.5% O. As a result, Dalton was led to believe the oxygen atom was only seven times as heavy as the hydrogen atom.

chemical properties are alike, they may *differ slightly in weight*. These different varieties of atoms of the same element are called *isotopes*. For example, natural oxygen consists of three isotopes, the heaviest of which is nine-eighths as heavy as the lightest and constitutes 0.20% of natural oxygen. Dalton would have been unaware of difficulties from the existence of isotopes because chemists generally work with enormous numbers of atoms having unvarying percentage of the various isotopes. Hence, an average is all that is needed.

In setting up a relative scale of atomc weights, one element is chosen as a reference standard, and all other elements are referred to it. The choice of standard is arbitrary, and for many years oxygen was used. At one time it was assigned a numerical value of 100; later, the value of 16 was used. In 1961, however, chemists and physicists the world over agreed that the reference standard would be the most common isotope of carbon; it was assigned a mass of exactly 12 atomic mass units (amu). By definition, then, one *atomic mass unit* is one-twelfth the mass of one atom of the most abundant variety of carbon atom. The mass of an "average" carbon atom, allowance being made for the percentage of each isotope, turns out to be 12.011 amu. Thus the atomic weight of carbon is given as 12.011 amu. Many of the other elements also have approximately integral atomic weights. For example, the atomic weight of hydrogen is 1.008 amu, sodium is 22.9898 amu, and sulfur is 32.06 amu. Chlorine, however, has an atomic weight of 35.453 amu. A complete list of the latest internationally accepted atomic weights is given inside the back cover of this book. Only small refinements in these values now occur from year to year.

In the chemical determination of atomic weights, there is a problem of deciding how many atoms of one element combine with one atom of another. An historically important way of solving this problem made use of the *law of Dulong and Petit*. This law, discovered in 1819, may be stated as follows: *For many solid elements the atomic weight multiplied by the heat capacity is approximately equal to 26 joules/deg*. The *heat capacity* (Section 1.3) can be measured experimentally for a given element. Some typical values are shown in Figure 2.3. For example, the heat capacity of silver is 0.236 joule per deg per g. We can use the law of Dulong and Petit to estimate its atomic weight by dividing 26 by the heat capacity:

$$\frac{26}{0.236} = 110$$

$$= \text{approximate atomic weight}$$

The use of this approximate atomic weight to get an exact atomic weight is illustrated in Example 2.

Figure 2.3
HEAT
CAPACITIES
AND ATOMIC
WEIGHTS

Element	Heat Capacity, joule/g · deg	Atomic Weight	Product, Heat Capacity × Atomic Weight
Aluminum	0.904	26.98	24.4
Calcium	0.657	40.08	26.3
Copper	0.386	63.54	24.5
Gold	0.128	196.97	25.2
Iodine	0.217	126.90	27.5
Iron	0.452	55.85	25.2
Lead	0.129	207.2	26.7
Magnesium	0.983	24.305	23.9
Nickel	0.439	58.71	25.8
Potassium	0.745	39.102	29.1
Silver	0.236	107.868	25.5
Sulfur	0.707	32.06	22.7
Tin	0.222	118.69	26.4
Zinc	0.388	65.37	25.4

EXAMPLE 2   In the formation of silver oxide, it is observed that 1.074159 g of this compound is formed from 1.000000 g of silver. From the heat capacity, it has been calculated that the *approximate* atomic weight of silver is 110 amu. The atomic weight of oxygen is 15.999 amu. What is the exact atomic weight of silver?

The information given above tells us that 0.074159 g of O combines with 1.000000 g of Ag. But what we want to know is what weight of *silver* combines with one atom (or 15.999 amu) of oxygen. First we determine how much silver combines with 1 g of O:

$$\frac{0.074159 \text{ g of O}}{1.000000 \text{ g of Ag}} = \frac{1.000000 \text{ g of O}}{x \text{ g of Ag}}$$

$$x \text{ g of Ag} = \frac{1.000000 \text{ g of O} \times 1.000000 \text{ g of Ag}}{0.074159 \text{ g of O}}$$

$$= 13.484539 \text{ g of Ag}$$

With 15.999 g of O, 15.999 times as much Ag will combine

$$13.484539 \text{ g of Ag} \times 15.999 = 215.74 \text{ g of Ag}$$

The amount of oxygen (15.999 g) was chosen, of course, because it is the number of grams equal to the atomic weight of oxygen. Therefore, if 15.999 g of O combine with 215.74 g of Ag, then 15.999 amu of O combine with 215.74 amu of Ag.

Now all that remains to be determined is how many atomic weights of silver the 215.74 amu represent—one, two, three, etc. If one oxygen atom is combined with *one* silver atom, the atomic weight of silver is 215.74 amu. If one oxygen atom is combined with *two* silver atoms, then the atomic weight of silver is half as much, or 107.87 amu. We need go no further however, because the approximate value of 110 from the law of Dulong and Petit indicates that the second value, 107.87, is correct.

Historically, atomic weights of elements were determined by experiments of the type illustrated in Example 2. Today there are alternative methods. In particular, the mass spectrometer, which, as discussed in the next section, is a device for measuring the magnetic deflection of charged atoms, has been especially useful in determining the mass of individual isotopes and their relative abundance. In the case of chlorine, for example, the natural isotope abundance is 75.77 percent of the light variety and 24.23 percent of the heavy variety. Of 1000 chlorine atoms selected at random, 758 have a mass of 34.97 amu,† and 2.42 atoms have a mass of 36.97 amu. The average atomic weight of chlorine is 758×34.97 plus 242×36.97 divided by 1000, or 35.45 amu.

2.3 The Mole and the Avogadro Number

Since atoms are extremely small, any laboratory experiment dealing with weighable amounts of chemicals must, of necessity, involve tremendous numbers of atoms. For example, in making carbon monoxide, it is not possible to weigh one carbon and one oxygen atom, because any weighable amount of carbon or oxygen contains an enormous number of atoms. However, it is possible to get equal numbers of carbon and of oxygen atoms by using the relative masses of these atoms. From the atomic weights (12 for C and 16 for O), we know that a carbon atom is twelve-sixteenths as heavy as an oxygen atom. Suppose we take any definite number of carbon atoms and an equal number of oxygen atoms. The weight of the entire collection of carbon atoms, then, is twelve-sixteenths as great as the weight of the collection of oxygen atoms. Conversely, any weight of carbon that is twelve-sixteenths as great as a weight of oxygen must contain just as many carbon atoms as there are oxygen atoms. For example, 12 g of carbon contains the same number of atoms as does 16 g of oxygen. In general, when we take amounts equal to the atomic weights of different elements, we always have the same number of atoms of each. We can take these amounts

---

†Note that the individual isotopes do not have whole-number atomic weights. We will see the reason for this in Section 2.8. *Nihil facilis.*

in grams, pounds, or any other convenient units of mass. In other words, 16 lb of oxygen contains the same number of atoms as 12 lb of carbon.

For convenience in specifying amounts, chemists define a mole of atoms as a collection of atoms whose total weight is the number of grams equal to the atomic weight.† Since sulfur has an atomic weight of 32.06 amu, a collection of sulfur atoms weighing 32.06 g is 1 mole of sulfur atoms. Since the atomic weight of iron is 55.85 amu, a collection of iron atoms weighing 55.85 g is 1 mole of iron. The collections have different weights, but each has the same number of atoms.

The concept of the mole enables us to choose the proper number of atoms for a reaction. Suppose we wish to make a compound, ferrous sulfide, in which there is one atom of iron for each atom of sulfur. If we take 1 mole of iron and 1 mole of sulfur, there are just enough iron atoms to match the sulfur atoms. Furthermore, the amounts taken are big enough to be handled with usual laboratory apparatus. As mentioned earlier, because equal numbers of moles of different elements contain equal numbers of atoms, it is convenient to refer to amounts of elements in terms of numbers of moles. For instance, 3.2 g of sulfur is 3.2/32, or 0.10, mole.

EXAMPLE 3   How many grams of sulfur are required to react with 336 g of iron in a reaction in which one atom of iron unites with one atom of sulfur? (The atomic weight of iron is 55.85 amu, and that of sulfur is 32.06 amu.)

The first thing to note about this problem is that it does not mention moles. The second thing to note is that this and similar problems, though they can be solved entirely in terms of grams and amu's, can be simplified by use of the mole.

$$\text{Moles of iron} = \frac{336 \text{ g}}{55.85 \text{ g/mole}} = 6.02 \text{ moles of Fe}$$

We need 6.02 moles of sulfur, or

$$(6.02 \text{ moles of S}) (32.06 \text{ g/mole}) = 193 \text{ g of S}$$

With modern techniques, it has been possible to determine the *number of atoms in 1 mole*. The value of this number is 6.02222×10^{23}, or 602,222,000,000,000,000,000,000. This number is referred to as the *Avogadro number* and should be memorized to at least three significant figures: 6.02×10^{23}. Detailed consideration of the methods used to determine the Avogadro number must be postponed until further principles have been discussed. Probably the most accurate determination of the Avogadro num-

---

†Historically, the concept of a mole has proved so useful that the original definition which referred only to molecules has been expanded to refer to atoms. In the recent past a mole of atoms was called a gram-atom, but since confusion resulted from a term which meant neither a gram nor an atom, that term is now dropped. The expanded use of "mole" can cause difficulty since some elements also form molecules and a mole of atoms will not be the same as a mole of molecules. However, the matter is usually settled by the context in which it appears.

ber is based on the study of solids. From the measured mass per unit volume (density) of a solid, the volume of 1 mole can be calculated. As discussed in Chapter 7, the spacing of atoms in a solid can be found by using X rays. This enables a precise determination of the number of atoms in the volume which contains 1 mole.

2.4 Divisible Atom

The Dalton theory went a long way toward answering the questions which had been raised by observations at that time. The simple, indivisible, structureless atom could account for all observations concerning weight relationships in chemical change. It confirmed Lavoisier's proposition that the principles of the matter before reaction remained as principles in the products of the reaction. Following Dalton, chemists merely substituted "atoms" for "principles." What the structureless atom could not account for was the fact that atoms seemed to show a definite combining ability such that a single atom of one kind combined with a definite number of atoms of another kind. Although attempts were made to explain this phenomenon by giving the atoms "valence hooks," real insight into combining capacity came as the result of studies of a different kind which forced scientists to conclude that atoms are in fact divisible.

Such studies, which revealed that atoms can be subdivided into electrically charged fragments, involve observation of the behavior of discharge tubes. A typical discharge tube is shown in Figure 2.4. It consists of a glass tube with two metal plates, or electrodes, sealed in at either end. The electrodes are connected to the positive and negative sides of a high-voltage source and are called *anode* and *cathode*, respectively. When the tube is full of air at normal pressure, nothing is observed even if 10,000 volts is applied across the electrodes. However, as air is pumped out of the tube, electricity starts to flow. When certain compounds, such as zinc sulfide, are

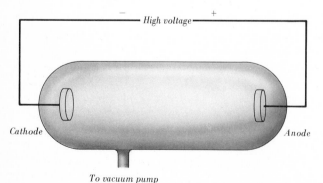

Figure 2.4
Simple discharge tube.

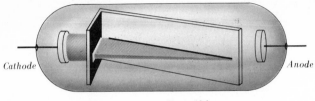

Cathode   Anode

Metal plate   *Zinc sulfide*
with slit   *screen*

Figure 2.5
Discharge tube to observe path of cathode particles.

placed between the electrodes, they give off an intense light and can serve as detectors.

More detailed information about the nature of the discharge can be obtained with the tube shown in Figure 2.5. This tube is fitted with a detecting screen covered with zinc sulfide and has a slotted metal plate near the cathode. When we observe the tube from the side, we see a line of light across the screen. We can explain this observation by assuming that electric beams, or rays, emanate from the cathode and move toward the anode. Most of these are stopped by the metal plate, but the slit lets through a narrow sheaf of them which hits the screen and produces a line of light. Because the beams seem to come from the cathode, they are called *cathode rays*. When the north pole of a magnet is brought up to the *side* of the tube, the line of light curves downward; when the south pole of the magnet is brought up, the line curves upward. The direction of the deflection indicates that the cathode rays are *negatively charged*. The same conclusion can be reached by noting that, if electrically charged plates are placed above and below the screen, the line curves toward the positive plate. Opposite charges attract each other; hence we can conclude cathode rays are negatively charged.

As described below, the quantitative determination of the electric charge and mass of the particles believed to make up cathode rays shows that the particles are identical, no matter what material the cathode is made of and no matter what gas is present in the tube. The particles are called *electrons* and are considered to be constituents of all matter. In the operation of the discharge tube, electrons appear to come from the cathode material.

Experiments first performed by Eugen Goldstein in 1886 suggested that *positive particles* are also formed in discharge tubes. The Goldstein tube is shown in Figure 2.6. The electrode on the right is positive; the electrode toward the left is negative and consists of a piece of metal with a hole bored in it. A detecting screen similar to that used in Figure 2.5 can be placed to the left of the negative electrode. A line of light appears on it and is deflected by a magnet in the direction *opposite* to the deflection in the experiment with cathode rays. The inference is that a positive beam exists to the left of the cathode. Its origin can be explained as follows: Electrons emitted from

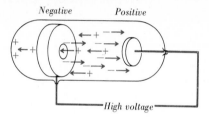

Figure 2.6
Discharge tube for studying positive particles.

the cathode are attracted to the anode. Since there is gas in the tube, these electrons collide with neutral atoms of the gas. If the electrons have enough energy, they can knock other electrons off the neutral atoms. This subtraction of negative electrons leaves residual positive particles, which are accelerated toward the negative electrode. Here most of them pick up electrons and are neutralized. Occasionally, a positive particle coasts through the hole, giving a beam of positive particles directed toward the left end of the tube. It is found that these particles are always more massive than electrons, with the mass dependent on the kind of gas that is in the tube.

The first quantitative study of the deflection of electron beams by electric and magnetic fields was made by J. J. Thomson in 1897. Figure 2.7 shows an arrangement which could be used for such a study. The negative electrode emits electrons, which are accelerated to the right. Some pass through the hole in the anode to give a narrow beam, which falls on the detecting screen at the face of the tube. The presence of the magnet causes the beam to be deflected. By suitably charging the plates above and below the beam, the deflection of the beam caused by the magnet can be counteracted by the deflection caused by the charged plates. The strength of the magnet and of the voltage on the plates required to produce no net deflection are measured. This information, when combined with a measurement of the

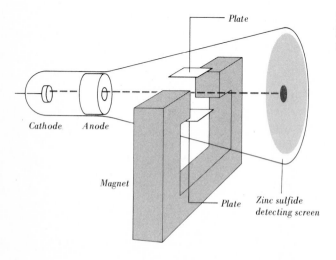

Figure 2.7
Deflection of cathode rays by electric and magnetic fields.

deflection observed when the magnet is removed or when the plates are uncharged, allows a calculation of the ratio of charge to mass for the particles. The amount of deflection depends on the charge of the particles, because the bigger the charge of a particle, the more it is attracted to (or repelled by) a charged plate. The deflection depends inversely on the mass of the particles, because the greater the mass of a moving particle, the more difficult it is to deflect the particle from its straight-line path.

Experiments such as that just described indicate that for an electron the ratio of charge to mass is equal to -1.76×10^8 coulombs/g. (The coulomb, a unit for measuring electric charge, is described in Appendix 3.5.) The minus sign of the ratio indicates the negative nature of the charge of the electron.

Beams of positive particles can also be studied by the above method by using a slightly modified apparatus. The mass spectrometer shown in Figure 2.8 is an example of such an instrument. It measures the charge-to-mass ratio of positive particles. Positive particles are produced by electron bombardment of neutral molecules near the filament. They are accelerated through the first slit and bent in a circular path by a magnetic field (not shown). Particles of different charge-to-mass ratio follow different paths, shown by the separation into two beams. When positive particles are produced from different gases, it is observed that the charge-to-mass ratio varies from one gas to another. Also it is observed that in all cases the charge-to-mass ratio (about 10^5 coulombs/g, or less) is considerably smaller than that for electrons. In other words, the positive particles are considerably more massive than the negative electrons.

A measurement of the charge-to-mass ratio of a particle tells nothing about the actual charge or the actual mass of the particle. However, once either of these quantities has been determined, the other can be calculated

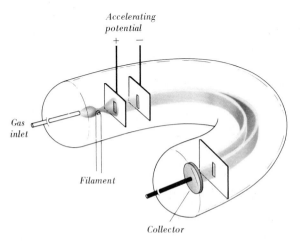

Accelerating potential

+ —

Gas inlet

Filament

Collector

Figure 2.8
Mass spectrometer (magnetic field not shown).

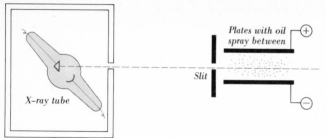

Figure 2.9
Oil-drop method for determining electronic charge.

from the ratio. The experiment of Thomson gave a value for the charge-to-mass ratio of the electron. The charge of the electron was measured in a classic experiment by R. A. Millikan in 1909. Figure 2.9 shows the essential features of the experiment.

An atomizer is used to spray oil droplets between two charged plates. Because of gravity, the droplets gradually settle. However, if the droplets can be given a negative charge, they can be made to rise because of attraction to the positive plate. (The negative charge can be imparted to the droplets by irradiation with X rays. The X rays presumably knock electrons off atoms in the air, just as the cathode rays do in the Goldstein tube. One or more of these electrons can be picked up by an oil droplet and thus make it negative.) The rate of rise of a charged oil droplet is measured by observing it with a telescope. From the rate of rise and the charge on the plate, the amount of charge on an individual droplet can be calculated. It is found that, although the charge on different droplets varies, the total charge on any one droplet is always a small whole-number multiple of -1.60×10^{-19} coulomb. Apparently, the smallest possible charge that any one droplet can pick up is -1.60×10^{-19} coulomb, and this is assumed to be the charge of an individual electron. Combining the charge of the electron (-1.76×10^8 coulombs/g) gives the mass of the electron as 9.1×10^{-28} g. For convenience, the charge of the electron is frequently referred to as $1-$.

The mass of positive particles can similarly be deduced from knowledge of their charge-to-mass ratio and their actual charge. Positive particles presumably result when electrons are pulled off neutral atoms. If a single electron is removed from a neutral atom, the positive particle left must have a charge exactly equal to but opposite in sign to the charge of the electron. For example, if an electron is pulled off a neutral hydrogen atom, the resulting positive particle (called the *proton*) has a charge of $+1.60 \times 10^{-19}$ coulomb. Its mass is found to be 1.67×10^{-24} g, which is about 1840 times the mass of the electron. If two electrons are removed from a neutral helium atom, the resulting particle (called the *alpha particle*) has a charge of $+3.20 \times 10^{-19}$ coulomb and a mass of 6.6×10^{-24} g. In terms of electronic charge units, the charge of the proton is $1+$ and that of the alpha particle is $2+$.

Comparison of the masses of electrons and protons, or similar positive particles, shows that electrons are much less massive. For example, the mass of an electron is only 5.5×10^{-4} that of the proton, the lightest of the positive particles. In terms of atomic mass units, the mass of an electron is 0.00055 amu.

2.5 · Discovery of the Nucleus

If atoms are complex as the discharge-tube experiments indicated, the question arises as to what their detailed structure is. J. J. Thomson in 1898 proposed that the atom be considered a sphere of positive electricity in which negative electrons are embedded like jelly beans in a ball of cotton. Most of the mass of the atom would have to be associated with the sphere of positive electricity, a conclusion drawn from the observation that the positive fragments of atoms are much heavier than the electrons. In 1911 Rutherford performed the classic experiment which tested the Thomson model. He was investigating the scattering of alpha particles by thin sheets of metal. According to the Thomson model, a metal consists of atoms which are spheres of positive electricity containing negative electrons; that is, the metal is essentially a sea of positive electricity containing negative charges. Since alpha particles are very energetic, they go right through metal foils. If the positive charge and the mass are distributed evenly throughout the metal, the alpha particle has little reason to swerve off its original path.

Figure 2.10 shows details of the Rutherford experiment. The alpha particles come from the radioactive element polonium. A thick lead plate, with a hole cut in it, serves to produce a beam of alpha particles. In the path of the beam is a metal foil. The alpha particles may be detected with a screen coated with zinc sulfide.

As expected, 99 percent of the alpha particles went through. However, some were deflected drastically. A few were actually reflected back along

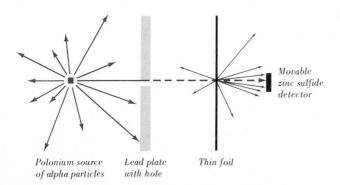

Movable zinc sulfide detector

Polonium source of alpha particles   *Lead plate with hole*   *Thin foil*

Figure 2.10
Rutherford's experiment for studying the scattering of alpha particles by metal foil.

their path. To Rutherford this was absolutely unbelievable. In his own words, "It was almost as incredible as if you fired a 15-inch shell at a piece of tissue paper and it came back and hit you." The Thomson model could not account for such large deflections. If mass and positive charge are uniformly spread throughout the metal, a positively charged alpha particle simply would not encounter a large repulsion or major obstacle anywhere in its path. According to Rutherford, the only way to account for the large deflection is to say that the positive electricity and mass in the metal foil are concentrated in very small regions. Although most of the alpha particles can go through without any deflection, occasionally one comes very close to the high concentration of positive charge. This high concentration of positive charge is essentially immovable because of its high mass. As the like charges get closer together, they repel each other, and the repulsion may be great enough to cause the alpha particle to swerve considerably from its original path. So Rutherford suggested that an atom has a *nucleus, or center, in which its positive charge and mass are concentrated.*

The quantitative results of scattering experiments such as Rutherford's indicate that the nucleus of an atom has a diameter of approximately 10^{-13} cm. Atoms have diameters about 100,000 times as great. In other words, the nucleus occupies an extremely tiny portion of the volume of the entire atom; practically all the volume of the atom is occupied by electrons. As an analogy, if an atom were magnified so that the nucleus were the size of the period at the end of this sentence, the whole atom would be bigger than a house.

2.6 Makeup of the Nucleus

After Rutherford had discovered the nucleus, appreciation of its complex character became apparent from the interpretation of earlier discoveries (1895) by Wilhelm Röntgen of X rays and Henri Becquerel of radioactivity. Röntgen found that highly penetrating radiation was produced when cathode rays bombarded metals; Becquerel discovered that apparently similar radiation was emitted spontaneously by uranium minerals, in the process now recognized as radioactive decay.

The nature of the radiation was studied by observing its deflection in a magnetic field. As shown in Figure 2.11, positive particles curve one way, negative particles curve the opposite way, and uncharged particles pass through the magnetic field undeflected. The paths of the particles can be detected by photographic film or ZnS screens or, in recent times, by a cloud chamber or a bubble chamber. In a cloud chamber, supercooled water vapor is caused to condense into droplets along the particle's trajectory; in a bubble chamber, decompressed liquid is caused to bubble along the trajectory. If,

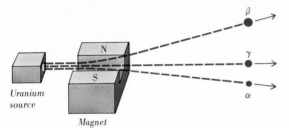

Uranium
source

Magnet

Figure 2.11
Radioactive-decay particles in a magnetic field.

as in the figure, a piece of uranium mineral is placed in the bottom of a hole drilled in a lead block, a beam of particles emerges and is separated by the magnetic field into three component beams. The positive component consists of alpha (α) particles, which are doubly charged helium nuclei; the negative component consists of beta (β) particles, which are electrons; the neutral component consists of gamma (γ) rays, which are like X rays. Other radioactive sources can give rise to a fourth kind of radiation consisting of *positrons*. Positrons are just like electrons except they carry a positive charge instead of a negative one. They are designated as β^+.

The emission of radiation by most radioactive elements is accompanied by the gradual disappearance of the original element and the appearance of a new element. Apparently, atoms of one element spontaneously change into atoms of the other element. This process is called *transmutation*. For example, uranium atoms divide to form atoms of thorium and alpha particles. The inference is that atomic nuclei are complex and in some cases break up into simpler fragments.

Direct evidence that the nucleus is composed in part of protons (nuclei of hydrogen atoms) was obtained by Rutherford in 1919. In studying the range of alpha particles in nitrogen gas, he observed at the ends of some alpha tracks the formation of new tracks, which he showed were not due to alpha particles. Instead, he reasoned that such a track arose from the capture of an alpha particle by a nitrogen nucleus to eject from it a proton and leave an oxygen nucleus . Here then was powerful support for believing that protons are constituent components of nuclei. But a question remained: How many protons are there in a given nucleus?

The answer came from the experimental investigations of Henry Moseley (1913) which led to the discovery of the *atomic number*. Moseley was studying the energy of the X rays emitted from various metals when bombarded by electrons. With an X-ray tube such as diagramed in Figure 2.12 various interchangeable metal target anodes (+) can be bombarded by electrons from the curved cathode (−) and their energies measured. A remarkable regularity was observed when the square root of the energy was plotted against the atomic weight of the element—the resulting graph was nearly a straight line. However, there were some gaps in the line (for ele-

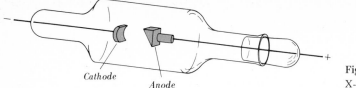

Cathode

Anode

Figure 2.12
X-ray tube.

ments not yet discovered at the time) and there were a few irregularities. To get rid of the apparent irregularities, Moseley suggested that atomic weight was not the fundamental property of the atom that decided the X-ray energy, but there was another parameter that was more fundamental.

By Moseley's time the idea had taken hold that the atom consisted of a positive center to which negative electrons were attracted by ordinary electrical attraction between opposite charges. To get a quantitative agreement between his X-ray observations and the energy that bound an electron to a nucleus, Moseley simply had to assume that the magnitude of the positive charge on the nucleus increased stepwise for all the elements. This positive charge is called the *atomic number of the element*. It is a simple next step to identify this number, usually designated by Z, as equal to the number of protons in the nucleus.

The atomic number, Z, is especially important in chemistry for the following reason. Because it gives the number of positively charged protons in the nucleus, the neutral atom must—to be neutral—have the same number of negatively charged electrons outside the nucleus. Thus Z, a nuclear property, gives us vital information about the rest of the atom.

Unsolved, however, was the problem of the mass of the nucleus. Furthermore, how is it possible to have isotopes? How is it possible to get different masses for different atoms of the same element if they all contain the same number of protons? There must be something else in the nucleus. The early theories resolved the problem, incorrectly as it turns out, by postulating that the nucleus consists of protons and electrons. Isotopes can then nicely be explained by saying that the number of protons and electrons can vary in the atoms of a given element and only the difference between them has to stay constant. As a specific example, there are two principal isotopes of neon both having atomic number Z = 10 but one having a mass of 20 amu and the other 22 amu. The old proton-electron model of the nucleus explained the mass-20 isotope as containing 20 protons plus 10 electrons and the mass-22 isotope as containing 22 protons plus 12 electrons. In both cases, the net nuclear charge would be 10+, and so the number of *extranuclear electrons* would be 10 and the atomic numbers would be identical. However, since the mass of the electron is negligible and only the proton masses need be added up, the atomic masses of the isotopes would differ. Unfortunately, the model does not stand closer scrutiny. It runs into an insuperable objection when the properties of small particles are examined. A particle as light as an electron simply cannot be confined in a volume as small as that of a nucleus. The kinetic energy of an electron increases as its confining volume decreases. In the case of a nucleus, the kinetic energy would become so enormous that the electron could not stay bound in the nucleus.

How, then, can isotopes be explained? The answer was not clear until 1932 when James Chadwick discovered another fundamental particle, the *neutron*. It has almost exactly the same mass as a proton but carries no electric charge. Because neutrons carry no charge, they are difficult to detect. They leave no track in a cloud chamber, for example. One indirect way to detect them is to aim the neutron beam at a piece of paraffin. The neutrons interact with the hydrogen in the paraffin so as to generate a beam of protons, which can be detected by the above-described conventional magnetic-deflection techniques.

At the present time the nucleus can be viewed as consisting of neutrons and protons. The neutron has a mass of 1.00866 amu and a charge of 0; the proton has a mass of 1.00727 amu and a charge of 1+. The charge of the nucleus is simply given by the number of protons in it. The mass of the nucleus is somewhat more complicated. It is *not* precisely equal to the total mass of the neutrons plus the total mass of the protons that make up the nucleus; rather, the observed mass is generally somewhat less than this. The reason, ramifications of which are further explored in Section 2.8, is that some of the total mass has been converted to binding energy for holding the nucleus together. The approximate mass of the nucleus is designated by what is called the *mass number, A,* which is equal to the number of protons plus the number of neutrons in the nucleus. Symbolically, a specific nucleus is designated by the symbol of the element with a subscript to designate the value of Z and a superscript to designate the value of A. Thus, for example, the two common isotopes of chlorine are denoted $^{35}_{17}\text{Cl}$ and $^{37}_{17}\text{Cl}$. The first isotope has 17 protons and 18 neutrons; the second, 17 protons and 20 neutrons. The number of neutrons is figured as $A - Z$, that is, the superscript minus the subscript.

2.7 Nuclear Stability

The difficult thing to understand about a nucleus is how the positive charges can be packed together into a region which is about 10^{-13} cm in radius without flying apart as a result of electric repulsion. Neutrons must be at least partly responsible for the binding, because, first, there is no nucleus consisting solely of several protons and, second, the more protons there are in a nucleus, the more neutrons are required *per proton* for stability.

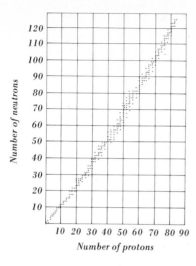

Figure 2.13
Stable nuclei.

The latter point is demonstrated by the *belt of stability* shown in Figure 2.13 which shows a plot of the stable (nonradioactive) nuclei. Each point corresponds to a known nucleus containing a given number of protons and a given number of neutrons. Whereas for the light elements nonradioactive nuclei contain approximately equal numbers of neutrons and protons, for the heavier elements nonradioactive nuclei contain considerably more neutrons than protons. Furthermore, the nuclei which do not fall within this belt of stability are radioactive; i.e., their neutron-to-proton ratios are either too high or too low for stability, and some kind of radioactive process must occur in order to bring the nucleus back to stability. The bigger question as to *how* neutrons act to bind the protons is still unanswered.

If a nucleus lies outside the belt of stability, it tends to reach a stable configuration by a radioactive process which changes the number of neutrons and/or protons to a more favorable value. There are two general cases which might be distinguished: *induced radioactivity* and *natural radioactivity*. Induced radioactivity results when stable nuclei are subjected to bombardment by other particles. If the energy of the incoming particles is of the proper magnitude, bombarded nuclei combine with incident particles to form new nuclei which, if unstable, undergo radioactive decay. An example of such a process occurs when $^{12}_{6}C$ nuclei are bombarded with protons which have been accelerated to high energies in a cyclotron. The bombardment adds a proton to the carbon nucleus to produce a nitrogen nucleus, $^{13}_{7}N$. This nucleus is unstable and lies below the belt of stability, with too few neutrons (six) for the number of protons (seven). A *positron*, or positive electron, is emitted. This has the effect of decreasing the number of protons to six, making the new nucleus a carbon nucleus, and increasing the number of neutrons to seven. $^{13}_{6}C$, the resulting nucleus, is stable.

The rate at which radioactive disintegration occurs gives a measure of the stability of a nucleus and is usually expressed in terms of the *half-life* of the nucleus, the time required for half of a given number of atoms to disintegrate. For the above decay of $^{13}_{7}N$, the half-life, usually designated as $t_{1/2}$, is 10.1 min. This means that, for any aggregation of $^{13}_{7}N$ nuclei, half of them will have disintegrated in 10.1 min; in another 10.1 min, half of the remainder will have disintegrated, etc. Since radioactive decay is a statistical process practically unaffected by changes in temperature or chemical environment, no one can predict which specific $^{13}_{7}N$ nucleus of a collection will disintegrate next; only the probability of decay can be stated, and this is done by specifying the half-life. The shorter the half-life, the more probable is the decay.

For certain other unstable nuclei, which like $^{13}_{7}N$ have too small a neutron-to-proton ratio and hence lie below the belt of stability, there is a different mechanism for increasing the ratio and reaching stability. It involves capture by the nucleus of an electron outside the nucleus. The electron is annihilated, and in the process its negative charge reduces the positive charge of the nucleus by 1 unit. An example of radioactive decay by electron capture is furnished by $^{37}_{18}Ar$, which decays with a 35-day half-life to $^{37}_{17}Cl$, which has a greater neutron-to-proton ratio and thus lies in the belt of stability.

Other nuclei are unstable for the opposite reason, too high a neutron-to-proton ratio (placing them above the belt of stability). For such nuclei, neutron emission may occur for certain cases such as $^{87}_{36}Kr$, but they are rare. More usual is beta (β) emission. It corrects a neutron-proton ratio that is too high by removing 1 unit of negative charge and thereby increasing the positive charge of the nucleus. Since a β particle (or electron) has very little mass, its emission does not change the weight of a nucleus appreciably. An example of β decay is furnished by radioactive carbon, $^{14}_{6}C$, which decays with emission of a β particle to $^{14}_{7}N$. The half-life of $^{14}_{6}C$ is 5580 years, a length of time that makes possible the use of this isotope in radioactive dating of objects from as far back as the last ice age.

Many nuclei are unstable because they lie *beyond* the belt of stability (too many protons). A nucleus having 84 or more protons is unstable and seems to contain too many charges crammed into one nucleus for stability, no matter how many neutrons are present. In such cases, a piece of the nucleus splits off. Most commonly, the piece split off is an alpha particle (two protons plus two neutrons), and in fact most heavy nuclei are alpha emitters. For example, $^{238}_{92}U$ emits an α to form $^{234}_{90}Th$ with a half-life of 4.5×10^9 years. This length of time is about equal to the age of the earth, and so nearly half the $^{238}_{92}U$ originally present at creation is still with us. For heavier elements, the nuclei are less stable and half-lives are shorter. They are therefore scarce or nonexistent in nature and can be formed artificially by bombardment of heavy nuclei with lighter ones.

Finally, to complete the discussion of nuclear stability and radioactive decay, mention must be made of γ rays (gamma rays). These are essentially bundles of energy, much like very high energy X rays, which frequently accompany beta emission and positron emission. They represent the principal way in which excited nuclei can get rid of excess energy.

2.8 Nuclear Energy

In the preceding section, we considered stability of nuclei with respect to radioactive decay. In this section, we consider stability of nuclei with respect to conversion of mass to energy. That such a conversion can occur is illustrated by the helium nucleus. A helium nucleus is thought of as containing two neutrons and two protons. Since the mass of a neutron is 1.00867 amu and since the mass of a proton is 1.00728 amu, we might expect that the mass of a helium nucleus would be

$$2 \times 1.00867 + 2 \times 1.00728 = 4.03190 \text{ amu}$$

However the experimentally observed mass of the helium nucleus is 4.0015 amu. What has happened to the missing mass of 0.0304 amu? It is believed that, in forming a helium nucleus from protons and neutrons, mass is converted to energy. From the Einstein relation ($E = mc^2$), the loss of 0.0304 amu of mass corresponds to the liberation of 2.72 trillion joules per mole of helium. Conversely, the same amount of energy is required to break up 1 mole of helium nuclei into protons and neutrons and thus it is a measure of the binding energy of the nucleus.

Similar calculations for other nuclei show that different nuclei have different binding energies. The binding energy per nuclear particle, i.e., the total binding energy of one nucleus divided by the number of protons plus neutrons in the nucleus, is plotted in Figure 2.14 for each of the different elements.† As can be seen, intermediate elements of mass number about 60 have the highest binding energies per particle and are the most stable. The other elements are unstable with respect to them. This means, for example, that, if a heavy element such as uranium is converted to iron, energy should be liberated. Similarly, if a light element such as hydrogen is converted to iron, energy should also be liberated. Such conversions are the basis for the utilization of nuclear energy.

†The different binding energies for different isotopes imply that in their formation different amounts of mass have been converted to energy. It therefore follows that individual isotopes cannot have whole-number atomic weights since they cannot be even multiples of any arbitrary standard.

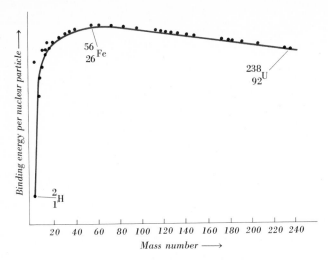

$\frac{56}{26}$Fe

$\frac{238}{92}$U

$\frac{2}{1}$H

Binding energy per nuclear particle $\longrightarrow$

20 40 60 80 100 120 140 160 180 200 220 240

Mass number $\longrightarrow$

Figure 2.14
Binding energies of the elements (energy per nuclear particle).

One method by which nuclear binding energy is made available is *nuclear fission,* the process in which a heavy nucleus breaks down to two approximately equal nuclei of intermediate mass. A typical fission process is shown in Figure 2.15, where a neutron striking a ^{235}U nucleus produces an unstable ^{236}U nucleus that splits into a ^{137}Te nucleus, a ^{97}Zr nucleus, and two neutrons. Both product nuclei have very high neutron-proton ratios and subsequently decay with the emission of beta particles. Actually, the fission shown is just one of the many ways in which a ^{236}U nucleus can split. Some of the ways of splitting result in nuclei which undergo decay by neutron emission. Since in any fission more neutrons are produced than are needed to initiate the fission, once it is started the fission process can become self-sustaining as a chain reaction.

For the very heaviest transuranium elements, fission is spontaneous, but for others, for example, $^{235}_{92}$U and $^{239}_{94}$Pu, fission can be initiated by exposure to neutrons. No matter how fission occurs, there is a change from a less stable nucleus (lower binding energy per nuclear particle) to more stable nuclei (higher binding energies per nuclear particle). In going from the less stable state to the more stable state, energy is liberated in large amounts. This release of energy by nuclear fission is the basis of nuclear reactors and atomic bombs. Nuclear reactors now account for a majority of the new

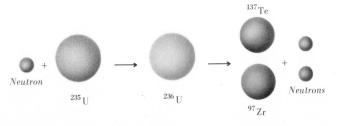

^{137}Te

Neutron

^{235}U

^{236}U

^{97}Zr

Neutrons

Figure 2.15
A typical fission.

power-generating capacities ordered by utilities in the United States. Reasons for this include the lower pollution problems associated with nuclear reactors as compared to coal-driven steam plants and the fact that $^{235}_{92}U$ liberates the heat equivalent of more than 2 million times its weight of coal.

Another process by which nuclear binding energy can be made available is *nuclear fusion,* the process in which two or more light nuclei combine to produce a single nucleus of heavier mass. Because of the steepness of the binding-energy curve (Figure 2.14) at the low end, greater energy can be liberated per nuclear particle by nuclear fusion than by nuclear fission. However, the process requires enormous temperatures (of the order of a million degrees) in order to occur. In the sun, temperatures are high enough so that nuclear fusion can take place to produce the energy that the sun pours out into space. It has been proposed that the chief source of this energy is the conversion of hydrogen to helium. In the overall reaction, which may occur in a number of steps, four ^1_1H nuclei produce a helium nucleus ^4_2He plus two positrons.

From the consumption of 1 kg of hydrogen by the nuclear reaction, the energy liberated is equivalent to the burning of about 20 million kg of coal. Since the mass of the sun is so large and since it is mostly hydrogen, it would take a conversion of only 1 percent of the sun's mass from hydrogen to helium to keep the sun shining at its present rate for another billion years.

Nuclear fusion, or *thermonuclear reaction,* as it is sometimes called, would seem impossible on the earth because of the high temperatures required. However, with new techniques, e.g., the release of energy by nuclear fission, temperatures have been produced on the earth high enough that thermonuclear reactions can occur. Hydrogen bombs illustrate the destructive violence that can be achieved when nuclear energy is rapidly released through nuclear fusion.

Despite the awful threat posed by hydrogen bombs, we seem to be unable or unwilling to create a political climate leading to rational control of nuclear weapons. It is unfortunate that the technological problems associated with the slow, controlled release of nuclear energy by nuclear fusion are so formidable that useful application of this energy has not been forthcoming to counterbalance its aspects of terror. Given our ingenuity for survival, it is possible that controlled nuclear fusion will open to us a source of energy that dwarfs any supply hitherto available.

QUESTIONS

2.1 *Dalton Theory* How does the Dalton atomic theory account for the fact that when 2.0 g of copper reacts with 2.0 g of sulfur, 1.5 g of sulfur is left unreacted? (Atomic weights: Cu = 63.5 amu, S = 32.1 amu.)

2.2 *Laws of Chemical Change* If elements did not form compounds, but only solutions, which of the three laws of chemical change would exist and which would be unknown? Explain.

2.3 *Dalton Theory* Using hypothetical reactions between elements X and Y, describe experiments illustrating the three laws of chemical change and show how Dalton's atomic theory would account for each.

2.4 *Atomic Weight* Distinguish clearly between (a) the atomic weight of sodium, (b) the weight of 1 mole of sodium, (c) the weight in grams of a sodium atom.

2.5 *Mole* For many years scientists made use of the concept of a mole without knowing the value of the Avogadro number. How is this possible?

2.6 *Atomic Weight* You wish to prepare a compound which contains equal numbers of lithium and aluminum atoms combined with four hydrogen atoms for each lithium atom. If you start with 10.0 g of Li, what is the minimum amount of Al and of H, you must use?

2.7* *Law of Dulong and Petit* 18.20 g of an element A combines with 5.21 g of oxygen to form an oxide. If the specific heat of element A is 0.452 joule/g $\cdot$ °C, what is its atomic weight?

2.8 *Half-life* 3.00 g of $^{132}_{55}Cs$ decays by β emission to produce stable xenon. If the half-life of $^{132}_{55}Cs$ is 55 days, how much $^{132}_{55}Cs$ will be left after 165 days?

2.9* *Moles* Potassium has atomic weight of 39.10 amu. (a) What is the weight in grams of 1 mole of potassium? (b) How many moles and how many atoms are there in 14.71 g of potassium?

2.10 *Symbols* (a) Indicate the number of protons and neutrons in each of the following nuclei: $^{13}_{6}C$, $^{16}_{8}O$, $^{64}_{30}Zn$, $^{238}_{92}U$. (b) Tell how many electrons are in the neutral atoms of each of the above.

2.11 *Isotopes* (a) How do isotopes of the same element differ? (b) What is the meaning of atomic weight for an element composed of two isotopes?

2.12 *Nucleus* If a particular atom whose nucleus contains equal numbers of neutrons and protons has a diameter 100,000 times the diameter of its nucleus, how many times as great is the density of the nucleus as the density of the atom?

2.13 *Radioactivity* Given four nuclei: $^{20}_{10}A$, $^{17}_{9}B$, $^{28}_{12}C$, $^{226}_{86}D$. One emits α, one β^-, and one β^+. Which is which?

2.14 *Radioactivity* Tell what is meant by each of the following: (a) gamma emission, (b) half-life, (c) electron capture.

2.15* *Mole* How many atoms are there in each of the following: (a) 100 moles of He, (b) 100 g of He, (c) 100 amu of He?

2.16 *Energy* What is the source of the energy produced by each of the following: (a) the sun, (b) burning coal, (c) fission of $^{235}_{92}U$?

2.17* *Avogadro Number* Determine the number of *atoms* in each of the following: (a) 5.00 g of KCl, (b) 2.10 moles of O_2, (c) 10.9 g of H_2SO_4, (d) 5.00 moles of SO_2.

2.18 *Moles* Compound C is 30% by weight oxygen and 70% by weight iron. How many moles of iron and oxygen would be needed to form 2 moles of compound C?

2.19* *Law of Dulong and Petit* 19.94 g of element B combines with oxygen to form 22.45 g of an oxide. If the specific heat of element B is 0.387 joule/g · °C, what is the name of element B?

2.20 *Moles* Potassium bromide (KBr) consists of 32.9% by weight potassium and 67.1% bromine. If 12.80 g of bromine is allowed to react with 7.21 g of potassium, how many moles of potassium combine with bromine to form KBr?

2.21* *Avogadro Number* An average weight of the ink used to print a page in this book is 0.721 mg. Assuming that 90 percent of the weight of the ink is carbon, calculate the average number of carbon atoms used to print a page.

2.22 *Half-life* 12.1 g of $^{37}_{18}$Ar decays by e^- capture to form stable chlorine. If the half-life of $^{37}_{18}$Ar is 35 days, how much chlorine will be present after 140 days?

2.23 *Atomic Particles* List the relative masses and charges of protons, neutrons, and electrons.

2.24 *Atomic Weight* What is the atomic weight of an element which, when 1.0 g combines with oxygen, produces 2.0 g of compound, and which has a heat capacity of 0.7 joule/g · °C?

2.25 *Multiple Proportions* The element barium forms two oxides. One contains 10.4% oxygen; the other contains 18.9% oxygen. Show that these compounds illustrate the law of multiple proportions.

2.26 *Isotopes* Atomic weight of an element is the result of the relative abundance of each isotope and the atomic weight of each isotope. Given that atomic weights of ^{79}Br and ^{81}Br are 78.92 and 80.92 respectively, determine the percent of natural abundance of each isotope. Atomic weight of bromine is 79.91.

2.27 *Properties of an Element* Suppose you discovered the new element pandemonium. Outline how you might proceed to determine each of the following experimentally: (a) its heat capacity, (b) its atomic number, (c) its atomic weight, (d) whether it consisted of more than one isotope, (e) whether radioactive and if so the type of radioactive particles emitted.

2.28 *Avogadro Number* Given the following facts and no others, calculate a value for both the mass of an H atom and the Avogadro number. (a) In a Millikan-type experiment you observe the following charges on different oil drops: -4.8×10^{-19}, -9.6×10^{-19}, -12.8×10^{-19} coulomb. (b) In a mass spectrometer you determine the ratio of charge to mass for a hydrogen ion to be 9.6×10^4 coulombs/g. (c) The atomic weight of H is 1.008 amu.

2.29* *Half-life* A lab desk is contaminated with radioactive $^{32}_{15}P$ which has a half-life of 14.30 days. How many half-lives must you wait until the radioactivity has dropped below 0.1 percent of its original value? How many days is this?

2.30 *Energy* From the Einstein relation, 1 g of mass is equivalent to 9×10^{13} joules of energy. When 12 g of carbon reacts with 32 g of oxygen, 392 kilojoules of energy is liberated. Calculate the corresponding decrease in mass to see whether it would make a change large enough for Dalton to have detected and hence caused him concern in using the law of conservation of mass.

2.31 *Science and Society* A county gets its electric power from a coal-fired steam plant owned by a private power company. The company wants to replace the plant with a nuclear reactor which will use a nearby lake as a supply of coolant water. Conversion will lower smoke pollution and ultimate power costs but will raise the temperature of the lake slightly and thereby change its ecology. Lakeside residents object. (*a*) Who should decide the issue, a vote of the entire county in which the lake residents are a minority, the state or federal power commission, the power company stockholders, the courts, or who? (*b*) What criteria should be used to decide?

2.32 *Atoms* Consider the period at the end of this sentence. Assume it consists only of carbon of density 2.2 g/cm³, is a circle of radius 0.10 mm, and has an average thickness of 0.010 mm. (*a*) How many C atoms does the period contain? (*b*) If all the volume is taken up by C atoms, how much does each appear to occupy? (*c*) What would be the radius of a sphere having the volume of your answer to (*b*)? (*d*) In your answer to (*b*), what fraction approximately is contributed by the nucleus if its diameter is 1×10^{-13} cm? (*e*) What fraction of the mass of the period is contributed by the electrons of the C atoms?

Electrons in Atoms

3

Dalton's atomic theory formed a basis for answering the chemical questions posed by Lavoisier and those who followed him. The mere concept of atoms is of great help in understanding observations concerning weight relations. However, deeper understanding of the reasons for the law of multiple proportions, for example, and for the differences in chemical behavior of different elements and their compounds requires a more detailed understanding of atomic structure. In particular, it is not primarily the atomic nucleus but the surrounding electrons which mainly determine the chemical behavior of materials. The reason for this belief is the convergence of information derived from two different sources: on the one hand, the systematic pattern of chemical properties of the elements; on the other, light emitted by heated atoms. In this chapter, we will explore the picture which has emerged from these two types of studies; it concerns the nature of the distribution of electrons within atoms.

A word of caution is in order. Scientists and philosophers have always suspected that there is a limit to the extent of our ultimate knowledge of the nature of matter. When drawing inferences from indirect evidence, as we must in describing the structure of atoms too small to be observed or sensed directly, we must reach a point at which we can no longer be certain. As recently as this present century it was not generally realized that the limit of certainty was so near at hand. However, it was reached when scientists tried to describe paths for electrons in atoms. A real triumph of modern science is that we have incorporated this limit of certainty into our fundamental concepts of nature and models for explaining observations. We have learned to accept models which are not quite as easy or as satisfying as we would have hoped. In certain cases we must even accept alternate models, each of which gives a partial explanation. Because of these necessary limitations on theory, it is vital to keep clear the distinction between observed fact and inferred explanation. We will start out with a discussion of observed facts, first those of a pattern of chemical properties of the elements, and then those of energy radiated by individual elements.

3.1 Periodic Law

Probably the single most useful generalization in chemistry is the observation that various *properties recur periodically when the chemical elements are ranged in sequence of increasing atomic number.* This generalization is called the *periodic law* and was discovered independently by Dmitri Mendeleev in Russia and Lothar Meyer in Germany in the late 1860s. At that time, the atomic-number concept was unknown and so elements were arranged according to increasing atomic weight, but this gives practically the same order. Furthermore, gaps had to be left for yet undiscovered elements and in a few cases the order based strictly on atomic weight had to be altered.

The best illustration of periodic behavior was unknown to Mendeleev because the key elements had not yet been discovered. We now know that, although most elements react with other elements, there are some, like helium and neon, which are relatively inert and have little tendency to react with any of the other elements. When all the elements are arranged in order of increasing atomic number, the "inert" elements are not bunched together but occur periodically throughout the sequence. This is shown in Figure 3.1. The relatively inert elements—helium, neon, argon, krypton, xenon, and radon, with atomic numbers 2, 10, 18, 36, 54, and 86, respectively—are all

Figure 3.1
Periodic occurrence of relative inertness in the elements.

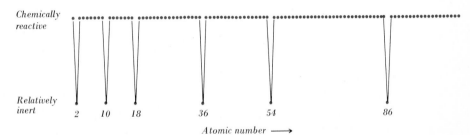

gases under usual conditions. For this reason, they are frequently called the *"inert" gases;* they are also and perhaps better known as the *noble gases.*

The elements that directly follow the noble gases—lithium, sodium, potassium, rubidium, cesium, and francium, with atomic numbers 3, 11, 19, 37, 55, and 87, respectively—are *metals;* i.e., they have a *shiny luster* and are *good conductors of heat and electricity.* As a group they are called the *alkali metals.* In their chemical properties, the alkali metals bear strong resemblance to each other. For example, they all react vigorously with water to liberate hydrogen and form basic solutions, which turn litmus paper blue and have a bitter taste and a slippery feel. Basic solutions can be neutralized by acids, which are substances that turn litmus red and have a sour taste. If the basic solutions made from the alkali metals are neutralized with hydrochloric acid and the water evaporated, a white salt is formed in each case. These salts, e.g., sodium chloride ($NaCl$) or potassium chloride (KCl), are quite similar to each other; for instance, all dissolve readily in water to give electrically conducting solutions. The salts can also be made by direct reaction between the alkali metals and chlorine gas. All the alkali metals form hydroxy compounds (for example, $NaOH$) which are basic.

The elements which directly precede the noble gases—fluorine, chlorine, bromine, iodine, and astatine, with atomic numbers 9, 17, 35, 53, and 85, respectively—also resemble each other. As a group, they are called the *halogens.* (The element hydrogen, which directly precedes helium, is not included in this group. As the first of all elements, hydrogen has unique properties, which do not resemble those of the halogens.) Unlike the alkali elements, the halogens are *nonmetals;* i.e., they are *poor conductors of heat and electricity.* Under usual conditions, fluorine and chlorine are gases; bromine is a liquid; iodine and astatine are solids. The halogens resemble each other in that all react with hydrogen to form compounds (e.g., hydrogen fluoride, HF, or hydrogen chloride, HCl) which dissolve in water to give acid solutions. Neutralization of these acid solutions with sodium hydroxide, followed by evaporation of the water, leads to the formation of white sodium salts. These salts, e.g., sodium fluoride (NaF) or sodium iodide (NaI), can also be prepared by the direct reaction of the halogens with sodium. With the exception of fluorine, halogens form hydroxy compounds, which are all acidic. An example is $HOCl$ (hypochlorous acid).

The elements that fall between an alkali metal and the next following halogen show a progressive gradation of properties between the two extremes. For example, the elements magnesium (atomic number 12), aluminum (13), silicon (14), phosphorus (15), and sulfur (16), which lie between sodium (11) and chlorine (17), represent such a gradation. In this sequence, there is a decrease in metallic character. Magnesium and aluminum are metals; phosphorus and sulfur are nonmetals; silicon is intermediate. Concurrently, there is a progressive change from basic to acidic character of the hydroxy compounds formed by these elements. The hydroxy compound

Figure 3.2
PERIODIC TABLE

Group	I	II											III	IV	V	VI	VII	0
Period 1	H 1																	He 2
2	Li 3	Be 4											B 5	C 6	N 7	O 8	F 9	Ne 10
3	Na 11	Mg 12				Transition elements							Al 13	Si 14	P 15	S 16	Cl 17	Ar 18
4	K 19	Ca 20	Sc 21	Ti 22	V 23	Cr 24	Mn 25	Fe 26	Co 27	Ni 28	Cu 29	Zn 30	Ga 31	Ge 32	As 33	Se 34	Br 35	Kr 36
5	Rb 37	Sr 38	Y 39	Zr 40	Nb 41	Mo 42	Te 43	Ru 44	Rh 45	Pd 46	Ag 47	Cd 48	In 49	Sn 50	Sb 51	Te 52	I 53	Xe 54
6	Cs 55	Ba 56	* 57–71	Hf 72	Ta 73	W 74	Re 75	Os 76	Ir 77	Pt 78	Au 79	Hg 80	Tl 81	Pb 82	Bi 83	Po 84	At 85	Rn 86
7	Fr 87	Ra 88	† 89–103	Ku 104	Ha 105													

*	La 57	Ce 58	Pr 59	Nd 60	Pm 61	Sm 62	Eu 63	Gd 64	Tb 65	Dy 66	Ho 67	Er 68	Tm 69	Yb 70	Lu 71
†	Ac 89	Th 90	Pa 91	U 92	Np 93	Pu 94	Am 95	Cm 96	Bk 97	Cf 98	Es 99	Fm 100	Md 101	No 102	Lr 103

of magnesium, $Mg(OH)_2$, or magnesium hydroxide, is basic; the hydroxy compounds of sulfur and phosphorus, for example, H_2SO_4 or $(HO)_2SO_2$, sulfuric acid, and H_3PO_4 or $(HO)_3PO$, phosphoric acid, are acidic; the hydroxy compounds of aluminum and silicon are intermediate.

In order to emphasize the periodic reappearance of properties, it is customary to lay out the elements, not in a long straight line as in Figure 3.1, but in what is known as a *periodic table*. There are many forms of the periodic table, one of which is shown in Figure 3.2. The number beneath the symbol of each element is the atomic number. The asterisk and the dagger represent the elements listed at the bottom.

The basic feature of the periodic table is the arrangement of the elements in order of increasing atomic number, with elements that are similar in properties placed under each other in a vertical column called a *group*. There are eight main groups, designated in Figure 3.2 as I, II, III, IV, V, VI, VII, and 0. Group I includes hydrogen plus the alkali metals; group VII, the halogens; and group 0, the noble gases. The elements intervening between groups II and III are called the *transition elements*. Each short vertical column of transition elements is called a *subgroup* and is named after the head element. Thus, Zn, Cd, and Hg make up the zinc subgroup.

A horizontal sequence of the periodic table is called a *period*. These are numbered from the top down. The first period contains two elements (H and He); the second and third, eight elements; the fourth and fifth, eighteen elements. The elements denoted by the asterisk are part of the sixth period; those by the dagger, the seventh period.

The periodic table is a useful device for organizing the chemistry of the elements. Furthermore, the fact that the elements can be arranged systematically in such a table indicates a periodic recurrence of detailed structure of individual atoms. We shall consider the periodic table in greater detail in Section 3.3.

3.2 Electronic Energy Levels

The picture of an atom as consisting of a positive nucleus with surrounding negative electrons presents a problem. Because of the opposite charges, electrons are attracted to the nucleus. They would be pulled into the nucleus if they were stationary, and so we must assume that the electrons are in some sort of motion which counteracts the pull of the nucleus. However, if they are in motion, they should radiate energy, since it is observed in all other cases that electric charges moving under the influence of attractive forces give off energy. Such a loss of energy would result in a slowing down of the electron, making it less able to withstand the attraction of the nucleus. Consequently, the electron would spiral down into the nucleus, and the atom should collapse. Since atoms do not collapse, there must be something wrong with the above argument.

A clue to solving the problem comes from the study of light emitted by substances when they are heated. It is a familiar fact that white light consists of different colors and is separated into its constituent colors when passed through a prism. Suppose that white light from a glowing solid, as, for example, the filament of a lamp, is passed through a prism, as diagramed in Figure 3.3. The screen shows a *continuous spectrum* of colors, a gradual blending from one color to the next. The colors correspond to light of different energies. On passing through the prism, the light of highest energy (violet) is bent most; the light of lowest energy (red) is bent least.

If the above experiment is repeated, with the light source a flame to which a vaporizable salt is added, the spectrum obtained is not continuous. As represented in Figure 3.4, the screen shows a *line spectrum*, narrow lines of colors. Since each of the lines corresponds to light of a definite energy,

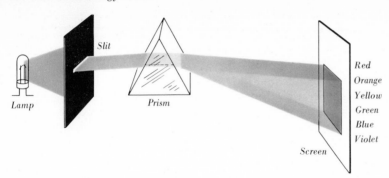

Figure 3.3
Continuous spectrum.

the existence of a line spectrum implies that atoms can radiate only certain energies. In other words, not just any energy is emitted but only definite, discrete values.

When compounds of the different elements are used as the light sources and the spectra are investigated as above, it is observed that each element contributes its own characteristic line spectrum. Furthermore, it is noted that there is a pattern of regularity in the lines of a given element. Also, it is found that the spectra of the elements are related.

Niels Bohr, a Danish physicist, in 1913 proposed a theory which not only accounted for the existence of line spectra but also suggested why atoms do not collapse. He made the revolutionary suggestion that the total energy (kinetic plus potential) of an electron in an atom is *quantized,* i.e., *restricted to having only certain values.* This amounts to assuming that in an atom an electron cannot have just any energy but only certain specific values. The only way an electron can change its energy is to shift from one discrete energy level to another. The transition cannot be gradual but must occur all at once. If no lower energy level is available, the electron cannot emit energy. For this reason, atoms do not collapse. If a lower energy level is available, the electron can radiate energy, but only a definite amount. This amount of energy has to be exactly equal to the difference between one

Figure 3.4
Line spectrum.

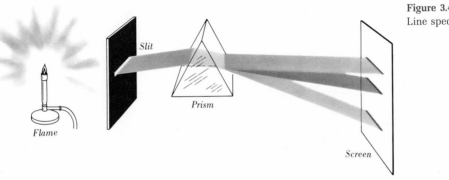

energy level and another. In the production of line spectra, electrons are presumably raised to higher energy levels by the heat energy of the flame. When the electrons drop back to lower energy levels, light of characteristic energy is emitted.

Bohr's assumption established the foundation for *quantum mechanics,* the study of the laws of motion that govern the behavior of small particles.[†] Particles of small mass, such as electrons, apparently do not follow Newton's laws of motion and the classis laws of electrodynamics which describe the interactions of moving charges. New principles are required. The basic principle is that only specified energy levels are possible for electrons in atoms. These energy levels are numbered, starting with the lowest as 1, the next higher as 2, the next higher as 3, etc. The number of the energy level, usually designated by n, is referred to as the *principal quantum number.* A second principle of quantum mechanics is that the electron population of any energy level in an atom is limited to $2n^2$. This means that, for the lowest energy level ($n = 1$), the maximum population is $2(1)^2$, or 2. For the second level ($n = 2$), the maximum population is $2(2)^2$, or 8; for the third level, $2(3)^2$, or 18; for the fourth level, $2(4)^2$, or 32; etc.

We can now draw what is known as an energy-level diagram. Figure 3.5 is such a diagram. The bottom line represents the lowest energy level—the level at which most energy must be added in order to remove the electron from the atom. Other lines represent levels of higher energy, requiring less energy to remove the electron. In principle, there are an infinite number of levels, but usually only the lowest seven or eight need to be considered. As shown in Figure 3.5, the energy difference between low energy levels is observed to be greater than that between high levels.

Electrons in the lowest energy level ($n = 1$) are referred to as being in the K shell, K orbit, or the innermost orbit. These electrons are the ones most tightly bound to the nucleus. The terms *shell* and *orbit* come from early models of the atom. Originally, Bohr suggested that the electrons in atoms move in curved orbits about the nucleus, tracing out a spherical shell. All electrons in a given shell were identified with one energy level. The shell closest to the nucleus corresponded to the lowest energy level; shells farther from the nucleus corresponded to higher energy levels. Although this picture of the atom is no longer acceptable, the terms shell and orbit are still sometimes used to refer to energy levels. Electrons in the second energy level ($n = 2$) are referred to as being in the L shell, or L orbit. The higher energy levels are numbered ($n = 3, 4, 5, . . .$) or lettered ($M, N, O, . . .$) consecutively from there on. Higher energy levels are referred to as outer energy levels.

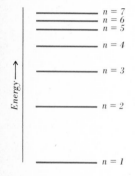

Figure 3.5
Energy levels of electrons in an atom.

[†] Although the laws of quantum mechanics are necessary for describing how small bodies move, they also apply to large bodies. In the latter case, the results from quantum mechanics are identical with those obtained more simply from Newton's laws of motion.

3.3 Energy Levels and the Periodic Table

The limitation of the number of electrons in a given energy level can be used to account for the periodic recurrence of properties in the elements, if it is assumed that the properties of atoms depend mainly on the number of electrons in the outermost energy level. Imagine the building up of an atom by addition of electrons to a nucleus of the proper atomic number. Each electron enters the lowest energy level available. In the case of hydrogen ($Z = 1$), the lone electron goes into the K shell. In helium, the nucleus contains two protons, and both electrons enter the K shell. For lithium, with $Z = 3$, the third electron has to go into the L shell since the maximum population in the K shell is two. Figure 3.6 lists the first 18 elements in order of increasing atomic number and shows the number of electrons in the various energy levels. Since the K shell can accommodate only two electrons, it becomes completely populated in the noble gas helium. Proceeding from helium, the L-shell population increases from one in lithium to eight in neon. In neon, the situation is like that of helium. With two electrons in the K shell and eight electrons in the L shell, the shells which are occupied are completely filled, and the shells which are empty are completely empty. Neon is a noble gas. In other words, after a period or a cycle of eight atoms, a repetition of the property of relative inertness appears. With the next eight elements, electrons add to the third, or M, shell, building it up gradually from one to eight electrons. The element argon, number 18, might not be expected to be "inert," because, according to the energy-level diagram, 10 more, or a total of 18, electrons can be put into the M shell. However, argon is observed to be relatively inert. It must be that eight electrons in the third shell behave like a full shell. This point will be considered in greater detail later in this section.

That the properties of atoms are closely tied to the number of electrons in the outermost energy level can be seen further from the following examples. In the case of lithium, there is one electron in the outermost energy

Figure 3.6
ELECTRONIC CONFIGURATIONS OF FIRST 18 ELEMENTS

| Atomic Number | 1 | 2 | 3 | 4 | 5 | 6 | 7 | 8 | 9 | 10 | 11 | 12 | 13 | 14 | 15 | 16 | 17 | 18 |
Element	H	He	Li	Be	B	C	N	O	F	Ne	Na	Mg	Al	Si	P	S	Cl	Ar
Electron population:																		
K level ($n = 1$)	1	2	2	2	2	2	2	2	2	2	2	2	2	2	2	2	2	2
L level ($n = 2$)			1	2	3	4	5	6	7	8	8	8	8	8	8	8	8	8
M level ($n = 3$)											1	2	3	4	5	6	7	8
		Noble gas								Noble gas								Noble gas

level (the L shell). Sodium also has one electron in its outermost energy level (the M shell). The properties of lithium and sodium are close to being identical, as already noted in Section 3.1. Likewise, beryllium ($Z = 4$) and magnesium ($Z = 12$) are similar. Each has two electrons in its outermost shell. In the periodic table, elements with similar properties are placed under each other. As it turns out, this results in the grouping together of atoms which have the same number of electrons in the outermost energy level.

In the periodic table (see Figure 3.2 or the back cover), the first period contains but two elements (H and He), a fact that is consistent with limiting the population of the K energy level to two electrons. The second period contains eight elements (lithium, beryllium, boron, carbon, nitrogen, oxygen, fluorine, and neon), a grouping consistent with the gradual filling of the L energy level to a maximum population of eight electrons. Since the L energy level is the outermost level occupied in these particular atoms, significant changes in properties are observed within the period.

The third period, covering sodium through argon, is more difficult to account for. On the basis of observed properties, it contains only 8 elements, whereas the energy-level picture suggests that there should be 18. (The reason for this apparent discrepancy is associated with the fact that, after eight electrons have been added to the third shell, the next two electrons go into the fourth shell, even though the third shell is not yet filled.) In the periodic table the element sodium, which has one electron in its outermost shell, is placed under lithium in group I; magnesium is placed under beryllium in group II; aluminum under boron in group III; silicon under carbon in group IV; phosphorus under nitrogen in group V; sulfur under oxygen in group VI; chlorine under fluorine in group VII; and argon under neon in group 0. Because they have the same number of electrons in the outermost shell, the elements of each pair just mentioned are chemically similar.

The fourth period, potassium through krypton, is even more complicated than the third period. As can be seen from the periodic table, there are 18 elements in the fourth period, ranging from atomic number 19 through atomic number 36. Of these 18 elements, the first two—K and Ca—and the last six—Ga, Ge, As, Se, Br, and Kr—correspond to addition of electrons to the outermost (fourth) shell. The 10 intervening elements—Sc, Ti, V, Cr, Mn, Fe, Co, Ni, Cu, and Zn—have no more than two electrons in the outermost shell. The buildup of the outermost shell is interrupted to allow for the belated filling of the next-to-outermost shell. Similar delayed filling of a next-to-outermost shell also occurs in the fifth and sixth periods.

The filling of shells in the fourth period occurs so that there are two fourth-period elements, potassium ($Z = 19$) and copper ($Z = 29$), both of which have one electron in the outermost, or fourth, energy level. Similarly, calcium ($Z = 20$) and zinc ($Z = 30$) have two electrons in the fourth energy level. Potassium and copper are somewhat similar in some properties, presumably because each has the same number of electrons in its outermost

OPTICAL SPECTRA

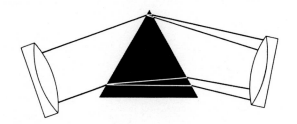

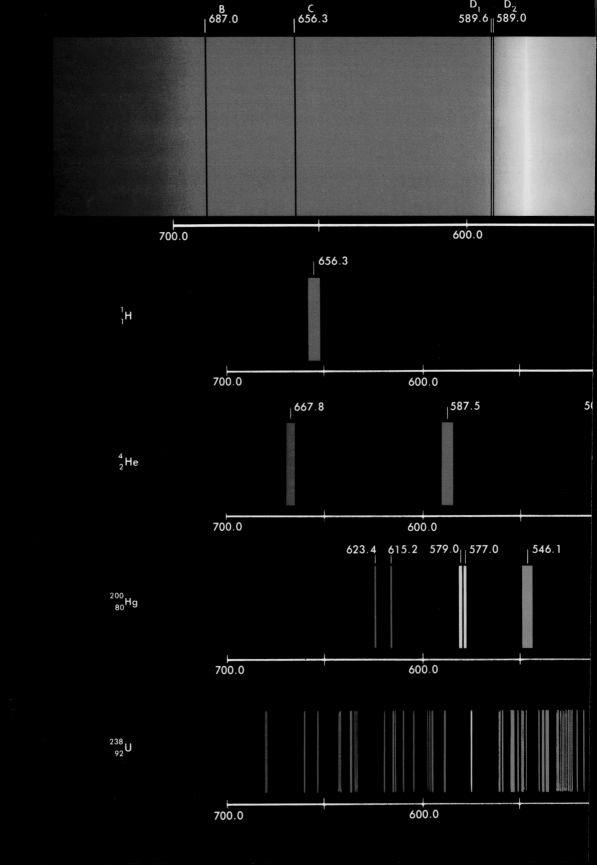

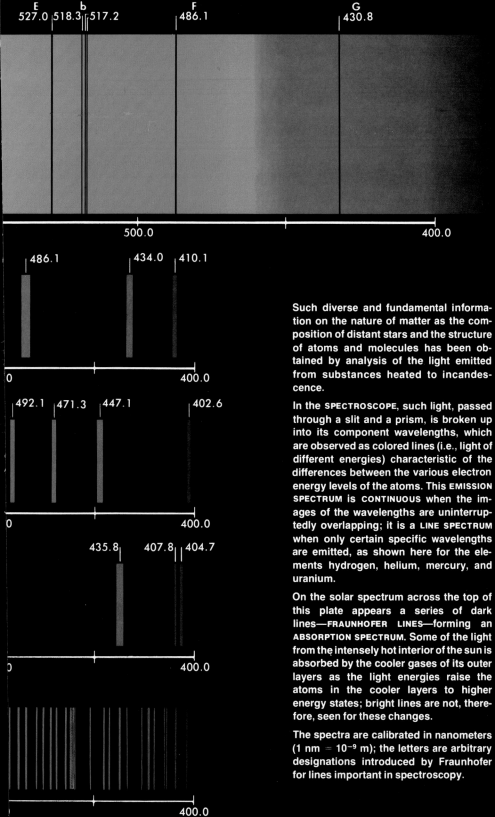

E 527.0 | b 518.3 | 517.2 | F 486.1 | G 430.8

500.0   400.0

486.1   434.0   410.1

0   400.0

492.1   471.3   447.1   402.6

0   400.0

435.8   407.8 | 404.7

0   400.0

400.0

Such diverse and fundamental information on the nature of matter as the composition of distant stars and the structure of atoms and molecules has been obtained by analysis of the light emitted from substances heated to incandescence.

In the SPECTROSCOPE, such light, passed through a slit and a prism, is broken up into its component wavelengths, which are observed as colored lines (i.e., light of different energies) characteristic of the differences between the various electron energy levels of the atoms. This EMISSION SPECTRUM is CONTINUOUS when the images of the wavelengths are uninterruptedly overlapping; it is a LINE SPECTRUM when only certain specific wavelengths are emitted, as shown here for the elements hydrogen, helium, mercury, and uranium.

On the solar spectrum across the top of this plate appears a series of dark lines—FRAUNHOFER LINES—forming an ABSORPTION SPECTRUM. Some of the light from the intensely hot interior of the sun is absorbed by the cooler gases of its outer layers as the light energies raise the atoms in the cooler layers to higher energy states; bright lines are not, therefore, seen for these changes.

The spectra are calibrated in nanometers (1 nm = 10^{-9} m); the letters are arbitrary designations introduced by Fraunhofer for lines important in spectroscopy.

Figure 3.7
ELECTRONIC
CONFIGURA-
TIONS OF
POTASSIUM
AND COPPER

Element	Z	K Level	L Level	M Level	N Level
Potassium	19	$2e^-$	$8e^-$	$8e^-$	$1e^-$
Copper	29	$2e^-$	$8e^-$	$18e^-$	$1e^-$

shell. However, K and Cu show major differences in other properties, attributable to the different numbers of electrons in the second-outermost shell (Figure 3.7). The second shell from the outside has an influence, sometimes quite large, on the chemical properties of an atom.

In succeeding periods, the electronic-configuration expansion proceeds in a similar but somewhat more complicated fashion. Before going on, we need to clear up one question raised previously. In the third period we found eight elements. From the energy-level diagram, we expected 18. Apparently the energy diagram as given is not satisfactory without some correction.

The previous discussion implied that all the electrons in a given shell are of the same energy. This is not quite true. Studies of the spectra of the different elements indicate that each energy level in Figure 3.5 actually consists of several energy levels closely bunched together. Technically, this is described by saying that each *main shell* consists of one or more *subshells*, or energy sublevels. The number of subshells in any main shell is equal to the principal quantum number n. Thus, the K shell ($n = 1$) consists of only one energy level. The L shell ($n = 2$) consists of two subshells. This means that not all the electrons in the L shell are of precisely the same energy. One set has an energy that is slightly higher than that of the other set. In the M shell ($n = 3$) there are three energy levels; in the N shell, four energy levels, etc. The subshells are designated by various devices. We shall find it most convenient to designate the lowest subshell of a given shell as an s subshell. The next higher subshell is labeled a p subshell; the next higher, a d subshell; the one above that, an f subshell.†

The energy-level diagram must now be redrawn. To the left of the broken lines in Figure 3.8 are shown the main shells. To the right of the broken lines are shown the component subshells. A distinctive feature is the overlapping of the higher-energy subshells, an overlapping which gets more complicated as the fifth and sixth main shells are added to the picture.

Just as the number of electrons that can be put in any main shell is limited, the population of a subshell is similarly limited. An s subshell can hold 2 electrons, a p subshell 6, a d subshell 10, and an f subshell 14. In

---

†The letters s, p, d, and f were originally chosen on the basis of observations of line spectra of elements such as sodium. Certain lines were observed to belong to a "sharp" series, and these were associated with energy transitions involving the s subshell; other spectrum lines were classified as belonging to a "principal," "diffuse," or "fundamental" series; hence the designations p, d, and f.

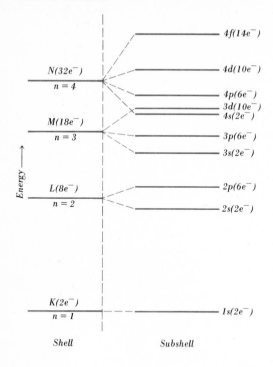

Shell　　　　*Subshell*

Figure 3.8

Energy-level diagram with component subshells.

Figure 3.8, the numbers in parentheses indicate the maximum population of the shells and subshells.

How does the existence of subshells affect the building of atoms from electrons and nuclei? So far as the first 18 elements are concerned, the number of electrons per main shell is as predicted before. As shown in Figure 3.9, element 18, argon, has two electrons in the 1s subshell, two in the 2s, six in the 2p, two in the 3s, and six in the 3p. Because the next subshell is so much higher in energy than the 3p (see Figure 3.8), argon behaves as an "inert" (noble-gas) atom. There seems to be special stability associated with eight electrons in any main shell everywhere in the periodic table.

Figure 3.9 ELECTRONIC CONFIGURA-TIONS AT START OF FOURTH SHELL	Element	Atomic Number	Electron Population									
			1s	2s	2p	3s	3p	3d	4s	4p	4d	4f
	Argon	18	2	2	6	2	6					
	Potassium	19	2	2	6	2	6		1			
	Calcium	20	2	2	6	2	6		2			
	Scandium	21	2	2	6	2	6	1	2			
	Titanium	22	2	2	6	2	6	2	2			

Z	Element	1	2		3			4				5				6				7
		s	s	p	s	p	d	s	p	d	f	s	p	d	f	s	p	d	f	s
55	Cs	2	2	6	2	6	10	2	6	10		2	6			1				
56	Ba	2	2	6	2	6	10	2	6	10		2	6			2				
57	La	2	2	6	2	6	10	2	6	10		2	6	1		2				
58	Ce	2	2	6	2	6	10	2	6	10	2	2	6			2?				
59	Pr	2	2	6	2	6	10	2	6	10	3	2	6			2?				
60	Nd	2	2	6	2	6	10	2	6	10	4	2	6			2				
61	Pm	2	2	6	2	6	10	2	6	10	5	2	6			2?				
62	Sm	2	2	6	2	6	10	2	6	10	6	2	6			2				
63	Eu	2	2	6	2	6	10	2	6	10	7	2	6			2				
64	Gd	2	2	6	2	6	10	2	6	10	7	2	6	1		2				
65	Tb	2	2	6	2	6	10	2	6	10	9	2	6			2?				
66	Dy	2	2	6	2	6	10	2	6	10	10	2	6			2?				
67	Ho	2	2	6	2	6	10	2	6	10	11	2	6			2?				
68	Er	2	2	6	2	6	10	2	6	10	12	2	6			2?				
69	Tm	2	2	6	2	6	10	2	6	10	13	2	6			2				
70	Yb	2	2	6	2	6	10	2	6	10	14	2	6			2				
71	Lu	2	2	6	2	6	10	2	6	10	14	2	6	1		2				
72	Hf	2	2	6	2	6	10	2	6	10	14	2	6	2		2				
73	Ta	2	2	6	2	6	10	2	6	10	14	2	6	3		2				
74	W	2	2	6	2	6	10	2	6	10	14	2	6	4		2				
75	Re	2	2	6	2	6	10	2	6	10	14	2	6	5		2				
76	Os	2	2	6	2	6	10	2	6	10	14	2	6	6		2				
77	Ir	2	2	6	2	6	10	2	6	10	14	2	6	7		2				
78	Pt	2	2	6	2	6	10	2	6	10	14	2	6	9		1				
79	Au	2	2	6	2	6	10	2	6	10	14	2	6	10		1				
80	Hg	2	2	6	2	6	10	2	6	10	14	2	6	10		2				
81	Tl	2	2	6	2	6	10	2	6	10	14	2	6	10		2	1			
82	Pb	2	2	6	2	6	10	2	6	10	14	2	6	10		2	2			
83	Bi	2	2	6	2	6	10	2	6	10	14	2	6	10		2	3			
84	Po	2	2	6	2	6	10	2	6	10	14	2	6	10		2	4?			
85	At	2	2	6	2	6	10	2	6	10	14	2	6	10		2	5?			
86	Rn	2	2	6	2	6	10	2	6	10	14	2	6	10		2	6			
87	Fr	2	2	6	2	6	10	2	6	10	14	2	6	10		2	6			1?
88	Ra	2	2	6	2	6	10	2	6	10	14	2	6	10		2	6			2
89	Ac	2	2	6	2	6	10	2	6	10	14	2	6	10		2	6	1		2?
90	Th	2	2	6	2	6	10	2	6	10	14	2	6	10		2	6	2		2
91	Pa	2	2	6	2	6	10	2	6	10	14	2	6	10	2	2	6	1		2?
92	U	2	2	6	2	6	10	2	6	10	14	2	6	10	3	2	6	1		2
93	Np	2	2	6	2	6	10	2	6	10	14	2	6	10	4	2	6	1		2?
94	Pu	2	2	6	2	6	10	2	6	10	14	2	6	10	6	2	6			2?
95	Am	2	2	6	2	6	10	2	6	10	14	2	6	10	7	2	6			2?
96	Cm	2	2	6	2	6	10	2	6	10	14	2	6	10	7	2	6	1		2?
97	Bk	2	2	6	2	6	10	2	6	10	14	2	6	10	8	2	6	1		2?
98	Cf	2	2	6	2	6	10	2	6	10	14	2	6	10	10	2	6			2?
99	Es	2	2	6	2	6	10	2	6	10	14	2	6	10	11	2	6			2?
100	Fm	2	2	6	2	6	10	2	6	10	14	2	6	10	12	2	6			2?
101	Md	2	2	6	2	6	10	2	6	10	14	2	6	10	13	2	6			2?
102	No	2	2	6	2	6	10	2	6	10	14	2	6	10	14	2	6			2?
103	Lr	2	2	6	2	6	10	2	6	10	14	2	6	10	14	2	6	1		2?
104	Ku	2	2	6	2	6	10	2	6	10	14	2	6	10	14	2	6	2		2?
105	Ha	2	2	6	2	6	10	2	6	10	14	2	6	10	14	2	6	3		2?

of which are important in understanding behavior of atoms and properties of matter.

Early attempts to describe energy levels in terms of paths of electron motion all ended in failure. At the time it seemed reasonable to believe that electrons in atoms could be described with the same relative certainty as could the planets in the solar system. Unfortunately, this is not possible. As indicated by the Heisenberg *uncertainty principle* (1927), it is impossible to know simultaneously the momentum and the position of an electron precisely enough to draw a picture of the path of an electron in a particular energy level. Any experiment, no matter how perfectly designed, to measure the location or the momentum of an electron must, in the very process of measuring, change either the momentum or the location. Since tracks cannot be drawn for electrons, the best we can do is to speak of the probability, or relative chance, of finding an electron at a given location within the atom.

The calculation of the probability of finding an electron at various points in an atom is an extremely involved mathematical problem. Quantum mechanics solves this problem by describing the electron as if it were a wave. Indeed, electrons do show wavelike properties under certain conditions; a beam of rapidly moving electrons, for example, can be made to exhibit diffraction, a property characteristic of wave motion. Consequently, quantum mechanics is frequently referred to as *wave mechanics*. What wave mechanics does is to take the mathematical equations which describe the motion of waves and use them to describe the probability of finding small particles such as electrons.

The probability of finding an electron, as calculated from wave mechanics, can be specified by a *probability distribution,* such as that given in Figure 3.11a. Here the probability of finding a 1s electron at a given location in space is plotted as a function of the distance of that location from the nucleus. The position of greatest probability is at the nucleus. Nowhere is the probability equal to zero. Even at points at very great distances from the nucleus there is some chance, although it is small, of finding the electron. Figure 3.11b is another way of representing the same electronic distribution. Here the intensity of the shading shows the relative probability of finding the 1s electron. Consistent with this picture, one can visualize an electron as forming a rather fuzzy charge cloud about a central nucleus. Sometimes it is convenient simply to indicate the shape of the charge cloud, as is done in Figure 3.11c. Remembering that atoms are three-dimensional, we may think of Figure 3.11c as a sphere within which the probability of finding the 1s electron approaches 90 percent. Thus we have in Figure 3.11 three different ways of representing the spatial distribution of an electron in a 1s energy level. Since these representations replace the Bohr idea of a simple orbit, they can properly be said to represent 1s orbits. In order to reduce any possible confusion between the old and new ideas, it has become customary to use the term *orbital* when referring to an energy level associated with a given electronic probability distribution.

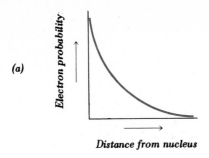

(a)

Electron probability

Distance from nucleus

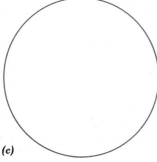

(b)

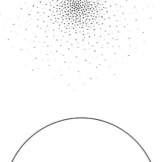

(c)

Figure 3.11
Representations of spatial distributions of 1s electron.

There is still another way of describing an electron in a 1s orbital, a way that serves to relate the idea of electronic "shells" to probability concepts. First we raise the question: If we imagine starting out from the nucleus and working our way along a straight line from the nucleus to the outside of the atom, how does the chance of finding the 1s electron change? Evidently, the chance decreases, consistent with Figure 3.11*a*. But now, suppose, as we work our way out of the atom, at each radial distance *r* from the nucleus, we investigate all the possible locations in three-dimensional space at that distance *r* from the nucleus and determine the chance of finding the 1s electron. Then, we move farther from the nucleus and investigate all the locations at a slightly bigger *r*. How does the chance of finding the 1s electron

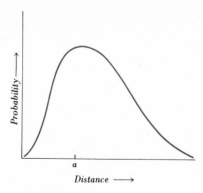

Figure 3.12

Probability plot for 1s electron.

change? The answer is not immediately obvious since we have to consider two factors. The chance of finding the electron at a given location decreases as we move away from the nucleus; but the number of locations to be investigated increases as we move away from the nucleus. Mathematically, this is equivalent to considering the atom to be divided into concentric layers and multiplying the probability per unit volume in a given layer by the volume of that layer. The result for a 1s electron is the probability curve shown in Figure 3.12. On this plot, the greatest probability of finding the 1s electron occurs at the distance a, which can be thought of as corresponding to the radius of an "electron shell."

Electrons that are in different energy levels differ from each other in having different probability distributions. For example, Figure 3.13 shows the probability that a 1s, a 2s, and a 2p electron are at various distances from a given nucleus. It should be noted that the distances of maximum probability for the 2s and 2p electrons are approximately the same and both are considerably larger than that for the 1s electron. This is consistent with the fact that the 2s and 2p electrons are of about the same energy but that this energy is considerably greater than that of the 1s electron. The peculiar little bump in the 2s distribution indicates that the 2s electron spends somewhat more of its time close to the nucleus than does the 2p electron. This can account for the fact that the 2s electron is bound a bit more tightly to

Figure 3.13

Probability plots for various electrons.

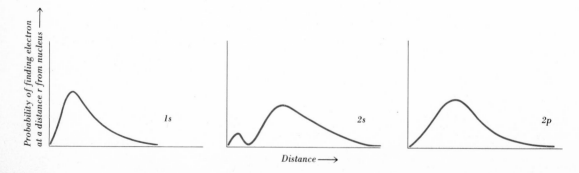

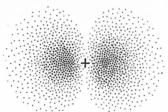

Figure 3.14
A 2p charge cloud (+ represents nucleus).

the nucleus (is of lower energy) than the 2p electron. Furthermore, it should be noted that all three of the distributions shown in Figure 3.13 overlap, implying that outer electrons penetrate the region occupied by inner electrons.

Actually, there is an essential difference between s and p electrons which is not evident from Figure 3.13. The spatial distribution of any s electron is spherically symmetrical; that is, its probability of being found is identical in all directions from the nucleus. On the other hand, p electrons are more probably found in some directions from the nucleus than in others. In fact, the probability distribution of a 2p electron can be thought of as forming two diffuse, somewhat flattened spheres, one on each side of the nucleus, as shown in cross section in Figure 3.14. This is called a 2p orbital, and the electron in a 2p orbital has equal probability of being found in either half of it. A 2p subshell is constructed of three such orbitals all perpendicular to each other, as shown in Figure 3.15. The three are distinguished from each other by the designations p_x, p_y, p_z according to the axis along which the probability is a maximum.

The d subshells consist of five orbitals and the f subshells of seven orbitals. Their spatial distribution is considerably more complicated. That of the 3d orbitals is shown in Figure 3.16.

Because the picture we get from the modern theory of electrons in atoms is less concrete than we might hope, let us summarize the essentials as a

Figure 3.15
Shapes of $2p_x$, $2p_y$, $2p_z$ orbitals.

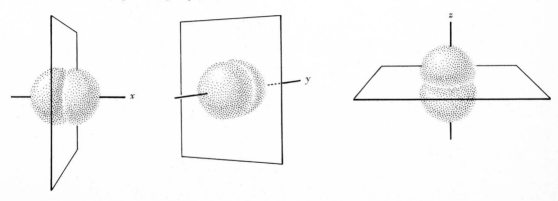

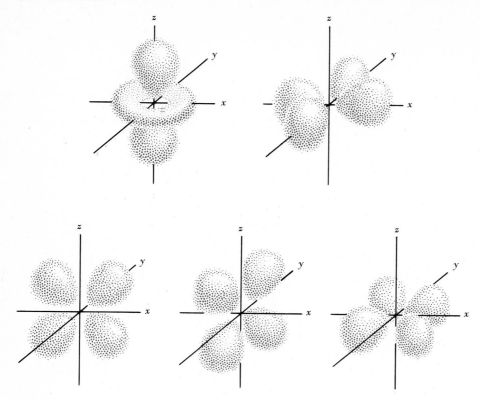

Figure 3.16
Shapes of 3d orbitals.

basis for our future use.

1 We cannot know the path taken by individual electrons in atoms.

2 We can picture a region in space (orbital) where an electron is most probably located.

3 The spatial regions (orbitals) for electrons differ for the different energy levels.

4 In a particular atom, the higher the principal quantum number, the farther from the nucleus will the bulk of its spatial region extend.

5 Electronic distributions (orbitals) for s subshells are spherically symmetrical about the nucleus.

6 Electronic distributions (orbitals) for p, d, and f subshells have preferred directions in space and so are not spherically symmetrical about the nucleus.

3.5 Electron Spin

In the preceding section it was noted that there is one orbital in an s subshell, three orbitals in a p subshell, five in a d, and seven in an f. Since these subshells can accommodate 2, 6, 10, and 14 electrons, respectively, it follows that any orbital can hold two electrons. The two electrons in the same orbital differ in one important respect, however: they have opposite "spin." The reason for talking about electron spin comes from observations on the magnetic behavior of substances.

It is a familiar observation that certain solids such as iron are strongly attracted to magnets. Such materials are called *ferromagnetic*. Other substances such as oxygen gas and copper sulfate are weakly attracted to magnets. They are called *paramagnetic*. Still other substances such as sodium chloride are very feebly repelled by magnets and are called *diamagnetic*. Ferromagnetism is exclusively a property of the solid state, but all three types of magnetic behavior just described are believed to arise from electrons in atoms.

Information about the magnetic behavior of individual atoms can be obtained from an experiment like the one first performed by Otto Stern and Walther Gerlach in 1921. In this experiment, shown in Figure 3.17, a beam of neutral silver atoms (from the vaporization of silver) was passed between the poles of a specially designed magnet. The beam was found to be split into two separate beams; i.e., half the atoms were deflected in one direction, and the rest in the opposite direction.

To account for this observation, it is assumed that each electron behaves like a tiny magnet. This magnetism of the electron can be thought of as coming from the spinning of the negative charge, since it is known that any spinning charge is magnetic. Two directions of spin are possible; an electron might spin about its axis in either a clockwise or a counterclockwise manner. These two directions of spin would correspond to two magnets oriented in opposite directions. If we have two electrons of opposite spin, we might expect them to attract each other, as two magnets would; but the electric

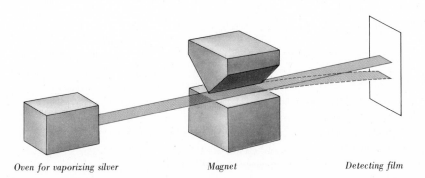

Figure 3.17
Stern-Gerlach experiment showing magnetic splitting of beam of silver atoms.

Oven for vaporizing silver   *Magnet*   *Detecting film*

repulsion due to like negative charges is very much greater than the magnetic attraction. Two electrons, forced to be in the same orbital, must have opposite spin. Thus, the electron pair in an orbital is nonmagnetic, because the magnetism due to one spin is canceled by the magnetism due to the opposite spin.

In silver atoms, as shown by the electronic configuration given in Figure 3.10, all the electrons are found in completed subshells except the one 5s electron. This electron obviously cannot be paired with another. Hence, its uncanceled spin gives magnetism to the silver atom. The two deflections observed in the Stern-Gerlach experiment presumably result from a separation of silver atoms into two types, which differ in the direction of spin of the unpaired electron, one having clockwise spin and the other, counterclockwise.

The discovery of electron spin completes the set of characteristics needed to describe an electron in an atom. There are four such characteristics. The first is the principal quantum number n which tells the order of grouping of electrons by increasing distance from the nucleus. The second characteristic describes the kind of orbital an electron occupies in terms of the symmetry of its spatial distribution (for example, s electrons have spherically symmetric distributions about the nucleus; p electrons have distributions along preferred directions in space). The third characteristic specifies which orbital of a given subshell the electron is in (e.g., an electron in a $2p_z$ orbital is most probably found along the z direction of space). The fourth and final characteristic specifies which of the two possible directions of spin the electron has.

Once the four characteristics are specified for an electron in a given atom, the electron is uniquely described and there can be no other electron in that same atom with an identical set of four characteristics. This fundamental limitation has come to be called the *Pauli exclusion principle*, after Wolfgang Pauli who first recognized its validity in 1925. It is a consequence of the Pauli principle that the population of atomic energy levels is limited and this in turn is the basis of the periodic law.

3.6 Electronic Symbols

In preceding sections we have described some of the complexities of the electronic structure of atoms. Before proceeding to a discussion of the relationship between properties of atoms and their electronic configurations, it will be convenient to set up a simplified way of expressing electronic configurations. As already pointed out, it is the number of electrons in the outermost shell which usually determines many properties of atoms. The number of outermost electrons in an atom is shown by its electronic symbol. Examples of electronic symbols are

Na·   Ca: :S̈·   H·   :F̈·   ·N̈·

As indicated, *electronic symbols* consist of the symbol of the element surrounded by dots. The letters represent the entire *core*, or *kernel*, of the atom. The core includes not only the *nucleus* but also the *electrons in inner shells*. The surrounding dots represent the outermost electrons. These are sometimes called the *valence electrons*, since they are thought to be the electrons that are mainly responsible for the *valence*, or *combining capacity*, of atoms.

The sulfur atom, for example, has a nucleus with a positive charge of 16. Its electron configuration is $1s^22s^22p^63s^23p^4$. In the electronic symbol only the six outermost electrons ($3s^23p^4$) are represented by dots. The rest of the atom makes up the core and is represented by the symbol S. Since the core contains 16 positive charges (protons in the nucleus) and 10 negative charges ($1s^22s^22p^6$), the net charge of the core is 6+. This 6+ charge is balanced in the neutral atom by the six negative charges of the outermost electrons. In writing electronic symbols, it does not matter where the dots are placed; *they have no significance so far as the actual positions of the electrons are concerned*. The dots are simply a convenient way of counting up the outermost electrons.

3.7   Atomic Sizes

The size of an atom is a difficult property to determine. For one thing, the electronic probability never becomes exactly zero, even at great distances from the nucleus. Therefore, the distance designated as the boundary of the atom is an arbitrary choice. For one thing, the electronic probability distribution is affected by neighboring atoms; hence the size of the atom changes somewhat in going from one condition to another, as, for example, in going from one compound to another. Therefore, in examining any table of atomic radii we must remember that the values listed may be meaningful only as a relative comparison of sizes. Figure 3.18 gives such a set of atomic radii (in nanometers, 10^{-9} m) deduced from the measured distances between centers of adjacent atoms in the pure elements. These interatomic spacings can be determined from X-ray and spectral studies of bound atoms.

In general, the atomic radii decrease in going from left to right across the periodic table and increase in going from top to bottom. How can we explain these trends?

Figure 3.19 shows the change of atomic radius going across the second period. It also shows the change in nuclear charge and electronic configuration. Within the second period, nuclear charge increases from 3+ to 9+. What effect might this have on the K electrons? In each of the elements, there are two K electrons, which are attracted to the nucleus by a force

Figure 3.18

ATOMIC RADII OF THE ELEMENTS (Number above symbol is atomic number; number below is radius in nm; 10^{-9} m)

1 H 0.037																		2 He —
3 Li 0.123	4 Be 0.089											5 B 0.080	6 C 0.077	7 N 0.074	8 O 0.074	9 F 0.072	10 Ne —	
11 Na 0.157	12 Mg 0.136											13 Al 0.125	14 Si 0.117	15 P 0.110	16 S 0.104	17 Cl 0.099	18 Ar —	
19 K 0.203	20 Ca 0.174	21 Sc 0.144	22 Ti 0.132	23 V 0.122	24 Cr 0.117	25 Mn 0.117	26 Fe 0.117	27 Co 0.117	28 Ni 0.116	29 Cu 0.115	30 Zn 0.117	31 Ga 0.125	32 Ge 0.122	33 As 0.121	34 Se 0.117	35 Br 0.114	36 Kr —	
37 Rb 0.216	38 Sr 0.191	39 Y 0.162	40 Zr 0.145	41 Nb 0.134	42 Mo 0.129	43 Tc 0.127	44 Ru 0.124	45 Rh 0.125	46 Pd 0.128	47 Ag 0.134	48 Cd 0.141	49 In 0.150	50 Sn 0.141	51 Sb 0.141	52 Te 0.137	53 I 0.133	54 Xe —	
55 Cs 0.235	56 Ba 0.198	*	72 Hf 0.144	73 Ta 0.134	74 W 0.130	75 Re 0.128	76 Os 0.126	77 Ir 0.126	78 Pt 0.129	79 Au 0.134	80 Hg 0.144	81 Tl 0.155	82 Pb 0.154	83 Bi 0.152	84 Po 0.153	85 At —	86 Rn —	
87 Fr —	88 Ra —	†	104 Ku —	105 Ha —														

*	57 La 0.169	58 Ce 0.165	59 Pr 0.165	60 Nd 0.164	61 Pm 0.163	62 Sm 0.166	63 Eu 0.185	64 Gd 0.161	65 Tb 0.159	66 Dy 0.159	67 Ho 0.158	68 Er 0.157	69 Tm 0.156	70 Yb 0.170	71 Lu 0.156
†	89 Ac —	90 Th 0.165	91 Pa —	92 U 0.142	93 Np —	94 Pu —	95 Am —	96 Cm —	97 Bk —	98 Cf —	99 Es —	100 Fm —	101 Md —	102 No —	103 Lr —

Figure 3.19 CHANGE OF ATOMIC RADIUS WITHIN A PERIOD	Li	Be	B	C	N	O	F
Atomic radius, nm	0.123	0.089	0.080	0.077	0.074	0.074	0.072
Nuclear charge	3+	4+	5+	6+	7+	8+	9+
K-level population	$2e^-$	$2e^-$	$2e^-$	$2e^-$	$2e^-$	$2e^-$	$2e^-$
L-level population	$1e^-$	$2e^-$	$3e^-$	$4e^-$	$5e^-$	$6e^-$	$7e^-$

proportional to the nuclear charge. As nuclear charge increases, the pull on the electrons is increased, and the maximum in the K probability-distribution curve gets closer to the nucleus.

What about the L electrons? Here the problem is complicated by the fact that the L electrons are screened from the nucleus by the K electrons, so that the attractive force of the nuclear positive charge is reduced by the intervening negative charges. In lithium, for example, the outermost electron is attracted not by a charge of 3+ but by a charge of 3+ screened by two intervening negative electrons. The net attractive charge is closer to a 1+ charge than to a 3+ charge. In the beryllium atom, the L electrons are attracted by a 4+ nucleus screened by two negative charges, or effectively a 2+ charge. Despite screening, from left to right across the period, the L electrons have a higher and higher positive charge attracting them to the center of the atom. Hence, just as the K shell becomes smaller because of this effect, the L shell gets smaller also.

How does the size of atoms change going down a group? Figure 3.20 gives the data for the alkali elements. There is an increase of size from top

Figure 3.20 CHANGE OF ATOMIC RADIUS WITHIN A GROUP	Element	Atomic Radius, nm	Nuclear Charge	Electronic Configuration				
	Li	0.123	3+	$2e^-$	$1e^-$			
	Na	0.157	11+	$2e^-$	$8e^-$	$1e^-$		
	K	0.203	19+	$2e$	$8e^-$	$8e^-$	$1e^-$	
	Rb	0.216	37+	$2e^-$	$8e^-$	$18e^-$	$8e^-$	$1e^-$
	Cs	0.235	55+	$2e^-$	$8e^-$	$18e^-$	$18e^-$	$8e^-$	$1e^-$

to bottom. In going down the sequence, the number of levels populated increases stepwise. The more levels used, the bigger is the atom. Because the nuclear charge progressively increases down the sequence, the individual shells get smaller, but adding a shell is such a big effect that it dominates. Similar behavior is found for many of the other groups of the periodic table. There are, however, some places in the periodic table where the size does not change much within the same group. This is particularly true when elements number 57 through 71 intervene between the two atoms compared; for example, 41 and 73 have the same radius.

3.8 Ionization Energy

The inference that the outer electrons of atoms are involved in chemical reactions focuses attention on the process whereby such electrons are completely removed from atoms. When an electron is pulled off a neutral atom, the particle which remains is a positively charged particle, or a *positive ion.* The process, called *ionization,* can be described by writing

$$\text{Na} \cdot \longrightarrow \text{Na}^+ + e^-$$

The electronic symbol on the left indicates the neutral sodium atom. In the process there is formed the positive sodium ion, shown on the right with a superscript $+$ to indicate a $1+$ charge. The electron is shown separately as e^-. The *ionization energy,* or *ionization potential,* is the work that is required to separate the negatively charged electron from the positively charged sodium ion that is attracting it. In other words, the ionization potential is the *energy required to pull an electron off an isolated atom.* Usually, the ionization potential is expressed in units of *electron volts,* one electron volt being a small amount of energy corresponding to 1.6×10^{-19} joule. One electron volt per atom is equivalent to 96.5 kilojoules per Avogadro number of atoms.

Figure 3.21 lists the values of the ionization potential for each atom of the second period. With some exceptions, there is a fairly steady increase from left to right.† Why is it harder to pull an electron off neon than off lithium? At least two factors must be considered. First, the nuclear charge increases from left to right across the period. By itself this predicts that the ionization potential increases from lithium to neon. Second, the size of the

Figure 3.21 IONIZATION POTENTIALS FOR SECOND-PERIOD ELEMENTS	Element	Li	Be	B	C	N	O	F	Ne
	Electron volts	5.4	9.3	8.3	11.3	14.5	13.6	17.4	21.6
	Nuclear charge	3+	4+	5+	6+	7+	8+	9+	10+
	K-level population	2e⁻	2e⁻	2e⁻	2e⁻	2e⁻	2e⁻	2e⁻	2e⁻
	L-level population	1e⁻	2e⁻	3e⁻	4e⁻	5e⁻	6e⁻	7e⁻	8e⁻

atoms decreases from left to right. The size effect by itself would also predict that the ionization potential should increase, since the closer an electron is to the nucleus, the harder it is to pull it off. The last element of the period, which has a shell of eight electrons (the so-called "octet"), has the highest ionization potential of any element in the period.

---

†Exceptions generally occur at elements following completion or half-completion of a subshell. With the completion of a subshell the next element must use a next higher orbital, in which the electron is necessarily less tightly bound. After half-completion of a subshell, the next element must accommodate two electrons in the same orbital, and their repulsion makes it easier to remove one.

How about the trend within a group? Figure 3.22 shows the values of the ionization potential for the alkali elements. There is a progressive decrease of the ionization potential from top to bottom. This is as predicted by the size change alone. The lithium atom is quite small; the electron which is being pulled off is, on an average, close to the nucleus. It is more firmly bound than in the case of cesium, where the electron is, on an average, much farther from the nucleus. The increase in nuclear charge, which we might

Figure 3.22 IONIZATION POTENTIALS FOR THE ALKALI ELEMENTS	Element	Ionization Potential, eV	Electronic Configuration				
	Li	5.4	$2e^-$	$1e^-$			
	Na	5.1	$2e^-$	$8e^-$	$1e^-$		
	K	4.3	$2e^-$	$8e^-$	$8e^-$	$1e^-$	
	Rb	4.2	$2e^-$	$8e^-$	$18e^-$	$8e^-$	$1e^-$
	Cs	3.9	$2e^-$	$8e^-$	$18e^-$	$18e^-$	$8e^-$	$1e^-$

expect to increase the ionization potential, essentially cancels out because of the screening effect of the intervening electrons.

Values of the ionization potentials of the elements are given in Figure 3.23. For each element, the value given refers to the first ionization, i.e., the removal of a single electron from the neutral atom.

3.9   Electron Affinity

Also important for determining chemical properties is the tendency of an atom to pick up additional electrons. This property can be measured by the *electron affinity, the energy released when an electron is added to an isolated neutral atom.* When a neutral atom picks up an electron from some source, it forms a negative ion, as indicated by writing

$$X + e^- \longrightarrow X^-$$

The amount of energy released in this process is the electron affinity. Thus, the electron affinity measures the tightness of binding of an additional electron to an atom. The values for the halogen elements are given in Figure 3.24.

Group VII elements are expected to have high electron affinity, because they are at the right of their periods where the size is small and the effective

Figure 3.23
IONIZATION POTENTIALS OF THE ELEMENTS (In eV)

1	2	3	4	5	6	7	8	9	10	11	12	13	14	15	16	17	18
1 H 13.6																	2 He 24.6
3 Li 5.4	4 Be 9.3											5 B 8.3	6 C 11.3	7 N 14.5	8 O 13.6	9 F 17.4	10 Ne 21.6
11 Na 5.1	12 Mg 7.6											13 Al 6.0	14 Si 8.1	15 P 11.0	16 S 10.4	17 Cl 13.0	18 Ar 15.8
19 K 4.3	20 Ca 6.1	21 Sc 6.6	22 Ti 6.8	23 V 6.7	24 Cr 6.8	25 Mn 7.4	26 Fe 7.9	27 Co 7.9	28 Ni 7.6	29 Cu 7.7	30 Zn 9.4	31 Ga 6.0	32 Ge 8.1	33 As 10	34 Se 9.8	35 Br 11.8	36 Kr 14.0
37 Rb 4.2	38 Sr 5.7	39 Y 6.6	40 Zr 7.0	41 Nb 6.8	42 Mo 7.2	43 Tc 7.3	44 Ru 7.5	45 Rh 7.7	46 Pd 8.3	47 Ag 7.6	48 Cd 9.0	49 In 5.8	50 Sn 7.3	51 Sb 8.6	52 Te 9.0	53 I 10.4	54 Xe 12.1
55 Cs 3.9	56 Ba 5.2	*	72 Hf 5.5	73 Ta 6	74 W 8.0	75 Re 7.9	76 Os 8.7	77 Ir 9.2	78 Pt 9.0	79 Au 9.2	80 Hg 10.4	81 Tl 6.1	82 Pb 7.4	83 Bi 8	84 Po 8.4	85 At —	86 Rn 10.7
87 Fr —	88 Ra 5.3	†	104 Ku —	105 Ha —													

*	57 La 5.6	58 Ge 6.9	59 Pr 5.8	60 Nd 6.3	61 Pm —	62 Sm 5.6	63 Eu 5.7	64 Gd 6.2	65 Th 6.7	66 Dy 6.8	67 Ho —	68 Er —	69 Tm —	70 Yb 6.2	71 Lu 5.0
†	89 Ac 6.9	90 Th —	91 Pa —	92 U 4	93 Np —	94 Pu —	95 Am —	96 Cm —	97 Bk —	98 Cf —	99 Es —	100 Fm —	101 Md —	102 No —	103 Lr —

Figure 3.24 ELECTRON AFFINITIES FOR GROUP VII

Element	Electron Affinity, eV	Electronic Configuration				
F	3.45	2e⁻	7e⁻			
Cl	3.61	2e⁻	8e⁻	7e⁻		
Br	3.36	2e⁻	8e⁻	18e⁻	7e⁻	
I	3.06	2e⁻	8e⁻	18e⁻	18e⁻	7e⁻

nuclear charge is great. The decrease of electron affinity observed from Cl to I is not unexpected, because the size increases in going down the group. In iodine, the electron to be added goes into the fifth shell. Being farther from the nucleus, the added electron is not so tightly bound as one added to the other elements of the group. The unexpectedly low value for fluorine must result from greater electron repulsion between the eight electrons in the outer shell of this small atom.

A knowledge of electron affinities can be combined with a knowledge of ionization potentials to predict which atoms can remove electrons from others. Unfortunately, the measurement of electron affinity is difficult and has been carried out for only a few elements. A different but related property of electron-attracting ability of atoms within a molecule is the electronegativity, which is discussed in the next chapter.

QUESTIONS

3.1 *Periodic Law* Graph the ionization potentials of the first 40 elements and show how your graph illustrates the periodic law.

3.2 *Periodic Table* (a) By reference to a periodic table, locate elements having each of the following atomic numbers by giving the number of its group and its period: 3, 10, 31, 37, 49, 84, 87. (b) Without reference to a periodic table, see if you can figure out which of the elements having the following atomic numbers are transition elements: 13, 36, 40, 55, 92.

3.3 *Periodic Behavior* In what group or groups would you expect to find elements which show the following properties: (a) generally unreactive, (b) form hydrogen compounds that give acid solutions, (c) form basic solutions and liberate hydrogen on reaction with water, (d) conduct electric current?

3.4 *Energy Levels* What is the maximum electron population of each for the following levels, sublevels, or orbitals: (a) $2p_x$, (b) $n = 2$, (c) $n = 4$, (d) $3d$?

3.5 *Line Spectrum* When elements are heated, line spectra can be observed. What purpose is served by heating?

3.6 *Energy Levels* (a) What is the maximum electronic population of the first four principal quantum levels? (b) Why are these not the numbers of elements in the first four periods of the periodic table?

3.7 *Electronic Structure* Describe in words, drawing, or both, the spatial distribution of each of the following: (a) an electron in a H atom, (b) two electrons in a 1s orbital, (c) three electrons in a Li atom, (d) one electron in a $2p_z$ orbital.

3.8 *Quantum Numbers* Complete specification of an electron in an atom requires use of four quantum numbers. What does each of these tell about the electron?

3.9 *Pauli Principle* Why does the Pauli principle account for the periodic law?

3.10 *Electronic Symbols* Write electronic symbols for each of the following elements after referring to Figure 3.2: He, O, Cl, Mg, P, Kr, Pb.

3.11 *Periodic Law* Using the periodic table, decide which of the following hydroxy compounds you would expect to be basic and which acidic: IOH, $Ca(OH)_2$, KOH, $ClO(OH)$, $NO_2(OH)$, $CO(OH)_2$.

3.12 *Atomic Structure* A neutral atom of an element has 2 K electrons, 8 L electrons, 10 M electrons, and 2 N electrons. Supply as many of the following quantities as possible from the information given above: (a) the total number of s electrons, (b) atomic weight, (c) total number of electrons, (d) atomic number.

3.13* *Ionization* How much energy in kilojoules is required to remove the two outermost electrons from each atom in 1 mole of calcium if the first and second ionization energies are 6.11 eV and 11.9 eV respectively?

3.14 *Charge Distribution* Which of the following would you expect to have nonspherical charge distribution: H, He, C, O, Cl, Li, N, P, Mg?

3.15 *Periodic Table* As we go from left to right through a group, the ionization potentials increase. Account for this change.

3.16 *Periodic Table* How is it possible for an atom to have electrons in the O shell even though the N shell is not filled?

3.17 *Atomic Size* Describe the effect of each of the following on the size of an atom: (a) gain of two electrons, (b) loss of an electron.

3.18 *Periodic Table* Explain the trend of atomic radii observed for elements 57 through 71.

3.19 *Energy Levels* Discuss reasons that the electronic configuration of Cr is $1s^2 2s^2 2p^6 3s^2 3p^6 3d^5 4s^1$ and not $1s^2 2s^2 2p^6 3s^2 3p^6 3d^4 4s^2$.

3.20 *Atomic Sizes* In going down a group, increase in nuclear charge and number of electrons causes a size increase. In going across a period, increase in nuclear charge and number of electrons causes a size decrease. Why the difference?

3.21* *Ionization Energy* How much energy (in kilojoules) is required to ionize each of the following: (a) 1.00 mole of Na, (b) 1.00 g of oxygen, (c) 1.00×10^{23} atoms of neon?

3.22 *Electron Affinity* (a) Is the energy associated with electron affinity liberated or required? (b) Give an explanation for the process being favorable or unfavorable as your answer to (a) indicates. (c) In most cases an element with a high ionization potential has a high electron affinity. Why is this? (d) Where would you expect there to be an exception to (c)?

3.23 *Periodic Law* Without reference to a periodic table, tell which of the elements having the following atomic numbers are in the same period and which in the same group with element $Z = 33$: $Z = 25, 15, 51, 41, 37$.

3.24 *Atomic Size* (a) Find the element in Figure 3.18 which has the largest atomic radius. (b) Which element would you expect to be even larger? (c) For your answer to (a), calculate the number of atoms, laid side to side, needed to stretch a distance of 1.00 m. (d) How many moles is this?

3.25 *Energy Levels* (a) Figure 3.8 might seem to be sufficient to account for elements as high as $Z = 60$. Why is this not the case? (b) What is the highest atomic number of an element which can be accounted for by Figure 3.8?

3.26 *Probability Distributions* Figure 3.11c is perhaps the easiest way to draw a representation of an orbital. Consider the $2p_x$, $2p_y$, and $2p_z$ orbitals and draw similar representations for each in each of the three planes xy, yz, and zx.

3.27 *Magnetism* (a) Tell whether each of the following atoms should be "paramagnetic," "diamagnetic," or "could be either": He, Li, Be, B, C. (b) Actually C is paramagnetic. By reference to Figures 3.10 and 3.15, see whether you can rationalize this fact. (c) By reference to Figures 3.10 and 3.16, what would you predict for Ni?

3.28 *Ionization* After the removal of one electron, the positive ion formed can be stripped of additional electrons by supplying additional energy. For a Ca atom the energies needed for the first three ionization steps are 6.11 eV, 11.9 eV, and 51.2 eV. (a) Why does the second require more than the first? (b) Why does the third require so very much more than the second?

3.29* *Atomic Size* (a) Why is it meaningless to talk about the size of a single atom? (b) Using Figure 3.18, calculate the volume apparently occupied by 1 mole of iron. (c) Calculate the apparent density from your answer to (b). (d) The density of iron is 7.86 g/cm^3. Why might this be different from your answer to (c)?

Electrons in Molecules: The Chemical Bond

4

Atomic theory came about as an explanation for the laws of chemical change. Its utility is determined in large part by its success in explaining chemical behavior. From our ideas concerning the structure of isolated atoms, we can hope to understand why certain kinds of atoms combine to form compounds and why other kinds of atoms do not. We can hope to learn why there is a definite number of atoms of one kind which combine with a definite number of atoms of another kind. We can hope to understand why certain discrete combinations of atoms lead to the properties observed. In short, we can hope to answer the questions raised by Lavoisier's writing nearly two centuries ago. Like Lavoisier, we will frame the answers in terms of the fundamental "principles" of the substances prior to reaction. In other words, chemists try to understand chemical reactions and complex products of reactions in terms of the properties of the atoms of the elements. As we shall see this is not always possible. Chemical reaction implies a change in the atoms, and after reaction the situation may be difficult to reconcile in terms of the starting atoms. However, understanding requires a frame of reference. For the chemist, the frame of reference is the atom.

The assumption that when atoms combine the nuclei remain unchanged is accurate. Because of this fact, mass is unchanged in chemical reactions. Other properties may depend on the extranuclear, or electronic, portion of atoms and so are altered by chemical reaction. Even in such cases, we will attempt to discuss reaction products in terms of the constituent atoms with the electronic structure altered as little as necessary. Our approach will be to discuss first the formation of simple substances, such as molecules containing only two atoms. Principles developed in such cases can be extended in our later discussions to complicated substances containing many atoms. The new principle needed is that of the chemical bond, which can then be envisioned as linking adjacent atoms in complex molecules. In fact we will need several types of chemical bonds to cover the many possibilities.

4.1 Ionic Bonds

One type of chemical bond is the *ionic bond* in which *electrons are completely transferred from one atom to another*. The formation of an ionic bond is favored in the reaction of an atom of low ionization potential with an atom of high electron affinity. An example of such a reaction is the one between sodium atoms and chlorine atoms. A sodium atom has a low ionization potential; i.e., not much energy is required to pull off the outer electron. A chlorine atom has a high electron affinity; i.e., considerable energy is released when an electron is added to its outer shell. Suppose these two atoms come together. As shown in Figure 4.1, sodium initially has one valence electron, and chlorine has seven. After electron transfer, chlorine

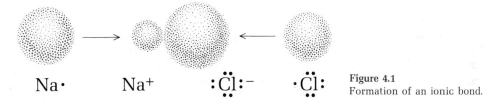

Na·   Na⁺   :C̈l:⁻   ·C̈l:

Figure 4.1
Formation of an ionic bond.

has all eight of the valence electrons. The sodium now has a positive charge because of the loss of a negative electron. The chlorine has a negative charge because of the gain of an electron. Thus, a positive ion and a negative ion are formed. Because the ions are of opposite electric charge, they attract each other to produce an ionic bond. The ionic bond is sometimes called the *electrovalent bond*.†

The formation of an ionic bond can be thought of in three steps:

Step 1   Sodium loses its outermost (3s) electron and becomes positively charged

$$Na(1s^22s^22p^63s^1) \longrightarrow Na^+(1s^22s^22p^6) + e^-$$

---

†A discrete unit consisting of one sodium ion and one chloride ion can properly be called a molecule only where such units exist, i.e., in gaseous sodium chloride. In the solid, each Na^+ is attracted to more than one Cl^- and each Cl^- to more than one Na^+. Thus, each crystal of salt is a "giant molecule" containing a huge number of ionically bound Na^+ and Cl^- ions.

Step 2 Chlorine gains an electron and becomes negatively charged

$$Cl(1s^22s^22p^63s^23p^5) + e^- \longrightarrow Cl^-(1s^22s^22p^63s^23p^6)$$

Step 3 Sodium and chloride ions combine

$$Na^+ + Cl^- \longrightarrow [Na^+][Cl^-]$$

Step 1 requires energy equal to the ionization potential of sodium, 5.1 eV. Step 2 releases energy equal to the electron affinity of chlorine, 3.61 eV. Step 3 releases energy of 5.2 eV, because of the attraction between positive and negative ions. Note that the ionic bond is formed only because the energy released in steps 2 and 3 is greater than that required in step 1.

In forming compounds by this process of electron transfer, it is necessary that there be a balance of electrons gained and lost. The reaction between sodium and chlorine requires one atom of sodium for every atom of chlorine. As a more complicated example, when calcium reacts with chlorine, each calcium atom loses its two valence electrons to form Ca^{2+}. Two chlorine atoms, each picking up one electron, are required to balance this. The compound formed, calcium chloride, contains one doubly positive calcium ion for every two singly negative chloride ions. The formula of the compound is $CaCl_2$. The subscript 2 for chlorine and subscript 1 (understood) for calcium indicate that in the compound there are two chlorine atoms for each calcium atom.

Since, in general, the elements on the left of the periodic table have low ionization potentials and the elements on the right have high electron affinity, ionic bonds are favored in reactions between these elements. Thus, alkali metals (group I) generally react with halogens (group VII) to form ionic compounds (for example, LiF). Similarly, most of the group II elements react with the halogens or with group VI elements to form ionic bonds (for example, MgO). In general, these ionic compounds resemble sodium chloride in that they are white, brittle solids at room temperature which dissolve in water to give conducting solutions. They melt at relatively high temperatures.

4.2 Covalent Bonds

Most bonds cannot be adequately pictured by assuming a complete transfer of electrons from one atom to another. For example, two H atoms react to form H_2, and in the hydrogen molecule it seems unreasonable that one hydrogen atom should pull an electron from the other, identical, hydrogen atom. In such cases, it is assumed that *electrons are shared between the atoms,* and the bond is called *covalent.* When two hydrogen atoms come together, as shown in Figure 4.2, each electron comes under the pull of both nuclei. Although this represents a net attraction, it is too small to account

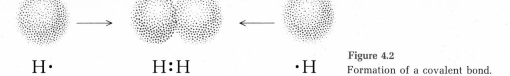

H·   H:H   ·H

Figure 4.2
Formation of a covalent bond.

for the entire attraction of the chemical bond. An additional contribution to the bonding comes about because of the increased space available to each electron. As already noted, it is a property of small particles that their kinetic energy increases as their confining volume decreases. In other words, if an electron is confined to an orbital of one atom, then its kinetic energy is higher than if it is spread out over an additional orbital of another atom. What this means is that it will take added energy to separate the atoms; i.e., the bond is more stable.

The electronic symbols at the bottom of Figure 4.2 show the conventional way of expressing the formation of a covalent bond. The formula of hydrogen is written as H_2, to indicate that there are two atoms per molecule; the molecule is *diatomic*. We emphasize that the pictures with dots and letters are only schematic representations of a complicated situation and tell little about the electron probability distribution. In the case of hydrogen, for instance, the charge distribution is such that the electrons occupy the whole molecule, spending equal time in the vicinity of each nucleus. In the resulting H_2 molecule, each electron is distributed over exactly the same region of space. Since the Pauli principle states that no two electrons can occupy the same region of space unless they differ in electron spin, the two electrons that constitute the shared pair have magnetic spins that cancel each other.

In the case of chlorine, the formation of a covalent bond is pictured as follows: Each atom has electron configuration $1s^2 2s^2 2p^6 3s^2 3p^5$, corresponding to seven valence electrons with a vacancy in one of the $3p$ orbitals. When two chlorine atoms come together, electron sharing occurs much as for H_2 except that the electrons involved occupy $3p$ orbitals instead of 1s. The $3p$ electron in the half-filled orbital of one atom pairs with the $3p$ electron in the half-filled orbital in the other atom. The pair constitutes a shared pair, the two electrons of which must differ in electron spin. In terms of electron symbols, the formation of the bond in Cl_2 can be expressed as follows:

$$\ddot{\underset{\cdot\cdot}{Cl}}\cdot \;+\; \cdot\ddot{\underset{\cdot\cdot}{Cl}}: \;\longrightarrow\; \ddot{\underset{\cdot\cdot}{Cl}}:\ddot{\underset{\cdot\cdot}{Cl}}:$$

The shared pair of electrons constitutes a *single bond* and is frequently designated, for example, as Cl—Cl. In other cases, *double* and *triple* bonds are made possible by sharing two or three pairs of electrons. A triple bond, for example, would be typified by N_2, where each N atom ($1s^2 2s^2 2p^3$) has three half-filled p orbitals that can share with a neighboring N atom. This may be represented by the "dot formula" $:N:::N:$, where each N atom is regarded as being surrounded by three pairs of shared electrons and one pair of unshared 2s electrons.

The formation of hydrogen chloride can be expressed as follows:

$$H\cdot + \cdot \overset{\cdot\cdot}{\underset{\cdot\cdot}{Cl}}: \longrightarrow H:\overset{\cdot\cdot}{\underset{\cdot\cdot}{Cl}}:$$

Although chlorine has a greater attraction for electrons than does hydrogen, the attraction is not great enough to result in complete transfer. In other words, HCl is *not* ionically bound. Instead, there is a covalent bond arising from electron sharing of the odd electrons of the two atoms, the 1s of the H and the 3p of the Cl. There is a fundamental difference between this single bond in HCl and that in H_2 or Cl_2. Because H and Cl are unlike atoms, the sharing of electrons is unequal. This means that part of the time both electrons will be on the Cl atom, so that, on an average, the electrons spend more time with Cl than with H.

4.3 Polarity of Bonds

Because electrons may be shared unequally between atoms, it is necessary to have some way of describing the electric-charge distribution in a bond. This is usually done by classifying bonds as *polar* or *nonpolar*. For example, the bonds in H_2 and Cl_2 are called nonpolar; the bond in HCl, polar.

Why are the covalent bonds in H_2 and Cl_2 called nonpolar? In both these cases, the "center of gravity" of the negative-charge distribution is at the center of the molecule, since the shared pair is distributed equally over the two atoms. The molecule is electrically neutral in two senses of the word. Not only does it contain an equal number of positive and negative charges (protons and electrons), but also the center of the positive charge coincides with the center of the negative charge. The molecule is a *nonpolar molecule*; it contains a *nonpolar bond* because an *electron pair is shared equally* between two atomic kernels (nucleus plus any inner electrons).

In the case of HCl, the bond is called polar because the center of positive charge does not coincide with the center of negative charge. The molecule *as a whole* is electrically neutral, because it contains an equal number of positive and negative charges. However, owing to the unequal sharing of the electron pair, the chlorine end of the molecule appears negative, and the hydrogen end positive. This arises because, as mentioned before, the bonding electrons spend more time on the chlorine atom than on the hydrogen atom. (Polarity does not arise simply from the fact that chlorine has more electrons than hydrogen; the charge of the unshared electrons is balanced by the greater positive charge of the chlorine nucleus.)

As another example of a polar covalent bond, consider the bond between chlorine and bromine in the molecule BrCl. Each of these atoms has one vacancy in its outermost p subshell. Sharing involving the p orbitals of the Br and Cl can produce a single covalent bond. However, the bromine atom is bigger than the chlorine atom, and it has a smaller attraction for electrons. In the covalent bond between Cl and Br, the pair of electrons is not shared equally but spends more of its time with the chlorine. The chlorine end of the molecule therefore appears negative with respect to the bromine end. In Figure 4.3 this polarity is indicated by a + at the center of positive-charge distribution and a − at the center of the negative charge. Again the molecule as a whole is electrically neutral—there are just as many positive charges as there are negative charges in the whole molecule—but there is a dissymmetry in the electrical distribution. Molecules in which the positive and negative centers of charge do not coincide are called *polar molecules,* and any bond in which the sharing between two kernels is *unequal* is a *polar bond.*

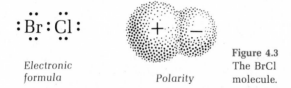

Electronic
formula

Polarity

Figure 4.3
The BrCl
molecule.

In the molecule BrCl there are two centers of charge. Such a molecule (or such a bond) is called a *dipole.* A dipole consists of a *positive and an equal negative charge separated* by some distance. Quantitatively, a dipole is described by giving its *dipole moment,* which is equal to the *charge times the distance* between the positive and negative centers. This number measures the tendency of the dipole to turn when placed in an electric field. As shown in Figure 4.4, each dipole turns because its positive end is attracted to the negative plate and its negative end to the positive plate. Since the positive and negative centers are part of the same molecule, the molecules can only turn; there is no migration toward the plates.

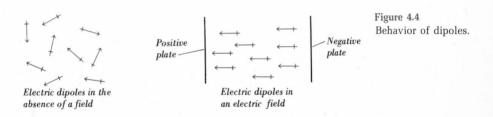

Positive
plate

Negative
plate

Figure 4.4
Behavior of dipoles.

*Electric dipoles in the
absence of a field*

*Electric dipoles in
an electric field*

The behavior of dipoles in an electric field provides an experimental means for distinguishing between polar and nonpolar molecules. The experiments involve the determination of a property called the *dielectric constant* (see Appendix 3.6). This property can be measured as follows: It is observed that an electric capacitor (two parallel metallic plates, like those shown in Figure 4.4) has the ability to store electric charge. The capacitance, i.e., the amount of charge that can be put on the plates for a given voltage, depends upon the material (dielectric) between the plates. The *dielectric constant* of a substance is defined as the ratio of the capacitance with that substance between the plates to the capacitance with a vacuum between them.

In general, a substance which consists of polar molecules has a high dielectric constant; i.e., a capacitor can store much more charge when such a substance is between its plates. This high dielectric constant can be thought of as arising in the following way: As shown in Figure 4.4, dipoles tend to turn in a charged capacitor, so that negative ends are near the positive plate and positive ends are near the negative plate. This partially neutralizes the charge on the plates and permits more charge to be added. Thus, measurement of the dielectric constant gives information about the polarity of molecules. The fact that hydrogen gas has essentially no effect on the capacitance (dielectric constant 1.00026, compared to 1.00000 for a vacuum) confirms the idea that H_2 molecules are nonpolar.

It is possible to predict whether a diatomic molecule is polar or nonpolar. If the two atoms are alike, the *bond* between them must be nonpolar, and therefore the *molecule* is nonpolar. If the two atoms are different, the *bond* is polar to some degree, and the *molecule* is also polar. The degree of polarity of diatomic molecules increases as the atoms bcome more unlike in electron-pulling ability. It is not so easy to predict the polar nature of a molecule containing more than two atoms. Such a *molecule* can be nonpolar even though all the *bonds* in the molecule are individually polar. Carbon dioxide (CO_2) is a good example. As shown in Figure 4.5, the two oxygen atoms are bonded to the carbon atom. Oxygen attracts the shared electrons more than carbon does so that each carbon-oxygen bond is

O   C   O

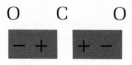

Figure 4.5
Nonpolar molecule containing polar bonds.

polar, with the shared electrons spending more time with the oxygen than with the carbon. The polarity of each bond is shown in the figure. Because the molecule is linear, however, when CO_2 molecules are placed in an electric field, they do not line up; any turning action of one bond is counteracted by the opposite turning action of the other bond. And, as we would expect carbon dioxide has a low dielectric constant.

Water (H_2O) is a triatomic molecule in which two hydrogen atoms are bonded to the same oxygen atom. There are two different possibilities for its structure. It could have a linear structure with the three atoms arranged in a straight line, or the atoms might be arranged in the form of a bent chain. The two possibilities are shown in Figure 4.6. The fact that water has a very high dielectric constant supports the bent structure, a structure which gives water its unique properties and makes possible life as we know it (Section 14.11).

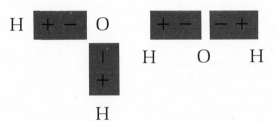

Figure 4.6
Possible structures of an H_2O molecule.

In the light of the foregoing discussion of polar bonds, it is interesting to note that *there is no sharp distinction between ionic and covalent bonds.* In the chemical bond between atoms A and B, all gradations of polarity are possible, depending on the nature of A and B. If A and B have the same ability to attract electrons, the bond is nonpolar. If the electron-pulling ability of B is increased (by taking different elements for B), the shared electrons spend more time on B, and the bond becomes more polar. In the limiting case, the electron pair is not shared at all but spends all its time on B. The result is a negative ion B^- and a plus ion A^+; the bond is ionic.

4.4 Electronegativity

In the preceding section we referred to the electron-pulling ability of atoms in molecules. A quantitative measure of this property could be obtained by considering various properties of molecules, such as dipole moments and energies required to break bonds. A listing of elements in the order of their *tendency to attract shared electrons* is called the scale of *electronegativity*. Numerical values are assigned for the various elements, as shown in Figure 4.7. These numbers describe the relative tendency of an atom, in forming a bond, to go to a negative condition, i.e., to attract a shared electron pair. Fluorine is assigned the highest electronegativity of any element in the periodic table. The noble-gas elements do not form many chemical bonds and their values, though low, have not been agreed upon. In general, from left to right across a period (i.e., with increasing nuclear charge), the electronegativity increases. The elements at the far left of the periodic table have low electronegativities. The elements at the far right, with the exception of

Figure 4.7
ELECTRONEGATIVITIES OF THE ELEMENTS

1 H 2.1																	2 He —
3 Li 1.0	4 Be 1.5											5 B 2.0	6 C 2.5	7 N 3.0	8 O 3.5	9 F 4.0	10 Ne —
11 Na 0.9	12 Mg 1.2											13 Al 1.5	14 Si 1.8	15 P 2.1	16 S 2.5	17 Cl 3.0	18 Ar —
19 K 0.8	20 Ca 1.0	21 Sc 1.3	22 Ti 1.5	23 V 1.6	24 Cr 1.6	25 Mn 1.5	26 Fe 1.8	27 Co 1.8	28 Ni 1.8	29 Cu 1.9	30 Zn 1.6	31 Ga 1.6	32 Ge 1.8	33 As 2.0	34 Se 2.4	35 Br 2.8	36 Kr —
37 Rb 0.8	38 Sr 1.0	39 Y 1.2	40 Zr 1.4	41 Nb 1.6	42 Mo 1.8	43 Tc 1.9	44 Ru 2.2	45 Rh 2.2	46 Pd 2.2	47 Ag 1.9	48 Cd 1.7	49 In 1.7	50 Sn 1.8	51 Sb 1.9	52 Te 2.1	53 I 2.5	54 Xe —
55 Cs 0.7	56 Ba 0.9	57–71 La–Lu 1.1–1.2	72 Hf 1.3	73 Ta 1.5	74 W 1.7	75 Re 1.9	76 Os 2.2	77 Ir 2.2	78 Pt 2.2	79 Au 2.4	80 Hg 1.9	81 Tl 1.8	82 Pb 1.8	83 Bi 1.9	84 Po 2.0	85 At 2.2	86 Rn —
87 Fr 0.7	88 Ra 0.9	89–103 *	104 Ku —	105 Ha —													

*	89 Ac 1.1	90 Th 1.3	91 Pa 1.5	92 U 1.7	93–103 Np–Lr 1.3

group 0, have high electronegativities. On the scale of electronegativity, the group VII elements are assigned the values F, 4.0; Cl, 3.0; Br, 2.8; and I, 2.5. The decreasing order is regular, unlike the order of electron affinities. In general, electronegativity decreases down a group (i.e., as the size increases).

Of what use are these values of electronegativity? For one thing, they can be used in predicting which bonds are ionic and which covalent. Since electronegativity indicates relative attraction for bonding electrons, two elements of very different electronegativity, such as Na (0.9) and Cl (3.0), are expected to form ionic bonds. Thus, electronegativities support the expectation that the alkali elements and the group II elements form essentially ionic bonds with the elements of groups VI and VII. Two elements of about equal electronegativity, such as Cl (3.0) and Br (2.8), are expected to form covalent bonds. In such cases, electronegativities can be used to predict polarity. The farther apart in electronegativity two elements are, the more polar the bond should be. Thus, the bond between H and Cl is more polar than that between Br and Cl. In both cases, the chlorine end should be more negative, since Cl has the higher electronegativity.

Support for the assignment of electronegativity values comes from measurements of dipole moments. For the hydrogen halides, the observed dipole moments are HF, 1.91; HCl, 1.03; HBr, 0.78; and HI, 0.38, expressed in Debye units (in honor of Peter Debye, who first studied them). The decreasing polarity from HF to HI indicates a trend toward equal sharing of electrons, consistent with decreasing electronegativity from F to I.

4.5 Saturation of Valence

Formulas of compounds are determined from experimental data. It is found that there is a limit to the combining ability atoms of one element have for atoms of another. For example, when calcium is combined with chlorine, no more than two chlorine atoms per calcium atom are found in the resulting compound; when carbon is combined with hydrogen, no more than four hydrogen atoms per carbon are found. If we use the term *valence* to describe the *ability of atoms to bind together,* we can summarize the above observations by saying that there is a *limit to the valence one atom shows for others;* i.e., there is a *saturation of valence.*

A common device for describing such saturation of valence is the use of electronic formulas written in terms of *octets* of electrons. The *octet rule* of classical chemistry states that, when atoms (other than H) combine, the bonds formed are such that each atom is surrounded by a complete octet of electrons. As the electron energy levels of atoms became better understood, the uniqueness of eight as a stable grouping disappeared. However, despite numerous exceptions, the octet rule remains as a useful generalization for most compounds. Its application is illustrated in the following examples.

Consider the compounds that fluorine forms with sodium, calcium, and aluminum. All three of these compounds are believed to be ionic, because fluorine is so much more electronegative than Na, Ca, or Al. In the combination of fluorine with sodium, there is only one fluorine atom per sodium atom. This can be accounted for by noting that each sodium atom (Na·) has one valence electron and each fluorine atom (:F̈·) has seven. If a fluorine atom takes one electron from a sodium atom, the result is fluoride ion (:F̈:⁻). The sodium ion (Na⁺) has no more valence electrons to lose to other fluorine atoms. Therefore, in sodium fluoride (NaF), only one fluorine atom is combined per sodium atom. (Note that the octet rule applies here.)

In the combination of fluorine with calcium, there are two fluorine atoms per calcium atom. The two valence electrons of the calcium atom (Ca:) can be lost to two fluorine atoms. When fluorine is combined with aluminum, no more than three fluorine atoms react per aluminum atom, because the aluminum atom (:Al·) has but three valence electrons.

It should be emphasized that the above discussion tells us nothing about the actual number of atoms in molecules. All three of the fluorides mentioned normally exist as white solids consisting of aggregates of tremendous numbers of positive and negative ions. For such ionic solids, saturation of valence simply indicates the relative numbers of positive and negative ions.

When covalent bonds are formed, not only is the relative number of atoms fixed, but also the actual number of atoms in the molecule may be limited. For example, in the combination of carbon with hydrogen, there are no more than four hydrogen atoms per carbon atom. Furthermore, the compound formed (methane) consists of discrete molecules, each of which has but one carbon atom and four hydrogen atoms. How can we account for this saturation of valence in methane? Each carbon atom (·Ċ·) has four valence electrons; each hydrogen (H·) has one. Since the electronegativities of C and H are similar (2.5 and 2.1, respectively), covalent rather than ionic bonds are expected. If a carbon atom contributes one electron to each covalent bond formed, four such bonds can be established. We can represent the formation of the compound as follows:

$$
\text{4H·} \; + \; \text{·C·} \; \longrightarrow \; \text{H:C:H with H above and H below}
$$

Other examples of valence saturation in covalent bonds are hydrogen compounds of fluorine, oxygen, and nitrogen. The formation of the compounds can be represented as follows:

$$
\text{H·} \; + \; \text{:F·} \; \longrightarrow \; \text{H:F:}
$$

Hydrogen fluoride

$$
\text{2H·} \; + \; \text{·O:} \; \longrightarrow \; \text{H:O: with H below}
$$

Water

$$
\text{3H·} \; + \; \text{·N·} \; \longrightarrow \; \text{H:N:H with H below}
$$

Ammonia

It should be noted that, in these compounds but not in methane, there are pairs of valence electrons which do not appear to be shared between atoms. One might imagine that these electron pairs could be used to bind to other atoms, but this can occur only if the additional atom has room for *two* more electrons in its valence shell. The unshared electron pairs cannot bind additional hydrogen atoms (H·), because three electrons cannot be accommodated by a single H atom.

Although hydrogen atoms cannot accommodate an additional pair of electrons, there are other species which can. For example, the hydrogen *ion* (H⁺), which has one electron less than a hydrogen atom, has room for a pair of electrons. Therefore, a hydrogen *ion* can bind to a molecule, such as ammonia, in which there is an unshared pair of electrons. Apparently, this is the explanation for the formation of ammonium salts. When ammonia (NH_3) is treated with hydrogen chloride (HCl), the white, solid ammonium chloride (NH_4Cl) is formed. The process can be pictured as follows:

$$\text{H:}\overset{\displaystyle ..}{\underset{\displaystyle \text{H}}{\text{N}}}\text{:H} + \text{H:}\overset{..}{\underset{..}{\text{Cl}}}\text{:} \longrightarrow \text{H:}\overset{\displaystyle \text{H}}{\underset{\displaystyle \text{H}}{\text{N}}}\text{:H}^+ \ \text{:}\overset{..}{\underset{..}{\text{Cl}}}\text{:}^-$$

We can imagine that first the HCl splits to give a hydrogen ion, leaving all the electrons on the chlorine, and then the hydrogen ion attaches to the NH_3. The hydrogen ion is bound to the NH_3 by sharing the nitrogen's extra pair of electrons. Since the original NH_3 molecule was electrically neutral and the hydrogen ion has a 1+ charge, the resulting ammonium ion (NH_4^+) carries a positive charge. Also, since the chlorine ends up in full possession of the hydrogen's electron, the resulting chlorine ion bears a negative charge. In ammonium chloride, positive ammonium ions and negative chloride ions attract each other. Once the ammonium ion is formed, all four nitrogen-hydrogen bonds are alike, even though the mode of formation of one of the bonds was different from that of the other three. Covalent bonds which are formed by donation of an electron pair from a single atom are sometimes called *donor-acceptor,* or *coordinate covalent,* bonds.

Saturation of valence is not restricted to bonding between unlike atoms. It also may occur when an atom forms covalent bonds with atoms of its own kind, as in the following examples:

$$:\overset{..}{\underset{..}{\text{F}}}:\overset{..}{\underset{..}{\text{F}}}: \qquad :\overset{..}{\underset{\text{H}}{\text{O}}}:\overset{..}{\underset{\text{H}}{\text{O}}}: \qquad \text{H:}\overset{..}{\underset{\text{H}}{\text{N}}}:\overset{..}{\underset{\text{H}}{\text{N}}}\text{:H} \qquad \text{H:}\overset{\text{H}}{\underset{\text{H}}{\text{C}}}:\overset{\text{H}}{\underset{\text{H}}{\text{C}}}:\overset{\text{H}}{\underset{\text{H}}{\text{C}}}\text{:H}$$

Fluorine   *Hydrogen*   *Hydrazine*   *Propane*
               peroxide

In addition to *saturation of valence,* the terms *saturated* and *unsaturated* are also used in describing carbon compounds. In such unsaturated compounds there are double or triple bonds between adjacent carbon atoms.

These compounds are called unsaturated because they can undergo chemical reaction in which atoms are added to the molecule. For example,

$$\underset{\text{H}}{\overset{\text{H}}{:}}\text{C}::\text{C}\underset{\text{H}}{\overset{\text{H}}{:}} + 2\text{H}\cdot \longrightarrow \text{H}:\underset{\text{H}}{\overset{\text{H}}{\text{C}}}:\underset{\text{H}}{\overset{\text{H}}{\text{C}}}:\text{H}$$

The compound on the left, ethylene, is unsaturated; each carbon atom shares two pairs of electrons with the other carbon atom. The compound on the right, ethane, is saturated; the carbon atoms are joined by sharing a single pair of electrons. If each electron pair is designated by a dash, the above equation is written

$$\underset{\text{H}}{\overset{\text{H}}{\diagdown}}\text{C}{=}\text{C}\underset{\text{H}}{\overset{\text{H}}{\diagup}} + 2\text{H} \longrightarrow \text{H}-\underset{\overset{|}{\text{H}}}{\overset{\overset{|}{\text{H}}}{\text{C}}}-\underset{\overset{|}{\text{H}}}{\overset{\overset{|}{\text{H}}}{\text{C}}}-\text{H}$$

Another unsaturated compound, acetylene, consists of molecules containing two carbon atoms and two hydrogen atoms. These atoms are arranged in a straight line with the carbons in the center. What is the electronic formula of this molecule? Each carbon atom makes available four valence electrons, and each hydrogen atom one valence electron, giving a total of 10 bonding electrons. Bonds between each hydrogen and its carbon take care of four electrons. There are six electrons left, which occur as three pairs of electrons shared between the two carbon atoms. This is a triple bond.

$$\text{H}:\text{C}:::\text{C}:\text{H} \qquad \text{or} \qquad \text{H}-\text{C}{\equiv}\text{C}-\text{H}$$

In general, for the same pair of bonded atoms, triple bonds are shorter than double bonds, and double bonds are shorter than single bonds. Experiment shows that the carbon-carbon distance (center to center) is 1.20×10^{-8} cm in acetylene, 1.33×10^{-8} cm in ethylene, and 1.54×10^{-8} cm in ethane.

Occasionally, no single electronic formula can be drawn for a molecule to account satisfactorily for its observed properties. Such a problem is encountered in the case of sulfur dioxide (SO_2). This molecule has a high dipole moment; hence we conclude that it is nonlinear, with the atoms arranged in a bent chain. Sulfur has six valence electrons, and oxygen also has six. There are thus a total of 18 valence electrons. These can be disposed in several ways:

$$\underset{(1)}{\overset{\text{S}}{\text{O} \quad \text{O}}} \qquad \underset{(2)}{\overset{\text{S}}{\text{O} \quad \text{O}}} \qquad \underset{(3)}{\overset{\text{S}}{\text{O} \quad \text{O}}} \qquad \underset{(4)}{\overset{\text{S}}{\text{O} \quad \text{O}}}$$

Neither (1) nor (2) is consistent with experimental fact, because each formula indicates that the SO_2 molecule has one double (short) sulfur-oxygen bond and one single (long) sulfur-oxygen bond. Experiments show the two bonds to be exactly the same length. The dilemma might be resolved by assuming that SO_2 is one of the compounds that violates the octet rule. In such a case, a possible formula would be (3). However, formula (3) is traditionally excluded because of the convenience of maintaining the sanctity of the octet rule. Formula (4) is excluded because it contains unpaired electrons. Molecules containing unpaired electrons are paramagnetic (Section 3.5); sulfur dioxide is not.

A situation in which *no single electronic formula conforms both to observed properties and to the octet rule* is described as *resonance*. In SO_2, the molecule is described by a combination of formulas (1) and (2). The actual electron distribution in the molecule is said to be a *resonance hybrid* of these contributing formulas. The choice of the word resonance for this situation is unfortunate, because it encourages people to think that the molecule resonates from one structure to the other, or that the extra electron pair jumps back and forth from one bond to the other. *Such is not the case.* The molecule has only one real electron structure. The problem is in describing it. For a fanciful analogy of resonance, we can recall the story of the medieval knight who encountered a rhinoceros in the forest and described it to his fellow knights back at the castle as something between a unicorn and a dragon. The rhinoceros (presumably) was real; it did not oscillate from being a unicorn at one instant to a dragon at another. In the same way, the properties of a resonance hybrid do not oscillate from those of one contributing structure to those of the other. The properties are fixed and are those of the actual resonance hybrid structure. Since resonance represents a problem in describing molecules, the difficulty lies in the description and not in the molecule itself. We will return to this point in Section 4.7.

4.6 Shapes of Simple Molecules

In the preceding sections, the formation of molecules has been represented in schematic fashion by using electron symbols. It has been emphasized that the dot representation of electrons says nothing about the spatial distribution of the electron clouds. Furthermore, the overall shape of a molecule is poorly presented in such electron-dot formulas. The theory of molecular shapes is a difficult subject but we might look at some of the simpler aspects of it.

Molecules which contain two atoms are necessarily linear but those containing three or more atoms present complications. For example, why is the water molecule nonlinear? To answer this question, we must consider the nature of the orbitals involved in bonding the hydrogen to the oxygen and specifically the spatial distribution of the electronic charge clouds about each of the nuclei. Imagine assembling the molecule H_2O from two H atoms and one O atom. Each H atom has originally a single electron in a 1s orbital, which is spherically symmetrical about the nucleus. The oxygen atom has originally in its outer shell two 2s electrons (spherically symmetrical) and four 2p electrons. Recalling the three p-type orbitals shown in Figure 3.15, we find two of these 2p electrons in one of the p orbitals and one electron in each of the other two p orbitals.

We have pictured the O—H bond as arising from the sharing of the 1s electron of the hydrogen with one of the unpaired 2p electrons of the oxygen. Such sharing favors bonding along the direction of the 2p orbital used. To tie on two H atoms requires use of two 2p orbitals, which are at right angles to each other (see Figure 3.15). Thus, on the basis of this simple picture we would expect the two O—H bonds in H_2O to be perpendicular to each other. Actually, they form the somewhat greater angle of 104°31′. We shall return to this discrepancy below.

Methane (CH_4) has a tetrahedral shape, as shown in Figure 4.8, with the carbon at the center of a tetrahedron and the four hydrogens at the corners. The angles between the C—H bonds are 109°28′. There are obvious problems with this molecule in that the bond angles are not 90° and furthermore there are four equivalent C—H bonds to be formed whereas we have just three p orbitals. Evidently, we need to use the 2s orbital of carbon as well. In fact, the formation of the CH_4 molecule can be pictured as involving the mixing of the orbitals with each other so that it is necessary to replace the one 2s and the three 2p orbitals on the carbon by a new set of four equivalent orbitals pointing to the corners of a tetrahedron. These new orbitals, which are called *hybrid orbitals,* partake simultaneously of the character of a 2s, $2p_x$, $2p_y$, and $2p_z$ orbital. They are called sp^3 orbitals; the superscripts, one (understood) on s and three on p, do not give electron population here but tell only how many orbitals have contributed to producing the final shapes. An sp^3 orbital is one that points from the nucleus in the direction of a tetrahedral vertex. The set of four sp^3 orbitals is called the *tetrahedral set.* (It is possible to have other hybrid orbitals. For example, sp hybrids, which have the combined character of an s orbital and one p orbital, are concentrated along a 180° line; they are appropriate for a linear

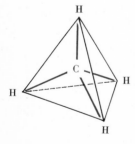

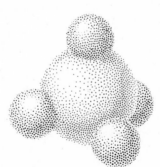

Figure 4.8
Tetrahedral CH_4 molecule.

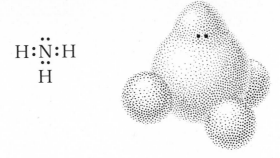

Figure 4.9
Ammonia molecule.

molecule such as Cl—Be—Cl. The three sp^2 hybrids point to the corners of a triangle and are appropriate for a planar molecule of the type BCl_3.) Electron sharing between each of the tetrahedral sp^3 orbitals and the 1s orbital of a hydrogen leads to the observed tetrahedral shape of the molecule CH_4.

The use of tetrahedral hybrid orbitals can account for the observed shapes of molecules other than methane even where there are *not* four attached atoms. For example, the NH_3 molecule can be imagined as having been built from a nitrogen atom ($1s^2 2s^2 2p^3$) with its five valence electrons distributed between four equivalent tetrahedral orbitals so that two of the electrons are paired and occupy one orbital. The other three electrons are shared with the H atoms using the other three tetrahedral orbitals. The result, as shown in Figure 4.9, is a pyramidal molecule in which the three hydrogens form the base, and the lone pair of electrons the apex. The observed angles between N—H bonds in NH_3 are 108°, which is very nearly that expected for a tetrahedron.

Tetrahedral orbitals can also help to explain the observed bond angle in H_2O. Following the reasoning of the preceding paragraph, we expect to find that H_2O is similar to NH_3 except for the two lone pairs of electrons in the case of H_2O. Figure 4.10 attempts to show that the two lone pairs of electrons and the two bound hydrogens are directed approximately to the corners of a tetrahedron. The resulting shape of the water molecule is

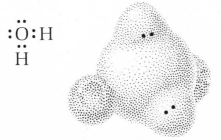

Figure 4.10
Water molecule.

the principal factor leading to the arrangement of H_2O molecules in ice that produces the low-density structure which causes ice to float on liquid water (Section 14.11).

4.7   Molecular Orbitals

In the above sections a simple picture of chemical bonding has been presented in which the bonded atoms retain the identity of isolated atoms except that electrons from an adjacent atom may occupy an unfilled orbital of the atom in question. Such a picture has the virtue of being about as simple as one can imagine, while being sufficiently detailed to account for practically all the observations we will be discussing in later chapters. However, we must note that nature is not quite this simple. Our picture does not hold for all cases. One example of its shortcomings which we have not yet mentioned is the O_2 molecule.

From what has been said thus far, we should expect the O_2 molecule to contain a simple double bond between the two O atoms to complete each O atom's octet of electrons, since each O atom has six valence electrons. Such is not the case. The O_2 molecule is known to contain unpaired electrons and so clearly violates the octet rule. Furthermore, many students find unsatisfying both the earlier resonance description of the SO_2 molecule (one of many such examples) and the lack of explanation as to why orbitals in atoms may hybridize, as C orbitals seem to in bond formation. All three of these kinds of problems can be handled in a more satisfying manner by a somewhat more complicated theory of chemical bonding. It may be well to discuss briefly some of the fundamentals of the more complicated theory of *molecular orbitals*. It should help us with these three problems and may give us a bit better understanding of what we mean when we say that electrons are "shared."

Basically, *molecular orbital theory* assumes that when molecules are formed the outer electrons no longer occupy the energy levels they occupied in the isolated atoms. Note that inner electrons and other unshared electrons may be considered to remain unchanged. However, new molecular energy levels may replace some of the old atomic levels. The new levels will have orbitals associated with them to contain the shared electron density. Figure 4.11 shows the formation of such a molecular orbital for the reaction of two H atoms to form H_2. The $+$ signs refer to the positions of the positively charged nuclei; the shaded area represents the electron distribution. There is a buildup of negative charge in the region where it is attracted by both nuclei, and bonding results. The orbital can hold no more than two electrons (because the Pauli exclusion principle applies to molecules as well as to atoms) with opposite spins. This filled orbital constitutes the "shared pair" chemical bond.

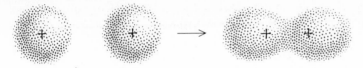

Thus far we have not really changed our picture of bonding. The differences come in when the resulting orbital bears less resemblance to the atomic orbitals than is true for H_2. An example of such a situation is that encountered earlier with methane. The point there is the fact that in the resultant molecule the lowest energy is achieved if the four molecular orbitals are arranged at tetrahedral angles even though similar orbitals were not present in the C atom prior to reaction.

For double and triple bonds, the molecular orbital description leads to some further insights. It is usual in such cases for one pair of electrons to be accommodated in a molecular orbital much like that formed by two H atoms as in Figure 4.11. Owing to the Pauli principle, the second shared pair must have a spatially different orbital. Frequently this orbital is formed by sharing electrons between two atomic p orbitals joined side to side as shown in Figure 4.12. The molecular orbital formed consists of two sausage-like electron clouds lying on opposite sides of the line connecting the two nuclei. This molecular orbital accommodates a shared pair of electrons and serves to concentrate electron density in a region where it is attracted to both nuclei. Hence, it helps to bond them together. Because its shape is different from that shown in Figure 4.11, it is called a π (pi) orbital, and the simpler orbital (of Figure 4.11) which has electron density concentrated along the line between nuclei is called a σ (sigma) orbital.

In the double bond of ethylene, $H_2C{=}CH_2$, one pair of shared electrons occupies a σ molecular orbital, and the other pair a π molecular orbital. What about the triple bond in acetylene, $HC{\equiv}CH$, or in nitrogen gas, $:N{\equiv}N:$? A second π orbital can be formed perpendicular to the first through the side-to-side combination of two more p atomic orbitals. Triple-bonded molecules contain a σ and two π orbitals. However, this represents the limit of bonding possibilities. In formation of a triple bond, the region between adjacent atoms is filled.

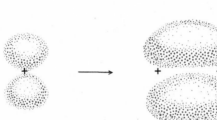

Figure 4.12
Molecular orbital formed by side-to-side p atomic orbitals.

For O_2 the molecular orbital picture can be briefly summarized as follows. The situation is like N_2, but in O_2 there are two more electrons. All the bonding molecular orbitals are filled and the two electrons remain. They are accommodated in other molecular orbitals which do not lie in the region between the atoms, because this has been filled by the triple bond. Actually each of the remaining electrons is in its own orbital and so can have the same spin quantum number (unpaired spin). Hence, O_2 is paramagnetic. Because these two electrons are not concentrated between the nuclei, they not only do not contribute to bonding, they fail to counteract the repulsion between the two positively charged nuclei. As a consequence, the bond in O_2 is weaker than that in N_2. Each molecule can be thought of as having three pairs of bonding electrons, but the greater unshielded nuclear charge of the O atoms weakens the bonding in O_2.

Finally, we might note that molecular orbitals may extend over more than two atoms. This is the basis for a number of failures of the simpler theory, and in particular those for which resonance was introduced. In the case of SO_2, we saw in Section 4.5 that structures (1) and (2) were both needed to describe the molecule. The two differ in that one shows a second pair of electrons in the left-hand bond and the other shows the pair in the right-hand bond. In molecular orbital terms, this pair occupies a single molecular orbital (π type) which extends over all three atoms. This π orbital comes about from the side-to-side overlap of one p atomic orbital from each of the three atoms, and it holds one pair of bonding electrons. We might think of each bond being worth, on average, $1\frac{1}{2}$ bonds, which is the result also of the resonance formulation.

An important example of either resonance or a multicentered molecular orbital is afforded by benzene, C_6H_6. As shown in Figure 4.13, the 12 atoms are arranged hexagonally in a plane. All C—C bonds are identical. The octet rule requires the two resonance formulations shown as (1) and (2). In molecular orbital terms, one p orbital (perpendicular to the plane of the molecule as shown) from each C atom forms molecular orbitals. It turns out that there are a total of three, multicentered orbitals that can contribute to the bonding. There are three pairs of electrons for these three orbitals and so each bond

Figure 4.13
Formulations for benzene.

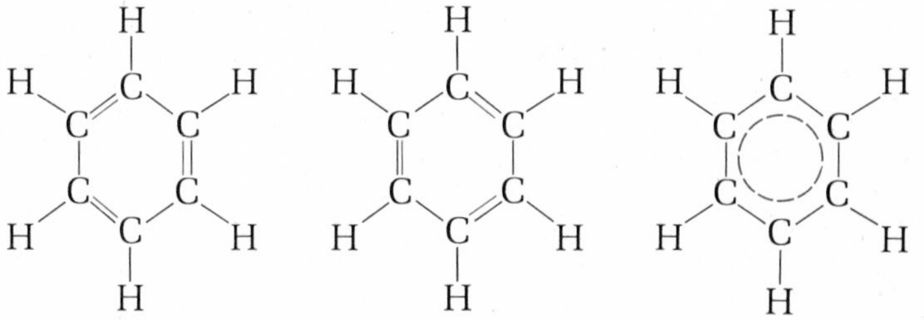

(1) (2) (3)

There are three pairs of electrons for these three orbitals and so each bond is classed as $1\frac{1}{2}$. This is precisely the answer indicated by the resonance formulas taken together. As with the resonance formulation, the problem comes in drawing the picture. One way the molecular orbital formation can be shown (not particularly successfully, but at least shown by a single formula) is given as (3) in Figure 4.13. Benzene's formula is often drawn either this way or with the inner circle solid.

In summary, attempts to picture bonds as resulting from sharing of electrons between unfilled atomic orbitals works for an amazingly large number of cases. In some cases however the bonding electrons occupy molecular orbitals which differ significantly from the orbitals of the isolated atoms. Such cases may be more satisfactorily described by the less simple molecular orbital theory. Often in science the simpler explanation is preferred. Consequently, more than one theory of the chemical bond remains in common use.

QUESTIONS

4.1 *Ionic Bonds* Predict for each of the following pairs of elements the relative numbers of atoms of each that should form the resultant ionic compound and draw dot formulas for the ions which compose it: (*a*) Li + O, (*b*) Ca + S, (*c*) Mg + Br, (*d*) Ba + H.

4.2 *Covalent Bonds* Through use of dot formulas, indicate the number of electrons shared in each of the following molecules: Br_2, N_2, CO_2, H_2O_2.

4.3 *Polarity of Bonds* (*a*) Tell whether each of the following molecules is "polar," "nonpolar," or "could be either": HI, N_2, CO, P_4, OF_2. (*b*) Actually, OF_2 is polar. What can you say about its shape?

4.4 *Electronegativity* (*a*) From Figure 4.7, what value would you predict for the electronegativity of element 105, hahnium? (*b*) From the table choose an element whose compound with hydrogen should contain nonpolar bonds. (*c*) Which pair of elements should have the greatest tendency to form ionic bonds?

4.5 *Saturation of Valence* With dot formulas show a compound of each of the following pairs of elements in which the valence is saturated: H + S; C + F; N + Cl; Na + S.

4.6 *Donor-Acceptor Bonds* By drawing dot formulas, show which of the following molecules might prove to be an electron-pair acceptor: CH_4, NI_3, BF_3, $AlCl_3$, N_2.

4.7 *Complex Ions* Draw dot formulas for each of the following chemical species which contains atoms covalently bonded to each other but has an electron surplus or deficiency to give the ionic charge shown: OH^-, H_3O^+, NH_2^-, NH_4^+, ClO_3^-, SO_4^{2-}.

4.8 *Multiple Bonds* Assuming the octet rule, which of the following should have multiple bonds: C_2Cl_6, CO, C_2F_4, N_2H_4, CH_4O?

4.9 *Types of Bonds* Describe the types of bonds in each of the following: (*a*) $NaCl$, (*b*) CH_4, (*c*) NH_4^+, (*d*) H_2O.

4.10 *Charge Distribution* Draw diagrams for the probable charge distribution in the nonpolar molecules O_2 and CCl_4, the ionic molecule KCl, and the polar molecules H_2O and CO.

4.11 *Polarity* Elements A, B, and C have one, five, and six valence electrons, respectively. (*a*) Write the electronic formula for the simplest hydrogen compound of each of these elements. (*b*) What shapes would you predict for these molecules? (*c*) Which of these molecules would you predict to be polar? Explain.

4.12 *Electronegativities* Tell why the electronegativities for noble gases are not given.

4.13 *Polarity* Draw the electronic formulas of H_2S and $CHCl_3$. Which of these two compounds has a dipole moment? Explain.

4.14 *Covalent Bonds* (*a*) What is the difference between a sigma bond and a pi bond? (*b*) Determine the number of sigma and pi bonds in the following molecules: CO_2, CH_4, C_2H_4, CH_3CN.

4.15 *Types of Bonds* Using electronegativities of atoms, predict whether bonds in the following would be ionic or covalent: KCl, PH_3, OsI_8, $AgBr$, CsF.

4.16 *Multiple Bonds* Each of the following molecules contains one or more double or triple bonds. For each molecule write an electronic formula consistent with the octet rule: C_4H_6, H_2CO_3, C_2H_4O, C_2Br_4.

4.17 *General* Given atoms A, B, C, and D in the same period, with one, four, six, and seven valence electrons, respectively. (*a*) What will be the formulas of the compounds formed between A and D, between B and A, and between C and D? (*b*) Compare the electronegativity of A with that of D. Will the bonding between A and D be ionic or covalent? Write the electronic formula of the compound. (*c*) Write the electronic formula of the compound formed between B and D. (*d*) Explain why you would expect the molecule in (*c*) to be either polar or nonpolar. (*e*) Which of these four atoms has the highest ionization potential?

4.18 *Molecular Orbitals* (a) Extend the reasoning of Section 4.7 to F_2. (b) Explain whether the bond should be stronger or weaker than that in O_2. (c) How is it that F_2 is not paramagnetic?

4.19 *Shapes of Molecules* (a) Show the dot formula for OF_2. (b) The molecule has a net dipole moment. What do you conclude about its shape? (c) Rationalize the fact that the bond angle is 103°.

4.20 *Molecular Orbitals* (a) How do the two electrons in a molecular orbital differ? (b) Why is a molecular orbital different from an atomic orbital? (c) How does a σ orbital differ from a π orbital?

4.21 *Resonance* Ozone, O_3, has a bent structure with two O's not bonded to each other. (a) Draw resonance structures and tell the meaning of the existence of more than one. (b) How would molecular orbital theory handle this case? (c) Show what the two formulations have in common.

4.22 *Ionic Bonds* As the data quoted in Section 4.1 show, when a sodium atom and chlorine atom are brought side by side it is energetically favorable for an electron to transfer from Na to Cl. (a) If the ions are now separated, should the electron hop back or stay with the Cl? (b) How much energy (in electron volts) would have to be supplied to give the opposite result to that which you predicted in (a)? (c) How much energy is this in kilojoules for a mole of Na and a mole of Cl?

4.23* *Covalent Bonds* When 2 moles of H atoms come together and form a mole of H_2 molecules, 431 kilojoules of energy is released. (a) Give two sources which contribute to this energy. (b) How many times as much energy would be needed to ionize 1 mole of H atoms (ionization potential of H is 13.6 eV)?

4.24 *Electronegativity* In addition to the scale of electronegativity given in Figure 4.7, there is an alternate scale which sets electronegativity equal to the average of the ionization potential and the electron affinity. Why does ionization potential get into the act?

4.25 *Polarity of Bonds* The table of electronegativities given in Figure 4.7 is based on the fact that a polar-covalent bond is stronger than a nonpolar-covalent bond. Recalling the three factors contributing to the energy of ionic-bond formation, which of these might justify the belief that polar bonds are stronger? Explain and outline the steps which could lead from bond strengths to electronegativities.

4.26 *Saturation of Valence* Assuming that in the compound formed, valence is saturated, and there is but one of the nonhydrogen atoms, how many grams of each of the following elements should react with 100 g of hydrogen: Cl, S, Si, P?

4.27 *Shapes of Molecules* Use hybrid orbitals to predict the shapes and sketch them, of the following molecules: C_2H_6, N_2H_4, H_2O_2, C_3H_8.

4.28 *Saturation of Valence* (a) In what way does the formation of NH_4^+ from NH_3 differ from the formation of C_2H_6 from C_2H_4? (b) Except for the charge of NH_4^+, is there any fundamental difference in the resulting bonds?

4.29 *Resonance* The nitrate ion consists of three O atoms (not bonded directly to each other) at the corners of an equilateral triangle with an N atom at the center. (a) Remembering the extra electron needed to make the negative charge on NO_3^- and the fact that the three bonds must be of equal length, draw resonance structures. (b) Describe a possible molecular orbital description of the molecule and show its similarity to the resonance formulation.

4.30 *Ionic Bonds* (a) Lithium is reacted with fluorine molecules to form a compound with the liberation of energy. Give the sources of this energy, noting that energy is required to break the bond in the fluorine molecule (75 kilojoules/mole of F atoms) and in removing an electron from Li. (b) Using data from Chapter 3, calculate the magnitude of these sources of energy if the total heat liberated is 610 kilojoules/mole of compound.

4.31 *Dipole Moments* Which should have the greater dipole moment, LiCl or CsBr? Give two distinct reasons for your answer.

4.32 *The Chemical Bond* A brash graduate student once told Robert M. Hutchins that his compilation of Great Ideas was incomplete because it failed to include the chemical bond. Dr. Hutchins asked: What is your concept of the chemical bond? What would you have answered in order to convince Dr. Hutchins that the chemical bond ranks as one of the Great Ideas?

Chemical Formulas and Equations

5

From the scientist's quest to understand the physical world, several simplifying concepts have emerged. The first of these is the fact that all the numberless different substances are combinations of less than 100 chemical elements. Next is the atomic theory of matter. Finally there is the important assumption that atoms undergo only limited changes during the course of chemical reaction. All these concepts are utilized by chemists in the shorthand notation they use to describe materials and their changes. In a sense the notation is a summary of centuries of observations and investigations which have made possible the simplifying formulas and equations on which future progress will be based. In earlier chapters we have developed the underlying principles and assumptions needed for understanding the formulas and equations which are the language tools of chemistry. In this chapter we will develop specifically these tools in order that we may use them throughout our subsequent discussions of the behavior of matter.

5.1 Simplest Formulas

The simplest formula, also called the empirical formula, gives the bare minimum of information about a compound, since it states only the *relative* number of atoms in the compound. The convention used in writing the simplest formula is to write the symbols of the elements with subscripts to designate the relative numbers of atoms of these elements. The formula A_xB_y represents a compound in which there are x atoms of A for every y atoms of B. Because of the relationship between moles and atoms, the simplest formula also gives information about the relative number of moles of atoms in the compound. In A_xB_y there are x moles of element A for every y moles of B.

The simplest formula is always the direct result of an experiment. The procedure is illustrated in the following example.

EXAMPLE 1

Suppose you are a chemist called on to determine the empirical formula of the hallucinogenic drug lysergic acid diethylamide (LSD, for short). You burn it and find that the combustion products indicate that LSD contains 74.27% carbon, 7.79% hydrogen, 12.99% nitrogen, and 4.95% oxygen. What is its simplest formula?

To solve the problem we first convert percentage to weight and then to moles. For convenience we assume that we have exactly 100 g of compound, in which case it will contain 74.27 g of carbon. The atomic weight of carbon is 12.01 amu, and so 1 mole of C weighs 12.01 g.

$$\frac{74.27 \text{ g of carbon}}{12.01 \text{ g/mole}} = 6.184 \text{ moles of C atoms}$$

Similarly

$$\frac{7.79 \text{ g of hydrogen}}{1.008 \text{ g/mole}} = 7.728 \text{ moles of H atoms}$$

$$\frac{12.99 \text{ g of nitrogen}}{14.007 \text{ g/mole}} = 0.927 \text{ mole of N atoms}$$

$$\frac{4.95 \text{ g of oxygen}}{15.999 \text{ g/mole}} = 0.309 \text{ mole of O atoms}$$

The empirical formula can be written

$$C_{6.184}H_{7.728}N_{0.927}O_{0.309}$$

Simplification is achieved by dividing all the subscripts by the smallest, 0.309, which gives

$$C_{20}H_{25}N_3O$$

5.2 Molecular Formulas

A second type of formula is the molecular formula. The amount of information provided by the molecular formula is greater than that given by the simplest formula. In the molecular formula the subscripts give the *actual* number of atoms of an element in one molecule of the compound. The molecule was defined previously as an aggregate of atoms bonded together tightly enough to be conveniently treated as a recognizable unit. In order to write the molecular formula, it is necessary to know how many atoms constitute the molecule. To find the actual number of atoms in a molecule, various experimental techniques can be used. For instance, X-ray determination of the positions of atoms in solids can give this information. Furthermore, some of the properties of gases and solutions depend on the number of atoms in each molecular aggregate.

A few molecular formulas thus determined are shown in Figure 5.1, where they are compared with the corresponding simplest formulas. In some cases, as for water and sucrose, the molecular and simplest formulas are identical. In other cases, they are not. We cannot tell from a formula whether it is molecular or simplest. However, if the subscripts given have a whole-number common divisor, it is a good bet that it is a molecular formula. The molecular formula gives all the information which is in the simplest formula, and more besides.

Figure 5.1 MOLECULAR AND SIMPLEST FORMULAS	Substance	Molecular Formula	Simplest Formula
	Benzene	C_6H_6	CH
	Acetylene	C_2H_2	CH
	Oxygen	O_2	O
	Water	H_2O	H_2O
	Sucrose (a sugar)	$C_{12}H_{22}O_{11}$	$C_{12}H_{22}O_{11}$
	Glucose (a sugar)	$C_6H_{12}O_6$	CH_2O
	Nicotine	$C_{10}H_{14}N_2$	C_5H_7N
	LSD	$C_{20}H_{25}N_3O$	$C_{20}H_{25}N_3O$

5.3   Moles and Molecules

The *formula weight* is the sum of all the atomic weights in the formula under consideration, be it simplest or molecular. For NaCl, the formula weight is the atomic weight of sodium, 22.9898 amu, plus the atomic weight of chlorine, 35.453 amu, a total of 58.443 amu. For $C_{12}H_{22}O_{11}$, the formula weight is equal to 12 times the atomic weight of carbon plus 22 times the atomic weight of hydrogen plus 11 times the atomic weight of oxygen, or 342.30 amu. In all cases, the formula weight depends on which formula is written. If the molecular formula is used, the formula weight is called the *molecular weight*. For example, 342.30 amu is the molecular weight of sucrose.

As noted above, the formula weight is given in atomic mass units. An amount of a substance whose mass in grams numerically equals the formula weight is called a *mole*. In the case of $C_{12}H_{22}O_{11}$, where the formula weight is 342.30 amu, 1 mole weighs 342.30 g. A pile of sugar weighing 342.30 g is 1 mole of sugar. We have already noted that for atoms, one mole contains the Avogadro number (6.02×10^{23}) of atoms. For molecules, one mole contains the Avogadro number of molecules. Thus 342.30 g of sugar contains the Avogadro number of $C_{12}H_{22}O_{11}$ molecules. For any substance whose molecular formula is known, 1 mole contains the Avogadro number of molecules. If the molecular formula is not known and only the simplest formula is given, then 1 mole contains the Avogadro number of such *formula-units* as are indicated by the formula.

Even though formulas are based on experimentally determined percentage composition, it is sometimes necessary to calculate percentage composition from a formula.

EXAMPLE 2   Cholesterol is an organic compound that is the main constituent of gallstones, is found in nearly all tissue, and is believed to be responsible for hardening of the arteries. Its molecular formula is $C_{27}H_{46}O$. Calculate the percentage composition.

One mole of $C_{27}H_{46}O$ contains

27 moles of C, or 27 × 12.01 g, or	324.27 g C
46 moles of H, or 46 × 1.008 g, or	46.37 g H
1 mole of O, or 1 × 16.00 g, or	16.00 g O
	386.64 g total

$$\%C = \frac{324.27 \text{ g C}}{386.64 \text{ g total}} \times 100 = 83.87\%$$

$$\%H = \frac{46.37 \text{ g H}}{386.64 \text{ g total}} \times 100 = 11.99\%$$

$$\%O = \frac{16.00 \text{ g O}}{386.64 \text{ g total}} \times 100 = 4.14\%$$

5.4   Chemical Reactions

It is possible to group chemical reactions into two broad classes: (1) reactions in which there is no electron transfer and (2) reactions in which there is electron transfer from one atom to another atom.

Reactions in which no electrons are transferred usually involve the joining or separating of ions or molecules. An example of a "no-electron-transfer" reaction occurs when a solution of sodium chloride is mixed with a solution of silver nitrate. The solution of sodium chloride contains sodium

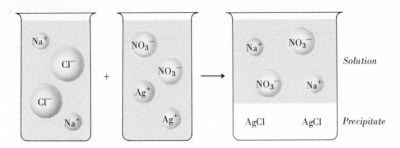

Figure 5.2
Chemical reaction without electron transfer.

ions and chloride ions. The solution of silver nitrate contains silver ions and nitrate ions. Figure 5.2 shows one beaker with sodium ions and chloride ions and another with silver ions and nitrate ions. On mixing the two solutions, a chemical reaction occurs, as shown by the formation of a white precipitate. This white solid consists of silver ions and chloride ions clumped together in large aggregates. In the final solution, sodium ions and nitrate ions remain dissolved as they were initially. In the chemical reaction, the only thing that has happened is that the silver ions have combined with chloride ions to form solid silver chloride, which is insoluble in the water. In shorthand form, the reaction is indicated as

$$Ag^+(soln) + \cancel{NO_3^-(soln)} + \cancel{Na^+(soln)} + Cl^-(soln) \longrightarrow$$
$$AgCl(s) + \cancel{Na^+(soln)} + \cancel{NO_3^-(soln)}$$

where the abbreviation "(soln)" indicates that the ion is in solution and the notation "(s)" emphasizes the fact that AgCl is formed as a solid; the strike-overs indicate cancellation of ions which do not change in the course of the reaction. The final net reaction is

$$Ag^+(soln) + Cl^-(soln) \longrightarrow AgCl(s)$$

Reactions in which electrons are transferred from one atom to another are known as *oxidation-reduction reactions*. Many of the most important chemical reactions fall into this class. For example, the combining of the sodium atom with a chlorine atom can be regarded as resulting from the transfer of an electron from the sodium to the chlorine, as shown schematically in Figure 5.3. A less obvious example of an oxidation-reduction reaction is that in which hydrogen and oxygen form water. In this case there is a change of the sharing of electrons during the course of the reaction:

Figure 5.3
Chemical reaction with electron transfer.

$$Na\cdot + \cdot\ddot{\underset{\cdot\cdot}{Cl}}: \longrightarrow Na^+ \left[:\ddot{\underset{\cdot\cdot}{Cl}}:\right]^-$$

$$H\!:\!H + :\overset{..}{O}: \longrightarrow \overset{..}{H\!:\!O}:$$
$$H$$

What has happened to the hydrogen in the course of this reaction? In the initial state, two hydrogen nuclei share a pair of electrons. Since the two hydrogen nuclei are identical, they share the pair equally and each hydrogen atom has a half-share of an electron pair. In the final state, the hydrogen shares a pair of electrons with oxygen. Since oxygen is the more electronegative, the electron pair is not shared equally but "belongs" more to the oxygen than to the hydrogen. In the course of the reaction, there is a change in the electron sharing, i.e., a partial transfer of electrons.

5.5 Oxidation Numbers

In order to keep track of electron shifts in oxidation-reduction reactions, it is convenient to use the *oxidation number,* sometimes called the *oxidation state.* The oxidation number is defined as the charge which an atom appears to have (with emphasis on the word *appears*) when electrons are counted according to some rather arbitrary rules. The first of these rules is that electrons shared between two unlike atoms are counted with the more electronegative atom. The second rule is that electrons shared between two like atoms are divided equally between the sharing atoms.

What is the oxidation number of hydrogen in the H_2 molecule? The electron pair is shared by two identical atoms, and so, according to the second rule, half of the electrons are counted with each atom, as shown by the line in Figure 5.4. Since the hydrogen kernel has a $1+$ charge and since one negative charge is counted with the kernel, the apparent charge of each hydrogen atom is zero. The oxidation number of hydrogen in H_2 is 0.

H ⌐:⌐ H   H kernel 1+   Oxidation number 0   **Figure 5.4** Assignment of oxidation number in H_2.

What are the oxidation numbers of hydrogen and oxygen in H_2O? Oxygen is the more electronegative, and so, according to the first rule, the shared electrons are counted with the oxygen, as shown by the line in Figure 5.5. The hydrogen therefore appears to have a charge of $1+$ and is assigned an oxidation number of $1+$. Since eight electrons are counted with the $6+$ oxygen kernel, the apparent charge of oxygen is $2-$. Oxygen has an oxidation number of $2-$ in H_2O.

In principle, the electronic pictures can be drawn for all molecules and electrons counted in this way to deduce the oxidation numbers of the various

H $\overset{\cdot\cdot}{\underset{\cdot\cdot}{:O:}}$ O kernel 6+   Oxidation number 2−   **Figure 5.5**
Assignment of electrons
H   H kernel 1+   Oxidation number 1+   for oxidation numbers in H_2O.

atoms. This is laborious. It is more convenient to use the following operational rules, which are derived from the above:

1. *In the free elements, each atom has an oxidation number of 0.* This is true no matter how complicated the molecule is. Hydrogen in H_2, sodium in Na, sulfur in S_8, and phosphorus in P_4 all have oxidation numbers of 0.

2. *In simple ions (ions containing one atom), the oxidation number is equal to the charge on the ion.* In these cases, the apparent charge of the atom is the real charge of the ion. In the tripositive aluminum ion, the oxidation number of the aluminum atom is 3+. Iron, which can form a dipositive or a tripositive ion, sometimes has an oxidation number of 2+ and sometimes 3+. In the dinegative oxide ion, the oxidation number of oxygen is 2−. It is useful to remember that elements of group I of the periodic table, lithium, sodium, potassium, rubidium, cesium, and francium, form only 1+ ions. Their oxidation number is 1+ in all compounds. The group II elements, beryllium, magnesium, calcium, strontium, barium, and radium, form only 2+ ions and hence always have oxidation numbers of 2+ in all compounds.

3. *In most compounds containing oxygen, the oxidation number of each oxygen atom is 2−.* This covers the vast majority of oxygen compounds. An important exception arises in the case of the peroxides, compounds of oxygen in which there is an oxygen-oxygen bond. In peroxides, e.g., hydrogen peroxide (H_2O_2), only seven electrons are counted with the 6+ kernel of oxygen. Figure 5.6 shows how the electrons are assigned. In the hydrogen-oxygen bond, the electrons are counted with oxygen,

$:\overset{\cdot\cdot}{\underset{\cdot\cdot}{O}}\overset{\cdot\cdot}{\underset{\cdot\cdot}{O}}:$ O kernel 6+   Oxidation number 1−   **Figure 5.6**
Assignment of electrons for oxidation
H   H   H kernel 1+   Oxidation number 1+   numbers in H_2O_2.

the more electronegative atom. In the oxygen-oxygen bond, the electron pair is shared between two like atoms and is split equally between the sharing partners. The apparent charge of the oxygen is thus 1−. Oxygen has an oxidation number of 1− in all peroxides.

4   *In most compounds containing hydrogen, the oxidation number of
hydrogen is 1+.* This rule covers practically all the hydrogen compounds. It fails in the case of the hydrides, in which hydrogen is bonded
to an atom less electronegative than hydrogen. For example, when
hydrogen is bonded to sodium in the compound sodium hydride (NaH),
the hydrogen is the more electronegative atom and two electrons are
counted with it. In hydrides, the oxidation number of hydrogen is 1−.

5   *All oxidation numbers must be consistent with the conservation of
charge.* Charge must be conserved in the sense that the sum of all the
apparent charges in a particle must equal the net charge of that particle.
This leads to the following conditions:

a   *For neutral molecules, the oxidation numbers of all the atoms must
add up to zero.*

b   *For complex ions (charged particles which contain more than one
atom), the oxidation numbers of all the atoms must add up to the
charge on the ion.*

As an example of a neutral molecule, consider the case of H_2O. The
oxidation number of hydrogen is 1+. There are two hydrogen atoms.
The total apparent charge contribution by hydrogen is 2+. The oxidation number of oxygen is 2−. The whole molecule appears to be
neutral. The neutrality rule enables us to assign oxidation numbers to
any single remaining atom. For example, what is the oxidation number
of sulfur in H_2SO_4? The oxidation number of hydrogen is 1+; the
oxidation number of oxygen is 2−. The two hydrogens give an apparent
charge of 2+; the four oxygens give an apparent charge of 8−. For
neutrality the sulfur must contribute 6+. Since there is but one sulfur
atom, the oxidation number of sulfur is 6+.

Since oxidation numbers are quite arbitrary, they may have values
which at first sight appear strange. For example, in sugar, $C_{12}H_{22}O_{11}$, the
oxidation number of carbon is 0. The total apparent charge of 22 hydrogen
atoms is canceled by that of 11 oxygen atoms. According to the oxidation
number, each carbon atom appears to contribute no charge to the molecule.
Fractional oxidation numbers are also possible, as in $Na_2S_4O_6$, where the
oxidation number of sulfur is $^{10}\!/_4+$.

In complex ions the apparent charges of all the atoms must add up to
equal the charge on the ion. This is true in hydroxide ion, OH^-, for example,
where the minus sign indicates that the ion has a net charge of 1−. Since
oxygen has an oxidation number of 2− and since hydrogen has an oxidation

number of 1+, the total apparent charge is $(2-) + (1+) = 1-$, which is the same as the actual charge of the ion. In $Cr_2O_7{}^{2-}$, a dinegative ion, the seven oxygen atoms contribute 14−. Chromium must contribute 12+ in order to make the ion have a net charge of 2−. Since there are two chromium atoms in the complex, each chromium atom has an oxidation number of 6+.

In order to avoid confusion with the actual charge on an ion, which is written as a superscript, the oxidation number of an atom, when needed, is written beneath the atom to which it applies. It should be emphasized strongly that oxidation numbers are not actual charges of atoms.

5.6 Oxidation-Reduction

Now the term *oxidation* refers to any chemical change in which there is an *increase in oxidation number, i.e., to a more positive value*. For example, when hydrogen (H_2) reacts to form water (H_2O), the hydrogen atoms change oxidation number from 0 to 1+. The H_2 is said to undergo oxidation. When sugar ($C_{12}H_{22}O_{11}$) is burned to give carbon dioxide (CO_2), carbon atoms increase in oxidation number from 0 to 4+. The sugar is oxidized. The term *reduction* applies to any *decrease in oxidation number*. For example, when oxygen (O_2) reacts to form H_2O, oxygen atoms change oxidation number from 0 to 2−; hence O_2 is said to undergo reduction. In oxidation and reduction the increase and decrease of oxidation numbers result from a shift of electrons. The only way by which electrons can be shifted away from an atom is for them to be pulled toward another atom. In this process the oxidation number of the first atom increases and the oxidation number of the second atom decreases. Oxidation and reduction must always occur together and must just compensate each other.

The *oxidizing agent* is, by definition, the *substance that does the oxidizing as it undergoes reduction*. It is that substance containing the atom which shows a decrease in oxidation number. For example, if in a reaction $KClO_3$ is converted to KCl, each chlorine atom decreases in oxidation number from 5+ to 1−. (This amounts to getting six electrons—six negative charges—from other atoms.) Thus, $KClO_3$ must cause oxidation and is acting as an oxidizing agent. Similarly, a *reducing agent* is the *substance that does the reducing as it undergoes oxidation*. It is the substance containing the atom which shows an increase in oxidation number. In the reaction of $C_{12}H_{22}O_{11}$ to give CO_2, $C_{12}H_{22}O_{11}$ is a reducing agent, because it contains carbon atoms which increase in oxidation number, from 0 in $C_{12}H_{22}O_{11}$ to 4+ in CO_2. Figure 5.7 summarizes the terms used to describe oxidation-reduction.

Listed in Figure 5.8 are some examples of oxidation-reduction processes. The numbers below the formulas indicate the oxidation numbers of interest.

Figure 5.7
OXIDATION-REDUCTION TERMS

Term	Oxidation-Number Change	Electron Change
Oxidation	Increase	Loss of electrons
Reduction	Decrease	Gain of electrons
Substance oxidized	Increase	Loses electrons
Substance reduced	Decrease	Gains electrons
Oxidizing agent	Decrease	Picks up electrons
Reducing agent	Increase	Supplies electrons

It must be emphasized that the term oxidizing agent or reducing agent refers to the entire substance and not to just one atom. For example, in the next-to-last reaction of the figure, the oxidizing agent is $KClO_3$ and not $5+Cl$. It can be shown that $KClO_3$ picks up electrons and therefore is an oxidizing agent, but it *cannot* be shown that it is the chlorine atom in $KClO_3$ that picks up electrons, because of the arbitrary rules for assigning oxidation numbers.

Figure 5.8
OXIDATION-REDUCTION REACTIONS

Oxidizing Agents	+	Reducing Agents	$\longrightarrow$	Products
O_2		H_2		H_2O
0		0		$1+\ 2-$
Cl_2		Na		$NaCl$
0		0		$1+\ 1-$
H^+		Mg		$Mg^{2+} + H_2$
$1+$		0		$2+\quad 0$
$KClO_3$		$C_{12}H_{22}O_{11}$		$KCl + CO_2 + H_2O$
$5+$		0		$1-\quad 4+$
H_2O_2		H_2O_2		$H_2O + O_2$
$1-$		$1-$		$2-\quad 0$

In the last reaction listed in Figure 5.8, H_2O_2 acts both as a reducing agent and an oxidizing agent. In oxidizing and reducing itself, it is said to undergo *auto-oxidation*, or *disproportionation*.

5.7 Chemical Equations

Chemical equations are shorthand designations which give information about a chemical reaction. We shall generally use *net equations, which specify only the substances used up and the substances formed* in the chemical reaction. Net equations omit anything which remains unchanged. The convention used in writing equations is to place what disappears (the *reactants*) on the left-hand side and what appears (the *products*) on the right-hand side. The reactants and products are separated by a single arrow $\longrightarrow$, an equal sign $=$, or a double arrow $\rightleftharpoons$, depending on what aspect of the chemical reaction is being emphasized. An example of a net equation is

$$Cl_2(g) + H_2O + Ag^+ \longrightarrow AgCl(s) + HOCl + H^+$$

The reactants and products are designated by appropriate formulas. The formula represents either one formula-unit (e.g., a molecule) or 1 mole. The notation (g) indicates the gas phase and (s) the solid phase. When no such phase notation appears, the liquid phase is understood.

To be valid, a chemical equation must satisfy three conditions.

1 It must be consistent with the experimental facts; that is, it must state what chemical species disappear and appear.

2 It must be consistent with the conservation of mass. (Since we cannot destroy mass, we must account for it. If an atom disappears from one substance, it must appear in another.)

3 The chemical equation must be consistent with the conservation of electric charge. (Since we cannot destroy electric charge, we must account for it.)

Conditions 2 and 3 are expressed by saying that the equation must be *balanced*. A balanced equation contains the same numbers of atoms of the different kinds on the left- and right-hand sides; furthermore, the net charge is the same on both sides.

How do we go about writing balanced equations? One method, useful only for simple reactions, is to balance the equation by inspection. For example, in the reaction between solid sodium and gaseous diatomic chlorine, solid sodium chloride is formed, and so we write first

$$Na(s) + Cl_2(g) \longrightarrow NaCl(s)$$

To balance this equation, we note that we have two chlorine atoms on the left, and so we ought to have two chlorine atoms on the right. We cannot change the subscript of Cl in the formula NaCl, because that would give the formula of a different compound. We can change only the coefficients; hence we put 2 in front of the NaCl. With two sodium atoms on the right we now need two sodium atoms on the left; therefore, we also place a 2 in front of the Na.

Chemical Formulas and Equations

The equation now reads

$$2Na(s) + Cl_2(g) \longrightarrow 2NaCl(s)$$

and has been balanced by inspection.

There are more complicated reactions involving electron transfer where balancing by inspection becomes quite a chore. For example, suppose that, in the reaction which occurs between potassium dichromate, sulfur, and water, the products are sulfur dioxide, potassium hydroxide, and chromic oxide.

$$K_2Cr_2O_7(s) + H_2O + S(s) \longrightarrow SO_2(g) + KOH(s) + Cr_2O_3(s)$$

Although the equation may be balanced by inspection, it is easier to balance it by matching up the electron transfer, i.e., the oxidation and the reduction. So far as electron transfer is concerned, we have to worry only about those atoms which change oxidation number. Applying the rules for assigning oxidation numbers, we see that sulfur changes oxidation number from 0 to 4+ and chromium from 6+ to 3+. As indicated below, each sulfur atom appears to lose four electrons and each chromium atom appears to gain three electrons:

$$K_2Cr_2O_7(s) + H_2O + S(s) \longrightarrow SO_2(g) + KOH(s) + Cr_2O_3(s)$$
6+ .... 0 .... 4+ .... 3+

3e⁻ per atom
6e⁻ per formula-unit | 4e⁻

Since each formula-unit of $K_2Cr_2O_7$ contains two chromium atoms, a formula-unit will pick up six electrons. These electrons must be furnished by the S. In order that the electron loss and the electron gain be equal, for every two $K_2Cr_2O_7$ formula-units that disappear (12 electrons picked up) three S atoms must be used up (12 electrons furnished). This is indicated by writing 2 in front of the $K_2Cr_2O_7$ and the Cr_2O_3 and 3 in front of the S and the SO_2 to give

$$2K_2Cr_2O_7(s) + H_2O + 3S(s) \longrightarrow 3SO_2(g) + KOH(s) + 2Cr_2O_3(s)$$

Although the difficult part is over, the equation is not balanced. To complete the job, the other coefficients must be made consistent with those already determined. We can do this by inspection. From the above equation we can see that we get four potassium atoms on the right, and so we place a 4 in front of the KOH. The result

$$2K_2Cr_2O_7(s) + H_2O + 3S(s) \longrightarrow 3SO_2(g) + 4KOH(s) + 2Cr_2O_3(s)$$

is still not balanced. Balance may be achieved by counting either the H atoms or the O atoms on the right. This shows that two molecules of H_2O are required. The balanced equation is

$$2K_2Cr_2O_7(s) + 2H_2O + 3S(s) \longrightarrow 3SO_2(g) + 4KOH(s) + 2Cr_2O_3(s)$$

Here in summary are the steps followed:

1   Assign oxidation numbers for those atoms which change.

2   Decide on the number of electrons to be shifted per atom.

3   Decide on the number of electrons to be shifted per formula-unit.

4   Compensate electron gain and loss by writing appropriate coefficients for the oxidizing agent and the reducing agent.

5   Insert other coefficients consistent with the conservation of matter.

A chemical equation is valuable from two standpoints. It gives information on an atomic scale and also on a laboratory scale. For example, for

$$8KClO_3(s) + C_{12}H_{22}O_{11}(s) \longrightarrow 8KCl(s) + 12CO_2(g) + 11H_2O(g)$$

on an atomic scale, the equation states that 8 formula-units of $KClO_3$ (each formula-unit containing a potassium atom, a chlorine atom, and three oxygen atoms) react with 1 formula-unit of $C_{12}H_{22}O_{11}$ to produce 8 formula-units of KCl, 12 formula-units of CO_2, and 11 formula-units of H_2O. Multiplying the equation through by the same number does not change its significance. Multiplying the equation through by the Avogadro number converts it from the atomic scale to something which is useful in the laboratory. The Avogadro number of formula-units in 1 mole, so that the equation signifies that 8 moles of $KClO_3$ reacts with 1 mole of $C_{12}H_{22}O_{11}$ to give 8 moles of KCl plus 12 moles of CO_2 plus 11 moles of H_2O. From the formula weights of the various compounds we can get further quantitative information from the equation. Eight moles of $KClO_3$ weighs 8×122.553 g, or 980.424 g; 1 mole of sucrose weighs 342.302 g; 8 moles of KCl weighs 8×74.555 g, or 596.44 g; 12 moles of CO_2 weighs 12×44.010 g, or 528.12 g; and 11 moles of H_2O weighs 11×18.015 g, or 198.17 g. The total mass on the left-hand side of the equation is 1322.73 g and that on the right-hand side 1322.73 g. Mass is conserved, as it must be.

Once a balanced chemical equation is obtained, it can be used for solution of problems involving mass relationships in chemical reactions. This is illustrated by the following examples.

EXAMPLE 3   The first chemical equation ever written, by Antoine Lavoisier for the fermentation of grape juice, can now be put in modern terms. In this reaction a sugar such as glucose, $C_6H_{12}O_6$, is converted to ethyl alcohol, C_2H_5OH, and carbon dioxide, CO_2. How many grams of ethyl alcohol can be made from 1 kg of glucose?

First write the balanced equation:

$$C_6H_{12}O_6 \longrightarrow 2C_2H_5OH + 2CO_2(g)$$

Now solve the problem in moles:

$$1 \text{ kg of } C_6H_{12}O_6 = \frac{1000 \text{ g}}{180 \text{ g/mole}} = 5.55 \text{ moles of glucose}$$

From the equation, 1 mole of $C_6H_{12}O_6$ gives 2 moles of C_2H_5OH. 5.55 moles of $C_6H_{12}O_6$ gives 5.55×2, or 11.1, moles of C_2H_5OH.

Finally, convert from moles to the answer required in grams: 11.1 moles of C_2H_5OH weighs

$$(11.1 \text{ moles})(46.0 \text{ g/mole}) = 510.6 \text{ g}$$

EXAMPLE 4   In the reaction of vanadium oxide (VO) with iron oxide (Fe_2O_3), the products are V_2O_5 and FeO. How many grams of V_2O_5 can be formed from 2.00 g of VO and 5.75 g of Fe_2O_3?

In solving this problem, we first write the balanced equation

$$2VO(s) + 3Fe_2O_3(s) \longrightarrow 6FeO(s) + V_2O_5(s)$$

Because there are two reactants we need to decide which reactant limits the amount of products and which is present in excess. To do this, we convert the data into moles. The formula weight of VO is 66.94 amu; the formula weight of Fe_2O_3 is 159.69 amu.

In 2.00 g of VO there is 2.00/66.94, or 0.0299, mole of VO.
In 5.75 g of Fe_2O_3 there is 5.75/159.69, or 0.0360, mole of Fe_2O_3.
According to the equation, 2 moles of VO require 3 moles of Fe_2O_3.
Therefore 0.0299 mole of VO requires (3/2)(0.0299), or 0.0449, mole of Fe_2O_3.

Since the 0.0360 mole of Fe_2O_3 is less than the 0.0449 mole required, there is not enough Fe_2O_3 to react with all the VO. The VO is present in excess, and the reaction is limited by the amount of Fe_2O_3.

Therefore we look at the equation again and note that
3 moles of Fe_2O_3 produce 1 mole of V_2O_5.
0.0360 mole of Fe_2O_3 produces (1/3)(0.0360), or 0.0120, mole of V_2O_5.
1 mole of V_2O_5 is 181.9 g.
0.0120 mole of V_2O_5 is 0.0120×181.9, or 2.18, g.

The above methods will suffice for obtaining solutions to chemical problems where the net equation is known or can be worked out. Historically, there were cases where chemical knowledge had not progressed to the point chemists could write complete equations. In a few applications that is still the case. Consequently an alternate method for handling chemical computations without the use of a balanced equation is still occasionally employed. This method, which we will but briefly touch upon, utilizes a quantity called the *equivalent*. One *equivalent of an oxidizing agent* is defined as *the weight of substance that picks up the Avogadro number of electrons*. One *equivalent of a reducing agent* is defined as *the weight of substance that releases the Avogadro number of electrons*. Equivalents are defined in this way so that 1 equivalent (equiv) of any oxidizing agent reacts exactly with 1 equivalent of any reducing agent. For example, when H_2 acts as a reducing agent, it forms two H^+ ions, and therefore each H_2 molecule releases two electrons. One mole of H_2 releases two moles of electrons or two times the Avogadro number of electrons. It follows that one mole of H_2 contains two equivalents of the reducing agent H_2 and one equivalent of H_2 is one-half mole of H_2.

EXAMPLE 5

How many grams of hydrogen sulfide (H_2S) react with 6.32 g of potassium permanganate ($KMnO_4$) to produce K_2SO_4 and MnO_2?

Mn changes oxidation number from 7+ to 4+ in this reaction.
1 mole of $KMnO_4$ picks up 3 moles of electrons.
1 equiv of $KMnO_4$ weighs 158.04/3, or 52.680, g.
6.32 g of $KMnO_4$ is 6.32/52.680, or 0.120, equiv.
1 equiv of $KMnO_4$ requires 1 equiv of H_2S.
0.120 equiv of $KMnO_4$ requires 0.120 equiv of H_2S.
S changes oxidation number from 2− to 6+ in this reaction.
1 mole of H_2S furnishes 8 moles of electrons.
1 equiv of H_2S weighs 34.076/8, or 4.26, g.
0.120 equiv of H_2S is 0.120 × 4.26, or 0.511, g.

It should be noted that the weight of an equivalent is defined for a particular reaction and will change if the products change. For example, $KMnO_4$ might also be reduced to Mn_2O_3, in which case the oxidation-number change would be from 7+ to 3+. Therefore, the electron change would be 4, and there would be 4 equiv/mole, leading to a weight per equivalent of 158.04/4, or 39.51, g.

QUESTIONS

5.1 *Simplest Formulas* Write simplest formulas for each of the following: (*a*) contains twice as many O atoms as N atoms (and no other elements), (*b*) contains 1 mole each of H atoms and N atoms with 3 moles of O atoms, (*c*) contains 7.0 g of N and 8.0 g of O, (*d*) N_2H_4.

5.2 *Percent Composition* Aspirin has a molecular formula $C_9H_8O_4$. What is its percentage composition?

5.3 *Simplest Formula* The plastic, polyethylene, is found to contain by weight 85.7% C and 14.3% H. What is its simplest formula?

5.4 *Molecular Formula* (a) What information if any does a simplest formula give that a molecular formula does not? (b) What information if any does a molecular formula give that a simpler formula does not? (c) How can you tell whether a particular formula as given is simplest or molecular? (d) For calculating the weights of various products which can be formed from weights of various reactants, does it matter whether you use simplest or molecular formulas? Explain.

5.5* *Moles* How many moles of H_2O are there in (a) 100 g of H_2O, (b) 1.00×10^{24} molecules of H_2O, (c) the weight of water which can be prepared from 1.00 g of H and 1.00 g of O?

5.6 *Moles* What is the weight in grams of each of the following: (a) 1 mole of CO_2, (b) one molecule of CO_2, (c) the total carbon in 1 mole of CO_2?

5.7 *Chemical Reactions* Without using examples given in Section 5.4, give equations for the two types of chemical change discussed there.

5.8 *Oxidation Numbers* (a) Show clearly how each of the five rules for assigning oxidation numbers (ignore the exceptions) arise from the two premises on which oxidation numbers are based. (b) Discuss briefly examples where H and O violate the usual rules for assigning oxidation numbers to compounds.

5.9 *Oxidation Numbers* Assign oxidation numbers to each atom in each of the following molecules: CO_2, $HClO$, $HClO_4$, K_2SO_4, $Na_2Cr_2O_7$.

5.10 *Oxidation Numbers* Assign oxidation numbers to each atom in each of the following complex ions: $CrCl_6^{3-}$, $S_2O_3^{2-}$, I_3^-, HSO_4^-, $H_3P_2O_7^-$.

5.11 *Oxidation-Reduction* Analyze each of the following by stating what is oxidized, what is reduced, what is the oxidizing agent, and what is the reducing agent.
(a) $CuS(s) + 2O_2(g) \longrightarrow CuSO_4(s)$
(b) $Cl_2(g) + 4NH_3(g) \longrightarrow N_2H_4(g) + 2NH_4Cl$
(c) $N_2H_4 + 2H_2O_2 \longrightarrow N_2(g) + 4H_2O$
(d) $C_2N_2(g) + 2NaOH(s) \longrightarrow NaCN(s) + NaOCN(s) + H_2O$

5.12 *Chemical Equations* Balance the following equations and tell which is the oxidizing agent in each case:
(a) $CO(g) + Fe_2O_3(s) \longrightarrow FeO(s) + CO_2(g)$
(b) $KNO_3(s) + CO(g) \longrightarrow CO_2(g) + NO_2(g) + K_2O(s)$
(c) $KClO_3(s) + C_{12}H_{22}O_{11}(s) \longrightarrow KCl(s) + CO_2(s) + H_2O$
(d) $Na_2CO_3(s) + 4C(s) + N_2(s) \longrightarrow 2NaCN(s) + 3CO(g)$

5.13* *Equivalents* Consider the reaction in which $KClO_3$ is reduced to KCl and the sugar $C_6H_{12}O_6$ is oxidized to CO_2. (*a*) What is the weight of 1 equiv of the oxidizing agent? (*b*) What is the weight of 1 equiv of the reducing agent? (*c*) How many grams of $KClO_3$ are needed to oxidize 10.0 g of the sugar?

5.14 *Simplest Formula* (*a*) 27.5 g of the element barium are heated in an excess of oxygen gas. A compound is formed which weighs 33.9 g. What is its simplest formula? (*b*) When the compound is heated at higher temperature under reduced oxygen, it loses 3.2 g of oxygen and weighs 30.7 g. What is this compound's simplest formula? (*c*) Show that these experiments illustrate all three laws of chemical change.

5.15 *Molecular Formula* 3.01×10^{24} molecules of a pure compound contain 10.0 moles of C atoms, 355 g of Cl, and 1.20×10^{25} H atoms. What is the compound's molecular formula?

5.16 *Moles* How many moles are there in each of the following: (*a*) 26.0 g of benzene, C_6H_6; (*b*) 1.00×10^{20} molecules of benzene; (*c*) that weight of benzene which contains 12.0 g of C; (*d*) that weight of benzene which can be prepared by complete reaction of 123 moles of acetylene, C_2H_2?

5.17* *Weight Relations* Given the equation

$$PCl_5 \longrightarrow PCl_3 + Cl_2$$

(*a*) How many grams of Cl_2 can be made by decomposing 121.2 g of PCl_5? (*b*) How many grams of PCl_3 can be made by decomposing 56.8 g of PCl_5?

5.18 *Weight Relations* The molecular formula of benzoic acid is $C_7H_6O_2$. How many grams of oxygen and carbon are there in (*a*) 221 g of benzoic acid? (*b*) 0.576 mole of benzoic acid?

5.19* *Equivalents* Consider the following reaction:

$$NiS(s) + O_2(g) \longrightarrow NiO(s) + SO_2(g)$$

(*a*) Calculate the weight of an equivalent of reducing agent. (*b*) Calculate the weight of an equivalent of oxidizing agent. (*c*) How many equivalents are there in 0.700 mole of reducing agent?

5.20 *Weight Relations* How many grams of carbon dioxide can be made from (*a*) 22.0 g of CO and 14.1 g of O_2, (*b*) 18.1 g of CO and 7.21 g of O_2?

5.21* *Moles* How many moles are there in (*a*) 5.71 g of oxygen, (*b*) 16.5 g of V_2O_5, and (*c*) 122.3 g of SO_2?

5.22 *Moles* How many moles of $KMnO_4$ are required to just react with (*a*) 58.7 g of $H_2C_2O_4$ and (*b*) 321 g of $H_2C_2O_4$ in the reaction

$$5H_2C_2O_4(s) + 2KMnO_4(s) \longrightarrow 4H_2O(g) + 10CO_2(g) + 2MnO(s) + 2KOH(s)$$

5.23* *Simplest Formula* Compound X is found to contain by weight 55.6% C, 3.3% H, 9.3% N, and 31.8% O. What is the simplest formula of compound X?

5.24 *Heat* Given the equation

$$2H_2(g) + O_2(g) \longrightarrow 2H_2O(l)$$

and that 0.565 kilojoule of heat is liberated when 1 mole of H_2O is formed, calculate the amount of heat that is liberated when (*a*) 1.00 g of H_2O is formed, (*b*) 12.51 g of O_2 reacts with excess H_2.

5.25* *Equivalents* Given the balanced equation

$$2Fe(s) + O_2(g) \longrightarrow 2FeO(s)$$

(*a*) How many equivalents are there in 1 mole of Fe? (*b*) What are the weights of an equivalent of Fe and an equivalent of O_2?

5.26 *Percent Composition* 1.51×10^{22} atoms of X weigh 2.94 g, and 0.112 mole of Y weighs 3.23 g. X and Y combine to form a compound which contains 3 atoms of X and 2 atoms of Y. (*a*) What is the simplest formula of the compound? (*b*) What is the atomic weight of X? (*c*) What is the percentage composition by weight of the compound?

5.27 *Chemical Equations* Balance the following equations and indicate which is the reducing agent in each case:
(*a*) $H_2(g) + Fe_3O_4(s) \longrightarrow Fe(s) + H_2O(g)$
(*b*) $P_4(g) + H_2(g) \longrightarrow PH_3(g)$
(*c*) $H_2(g) + V_2O_3(s) \longrightarrow VO(s) + H_2O(g)$
(*d*) $NH_4NO_3(l) \longrightarrow N_2O(g) + H_2O(g)$

5.28 *Oxidation-Reduction* (*a*) Why, in any chemical reaction, is there a fixed ratio of weight of reducing agent to weight of oxidizing agent? (*b*) Show with a specific example that the classification of reactions as being either electron transfer or nonelectron transfer is a simplification of what may actually occur. (*c*) Why don't we assign oxidation numbers on a more realistic basis than "most electronegative takes all?"

5.29 *Oxidation Numbers* In assigning oxidation numbers according to the five rules of Section 5.5, no account is taken of resonance. (*a*) Show that resonance can be ignored in assigning oxidation numbers to SO_2, both through use of the rules and by working with the resonance formulas. (*b*) If S were more electronegative than O, show what oxidation numbers you would get. (*Note:* Both O's must have same oxidation number.)

5.30 *Chemical Equations* (*a*) By assuming that iron (Fe) does not change oxidation number in the following equation, assign oxidation numbers to all the atoms, tell what is oxidized, what reduced, what is the oxidizing agent, what is the reducing agent, and balance the equation:

$$Na_2CO_3(s) + FeCr_2O_4(s) + O_2(g) \longrightarrow Fe_2O_3(s) + Na_2CrO_4(s) + CO_2(g)$$

(b) Now assume that Fe is initially in the 2+ oxidation state (a more usual assumption for this reaction) and show that you get the same equation, and label the same substances as oxidizing agent and reducing agent.

5.31 *Chemical Equations* In balancing a chemical equation, you multiply an entire formula-unit by a factor. Often balance can be achieved more simply by changing the subscripts within the formula-units. Why is this practice wrong?

5.32 *Chemical Reactions* In the preparation of the element phosphorus, phosphate rock, $Ca_3(PO_4)_2$, is heated with sand, SiO_2, and coke, C. The products are calcium silicate, $CaSiO_3$, carbon monoxide, and phosphorus gas, P_2. (a) Write a balanced equation by assigning oxidation numbers to all atoms. (b) How many moles of P_2 can be prepared from a mixture containing 1 mole of oxidizing agent and 1 mole of reducing agent and an excess of the other reactant? (c) How many grams of P_2 can be obtained from a mixture containing 100 g of phosphate rock, 50.0 g of sand, and 10.0 g of coke?

5.33 *General* A pure, newly discovered element is heated in oxygen until reaction is complete to form a compound. You know the weight of element and the weight of compound. (a) What additional piece of information do you need to write an equation for the reaction? (b) What additional piece of information, in addition to your answer to (a), do you need to have to specify the oxidation state of the metal in the compound?

5.34 *General* A platinum crucible containing iron powder is heated until all the iron is oxidized to Fe_2O_3. If the original weight of crucible plus iron was 74.4 g and the final weight was 86.4 g, what was the weight of the crucible?

Gases

We are now equipped with sufficient background and chemical terminology to begin a systematic study of the materials that compose our world. For the study of any material we must first note its *state,* that is, whether it is solid, liquid, or gas. Such classification is important since many properties exhibited by a substance are common to all substances in that state of matter. Thus, we achieve a tremendous simplification by knowing the properties that can be expected for substances in each of the three states. Of the three, the gaseous state, although historically the last to be recognized, is actually the simplest and best understood. The reason for this simplicity is the fact that molecules in the gaseous state are little influenced by their neighbors. Strictly speaking molecules are best defined in the gaseous state, and much of what we understand concerning molecules is exact only for gases. For these reasons we will discuss in some detail the general observations concerning the properties of the gaseous state, the theory for understanding these observations, and finally the deviations from the theory and idealized behavior of gases.

6.1 Volume

The behavior of any gas can be described in terms of three variables—volume, temperature, and pressure. The volume of any substance is the space occupied by that substance. For gases, the volume of a sample is the same as the volume of the container in which it is held. Ordinarily, this volume is specified in units of liters (l), milliliters (ml), or cubic centimeters (cm³ or cc). As the name implies, one cubic centimeter is the volume of a cube one centimeter on an edge. One liter is exactly 1000 times as great as one cubic centimeter. It follows that one milliliter, which is one-thousandth of a liter, is equal to one cubic centimeter.

For liquids and solids, volume does not change much with a change of pressure or temperature. Consequently, to describe the amount of a solid or liquid being handled, e.g., the number of moles, it is usually sufficient to specify only the volume of the sample. For gases, this is not enough. As an example, 1 ml of hydrogen at a certain pressure and temperature will contain a different number of moles and have a different mass from 1 ml at some other pressure and temperature. In order to fix the number of moles in a given volume of gas, it is necessary to know its pressure and temperature.

When solids or liquids are mixed together, the total volume is roughly equal to the sum of the original volumes. However, this is not necessarily true for gases. For example, if gaseous bromine is added to a bottle full of air, the brown bromine gas spreads through the whole bottle, so that both the air and the bromine now occupy the same volume which originally contained only air. Since all gases can *mix in any proportion,* they are said to be *miscible.*

6.2   Temperature

We have already referred (Section 1.3) to temperature as a measure of the degree of hotness of a substance. It is a familiar observation that a hot and cold substance placed in contact with each other change so that the hot substance gets colder and the cold substance hotter. This is interpreted as resulting from a flow of heat energy from the hot body to the cold body. The hot body is said to have a higher temperature; the cold body, a lower temperature. Therefore, temperature determines the direction of heat flow: heat always flows from a region of higher temperature to one of lower temperature.

The international scale for measuring temperature is an absolute scale in the sense that it starts with the absolute zero. Absolute zero is the lower limit of temperature; or, to put it another way, no temperatures lower than absolute zero are attainable. The scale is called the kelvin scale after the English scientist Lord Kelvin, who proposed it in the year 1848. It is now defined by assigning the value 273.16 K (kelvins) to the temperature at which H_2O coexists in the liquid, gaseous, and solid states, i.e., the triple point. Thus, the size of the kelvin is defined as 1/273.16 of the temperature difference between absolute zero and the triple point of H_2O. The triple point of H_2O is 0.01 K higher than the normal† freezing point of H_2O, so that on the kelvin scale the normal freezing point of H_2O is 273.15 K.

Outside the laboratory temperatures below 200 K are rarely encountered and other temperature scales are used. On the Celsius scale the normal freezing point of H_2O (273.15 K) is set as zero. The size of the Celsius degree (°C) is the same as the kelvin. The normal boiling point of H_2O on the Celsius scale is 100°C. Because there are 100 degrees between the normal freezing and boiling points of H_2O, this temperature scale is sometimes called the centigrade scale. A comparison of the kelvin scale with the Celsius and Fahrenheit scales is shown in Figure 6.1. Although the size of one degree is the same on the Celsius and kelvin scales, the Fahrenheit degree is only

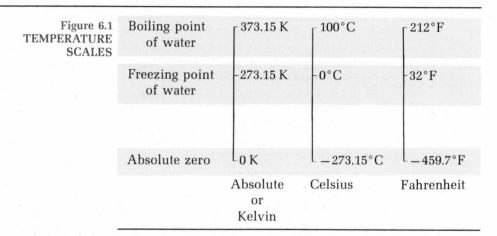

Figure 6.1 TEMPERATURE SCALES		Absolute or Kelvin	Celsius	Fahrenheit
Boiling point of water		373.15 K	100°C	212°F
Freezing point of water		273.15 K	0°C	32°F
Absolute zero		0 K	−273.15°C	−459.7°F

five-ninths as large. Temperature on the Celsius scale is converted to temperature on the kelvin scale by adding 273.15°:

$$°C + 273.15 = K$$

To convert Fahrenheit temperature to kelvin temperature, it is also necessary to correct for the difference in the size of the degree:

$$(°F - 32) \times \tfrac{5}{9} + 273.15 = K$$

6.3 Pressure

Just as temperature determines the direction of heat flow, pressure is a property which determines the direction of mass flow. Unless otherwise constrained, matter tends to move from a place where it is at higher pressure

†The *normal freezing point* is the temperature at which H_2O freezes under an atmosphere of pressure; the *triple point* is the temperature at which H_2O freezes under the pressure of its own vapor alone.

to a place of lower pressure. For example, when air escapes from an automobile tire, it moves from a region of higher pressure to one of lower pressure. Quantitatively, *pressure* is defined as *force per unit area.*

In *fluids,* a general term which includes *liquids and gases,* the pressure at a given point is the same in all directions. This can be visualized by considering a swimmer under water. At a given depth, no matter how he turns, the pressure exerted on him by the water is always the same. However, as he increases his depth, the pressure increases. This comes about because of the pull of gravity on the water above him. We can picture his body as being compressed by the weight of the column of water directly above him. In general, for all fluids, the greater the depth of immersion, the greater is the pressure.

The earth is surrounded by a blanket of air approximately 500 miles thick. In effect, we live at the bottom of an ocean of fluid, the atmosphere, which exerts a pressure. The existence of this pressure can be shown by evacuating a tin can. As the air is pumped out, atmospheric pressure crushes the can. A more subtle indication of atmospheric pressure can be obtained by filling a long test tube with mercury and inverting it in a dish of mercury. (Any other liquid would do, but mercury has the advantage of not requiring too long a test tube.) Some of the mercury runs out of the tube, but the important thing we observe is that not all of it runs out. The experiment is represented in Figure 6.2. No matter how large the diameter of the tube and no matter how long the tube, the difference in height between the mercury level inside and outside the tube is always the same. The fact that all the mercury does not run out shows that there must be a pressure exerted on the surface of the mercury in the dish sufficient to support the column of mercury. There is essentially nothing in the space above the mercury level in the tube, because at room temperature mercury does not evaporate much. To a good approximation, the space is a vacuum and exerts no pressure on

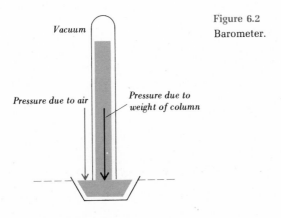

Vacuum

Pressure due to air

Pressure due to weight of column

Figure 6.2
Barometer.

the upper mercury level. The pressure at the bottom of the mercury column is therefore due only to the weight of the mercury column.

As noted, it is a general property of liquids that at any given level in the liquid the pressure is constant. In Figure 6.2 the broken line represents the level which is of interest. At this level, outside the tube, the force per unit area is due to the atmosphere and can be labeled P_{atm}. The pressure inside the tube at this level is due to the pressure of the column of mercury and can be labeled P_{Hg}. The equality $P_{atm} = P_{Hg}$ provides a method for measuring the pressure exerted by the atmosphere. The device shown in Figure 6.2 is called a *barometer*.

The atmospheric pressure changes from day to day and from one altitude to another. A *standard atmosphere* is defined as the pressure which supports a column of mercury that is 760 mm high at 0°C at sea level.† A standard atmosphere is referred to as 1 atm. We will use the standard atmosphere as our principal unit of pressure. However, because we often measure atmospheric pressure with a mercury barometer, it is necessary to remember that 760 mm of mercury equals one standard atmosphere.

We can also measure pressure by the height of a water column. Since water has a density of 1 g/ml, whereas mercury has a density of 13.6 g/ml, a given pressure supports a column of water that is 13.6 times as high as one of mercury. One atmosphere pressure supports 760 mm of mercury, or 760 × 13.6 mm of water, the latter being roughly 34 ft. Finally, because pressure is really force per unit area, there are other units for expressing pressure. In such units, a standard atmosphere equals 14.7 lb per square inch or 1.013×10^5 newtons/meter². However, for simplicity we will express pressure in units of atmospheres (meaning, the standard atmosphere) wherever possible.

The device shown in Figure 6.3 is a *manometer*, used to measure the pressure of an enclosed sample of gas. The manometer is constructed by placing a liquid in the bottom of a U tube with the gas sample in one arm of the U. If the right-hand tube is open to the atmosphere, the pressure which is exerted on the right-hand liquid surface is atmospheric pressure P_{atm}. At the same liquid level in the other arm of the tube, the pressure must be equal; otherwise, there would be a flow of liquid from one arm to the other. At the level indicated by the broken line in Figure 6.3, the pressure in the left arm is equal to the pressure P_{gas} due to the trapped gas plus the pressure P_{liq} exerted by the column of liquid above the broken line. We can therefore write

---

† If we think of pressure as weight per unit area, we can see why it is necessary that both 0°C and sea level be specified in defining the standard atmosphere. The density of liquid mercury changes with temperature and therefore the weight of a 760-mm-high Hg column of fixed cross section changes with temperature. Hence the temperature must be specified. Similarly, the force of gravity changes slightly with altitude; hence the weight of the Hg column changes when moved away from sea level.

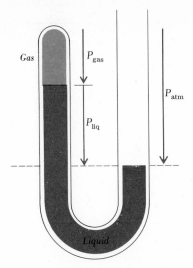

Figure 6.3
Manometer.

$$P_{atm} = P_{gas} + P_{liq}$$

or

$$P_{gas} = P_{atm} - P_{liq}$$

The atmospheric pressure can be measured by a barometer, and P_{liq} can be obtained from the difference in height between the liquid level in the right and left arms. P_{atm} and P_{liq} must be expressed in the same units. For example, if P_{atm} is measured in millimeters of mercury and the manometer liquid is not mercury, the difference in height must be converted to its mercury equivalent. If the bottom of the U tube consists of flexible rubber tubing, the right arm can be raised with respect to the left arm until the two liquid levels are the same height; in such case $P_{liq} = 0$, and $P_{gas} = P_{atm}$.

6.4 Boyle's Law

A characteristic property of gases is their great compressibility. This behavior is summarized quantitatively in Boyle's law (1662). *Boyle's law* states that *at constant temperature a fixed mass of gas occupies a volume inversely proportional to the pressure exerted on it.* If the pressure is doubled, the volume becomes one-half as large. Figure 6.4 shows a sample of gas trapped in a cylinder with a movable piston. When the weight on the piston is doubled, the pressure exerted on the gas is doubled and the gas volume shrinks to half its original volume. Boyle's law can be summarized by a pressure-volume, or *P-V*, plot like that shown in Figure 6.5. In this graph

Gases

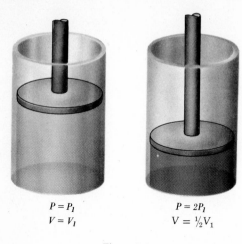

$P = P_1$
$V = V_1$

$P = 2P_1$
$V = \frac{1}{2}V_1$

Figure 6.4
Boyle's law experiment.

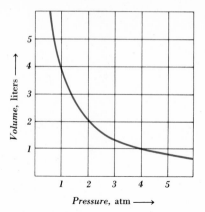

Figure 6.5
P-V plot for a gas.

the horizontal axis represents the pressure of a given sample of gas, and the vertical axis, the volume occupied. The curve is a hyperbola, the equation for which is PV = constant, or V = constant$/P$. (The size of the constant is fixed once the mass of the sample and its temperature are specified.) If at 4 atm the volume is 1 liter, then at 1 atm the volume is 4 liters. This can be seen from either the graph or the equation.

The behavior specified by Boyle's law is not always observed. For any gas, the law is most nearly followed at lower pressures and at higher temperatures, but as the pressure is increased or as the temperature is lowered, deviations may occur. This can be seen by considering the experimental data in Figure 6.6. In each of these experiments, the quantity of gas is fixed at 39.95 g, and the temperature is fixed either at 100°C or at −50°C. The pressure is measured when the given mass of gas is contained in different volumes. The PV products in the last column, obtained by multiplying the values in the second and third columns, should be, according to Boyle's law, constant at a constant temperature. The data shown indicate that, at the high

Figure 6.6 PRESSURE-VOLUME DATA FOR 39.95 g OF ARGON GAS	Temperature, °C	V, liters	P, atm	PV
	100	2.000	15.28	30.560
		1.000	30.52	30.520
		0.500	60.99	30.500
		0.333	91.59	30.530
	−50	2.000	8.99	17.980
		1.000	17.65	17.650
		0.500	34.10	17.050
		0.333	49.50	16.500

Figure 6.7	P, atm	0.5	1.0	2.0	4.0	8.0
PV PRODUCTS FOR A SAMPLE OF ACETYLENE	PV	1.0057	1.0000	0.9891	0.9708	0.9360

temperature, Boyle's law is closely obeyed. However, at the low temperature, the PV product is not constant but drops off significantly as the pressure increases; Boyle's law is not obeyed. In other words, as the temperature of argon is decreased, its behavior deviates from that specified in Boyle's law.

The fact that deviations from the law increase at higher pressures can be seen from the experimental data for acetylene given in Figure 6.7. When the pressure is doubled from 0.5 to 1.0 atm, the PV product is essentially unchanged, so that in this pressure range acetylene follows Boyle's law reasonably well. However, when the pressure is doubled from 4.0 to 8.0 atm, the PV product decreases by more than 3 percent; in this pressure range, Boyle's law is not followed so well. For any gas, the lower the pressure, the closer is the approach to Boyle's law behavior. When the law is obeyed, the gas is said to show *ideal behavior*.

The following example shows how Boyle's law is used. It is assumed that the temperature remains constant and that the behavior is ideal.

EXAMPLE 1   To what pressure must a gas be compressed in order to get into a 3.00-liter tank all the gas that occupies 400 liters at atmospheric pressure?

Volume changes to 3.00/400 of the original volume.
Pressure changes inversely.
Pressure changes to 400/3.00 of original

(400/3.00)(1.00 atm) = 133 atm

6.5 Charles' Law

Another characteristic property of gases is their thermal expansion. Like most other substances, all gases increase in volume when their temperature is raised. Experimentally, the increase of volume with increasing temperature can be measured by confining a fixed mass of gas in a cylinder fitted with a sliding piston, shown in Figure 6.8. The mass on top of the piston is constant, and so the gas sample remains at constant pressure. It is observed that, as the gas is heated, the piston moves out and the volume increases.

Typical numerical data are plotted in Figure 6.9. The points fall on a straight line, indicating that the volume varies linearly with temperature. If the temperature is lowered sufficiently, the gas liquefies and no more

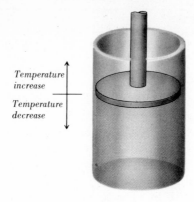

Figure 6.8
Gas cylinder with movable piston.

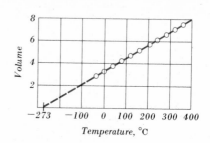

Figure 6.9
Volume of a gas as a function of Celsius temperature.

experimental points can be obtained. However, if the straight line is extended, or extrapolated, to lower temperatures, as shown by the broken line, it reaches a point of zero volume. The temperature at which the broken line reaches zero volume is −273.15°C, no matter what the kind of gas used or the pressure at which the experiment is performed. Designating −273.15°C as absolute zero (0 K) is reasonable, since temperatures below this would correspond to negative volume.

If volume-temperature data like those plotted in Figure 6.9 are given in terms of absolute temperature (K), it is found that, *at constant pressure, the volume occupied by a fixed mass of gas is directly proportional to the absolute temperature.* This summarization of gas behavior is called *Charles' law* (1787) and can be expressed mathematically as $V = $ constant $\times T$, where T is in kelvins. The value of the constant depends on pressure and on the quantity of gas.

Charles' law, like Boyle's law, represents the behavior of an *ideal,* or *perfect,* gas. For any *real* gas at high pressures and at temperatures near the liquefaction point, deviations from Charles' law are observed. Near the liquefaction point the observed volume is less than that predicted by Charles' law. For simplicity, in solving gas problems, we shall assume ideal behavior, which is usually a good approximation.

EXAMPLE 2   A sample of nitrogen gas weighing 9.3 g at a pressure of 0.988 atm occupies a volume of 12.3 liters when its temperature is 450 K. What is its volume when its temperature is 300 K?

Absolute temperature changes to 300/450 of its original value.
Volume changes proportionally.
Volume changes to 300/450 of its original value.
Final volume is (300/450)(12.3), or 8.20, liters.

Because of the Charles' law relation of volume to absolute temperature, calculations involving gases require conversion of temperatures to the kelvin scale. It is also convenient in working with gases to have a reference point. The customary reference point for gases is 273.15 K (0°C) and 1 standard atmosphere pressure. These conditions are called *standard temperature and pressure* (STP).

6.6   Dalton's Law of Partial Pressures

The behavior observed when two or more gases are placed in the same container is summarized in Dalton's law of partial pressures (1801). *Dalton's law* states that the *total pressure exerted by a mixture of gases is equal to the sum of the partial pressures* of the various gases. The *partial pressure of a gas in a mixture* is defined as the *pressure the gas would exert if it were alone in the container*. Dalton's law can be illustrated with the aid of Figure 6.10. Each of the boxes is of the same volume, and each has a manometer for measuring pressure. Suppose a sample of hydrogen is

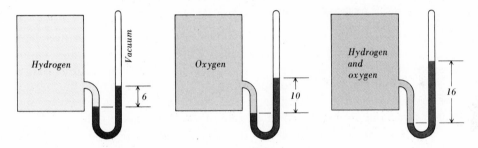

Figure 6.10
Dalton's law of partial pressures.

pumped into the first box, and its pressure is measured as 6 cmHg; a sample of oxygen is pumped into the second box, and its pressure is measured as 10 cmHg. If now both samples are transferred to the third box, the pressure is observed as 16 cmHg. For the general case, Dalton's law can be written

$$P_{total} = P_1 + P_2 + P_3 + \cdots$$

where the subscripts denote the various gases occupying the same volume. Actually, Dalton's law is an idealization but is closely obeyed by most mixtures of gases.

In many laboratory experiments dealing with gases, the gases are collected above water, and water vapor contributes to the total pressure measured. Figure 6.11 illustrates an experiment in which oxygen gas is collected

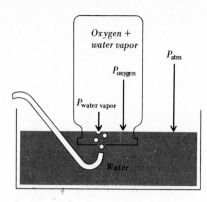

Figure 6.11
Oxygen collected over water.

by water displacement. If the water level is the same inside and outside the bottle, we may write

$$P_{atm} = P_{oxygen} + P_{water\ vapor}$$

or

$$P_{oxygen} = P_{atm} - P_{water\ vapor}$$

P_{atm} is obtained from a barometer. As we shall see later (Section 8.2), $P_{water\ vapor}$ depends only on the temperature of the water. This so-called vapor pressure of water has been measured at various water temperatures and is recorded in tables such as the one given in Appendix 5. Thus, the partial pressure of oxygen can be determined from an observed pressure and temperature and by reference to a table of vapor-pressure data.

The following example shows how Dalton's law of partial pressures enters into calculations involving gases.

EXAMPLE 3   If 40.0 liters of nitrogen is collected over water at 22°C when the atmospheric pressure is 0.956 atm, what is the volume of the dry nitrogen at standard temperature and pressure, assuming ideal behavior?

	Initial	Final
Volume, liters	40.0	?
Pressure, atm	0.956 − 0.026 = 0.930	1.000
Temperature, K	295	273

The initial volume of the nitrogen is 40.0 liters. The final volume is unknown. The initial pressure of the nitrogen gas is the atmospheric pressure, 0.956 atm, minus the vapor pressure of water. From the table of water pressures in Appendix 5, it can be seen that, at 22°C, water has a vapor pressure of 0.026 atm. The initial temperature of the nitrogen is 22°C, or 273 + 22 = 295 K. Final conditions are standard; i.e., the final pressure is 1.000 atm, the final temperature is 273 K. The problem is solved by considering separately how the volume is affected by a change in pressure and a change in temperature.

Pressure changes to 1.000/0.930 of its original value.
Volume changes inversely.
Volume changes to 0.930/1.000 of its original value.
Temperature changes to 273/295 of its original value.
Volume changes proportionally.
Volume changes to 273/295 of its original value.

$$V_{final} = V_{initial} \times \begin{array}{c}\text{(correction for}\\ \text{pressure change)}\end{array} \times \begin{array}{c}\text{(correction for}\\ \text{temperature change)}\end{array}$$

$$= 40.0 \text{ liters} \times (0.930/1.000) \times (273/295)$$

$$= 34.4 \text{ liters}$$

6.7 Gay-Lussac's Law of Combining Volumes

In the previous section we assumed that, when gases are mixed, they do not react with each other. However, sometimes they do react. For example, when a spark is passed through a mixture of hydrogen and oxygen gas, reaction occurs to form gaseous water. Similarly, when a mixture of hydrogen and chlorine gas is exposed to ultraviolet light, reaction occurs to form the gas hydrogen chloride. In any such reaction involving gases, it is observed that at constant temperature and pressure the volumes of the individual gases which actually react are simple multiples of each other. As a specific example, in the reaction of hydrogen with oxygen to form water, 2 liters of hydrogen are required for every 1 liter of oxygen. In the reaction of hydrogen with chlorine, each liter of hydrogen requires 1 liter of chlorine, and 2 liters of hydrogen chloride gas is formed. These observations are summarized in *Gay-Lussac's law of combining volumes* (1809), which states that, *at a given pressure and temperature, gases combine in a simple proportion by volume, and the volume of any gaseous product bears a whole-number ratio to that of any gaseous reactant.*

6.8 The Avogadro Principle

In the law of multiple proportions (Section 2.1), the observation of simple ratios between combining masses of elements implies that matter is atomic. Similarly, the occurrence of simple ratios between combining volumes of gases suggests that there is a simple relation between gas volume and number of molecules. Avogadro, in 1811, was the first to propose that equal volumes of gases at the same temperature and pressure contain equal numbers of molecules. That this principle accounts for Gay-Lussac's law can be seen from the following example:

When hydrogen combines with chlorine, the product, hydrogen chloride, can be shown by chemical analysis to contain equal numbers of hydrogen and chlorine atoms. These equal numbers of H and Cl atoms come from the original molecules of hydrogen gas and chlorine gas. If we assume that both hydrogen and chlorine molecules are diatomic, then equal numbers of hydrogen and chlorine molecules are required for reaction. According to the Avogadro principle, these occupy equal volumes, consistent with the observation that the combining volumes of hydrogen and chlorine gas are equal.

The assumption that hydrogen and chlorine molecules are diatomic rather than monatomic can be justified as follows: If hydrogen were monatomic, i.e., consisted of individual H atoms, and if chlorine were also monatomic, then 1 liter of hydrogen (n atoms) would combine with 1 liter of chlorine (n atoms) to give 1 liter of HCl gas (n molecules). This is contrary to the observation that the volume of HCl formed is *twice* as great as the volume of hydrogen or of chlorine reacted. It must be that the hydrogen and chlorine molecules are more complex than monatomic. If hydrogen and chlorine are diatomic, then 1 liter of hydrogen (n molecules, or $2n$ atoms) will combine with 1 liter of chlorine (n molecules, or $2n$ atoms) to form 2 liters of hydrogen chloride ($2n$ molecules). This agrees with experiment.†

As first shown by Stanislao Cannizzaro (1858), the Avogadro principle can be used as a basis for the determination of molecular weights. If two gases at the same temperature and pressure contain the same number of molecules in equal volumes, the masses of equal volumes give directly the relative masses of the two kinds of molecules. For example, at STP 1 liter of gas X is observed to weigh 0.0900 g, whereas 1 liter of oxygen weighs 1.43 g. Since the number of molecules is the same in both samples, according to the Avogadro principle, each X molecule must be 0.0900/1.43, or 0.0630, times as heavy as each oxygen molecule. Since the diatomic oxygen molecule has a molecular weight of 32.00 amu, the molecular weight of gas X is 0.0630 times 32.00, or 2.016, amu.

The volume occupied at STP by 28.01 g of nitrogen (1 mole) has been

---

† A footnote to Section 2.2 indicates that Dalton believed water to contain one H for each O. This error could have been corrected by the following reasoning: Two volumes of hydrogen react with one volume of oxygen to form two volumes of gaseous water. Since one volume of oxygen gives two volumes of water, the oxygen molecule must contain an even number of oxygen atoms. If oxygen, like hydrogen, is diatomic, the fact that two volumes of hydrogen are needed per volume of oxygen implies that the water molecule contains twice as many H atoms as O atoms.

Figure 6.12 MOLAR VOLUMES AT STP	Gas	Molar Volume, liters
	Hydrogen	22.432
	Nitrogen	22.403
	Oxygen	22.392
	Carbon dioxide	22.263
	Ideal gas	22.414

determined by experiment to be 22.4 liters. This is called the *molar volume of nitrogen at STP,* and should be the volume occupied by 1 *mole* of any so-called ideal gas at STP. Figure 6.12 shows the observed molar volumes of some gases. The value for the ideal gas is obtained from measurements made on gases at high temperatures and low pressures (where gas behavior is more nearly ideal) and extrapolated to STP by using Boyle's and Charles' laws. For the first three gases, agreement with ideality is quite satisfactory. Even for the fourth, carbon dioxide, the agreement is better than 1 percent. Consequently, in the future, we shall assume that at STP the molar volume of any gas is 22.4 liters. The following example shows how the molar volume can be used to determine molecular weight and molecular formula.

EXAMPLE 4   Chemical analysis shows that acetylene has a simplest formula corresponding to one atom of carbon for one atom of hydrogen. It has a density of 1.16 g/liter at standard temperature and pressure. What is the molecular weight and the molecular formula of acetylene?

At STP, 1 mole of gas (if ideal) has a volume of 22.4 liters.
Each liter of acetylene weighs 1.16 g, and so 1 mole of acetylene weighs 22.4 times 1.16 g, or 26.0 g.
Since the simplest formula is CH, the molecular formula must be some multiple of that, or $(CH)_x$.
Now the weight of CH is the atomic weight of carbon plus the atomic weight of hydrogen, or 13.0 amu.
For $(CH)_x$, the molecular weight is equal to x times 13.0.
Since 1 mole of acetylene weighs 26.0 g, x = 2.
The molecular formula of acetylene is $(CH)_2$, or C_2H_2.

6.9 Equation of State

Boyle's law, Charles' law, and Avogadro's principle can be combined to give a general relation between the volume, pressure, temperature, and number of moles of a gas sample. Such a general relation is called an *equation of* state, and it tells how, in the change from one set of conditions to another, the four variables V, P, T, and n (the number of moles) change. For an ideal gas, the equation of state can be deduced as follows: According to Boyle's law, V is inversely proportional to P; according to Charles' law, V is directly proportional to T; according to Avogadro's principle, V is directly proportional to n. With the symbol $\propto$ used for "is proportional to," this can be written

$$V \propto \frac{1}{P} \qquad \text{at constant } T \text{ and } n$$

$$V \propto T \qquad \text{at constant } P \text{ and } n$$
$$V \propto n \qquad \text{at constant } T \text{ and } P$$

or, in general,

$$V \propto \frac{1}{P}(T)(n)$$

(That this last relation embodies each of the other three can be seen by imagining any two of the variables, such as T and n, to be held constant and noting the relation of the other two.) Written as a mathematical equation, the general relation becomes

$$V = R\frac{1}{P}(T)(n) \qquad \text{or} \qquad PV = nRT$$

R, inserted as the constant of proportionality (Appendix 2.2), is called the *universal gas constant*. The equation $PV = nRT$ is the *equation of state for an ideal gas*, or the *perfect-gas law*.

The numerical value of R can be found by substituting known values in the equation. At STP, $T = 273.15$ K, $P = 1$ atm, and for 1 mole of ideal gas $(n = 1)$, $V = 22.414$ liters. Consequently

$$R = \frac{PV}{nT} = \frac{1 \times 22.414}{1 \times 273.15} = 0.082057$$

The units of R in this case are liter · atmosphere per degree per mole. In order to use this value of R in the equation of state, P must be expressed in atmospheres, V in liters, n in number of moles, and T in kelvins. (Other units could be used, of course, and the numerical value of R would be different.)

EXAMPLE 5   The density of an unknown gas at 98°C and 740 mmHg pressure is 2.50 g/liter. What is the molecular weight of this gas, on the assumption of ideal behavior?

Temperature is 98 + 273 = 371 K.
Pressure is 740/760 = 0.974 atm.
From the equation of state, $PV = nRT$, we can calculate the number of moles in 1 liter:

$$\frac{n}{V} = \frac{P}{RT} = \frac{0.974}{0.0821 \times 371} = 0.0320$$

Since 0.0320 mole weighs 2.50 g, 1 mole weighs 2.50/0.0320, or 78.1 g.

6.10 Graham's Law of Diffusion

As already noted, a gas spreads to occupy any volume accessible to it. This spontaneous spreading of a substance throughout a phase is called *diffusion*. Diffusion can readily be observed by liberating some ammonia gas in a room. Its odor soon fills the room, indicating that the ammonia has become distributed throughout the entire volume of the room. Furthermore, for a series of gases it is found that the lightest gas (i.e., the one of the lowest molecular weight) diffuses most rapidly. Quantitatively, under constant conditions the *rate of diffusion of a gas is inversely proportional to the square root of its molecular weight*. This is *Graham's law of diffusion* (1829) and in mathematical form is written

$$\text{Rate} = \frac{\text{constant}}{\sqrt{m}} \qquad \text{or} \qquad \frac{\text{rate}_1}{\text{rate}_2} = \frac{\sqrt{m_2}}{\sqrt{m_1}}$$

Rate$_1$ and rate$_2$ are the rates of diffusion of gases 1 and 2, and m_1 and m_2 are their respective molecular weights. In the case of oxygen gas and hydrogen gas

$$\frac{\text{Rate}_{H_2}}{\text{Rate}_{O_2}} = \frac{\sqrt{m_{O_2}}}{\sqrt{m_{H_2}}} = \sqrt{\frac{32}{2}} = \sqrt{16} = 4$$

The fact that heavier gases diffuse more slowly than light gases has been applied on a mammoth scale to effect the separation of uranium isotope ^{235}U (which undergoes nuclear fission) from ^{238}U (which does not). Natural uranium is converted to the gas UF_6 and passed at low pressure through a porous solid. The heavier $^{238}UF_6$ diffuses less rapidly than $^{235}UF_6$; hence the gas mixture which first emerges from the solid is richer in ^{235}U than is the starting mixture. Since the square root of the ratio of molecular weights is only 1.0043, the step must be repeated thousands of times, but eventually substantial enrichment of the desired 235 isotope is obtained.

6.11   Brownian Motion

One aspect of observed gas behavior which gives the strongest clue to the nature of gases is the phenomenon known as *Brownian motion*. This motion, first observed by the Scottish botanist Robert Brown, in 1827, is the *irregular zigzag movement of extremely minute particles when suspended in a liquid or a gas.* Brownian motion can be observed by focusing a microscope on a smoke particle illuminated from the side. The particle does not settle to the bottom of its container but moves continually to and fro and shows no sign of coming to rest. The smaller the suspended particle observed, the more violent is this permanent condition of irregular motion. The higher the temperature of the fluid, the more vigorous is the movement of the suspended particle.

The existence of Brownian motion contradicts the idea of matter as a quiescent state and suggests rather that the molecules of matter are constantly moving. A particle of smoke appears to be jostled by molecules of the air, and thus indirectly the motion of the smoke particle reflects the motion of the submicroscopic, invisible molecules of matter. Here then is powerful support for the suggestion that gases consist of minute bits of matter which are ever in motion.

6.12   Kinetic Theory

The "moving-molecule" theory is known as the kinetic theory of matter. Its two basic assumptions are that molecules of matter are in motion and that heat is a manifestation of this motion.

Like any theory, the kinetic theory represents a model which is proposed to account for an observed set of facts. In order that the model be useful, certain simplifying assumptions must be made about its properties. The validity of each assumption and the reliability of the whole model can be checked by how well the facts are explained. For an ideal gas, the following postulates are made:

1   Gases consist of tiny molecules, which are so small and so far apart, on an average, that the actual volume of the molecules is negligible compared to the empty space between them.

2   In the perfect gas, there are no attractive forces between molecules. The molecules are completely independent of each other.

3   The molecules of a gas are in rapid, random, straight-line motion, colliding with each other and with the walls of their container. In each collision, it is assumed that there is no *net* loss of kinetic energy, although there may be a transfer of energy between the partners in the collision.

4   At a particular instant in any collection of gas molecules, different molecules have different speeds and, therefore, different kinetic energies. However, the average kinetic energy of all the molecules is assumed to be directly proportional to the absolute temperature.

Before discussing each of these assumptions, we might ask how the model is related to the observable quantities V, P, and T. The accepted model of a gas is that it consists mostly of empty space in which billions of tiny points representing molecules move in violent motion, colliding with each other and with the walls of the container. Figure 6.13 shows a schematic version of this model. The *volume* of a gas is mostly empty space but is *occupied,* in the sense that moving particles occupy the entire region in which they move. *Pressure,* defined as force per unit area, is exerted by gases because molecules collide with the walls of the container. Each collision produces a tiny push, and the sum of all the pushes per second on 1 cm^2 (sq cm) of wall is the pressure. *Temperature* gives a quantitative measure of the average motion of the molecules.

That the first of the four postulates listed above is reasonable can be seen from the fact that the compressibility of gases is so great. Calculations show that, in oxygen gas, for example, at STP, 99.96 percent of the total volume is empty space at any instant. Since there are 2.7×10^{19} molecules per milliliter of oxygen gas at STP, the average spacing between molecules

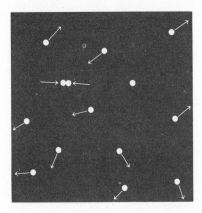

Figure 6.13
Kinetic model of a gas.

is about 3.7×10^{-7} cm, which is about 12 times the molecular diameter. When oxygen or any other gas is compressed, the average spacing between molecules is reduced; i.e., the fraction of free space is diminished.

The validity of the second postulate is supported by the observation that gases spontaneously expand to occupy all the volume accessible to them. This behavior occurs even for a highly compressed gas, where the molecules are fairly close together and any intermolecular forces should be greatest. It must be that there is no appreciable binding of one molecule of a gas to its neighbors.

As already indicated, the observation of Brownian motion implies that molecules of a gas move, in agreement with postulate 3. Like any moving body, molecules have an amount of kinetic energy equal to $\frac{1}{2} ms^2$, where m is the mass of the molecule and s is its speed. That molecules move in straight lines follows from the postulate of no attractive forces. Only if there were attractions between them could molecules be bent from straight-line paths. Because there are so many molecules in a gas sample and because they are moving so rapidly (at $0°C$ the average speed of oxygen molecules is about 1000 mph), there are frequent collisions. It is necessary to assume that the collisions are perfectly elastic in that no kinetic energy is lost by conversion to potential energy (as by distorting molecules). If this were not true, motion of the molecules would eventually stop, and the molecules would settle to the bottom of the container.

The fourth postulate has two parts: (1) that there is a distribution of kinetic energies and (2) that the average kinetic energy is proportional to the absolute temperature. The distribution, or range, of energies comes about as the result of molecular collisions, which continually change the speed of a particular molecule. A given molecule may move along with a certain speed until it hits another, to which it loses some of its kinetic energy; perhaps, later, it gets hit by a third and gains kinetic energy. This exchange of kinetic energy between neighbors is constantly going on, so that it is only the total kinetic energy of a gas sample that stays the same, provided, of course, that no energy is added to the gas sample from the outside, as by heating (or removed, as by cooling). The total kinetic energy of a gas is made up of the kinetic energies of all the molecules, each of which may be moving at a different speed. At a particular instant, a few molecules may be "standing still," with no kinetic energy; a few may have high kinetic energy; most will have kinetic energies near the average. The situation is summarized in Figure 6.14, which indicates the usual distribution of kinetic energies in a gas sample. Each point on the curve tells what fraction of the molecules have the specified value of the kinetic energy.

The temperature of a gas may be raised by the addition of heat. What happens to the molecules as the temperature is raised? The heat which is added is a form of energy and so can be used to increase the speed of the molecules and, therefore, the average kinetic energy. This is shown in Figure

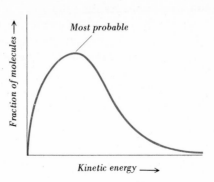

Figure 6.14
Energy distribution in a gas.

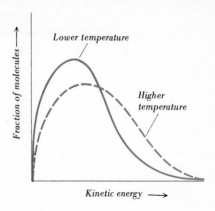

Figure 6.15
Energy distribution in a gas at two temperatures.

6.15, where the dashed curve describes the situation at higher temperature. At the higher temperature the molecules have a higher average kinetic energy than at the lower temperature. Thus, temperature serves as a measure of the average kinetic energy.

We can now also see how the kinetic theory accounts for the observed behavior of gases as expressed in the gas laws. It is worth remembering that these laws only describe gas behavior as kinetic theory explains it.

Boyle's law   The pressure exerted by a gas depends only on the number of molecular impacts per unit wall area per second, if the temperature is kept constant so that the molecules move with the same average speed. As shown in Figure 6.16, when the volume is reduced, the molecules do not have as much volume in which to move. They must collide with the walls more frequently; hence the walls receive more impacts per second, and the observed pressure is greater in the smaller volume.

Figure 6.16
Kinetic-theory explanation of Boyle's law.

Charles' law   The effect of raising the temperature of a gas is to raise the average kinetic energy of the molecules. As the molecules move more energetically, they collide with the walls of the container more frequently and more vigorously, thus producing a greater pressure. As shown in Figure 6.17, if the external pressure on a balloon is constant but the temperature is raised, the gas expands the balloon to a larger volume in which the more vigorous molecular motion is compensated for.

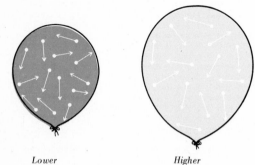

*Lower
temperature*

*Higher
temperature*

Figure 6.17
Kinetic-theory explanation of Charles' law.

Dalton's law   According to the kinetic theory, there are no attractive forces between the molecules of an ideal gas. In a mixture of gas molecules, each molecule strikes the walls the same number of times per second and with the same force as if no other molecules were present. Therefore, the partial pressure of a gas is not changed by the presence of other gases in the container.

Brownian motion   When a particle is suspended in a gas, gas molecules collide with it. If the particle is very large, the number of bombarding molecules on one side is about equal to the number of bombarding molecules on the other side. However, if the particle is small, so that the number of bombarding molecules at any instant is small, collisions on one side of the particle may predominate, so that the particle experiences a net force which causes it to move. This is illustrated in Figure 6.18. An analog of Brownian

Figure 6.18
Kinetic-theory explanation of Brownian motion.

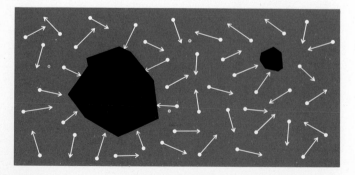

motion is observed when a small chunk of bread is thrown on the surface of a pool in which there are many small fish. The bread darts to and fro as if propelled by some unseen force, the invisible force being due to the bumping of the nibbling fish. The larger the piece of bread, the less its erratic motion.

Graham's law   This law can be deduced directly from the fourth postulate of the kinetic theory, that the average kinetic energy of molecules is fixed for a given temperature. When gas A and gas B are compared at the same temperature,

Average kinetic energy of A = average kinetic energy of B

$$\tfrac{1}{2}\,m_A s_A{}^2 = \tfrac{1}{2}\,m_B s_B{}^2$$

where m_A and m_B are the masses of the different molecules and s_A and s_B are their average speeds.

$$\frac{s_A{}^2}{s_B{}^2} = \frac{\tfrac{1}{2}\,m_B}{\tfrac{1}{2}\,m_A} = \frac{m_B}{m_A}$$

$$\frac{s_A}{s_B} = \sqrt{\frac{m_B}{m_A}}$$

Qualitatively, this last equation states that the speeds of molecules depend inversely on their mass, i.e., that heavier molecules must move more slowly than light ones because their kinetic energies are the same. The relative rates of diffusion of molecules should be measured by their relative average velocities, so that

$$\frac{\text{Rate of diffusion of gas A}}{\text{Rate of diffusion of gas B}} = \sqrt{\frac{m_B}{m_A}}$$

This is Graham's law.

6.13   Deviations from Ideal Behavior

We have seen that, at high pressures and at low temperatures, gases deviate from ideal behavior. For example, although a gas compressed at constant temperature from 1 to 2 atm changes its volume to one-half the original, the same gas compressed from 1000 to 2000 atm may change its volume to something other than one-half the original. Similarly, at constant low pressure a gas cooled from 300 to 200 K changes its volume to two-thirds the original volume, but the same gas cooled from 30 to 20 K may change its volume to something other than two-thirds the original. In fact, at sufficiently high pressures and sufficiently low temperatures, the volume of the gas phase may "disappear." The gas liquefies, a phenomenon which is the extreme case of nonideal behavior.

What are the sources of deviations from ideal behavior? In the kinetic theory of gases, two postulates can be questioned: (1) that the actual volume of the molecules is negligible compared to the empty space between them; (2) that there are no attractive forces between the molecules. Under what conditions would these assumptions break down? The volume of the molecules is not negligible when the molecules are close together, as they are at high pressure. Further compression is resisted by the impenetrability of molecules. Eventually the volume cannot be reduced any more, because all the free space is gone. Considering this molecular-volume effect alone, volumes observed at high pressure would be greater than predicted by the ideal-gas law. Assumption 2 is not strictly true, since it is known that there *are* attractive forces between molecules. However, at high temperatures, molecules are in violent motion, so that these attractions are not important. At low temperatures, on the other hand, the attractions are important, and they tend to make the observed pressure less than that predicted by the ideal-gas law.

6.14 Attractive Forces

In some cases it is easy to see the reason for attractive forces between molecules. For example, in polar molecules, the positive end of one molecule attracts the negative end of another molecule. It is not surprising, therefore, that polar substances deviate markedly from ideal behavior. Water vapor, as an illustration, is so nonideal that even at room temperature it liquefies under slight pressure. It is not so easy to see the reason for attractive forces between nonpolar molecules. That these forces do exist was suggested by the Dutch physicist Johannes van der Waals, for whom they are named. The *van der Waals forces* arise from the motion of electrons in atoms and molecules. They are present in all matter.

The van der Waals forces can be described in the following way: Suppose we consider two neon atoms extremely close together, as shown in Figure 6.19. We can imagine that *instantaneously* the electron distribution

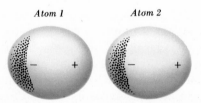

Atom 1   *Atom 2*

Figure 6.19
Model of van der Waals attraction.

in atom 1 becomes unsymmetrical, with a slight preponderance on one side. For a fraction of a microsecond, the atom is in a state in which one end appears slightly negative with respect to the other end; i.e., the atom is momentarily a dipole. As a result the neighboring atom is distorted, because the positive end of atom 1 displaces the electrons in atom 2. As shown in the figure, there is an instantaneous dipole in each of the neighboring atoms, with a consequent attraction. This picture persists only for an extremely short time, because the electrons are in motion. As electrons in atom 1 move to the other side, electrons in atom 2 follow. In fact, we can think of van der Waals forces as arising because of electron motion in adjacent molecules, so as to produce fluctuating dipoles which give rise to an instantaneous attraction. The attraction is strong when particles are close together but rapidly weakens as they move apart. The more electrons there are in a molecule and the less tightly bound these electrons are, the greater are the van der Waals forces.

Under what conditions do the forces of attraction produce the biggest effect? The closer the molecules are together, the greater is the attraction. This means that the attractive forces become more important as the pressure increases and the molecules are crowded together. The attractive forces become less important as the temperature increases, because a rise in temperature produces an effect that opposes the attractive forces. This effect is a disordering one due to the molecular motion, which increases in speed as the temperature rises. Disordering arises because the molecules of a gas move in random fashion. The attractive forces try to draw the molecules together, but the latter, because of their motion, stay apart. As the temperature is lowered, the molecules become sluggish. They have less ability to overcome the attractive forces. The attractive forces are unchanged, but the motion of the molecules decreases; hence the attraction becomes relatively more important. At sufficiently low temperature the attractive forces, no matter how weak they are, take over and draw the molecules together to form a liquid. The *temperature at which gas molecules coalesce to form a liquid* is called the *liquefaction temperature*. Liquefaction is easier at high pressures, where distances between molecules are smaller and hence intermolecular forces are greater. The higher the pressure of a gas, the easier it is to liquefy it and the less it needs to be cooled to accomplish liquefaction. Thus, the liquefaction temperature increases with increasing pressure.

6.15   Critical Temperature

There is for each gas a temperature above which the attractive forces are not strong enough to produce liquefaction, no matter how high the pressure. This temperature is called the *critical temperature* of the substance and is designated by T_c. It is defined as the *temperature above which the substance*

Figure 6.20 CRITICAL CONSTANTS	Substance	Critical Temperature, K	Critical Pressure, atm
	Water, H_2O	647	217.7
	Hydrogen chloride, HCl	324	81.6
	Carbon dioxide, CO_2	304	73.0
	Oxygen, O_2	154	49.7
	Nitrogen, N_2	126	33.5
	Hydrogen, H_2	33	12.8
	Helium, He	5.2	2.3

can exist only as a gas. Above the critical temperature the motion of the molecules is so violent that, no matter how high the pressure, the molecules occupy the entire available volume as a gas. The critical temperature depends on the magnitude of the attractive forces between the molecules.

Figure 6.20 contains values of the critical temperature for some common substances. Listed also is the *critical pressure, the pressure which must be exerted to produce liquefaction at the critical temperature.* Above the critical temperature, no amount of pressure can produce liquefaction. For example, above 647 K (374°C), H_2O exists only in the gaseous state. The high critical temperature indicates that the attractive forces between the polar water molecules are so great that, even at 647 K, they can produce coalescence. The attractive forces between CO_2 molecules are less than those between water molecules; hence the critical temperature of CO_2 is lower than that of water, and liquefaction cannot be achieved above 304 K (31°C). In the extreme case of helium the attractive forces are so weak that liquid helium can exist only below 5.2 K (−267.9°C). At this very low temperature the molecular motion is so slow that the weak van der Waals forces can hold the atoms together in a liquid. In Figure 6.20, the order of decreasing critical temperature is also the order of decreasing attractive forces, and we can think of the critical temperatures as giving a measure of the attractive forces between molecules.

QUESTIONS

6.1 *Volume* Why is the specification of the volume of a gas not very helpful unless its pressure and temperature are also given?

6.2 *Temperature* The kelvin scale is defined in terms of only a single temperature (triple point of water is 273.16 K), yet the Celsius and Fahrenheit scales are defined by two temperatures (normal freezing and boiling points of water). Why the difference?

6.3 *Pressure* How can you demonstrate that the atmosphere exerts a pressure?

6.4 *Boyle's Law* The air in a room of volume 50.0 m^3 is at a pressure of 0.950 atm. It is pumped into a tank of volume 10.0 liters without change of temperature. What should the air pressure in the tank be, assuming Boyle's law behavior?

6.5* *Charles' Law* A sample of gas is placed in a cylinder with a movable piston as shown in Figure 6.8. At 25°C, the bottom of the piston is 20.0 cm from the bottom of the cylinder. (*a*) At what temperature should the piston be raised 10.0 cm? (*b*) At what temperature should the piston be lowered 10.0 cm from its original position?

6.6 *Dalton's Law* A particular sample of moist, warm air contains oxygen at a partial pressure of 0.180 atm, nitrogen at 0.680 atm, and water vapor at 0.100 atm. The water vapor is removed, and the gas expanded to twice its volume. What should the partial pressures of oxygen and nitrogen become according to Dalton's law?

6.7 *Combining Volumes* Two volumes of hydrogen and one volume of oxygen react to form two volumes of water vapor. Suppose that is all you knew about hydrogen, oxygen, and water. According to the Avogadro principle, which of the following formulations for the three molecules account for the observed combining volumes? (*a*) H_2, O_2, HO; (*b*) H_2, O, H_2O; (*c*) H, O_2, HO; (*d*) H_3, O_2, H_3O; (*e*) H_2, O_4, H_2O_2; (*f*) H, O_2, HO_2.

6.8 *Molar Volume* The rocket fuel hydrazine has simplest formula NH_2. At STP its density is 1.43 g/liter. What is its molecular weight and molecular formula?

6.9* *Equation of State* (*a*) Derive an equation expressing the density of an ideal gas in terms of its pressure, temperature, and molecular weight. (*b*) How will each of the following separate changes affect the density of the gas: halving the pressure; doubling the kelvin temperature?

6.10 *Diffusion* Which should diffuse faster, $SO_2(g)$ or $CH_4(g)$? How many times as fast?

6.11 *Brownian Motion* Why does Brownian motion appear more violent at high temperature? Why does it appear more violent the smaller the smoke particle?

6.12 *Kinetic Theory* How does kinetic theory account for each of the following? (*a*) Gases exert pressure. (*b*) The pressure of a gas increases as its temperature is raised. (*c*) The partial pressure exerted by a gas sample is independent of the presence of other gases. (*d*) Gaseous diffusion rates are proportional to the square root of molecular weight.

6.13 *Deviations from Ideal Behavior* In terms of the assumptions basic to kinetic theory, tell why the PV product increases as V increases at low temperature but may increase as P increases at high P and high T.

6.14 *Attractive Forces* At atmospheric pressure neon gas liquefies at $-246°C$ and argon gas at $-186°C$. (a) What is the explanation for molecular attraction which allows liquefaction for these gaseous elements? (b) Account for the direction of the apparent difference in strength of molecular attractions in the two cases.

6.15 *Critical Temperature* (a) Which are moving faster at 40°C, $HCl(g)$ molecules or $CO_2(g)$ molecules? (b) At this temperature HCl can be liquefied but $CO_2(g)$ cannot. Why the difference?

6.16 *Gas Calculations* What volume should be occupied by 3.20 g of O_2 gas under each of the following conditions assuming ideal behavior: (a) STP, (b) 0°C and 0.500 atm, (c) 273°C and 2.00 atm, (d) when mixed with an equal number of moles of N_2 and placed at 0°C and a total pressure of 1.50 atm?

6.17* *Gas Calculations* How many moles of gas are there in a sample which occupies 700 ml at a pressure of 0.800 atm and temperature of 27°C?

6.18 *Gas Calculations* Suppose you have collected 300 ml of oxygen gas as in Figure 6.11. The atmospheric pressure is 0.950 atm and the temperature is 25°C, which means the water vapor pressure is 0.031 atm. How many grams of oxygen have you collected?

6.19 *Kinetic Theory* In terms of kinetic theory compare 1-liter samples of H_2 and O_2 at STP with respect to each of the following: (a) average kinetic energy of the molecules, (b) average speed of the molecules, (c) number of collisions with the walls of the container. (d) Now rationalize the fact that the total pressure is the same in both containers.

6.20 *Avogadro Principle* When a 3.21-g sample of an unknown gas is collected over water, the volume occupied at 0.976 atm atmospheric pressure and 30°C is 0.850 liter. Assuming ideal behavior of the gas, what is the molecular weight of the gas?

6.21* *Boyle's Law* Given that 4.2 g of an ideal gas occupies 7.2 liters at STP. What is its volume at 0°C and a pressure of 92 cmHg?

6.22 *Charles' Law* Given that 5.6 g of an ideal gas occupies 6.9 liters at STP. What is its volume at 100°C and a pressure of 76 cmHg?

6.23* *Gas Laws* Given that 5.6 g of an ideal gas occupies 6.9 liters at STP. What is its volume at 80°C and a pressure of 83 cmHg?

6.24 *Diffusion* Given four gases HCl, H_2O, SO_2, and NH_3. (a) Which of these gases will diffuse the fastest? (b) Calculate the ratio of diffusion of fastest to slowest.

6.25* *Avogadro Principle* Suppose you are given a gaseous compound for which the formula is C_xH_{2x-2}. At the same temperature and pressure at which oxygen weighs 4.80 g/liter, the unknown gas weighs 8.10 g/liter. What is the value of x?

6.26 *Partial Pressure* A tank was filled with 10.20 moles of N_2, 2.30 moles of H_2O, and 5.60 moles of O_2 at 120°C and 2.50 atm. What is the partial pressure of each gas in the tank?

6.27* *Avogadro Principle* Calculate the number of molecules in (a) 12.2 liters of oxygen at STP, (b) 12.2 liters of HCl at 70°C and 1.20 atm, (c) 5.60 liters of N_2 at 400°C and 17.2 atm.

6.28 *Stoichiometry* Sulfur dioxide in the amount of 12.1 g was produced by the reaction

$$2Fe_2S_3 + 9O_2 \longrightarrow 2Fe_2O_3 + 6SO_2$$

(a) How many liters of O_2 at STP were used? (b) How many liters of O_2 at 900°C and 2.20 atm were used?

6.29* *Gas Laws* In an experiment, 24.8 g of O_2 is to be collected over water at 30.0°C and a barometer reading 736 mmHg. What volume will be occupied by this oxygen?

6.30 *Partial Pressure* A 32.0-liter tank was filled with 7.86 moles of HCl, 11.2 moles of O_2, and 3.52 moles of H_2 at 56.0°C. What is the partial pressure of each gas in the tank?

6.31* *Avogadro Principle* A 10.0-liter box contains 41.4 g of a mixture of gases C_xH_8 and C_xH_{12}. At 44°C, the total pressure is 1.56 atm. Analysis of the gas mixture shows 87.0% carbon and 13.0% hydrogen. (a) What gases are in the box? (b) How many moles of each gas are in the box?

6.32 *Attractive Forces* Making gasoline from natural petroleum involves separating nonpolar hydrocarbon molecules from each other based on differences of boiling point (liquefaction temperatures). It is observed that lower molecular weight hydrocarbons have lower boiling points. What does this indicate concerning forces of attraction, and why should they differ in this manner?

6.33* *Reaction of Gases* Equal moles of H_2 and O_2 are injected into a steel cylinder until the pressure at 27°C is 2.00 atm. A spark plug ignites the mixture to form $H_2O(g)$ and the final temperature is raised to 127°C. What is the final pressure if behavior is ideal?

6.34 *Kinetic Theory* Dry air contains, on a mole basis, 78% N_2 and 21% O_2. On the average how many times as frequently does an N_2 molecule strike your forehead as does an O_2 molecule?

6.35* *General* What weight of air does an automobile tire hold under the following conditions: volume = 4.00 liters, temperature = 27°C, atmospheric pressure = 14.7 lb/in.^{2} (sq in.) plus gauge pressure of 28 lb/in.^{2}? Assume air is 79% N_2 and 21% O_2.

6.36 *Kinetic Theory* As noted in Chapter 4 there is a problem concerning the size of an isolated atom (or molecule) in that the electron probability distribution never becomes zero. However, on the assumption that all the electrons are accommodated to the same degree in gaseous and liquid water (density 1.00 g/ml), calculate the apparent fraction of empty space in the gas phase of water vapor at its vapor pressure at 0°C and at 100°C.

The Solid State 7

In some ways the solid state seems simpler than the gaseous state. In the first place the volume and even shape of a solid is much less dependent on external conditions of temperature and pressure than is true for gas. Second, whereas the gaseous state is characterized by disorder, in that the molecules are in constant motion, the solid state consists of atoms arranged in fixed, ordered positions. This regularity of arrangements of atoms makes the detailed study of solids possible. It also presents the challenge of accounting for observed properties in terms of these arrangements. However, the fact that there are many different solid structures and the fact that the total volume is pretty well filled with atoms whose differences influence observed properties, means that different solids have fewer properties in common than do gases. In this sense the solid state is not as simple as the gaseous state. Not only is there not a single equation of state or laws of solid behavior as there are for gases, but the kinetic theory, which is so successful for gases, provides only a qualitative understanding of solids.

7.1 Properties of Solids

Unlike a gas, which expands to occupy any accessible volume, a solid sample has a characteristic volume that does not change with the volume of the container or with any but quite large changes of pressure or temperature.

In other words, compared with gases, solids are nearly incompressible and do not expand much with an increase of temperature. This results from the existence of strong attractive forces between closely spaced atoms or molecules, But, there are also strong repulsive forces which keep electron clouds of atoms from penetrating each other significantly.

Another marked difference between the solid and gaseous states is found in the diffusion rates. Whereas any gas can diffuse through another in a time short enough to allow easy observation, many solids diffuse so slowly as to make changes essentially imperceptible. For example, rock layers have been in contact with each other for millions of years and still retain sharp boundaries. However, in some solids considerably more rapid diffusion has been demonstrated: metals interpenetrate to a depth of 0.1 mm in a matter of hours at elevated temperatures that are still below the melting point. If it is assumed that solids consist of atoms in virtual contact, it is reasonable that diffusion should be a slow process in solids. In fact, from this view one might ask why diffusion occurs at all. One principal reason is that solids are generally imperfect, having, for example, occasional vacancies where atoms or molecules should reside. Movement into these vacancies permits diffusion to occur. For many purposes, the existence of these defects can be ignored, but for certain properties such as diffusion, conductivity, and mechanical strength, their presence can be decisive.

Solids form *crystals,* definite geometric forms which are characteristic of the substance in question. The crystals of a given substance are bounded by plane surfaces called *faces.* The faces always intersect at an angle characteristic of the substance. For example, sodium chloride crystallizes in the form of cubes with faces which intersect at an angle of 90°. When a crystal is broken, it splits, or shows *cleavage,* along certain preferred directions, so that the characteristic faces and angles result even when the material is ground to a fine powder. Figure 7.1 shows the usual crystal shapes of some common chemicals. It should be pointed out that the same chemical substance can, under different conditions, form different kinds of crystals. For instance, although NaCl almost always crystallizes as cubes, it can also

Sodium chloride

Quartz

Alum

Figure 7.1
Crystals.

be made to crystallize in the shape of octahedral crystals like those shown for alum. The occurrence of different crystalline modifications of the same chemical is called *polymorphism.* The existence of crystals can be accounted for by assuming that the molecules or atoms which make up a crystal are arranged in orderly patterns characteristic of the substance.†

7.2 Determination of Structure

Information about the arrangement of fundamental particles in solids can be gained from the external symmetry of crystals. However, much more information is obtained from X-ray diffraction. X rays are radiant energy much like light but more energetic and of greater penetrating power. They can be produced by bombarding a metal with energetic electrons. The energy of the electrons is transferred to the atoms of the metal and excites the atoms to a higher energy level. When the atoms return from a high energy level to a low energy level, they emit energy in the form of X rays. The target metal is usually copper or molybdenum.

Figure 7.2 shows a simple X-ray tube. It is a gas discharge tube, with the cathode connected to the negative side of a high voltage and the anode to the positive side. The cathode is concave, so that the electrons which come from it are focused to a point on the metal target. The resultant X rays radiate from the target in all directions, and present a hazard that requires shielding.

For the study of crystalline materials, X rays are collimated into a beam by a lead shield with a hole in it. If sufficiently thick, the shield stops all

† Not all solids are crystalline in form. There are some substances, such as glass, which have the solid-state properties of extremely slow diffusion and virtually complete maintenance of shape and volume but do not have the ordered crystalline state. As a result, glass flows extremely slowly as by the distortion that can be seen in very old window panes. These substances are sometimes called amorphous solids and will be discussed in Chapter 8.

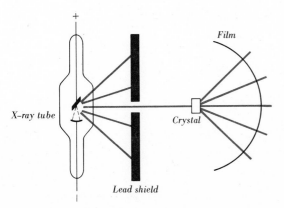

Figure 7.2
X-ray determination of structure.

X rays except the beam which comes through the hole. A well-formed crystal is mounted in the path of the X-ray beam, as shown in Figure 7.2. As the X rays penetrate the crystal, the atoms which make up the crystal scatter or deflect some of the X rays from their original path. The X rays can be detected by a piece of photographic film at some distance from the crystal. The X-ray beam exposes the film, and when the film is developed, a spot appears at each point where the X rays struck it. The developed film shows not just one spot, but a pattern, which is uniquely characteristic of the crystal investigated.

Figure 7.3 shows the pattern for a typical crystal. The big spot in the center corresponds to the main unscattered beam. The other spots represent a scattering of part of the original beam through various characteristic angles. The creation of many beams from one is like the effect observed when a light beam falls on a diffraction grating—a piece of glass on which are scratched thousands of parallel lines that are opaque and leave, in effect, many narrow slits of unscratched surface for light to pass through. The spreading of the light wave from each slit results in interference between waves from different slits, giving rise to alternate regions of light and dark. The diffraction of X rays by crystals is similar to diffraction by a grating and suggests that crystalline materials consist of regular arrangements of atoms in space in which the lines of atoms act like tiny slits for the X rays.

The diffraction of X rays by a crystal is a complex phenomenon involving the interaction between incoming X rays and the electrons that make up atoms. We can consider the X rays to consist of electrical pulsations, and their interactions with atoms as producing a corresponding pulsation of the electrons in each atom. Thus, each atom in the path of an X-ray beam receives pulsations and regenerates them in all directions. Consider (as shown in Figure 7.4) two atoms, side by side in an X-ray beam, set into electrical pulsation at the same frequency. Viewed from head on (position A),

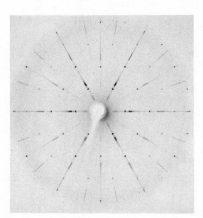

Figure 7.3
X-ray pattern of NaCl.

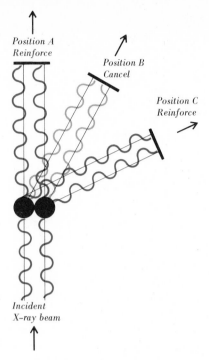

Position A
Reinforce

Position B
Cancel

Position C
Reinforce

Incident
X-ray beam

Figure 7.4
Two-atom model for X-ray scattering.

the signals from the two atoms are received "in step" so that the signal from one reinforces that from the other. This need not be true when viewed at some general angle to the incident beam. In general, the signals will be out of step with each other (position B), because one signal has to travel slightly farther than the other. If, however, the path difference is just exactly equivalent to a whole pulsation (position C), there will again be reinforcement of the signals. For a real crystal consisting of many atoms regularly arranged, there will be certain angles (relative to the incident X-ray beam) at which there will be reinforcement in the emergent beam so as to produce spots on a photographic film.

7.3 Space Lattice

A careful mathematical analysis of a spot pattern resulting from X-ray diffraction enables X-ray crystallographers to calculate the positions particles might occupy in order to produce such a pattern. The process of calculation is an indirect one which involves guessing probable structures, calculating the X-ray patterns they would produce, and comparing them with experiment. These previously laborious calculations have been immensely facilitated by high-speed computers.

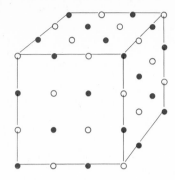

Figure 7.5
Space lattice of NaCl.

The pattern of points which describes the arrangement of molecules or atoms in a crystal is known as a *space lattice*. Figure 7.5 shows the space lattice of NaCl.† Each of the points corresponds to the position of the center of an ion. The solid circles locate positive sodium ions and the open circles negative chloride ions. The circles do not represent sodium ions and chloride ions but only the positions occupied by their centers. In fact, in sodium chloride the ions are of different sizes and are practically touching each other, as shown in Figure 7.6.

The space lattice has to be thought of as extending in all directions throughout the entire crystal. In discussing the space lattice, it is sufficient to consider only enough of it to represent the order of arrangement. This small fraction of a space lattice, which shows the pattern of the whole lattice, is called the *unit cell*. It is defined as the *smallest portion of the space lattice which, moved a distance equal to its own dimensions in various directions,*

† In the strictest sense, a space lattice is concerned only with points and, hence, all points in a space lattice must be identical. In this sense, NaCl can be represented by two identical interpenetrating space lattices, one for the positions of the sodium ions and one for those of the chloride ions.

Figure 7.6
Model of NaCl.

Chloride ion

Sodium ion

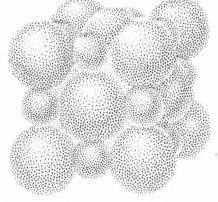

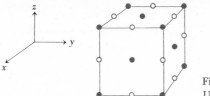

Figure 7.7
Unit cell of NaCl.

generates the whole space lattice. A unit cell of sodium chloride is the cube shown in Figure 7.7. If this cube is moved through its edge length in the x direction, the y direction, and the z direction, many times, eventually the whole space lattice is reproduced.

There are several different kinds of symmetry patterns which occur in crystalline substances. The simplest is known as *simple cubic*. The unit cell is shown in Figure 7.8a. Each point at the corner of the cube represents a position occupied by an atom or a molecule. The arrows represent the three characteristic directions of space, or *axes*, along which the structure must be extended to reproduce the entire space lattice. Closely related to the simple cubic symmetry is *body-centered cubic,* the unit cell for which is shown in Figure 7.8b. It is made up of points at the corners of a cube, with an additional point in the center. In *face-centered cubic* symmetry there are points at the corners of the cube with additional points in the middle of each face, as shown in Figure 7.8c.

Figure 7.8
Unit cells: (*a*) simple cubic, (*b*) body-centered cubic, (*c*) face-centered cubic.

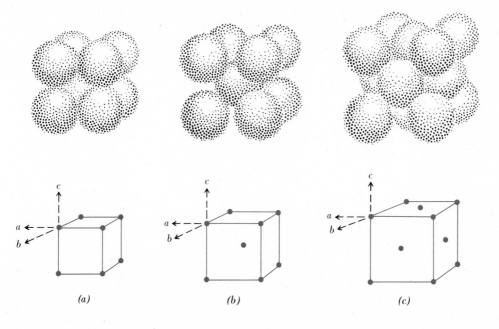

(a)   (b)   (c)

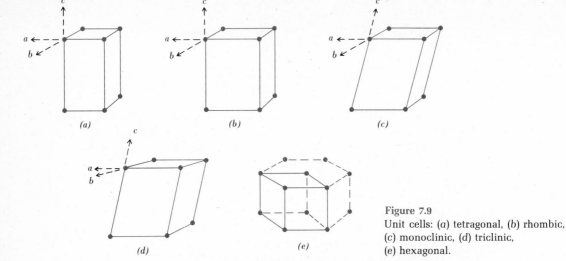

Figure 7.9
Unit cells: (*a*) tetragonal, (*b*) rhombic, (*c*) monoclinic, (*d*) triclinic, (*e*) hexagonal.

In addition to cubic symmetry there are five other basic types, as shown in Figure 7.9. In *tetragonal* symmetry, the cube is elongated in one direction. The lines of atoms still form right angles with each other, but the distance between points along one axis differs from that along the other two. Figure 7.9*a* shows a tetragonal unit cell. The separation of points along the *a* axis is the same as that along the *b* axis, but that along the *c* axis is different. In *rhombic,* or *orthorhombic,* symmetry (Figure 7.9*b*), the unit cell retains mutually perpendicular edges, but the point separation is unequal in the *a*, *b*, and *c* directions. In *monoclinic* crystals the three axes *a*, *b*, and *c* are no longer perpendicular to each other. Monoclinic symmetry differs from rhombic symmetry in that the *c* axis does not make a right angle with the *ab* plane. An example is shown in Figure 7.9*c*. In *triclinic* symmetry none of the three axes *a*, *b*, and *c* is perpendicular to any of the others (Figure 7.9*d*). In the *hexagonal* type of symmetry, as shown in Figure 7.9*e*, the atoms or molecules are arranged in the form of hexagons, with the hexagons stacked on top of each other.

7.4 Packing of Atoms

The unit cells just discussed concern points that locate atomic or molecular centers. Atoms are space-filling entities, and structures can be described as resulting from the packing together of representative spheres. The most efficient packing together of equal spheres, called *close-packing,* can be achieved in two ways, each of which utilizes the same fraction (0.74) of total space. One of these close-packing arrangements is called *hexagonal close-*

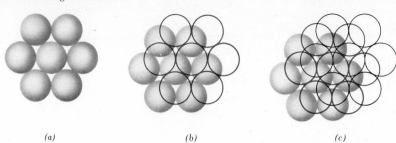

(a)   *(b)*   *(c)*

Figure 7.10
Close-packing of spheres.

packed. It can be envisioned as being built up as follows: Place a sphere on a flat surface. Surround it with six equal spheres as close as possible to the central sphere and in the same plane. If one looks down at this plane, the projection is as shown in Figure 7.10*a*. Now form a second layer of equally bunched spheres, staggered as shown in Figure 7.10*b* so that the second-layer spheres nestle into the depressions formed by the first-layer spheres. A third layer can now be added with each sphere directly above a sphere of the first layer. The fourth layer lies directly above the second layer, and so forth in alternating fashion until the hexagonal close-packed structure has been generated.

The other close-packed structure, called *cubic close-packed,* results if the buildup of layers *a* and *b* is the same as that described above but then a third layer *c* is added, as shown in Figure 7.10*c*. The spheres of layer *c* are not directly above those of either layer *a* or *layer b*. In generating this cubic close-packed structure the sequence of layers is *abcabc* . . . as contrasted to the sequence *ababab* . . . for hexagonal close-packing.

These two structures are represented in terms of unit cells. For hexagonal close-packing, a unit cell is like that of Figure 7.9*e* except that the *b* layer is inserted between the top and bottom faces so as to add three lattice points at mid-height in the hexagonal prism shown. For cubic close-packing the unit cell is a face-centered cube; the body diagonal is perpendicular to the stacking layers. The difference between the two kinds of close-packing may be seen in Figure 7.11.

Figure 7.11
Hexagonal and cubic close-packing.

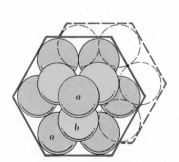

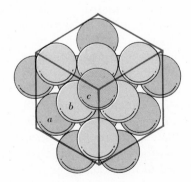

Figure 7.12
Tetrahedral and octahedral holes between layers of close-packed spheres.

Among the common materials that crystallize with hexagonal close-packing are many of the metals, such as magnesium, zinc, and titanium. Also showing hexagonal close-packing is solid H_2, where gyrating H_2 molecules are equivalent to the close-packing spheres. Cubic close-packing is shown by other metals, such as aluminum, copper, silver, and gold, as well as by simple rotating molecules such as CH_4 and HCl.

Close-packing of spheres can also be used to describe many ionic solids. For example, NaCl can be viewed as a cubic close-packed array of chloride ions with the sodium ions fitting into interstices between the chloride layers (see Figure 7.6). Interstices between layers of close-packed spheres are of two kinds, one of which (called a tetrahedral hole) has four spheres adjacent to it and the other of which (called an octahedral hole) has six spheres adjacent to it. The difference between the two kinds is illustrated in Figure 7.12. The a layer of close-packed spheres is represented by filled balls, and the b layer above it by open balls. On the left, marked by color, is a grouping of four adjacent balls (three from layer a and one from layer b) surrounding a tetrahedral hole. This is called a tetrahedral hole because a small atom inserted in the hole would have four neighboring atoms arranged at the corners of a regular tetrahedron. Such a grouping of four spheres and its relation to a tetrahedron are shown at the bottom of the figure. It should be recalled that a regular tetrahedron is a triangle-based pyramid in which each of the four faces is an equilateral triangle.

On the right in Figure 7.12, also marked in color, is a grouping of six adjacent balls (three from layer a and three from layer b) surrounding an octahedral hole. The octahedron, as indicated at the bottom of the figure, is an eight-faced figure all of whose faces are equilateral triangles. In the NaCl structure (called the *rock-salt structure*), the Na^+ ions are regarded as being in the octahedral holes created between layers of Cl^- ions. The number of octahedral holes is just equal to the number of packed spheres, so that the number of Na^+ ions accommodated just equals the number of Cl^- ions.

Another important structure is the *zinc blende structure,* named after the mineral ZnS. It can be visualized as consisting of a cubic close-packed array of sulfur atoms with zinc atoms disposed in tetrahedral holes. Examples of the zinc blende structure include the technologically important group III–V compounds, which are formed from an element of group III (for example, Al, Ga, In) plus an element of group V (for example, P, As, Sb). They are important in solid-state electrical devices such as transistors and rectifiers.

In the compounds thus far discussed in this section, only the holes of one kind (i.e., all tetrahedral or all octahedral) have been utilized. More complicated structures can be built up by using both kinds of holes simultaneously. A most important example is the *spinel structure,* named after the mineral spinel, $MgAl_2O_4$. This consists of a cubic close-packed array of oxide ions with Mg^{2+} in tetrahedral holes and Al^{3+} in octahedral holes. There are many compounds with the spinel structure (for example, $ZnFe_2O_4$, used in making memory cores for computers) in which the dipositive ions (Zn^{2+}) reside in tetrahedral holes and the tripositive ions (Fe^{3+}) in octahedral holes. Such compounds are referred to as *normal spinels.*

In *inverse spinels* the dipositive ions go to octahedral holes and the tripositive ions are distributed half and half between tetrahedral and octahedral holes. Fe_3O_4, the mineral magnetite (also called lodestone), is an interesting example of an inverse spinel. It can be written $Fe^{2+}Fe_2^{3+}O_4$ to indicate the presence of Fe^{2+} and Fe^{3+}. The Fe^{2+} ions occupy octahedral holes, and the Fe^{3+} ions are divided between tetrahedral and octahedral interstices. An electron can jump from an Fe^{2+} to an Fe^{3+} that is also in an octahedral hole, thereby giving rise to electric conductivity, intense light absorption responsible for the black color, and ferromagnetism resulting from magnetic interaction between structurally equivalent ions.

7.5   Types of Solids

Instead of classifying solids by their geometry, it is often more useful to classify them by the nature of the units that occupy the lattice points. On this basis, there are four types of crystals: *molecular, ionic, covalent,* and *metallic.* Figure 7.13 lists for each type the units that occupy lattice points, the forces that bind these units together, characteristic properties, and some typical examples.

Molecular solids are those in which the lattice points are occupied by molecules. In a molecular solid the bonding within the molecule is covalent and, in general, is much stronger than the bonding between the molecules. The bonding between molecules can be of two types, dipole-dipole inter-

Figure 7.13
TYPES OF SOLIDS

	Molecular	Ionic	Covalent	Metallic
Units that occupy lattice points	Molecules	Positive ions Negative ions	Atoms	Positive ions in electron gas
Binding force	Van der Waals Dipole-dipole	Electrostatic attraction	Shared electrons	Electrical attraction between + ions and − electrons
Properties	Very soft	Quite hard and brittle	Very hard	Hard or soft
	Low melting point Volatile	Fairly high melting point	Very high melting point	Moderate to very high melting point
	Good insulators	Good insulators	Noncon-ductors	Good con-ductors
Examples	H_2 H_2O CO_2	NaCl KNO_3 $CaCO_3$	Diamond, C Carborundum, SiC Quartz, SiO_2	Na Cu Fe

action or van der Waals attraction. Dipole-dipole attraction is encountered in solids consisting of polar molecules. As in the case of water, the negative end of one molecule attracts the positive end of a neighboring molecule. Van der Waals attractions (Section 6.14) are present in all molecular solids. Because the total intermolecular attraction is small, molecular crystals usually have low melting temperatures. Not surprisingly, molecular substances are usually quite soft, since the molecules can be easily pushed around from one place to another. Finally, they are nonconductors of electricity, because there is no easy way for an electron associated with one molecule to get over to another molecule. Most substances which exist as gases at room temperature form molecular solids at lower temperatures.

In an *ionic solid*, the units that occupy the lattice points are positive and negative ions. For example, in $CaCO_3$ some of the lattice points are occupied by calcium ions, Ca^{2+}, and the others by carbonate ions, CO_3^{2-}. The forces of attraction are those between a positive and a negative charge and are high. Hence ionic solids usually have fairly high melting points, well

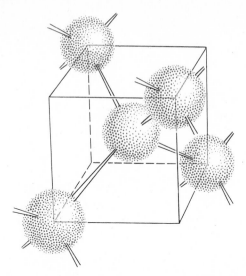

Figure 7.14
Diamond structure.

above room temperature. Calcium carbonate, for example, melts at 1339°C. Also, ionic solids tend to be brittle and fairly hard, with great tendency to fracture by cleavage. In the solid state, the ions are not free to move; therefore these ionic substances are poor conductors of electricity. However, when melted, they become good conductors.

In a *covalent solid,* the lattice points are occupied by atoms that share electrons in forming bonds with their neighbors. The covalent bonds extend in fixed directions, so that the result is a giant interlocking structure. The classic example of a covalent solid is diamond, in which each carbon atom is joined by pairs of shared electrons to four other carbon atoms, as shown in Figure 7.14. Each of these carbon atoms in turn is bound to four carbon atoms, etc., giving a giant three-dimensional molecule. In any solid of this type the bonds between the individual atoms are covalent and, usually, are quite strong. Substances with covalent structures have high melting points, are quite hard, and, in general, are poor conductors of electricity.

In a *metallic solid,* the points of the space lattice are occupied by positive ions. This array of positive ions is immersed in a cloud of highly mobile electrons derived from the outer atomic shells. In solid sodium, sodium ions are arranged in a body-centered cubic pattern. The cloud of electrons, or *electron gas* as it is often called, arises from the contribution by each neutral sodium atom of its lone outermost electron. The electron cloud belongs to the whole crystal. A metal like sodium is held together by the attraction between the positive ions and the cloud of negative electrons. In other cases, e.g., tungsten, there is covalent binding between the positive ions superimposed on the ion-to-electron-gas attraction. Because electrons can wander at will throughout the metal, a metallic solid is characterized by high electric conductivity. It is difficult to specify the other properties of metallic solids, because they vary widely. Sodium, for example, has a low melting point; tungsten has a very high melting point. Sodium is soft and can be cut with a knife; tungsten is very hard.

What does the term molecule mean in these various solids? In a molecular crystal, e.g., solid CO_2 (dry ice), it is possible to distinguish discrete molecules. Each C atom has two relatively close O atoms as neighbors, and all other atoms are at considerably greater distances. In an ionic substance like sodium chloride this is not true. Each sodium ion is equally bound to its neighboring six chloride ions, as shown in Figure 7.6. These six chloride ions must be considered as belonging to the same aggregate as the original sodium ion. But each chloride ion in turn is bonded to six sodium ions, which must also be counted as part of the aggregate. Actually, all the ions in the whole crystal belong to the same aggregate, or giant molecule. A similar situation occurs in metallic and covalent crystals, in which all the ions or atoms are bound together as one giant aggregate. The concept of the molecule is not useful for describing ionic, metallic, or covalent solids.

7.6 Solid-State Defects

The preceding discussion concerned only ideal crystals. An ideal crystal is one which can be completely described by the unit cell; i.e., it contains no *lattice defects*. There are several important kinds of lattice defects. One, called *lattice vacancies*, arises if some of the lattice points are unoccupied. Another, called *lattice interstitials*, arises if atoms occupy positions between lattice points. All crystals are imperfect to a slight extent and contain lattice defects. For example, in NaCl some of the sodium ions and chloride ions are missing from the regular pattern (Figure 7.15a). In silver bromide (AgBr) some of the silver ions are missing from their regular positions and are found squeezed in between other ions (Figure 7.15b). Lattice vacancies occur to

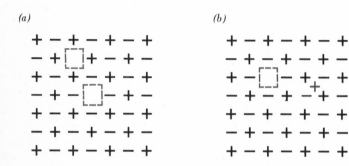

(a)

(b)

Figure 7.15
Lattice defects: (a) vacancies in NaCl, (b) misplaced Ag^+ in AgBr.

some extent in all crystals; their presence helps to explain how diffusion and ionic conductivity can occur in the solid state. Lattice interstitials are considerably less probable; they occur when small positive ions have moved into positions between the normal planes of large negative ions.

The presence of interstitial Ag^+ in AgBr is believed to be important for the formation of a photographic image when AgBr crystals are exposed to light. It is postulated that the quantum of light that strikes a grain of a photographic AgBr emulsion causes an electron to move from a bromide ion to a defect on the surface of the grain. The negative charge attracts an interstitial Ag^+ to the surface. A clump of neutral silver atoms so formed acts as a catalyst for the reduction of other Ag^+ ions in the grain when a photographic developer (a mild reducing agent) is subsequently added. The result is that black areas appear on the film where the light was strongest. Since AgBr slowly turns black when exposed to light, the whole film would turn black eventually. However, the photographic image can be fixed by washing out with "hypo" ($Na_2S_2O_3$) unexposed AgBr grains.

In addition to the defects arising from structural imperfections are defects of a more chemical nature in that they are associated with the presence of chemical impurities. Such impurities can drastically change the properties of materials; hence their controlled introduction is being exploited in producing new materials with desirable combinations of properties. As an example of how properties can be modified by impurities, it might be noted that the addition of less than 0.1% $CaCl_2$ to NaCl can raise the conductivity by 10,000 times. This comes about as follows: In the mixed crystal the Cl^- ion lattice is unchanged, but the Ca^{2+} ions occupy positions in the Na^+ lattice. Because of the requirement for electrical neutrality, each insertion of Ca^{2+} for $2Na^+$ leads to the creation of a lattice vacancy. The vacancy allows Na^+ ions to move and hence results in increased conductivity.

A more practical utilization of impurity defects involves germanium and silicon crystals. The electric conductivity of these group IV elements, in the pure state, is extremely low. However, on addition of trace amounts of elements from either group III or group V, the conductivity is greatly enhanced. Both germanium and silicon have the diamond structure, shown in Figure 7.14. Each atom is bonded to four neighbors by four covalent bonds. These require all four outer electrons of each group IV atom. In the pure elements Ge and Si there are no conduction electrons (except at very high temperatures). But when, for example, a group V element—P, As, Sb, or Bi—is substituted for a Ge atom, an extra electron above that required for forming four covalent bonds is introduced. This extra electron can act something like a conduction electron in metals; hence an arsenic-containing germanium crystal exhibits marked conductivity. If, on the other hand, a group III element—B, Al, Ga, or In—is substituted for a Ge atom, an electron deficiency in the covalent-bonding network is introduced. Such an electron vacancy is not confined to the impurity atom site but can "move" through the structure as other electrons move in to fill it. Thus, the introduction of electron vacancies, or "holes" as they are sometimes called, by substituting ("doping") Ge with traces of group III atoms allows the electron motion necessary for increased conductivity. Such materials are called *semiconductors,* because their conductivity is small and generally increases as the temperature is raised, which is opposite to the behavior of metals.

Impurity-doped germanium and silicon act as semiconductors because there is a weak binding of the excess electron or the hole to the impurity center responsible for it. Additional energy, as supplied by a temperature rise, is needed to free the electron or hole sufficiently for conductivity. Semiconductors in which electric transport is mainly by "excess" electrons are called *n* type (*n* for negative), and those in which electric transport is mainly by "holes" are called *p* type (*p* for positive). It must be emphasized that both *n*- and *p*-type semiconductors are electrically neutral. The electron "excess" or "deficiency" is with respect to that required for covalent bonding, not with respect to atom neutrality. To illustrate, substitution of As for Ge introduces not only an additional electron but also an additional unit of positive nuclear charge.

An interesting application of impurity semiconductors results from combination of *n*- and *p*-type materials to form a junction, the so-called "*n-p* junction." This device can pass electric current more easily in one direction than in the reverse; hence it can act as a rectifier for converting alternating current to direct current. Figure 7.16 shows schematically why there should be a difference in the ease of passing current in the two directions. The left side of the junction is *n* type, and the minuses represent the extra electrons arising from the group V element impurity centers; the right side of the junction is *p* type, and the pluses represent the electron holes arising from the group III element impurity centers. When an external voltage is applied (as shown in Figure 7.16*a*) so as to favor motion of electrons from left to right and motion of holes from right to left, conductivity readily occurs. At the

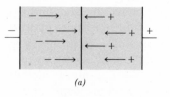

(a)

Figure 7.16
The *n-p* function with two directions, (*a*) and (*b*), of external voltage.

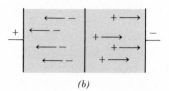

(b)

interface between the n- and p-type zones, electrons coming from the left annihilate holes coming from the right. Conductivity does not stop, because the external voltage acts to supply more n carriers on the left and p carriers on the right. On the other hand, when the external voltage is reversed (as shown in Figure 7.16b), so as to separate the n and p carriers, conductivity stops. There is no way in which new carriers can be regenerated at the n-p interface.

QUESTIONS

7.1 *Properties of Solids* List three properties which differentiate the solid state from gases and account for the differences in terms of kinetic theory.

7.2 *Structure Determination* In terms of Figure 7.4, would it seem to be necessary that the wavelength of X rays is roughly equal to atomic dimensions? Tell what you think might happen if radiation were used which had wavelength 100 times longer and then if radiation were used that was 100 times shorter.

7.3 *Space Lattice* Imagine a space lattice based on a simple cubic unit cell. Account for the fact that there are equal numbers of atoms and unit cells despite the fact that each unit cell locates eight atoms.

7.4 *Unit Cells* For NaCl, why is the unit cell that which is shown in Figure 7.7 and not a simple cube with Na^+ and Cl^- ions at alternate corners?

7.5 *Packing of Atoms* (a) Which should occupy a smaller volume, a million marbles in cubic close-packing or a million marbles in hexagonal close-packing? Explain. (b) Which should have the greater radius, a tetrahedral hole or an octahedral hole? Explain.

7.6 *Types of Solids* Tell which of the four types of solids each of the following probably is: (a) soft and nonconducting, (b) soft and high melting, (c) insulator that conducts on melting.

7.7 *Defects* (a) What type of solid could show the defects described by Figure 7.15? (b) Show for both types of defects.

7.8 *Properties* Unlike gases, two solids composed of molecules of identical molecular weight may have different densities. Give two reasons why this might be.

7.9 *Terms* Differentiate between unit cell and space lattice.

7.10 *Crystal Forms* Explain why you would expect the melting points of ionic solids to be higher than the melting points of molecular solids.

7.11* *Unit Cell* Potassium metal crystallizes in a face-centered arrangement of atoms where the edge of the unit cell is 0.574 nm. What is the shortest separation of any two potassium nuclei?

7.12 *Unit Cell* Working from the unit cells shown in Figure 7.9, decide the number of *equidistant* closest neighbors of (a) any atom in tetragonal symmetry, (b) center atom in hexagonal symmetry.

7.13 *Types of Solids* Given three solids A, B, and C. Supply, wherever possible, the information missing from the table.

Characteristic	A	B	C
Type of lattice	Molecular		
Melting point			High
Hardness	Soft	Very hard	Soft
Electrical properties		Nonconductor; melt conducts	

7.14 *Nonstoichiometry* One way to explain nonstoichiometric compounds such as $Fe_2Se_{2.37}$ is to say that the selenium lattice is normal, as it would be in FeSe, but that some of the iron atoms are missing. For electrical neutrality, the remaining iron atoms would have to be partly in the $3+$ state as well as the $2+$ state. To explain $Fe_2Se_{2.37}$, what ratio of Fe^{2+}/Fe^{3+} would be needed?

7.15 *Semiconductors* (a) What is the fundamental distinction between an n-type and a p-type impurity? (b) Give specific examples of an n-type and a p-type impurity for inclusion in Ge. (c) Would your answers to (b) be changed if B were the host lattice instead of Ge? Explain.

7.16 *Types of Solids* Tungsten is a solid element which seems to consist of an electron gas plus ions which are covalently bonded to each other through sharing electrons from d atomic orbitals. Predict its properties.

7.17* *Unit Cell* Copper metal has a face-centered cubic structure with the unit-cell length equal to 0.361 nm. Picturing copper ions in contact along the face diagonal, what is the apparent radius of a copper ion?

7.18 *Types of Solids* Consider the solid elements potassium and iodine and the solid compound they form. Describe these three solids in terms of (a) units that occupy the lattice points, (b) binding force holding the solid together, (c) properties you would expect.

7.19 *Ionic Radii* From X-ray studies of NaCl and similar compounds, it is possible to determine the distances between adjacent ionic centers. It is not possible to determine ionic size, or in other words, how much of this distance to attribute to each of the two ions. One method for getting at individual ionic radii is to look at a series of alkali halides with very different ionic radii and pick one where the anions are so much larger than the cations that the anions should be in contact. This spacing can then be divided by 2 and called the *anion radius*. Show whether this condition should be met for LiI which has the NaCl structure with Li^+ assigned a radius of 0.068 nm and I^- a radius of 2.20 nm. Would it be met for NaI, where Na^+ is assigned a radius of 0.097 nm?

7.20 *Lattice Defects* How might you be able to measure quantitatively how many of the vacancies described by Figure 7.15a a particular sample of NaCl apparently has?

7.21* *Packing of Atoms* (a) By reference to Figure 7.12 tell which will accommodate the larger atoms, tetrahedral or octahedral holes. (b) There are as many octahedral holes as atoms in the lattice. How many tetrahedral holes are there?

7.22 *Space Lattice* (a) Tell how many nearest neighbors each atom has in a simple cubic lattice. (b) How many in a close-packed lattice?

7.23* *Unit Cell* What is the simplest formula of a solid whose cubic unit cell has an A atom at each corner, a B atom at each face center, and a C atom at the body center?

7.24 *Packing of Atoms* Gold has a close-packed structure which can be viewed as spheres occupying 0.74 of the total volume. If the density of gold is 19.3 g/cm³, what is the apparent radius of a gold ion in the solid?

7.25* *Avogadro Number* Given: Density of solid NaCl is 2.165 g/cm³ and the distance between centers of adjacent Na^+ and Cl^- is 0.2819 nm. (a) Calculate the edge length of a cube containing 1 mole of NaCl. (b) Calculate the number of Na^+ plus Cl^- ions along one edge of the cube. (c) Calculate the Avogadro number.

7.26 *Unit Cell* Show which of the three unit cells in Figure 7.8 is most densely packed by calculating the fraction of empty space contained in each if the atoms are considered as hard spheres in contact. *Hint*: Assume unit radius for spheres and calculate cube edge length in terms of this radius by knowing that the spheres are in contact (a) along the edge, (b) along the body diagonal, and (c) along the face diagonal. Then calculate the number of spheres or fractions of spheres which lie *completely* within the cube.

Liquids, Amorphous Materials, and Colloids

8

In the gaseous state, molecules are independent of each other and the collection of random particles represents a highly disordered condition. At the other extreme, in the solid state, atoms and molecules are highly organized into well-defined patterns. The randomness of gases and the order of solids make possible mathematical simplifications that have helped in understanding these two states of matter. Between the extremes of randomness and order are liquids and various amorphous materials, all of which show partial ordering of atoms and molecules. Although amorphous materials, especially liquids, share some of the properties of gases and solids, the complexity of mathematically describing partially ordered states is so great that our knowledge of them is still fragmentary. Vigorous research, aided by electronic computers, is currently under way because amorphous substances include many materials of great technological and biological importance.

8.1 Properties of Liquids

When a sample of gas is cooled or compressed, or both, it liquefies. In the process the molecules of the gas, originally far apart on the average, are slowed down and brought close enough together that attractive forces become appreciable. The individual molecules coalesce into a cluster which settles to the bottom of the container as liquid. What are the general properties of liquids? How can they be accounted for by the kinetic theory?

Liquids are practically incompressible   Unlike gases but like solids, a liquid does not change much in volume when the pressure on it is changed, even when pressures of thousands of atmospheres are involved. The kinetic theory accounts for this by saying that the amount of free space between the molecules of a liquid has been reduced almost to a minimum. Any attempt to compress the liquid meets with resistance as the electron cloud of one molecule repels the electron cloud of an adjacent molecule.

Liquids maintain their volume   No matter what the shape or size of the container, a 10-ml sample of liquid occupies a 10-ml volume, whether it is placed in a small beaker or in a large flask, whereas a gas spreads out to fill the whole volume accessible to it. Gases do not conserve their volume, because the molecules are essentially independent of each other and can move into any space available. In liquids, the molecules are close together, and mutual attractions are strong. Consequently, the molecules are clustered together.

Liquids have no characteristic shape   A liquid sample assumes the shape of the bottom of its container. The kinetic theory explains this property by saying that there are no fixed positions for the molecules. The molecules are free to slide over each other in order to occupy positions of the lowest possible potential energy. On earth, gravity pulls the liquid specimen to the bottom of its container: in an orbiting satellite, intermolecular forces pull the specimen into a spherical glob.

Liquids diffuse slowly   When a drop of ink is carefully released in water, there is at first a rather sharp boundary between the ink cloud and the water. Eventually the color diffuses throughout the rest of the liquid. In gases, diffusion is much more rapid. Diffusion occurs because molecules have kinetic energy and move from one place to another. In a liquid, molecules do not move very far before they collide with neighboring molecules. Eventually each molecule of a liquid does migrate from one side of its container to the other, but it undergoes many billions of collisions in doing so. By contrast, a gas is mostly empty space, and the molecules of one gas can rather quickly mix with those of another.

Liquids evaporate   Although there are attractive forces which hold the molecules together in a clump, it is evidently possible for molecules to escape from open containers. The molecules with kinetic energy great enough to overcome the attractive forces can escape into the gas phase. In any

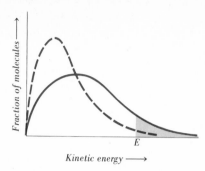

Figure 8.1
Energy distribution in a liquid.

collection a given molecule does not have the same energy all the time. There is continual exchange of energy in collisions. The collection might start with all molecules of the same energy, but this situation does not last long. Two or more molecules may simultaneously collide with a third. Molecule 3 now has not only its initial energy but possibly some extra energy received from its neighbors. Molecule 3 now has higher than average kinetic energy. If it happens to be near the surface of the liquid, it may be able to overcome the attractive forces of neighbors and go off into the gas phase.

Figure 8.1 shows a typical energy distribution for the molecules of a sample of liquid at a given temperature. The value marked E represents the minimum kinetic energy required by a molecule to overcome attractive forces and escape from this liquid. All the molecules in the shaded area under the curve have enough energy to overcome attractive forces. These are the molecules that have the possibility of escaping, provided that they are close enough to the surface. If these highly energetic molecules leave the liquid, the average kinetic energy of those left behind is lower. That is, each molecule that escapes carries along with it more than an average amount of energy, part of which it uses in working against the attractive forces. Since the average kinetic energy of the remaining molecules is lower, their temperature drops. Evaporation is therefore accompanied by cooling.

When a liquid evaporates from an *uninsulated* container such as a beaker, the temperature of the liquid cannot fall very far before there is an appreciable heat flow from the surroundings into the liquid. If the rate of evaporation is not too great, this flow of heat is sufficient to compensate for the energy required for evaporation. As a consequence, the temperature of a liquid remains near room temperature, even though the liquid is evaporating. Eventually, the liquid disappears.

When evaporation proceeds from an open container *insulated* to reduce heat flow, the heat flow from the surroundings is slower, and the temperature of the liquid drops. The flow of heat from outside is not fast enough to compensate for the evaporation. The average kinetic energy of the molecules decreases, and the rate of evaporation diminishes until the heat flow inward just equals the heat required for evaporation. At the lower temperature the distribution of the kinetic energies is shifted to the left, as shown by the dashed line in Figure 8.1. As the temperature decreases, the fraction of energetic molecules decreases, and the rate of evaporation decreases. The temperature now stays constant but at a value which may be considerably lower than the original temperature. Even liquid air, an extremely volatile liquid, remains in an open insulated flask at roughly $-190°C$ for many hours.

8.2 Equilibrium Vapor Pressure

When a lid is placed on a beaker of evaporating liquid, as shown in Figure 8.2, the liquid level drops for a while but then becomes constant. This can be explained as follows: Molecules escape from the liquid into the gas, or vapor, phase. After escaping, they are confined to a limited space. As the molecules accumulate in the space above the liquid, there is an increasing chance that in their random motion some of them go back to the liquid. The longer the experiment proceeds, the better is the chance that the molecules return. Eventually, a situation is established in which molecules return to the liquid just as fast as other molecules leave it. At this point the liquid level no longer drops, because the number of molecules evaporating per second is equal to the number of molecules condensing per second. A condition in which *changes exactly oppose each other* is referred to as *dynamic equilibrium*. Although the system is not at a state of rest, there is no net change in the system. The level of liquid in the beaker stays constant; the concentration of molecules in the vapor above the liquid is constant. A particular molecule spends part of its time in the liquid and part in the vapor phase. As molecules pass from liquid to gas, other molecules move from gas to liquid, keeping the number of molecules in each phase constant.

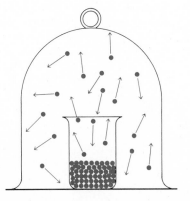

Figure 8.2
Evaporation in a confined space.

The molecules which are in the vapor exert a pressure. At equilibrium, this pressure is characteristic of the liquid at a given temperature and is known as the *equilibrium vapor pressure*. As the term implies, it is the *pressure exerted by a vapor when in equilibrium with its liquid*. The magnitude of the equilibrium vapor pressure depends on the nature of the liquid and on its temperature.

The nature of the liquid is involved, since each liquid has characteristic attractive forces between its molecules. Molecules which have large mutual attraction have a small tendency to escape into the vapor phase. Such a liquid has a low equilibrium vapor pressure because there will be fewer molecules in the gas phase. Liquids composed of molecules with small mutual attraction have a high escaping tendency and therefore a high equilibrium vapor pressure.

So far as temperature is concerned, a temperature rise means that the average kinetic energy of molecules is increased. The number of high-energy molecules capable of escaping becomes larger; hence the equilibrium vapor pressure increases.

There are various devices for measuring equilibrium vapor pressure, one of which is shown in Figure 8.3. It consists of a barometer set up in the usual manner. The difference between the upper mercury level and the lower mercury level represents the atmospheric pressure. Above the mercury there is a vacuum. By squeezing the medicine dropper, a drop of liquid can be ejected into the mercury. Since practically all liquids are less dense than mercury, they float to the top of the mercury, where enough liquid evaporates to establish its equilibrium vapor pressure. This vapor pressure pushes down the mercury column. (The weight of the excess liquid also helps to push down the mercury column, but this is a negligible effect.) The extent to which the mercury level is depressed gives a quantitative measure of the vapor pressure of the liquid. At 20°C, water has an equilibrium vapor pressure of 17.5 mmHg (0.023 atm) and chloroform ($CHCl_3$), 160 mmHg

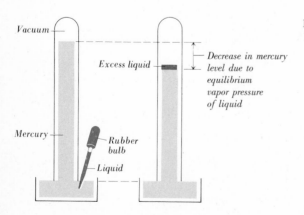

Figure 8.3
Measurement of vapor pressure.

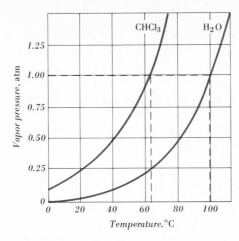

Figure 8.4

Change of vapor pressure with temperature.

(0.21 atm). These values of the vapor pressure give an idea of the escaping tendencies of molecules from the various liquids. In water, the attractive forces between molecules are apparently quite large; hence relatively few water molecules escape. In chloroform, the attractive forces between molecules are small, and it evaporates readily to give a high vapor pressure.

By repeating the above experiment at different temperatures, it is possible to determine vapor pressure of liquids as a function of temperature. Appendix 5 is a table showing the results for water in great detail. The general behavior of water and chloroform is shown in Figure 8.4. The vertical scale represents the vapor pressure and the horizontal scale the temperature. As the temperature increases, the vapor pressure rises, first slowly and then more steeply, until at high temperature it is rising almost vertically. The curve continues to the critical temperature. It does not go beyond the critical temperature, because above the critical temperature the liquid cannot exist.

8.3 Boiling

Boiling is a special case of vaporization and is the passage of a liquid (from an open vessel) into the vapor state through the formation of bubbles.† A liquid is said to boil at its *boiling point* (abbreviated bp), the *temperature at which the vapor pressure of the liquid is equal to the prevailing atmospheric pressure*. At the boiling point, the vapor pressure of the liquid is high enough that it simply pushes the atmosphere aside. Therefore, bubbles of

† When water is heated in an open container, it is usually observed that, as the liquid is warmed, tiny bubbles gradually form at first, and then, at a higher temperature, violent bubbling commences. The first bubbling should not be confused with boiling. The first tiny bubbles are due to the expulsion of the air usually dissolved in water.

vapor can form in the interior of the liquid, allowing vaporization to occur at any point in the liquid. In general, a molecule can evaporate only if two requirements are met: it must have enough kinetic energy, and it must be close enough to a liquid-vapor boundary. At the boiling point, bubbling infinitely increases the liquid-vapor boundary, and therefore it is necessary only that molecules have enough kinetic energy to escape from the liquid. All of the heat added to a liquid at its boiling point gives more molecules sufficient energy to escape; hence the average kinetic energy of molecules remaining in the liquid cannot increase. The temperature of a pure boiling liquid is therefore constant.

The boiling point of a liquid depends on pressure. For instance, when the atmospheric pressure is 0.921 atm, water boils at 97.7°C; at 1.00 atm, it boils at 100°C. To avoid ambiguity, it is necessary to define a *standard*, or *normal, boiling point*. The normal boiling point is the temperature at which the vapor pressure of a liquid is equal to one standard atmosphere (760 mmHg). The normal boiling point is the one usually listed in handbooks and textbooks. It can be determined from the vapor-pressure curve by finding the temperature which corresponds to 1 atm pressure. Figure 8.4 shows that the normal boiling point of water is 100°C and that of chloroform is 61°C. In general, the higher the normal boiling point, the greater must be the attractive forces between the molecules of a liquid.

It is usually observed that in a series of similar compounds, such as CH_4, C_2H_6, and C_3H_8, the normal boiling point is higher for the compound of greatest molecular weight. Although it is tempting to explain a high boiling point in terms of large gravitational attraction between heavy molecules, gravitational attraction in reality is small. A more reasonable explanation is that heavy molecules usually contain more electrons than light molecules and hence have greater van der Waals attractions.

8.4 Amorphous Solids

The important difference between liquids and crystalline solids is that liquids do not show the *extended* ordering of molecules which is characteristic of crystals. However, the strong intermolecular forces that exist in liquids cause molecules to pull together into a *locally* ordered array. In other words, the near-neighbor molecules arrange themselves into a more or less definite pattern, but the pattern does not extend very far. As an example, in liquid waters it is believed that about four H_2O molecules arrange themselves approximately tetrahedrally about each H_2O molecule. However, the network formed is not perfect. Some scientists suggest that some of the H_2O molecules have other than four neighbors; others suggest that the tetrahedral arrangement is randomly distorted. In any case, the imperfections of arrangement prevent the molecules of H_2O in the liquid from having the long-range ordering characteristic of ice.

Amorphous solids resemble liquids in that they lack the extended ordering characteristic of crystalline solids. On the other hand, they differ from liquids in not being fluid, presumably because the intermolecular forces are so strong that the imperfect network persists. A good example of an amorphous solid is glass, the important constituents of which are silicon and oxygen together with other elements such as sodium and calcium. In the structure there seem to be tetrahedra of four oxygen atoms about the silicon atom with each of the oxygen atoms joined to a second silicon atom at the center of its own tetrahedron. Thus, the tetrahedra are connected to each other through a shared oxygen atom. The angles that various oxygens make with their two silicon atoms are not all the same, so that the result is a disordered array of tetrahedra. The sodium and calcium atoms, in this picture, are stuck in holes of the network structure. Because the network contains strong Si—O—Si bonds, it is quite rigid and glass does not flow appreciably except at high temperatures.

The word *glass* is a general term used to describe many amorphous solids. It describes a rigid noncrystalline state. As might be expected, the X-ray picture of a glass is quite different from that of a crystal and resembles that of a liquid. Instead of a spot pattern, the X-ray picture shows concentric rings. The existence of these rings indicates that there is a certain amount of order, but it is far from perfect. Another indication that glasses do not have a long-range ordering of atoms is their behavior on being broken. Instead of showing cleavage with formation of flat faces and characteristic angles between faces, glasses break to give conchoidal or shell-like depressions such as are observed in the broken edge of glass.

An interesting class of amorphous solids are the plastics, such as Bakelite, polyethylene, polystyrene, and nylon. These consist of molecules (hydrocarbons or hydrocarbon derivatives) cross-linked together so as to establish giant random networks. Depending on the stiffness of the component molecules and the number of cross-links between molecules, the resulting material may be rubberlike, fibery, or brittle. The pervasion of modern living by plastic objects of all forms, shapes, and colors has come about because of the ease of fabricating objects from plastics. Fabrication can be done either by forming the cross-linked material inside a mold or by cooling the fluid plastic. The object so formed has for its weight considerable strength owing to the cross-linkage of molecules into the network structure.

8.5 Colloids

There are numerous important substances including milk, beer, blood, and jelly that are not simply classified as solid, liquid, or gas. These substances constitute a class called *colloids,* from the Greek word *kolla* meaning glue.

In order to get an idea of what a colloid is, imagine a process in which a sample of solid is placed in a liquid and subdivided. So long as distinct particles of solid are visible to the naked eye, there is no question that the system is heterogeneous. On standing, these visible particles separate out. Depending on the relative densities of the solid and the liquid, the solid particles float to the top or settle to the bottom. They can be separated easily by filtration. Eventually, however, as the solid is progressively subdivided, a state is reached in which the dispersed particles have been broken down to individual molecules or atoms. In this limit, a solution is produced in which two phases can no longer be distinguished. No matter how powerful a microscope is used, the solution appears uniform throughout, and individual molecules cannot be seen. On standing, the dispersed particles do not clearly separate out, nor can they be separated by filtration.

Between coarse suspensions and true solutions, there is a region of change from heterogeneity to homogeneity. In this region, dispersed particles are so small that they do not form an obviously separate phase, but they are not so small that they can be said to be in true solution. This state of subdivision is called the *colloidal state*. On standing, the particles of a colloid do not separate out at any appreciable rate; they cannot be seen under a microscope; nor can they be separated by filtration. The dividing lines between solutions, colloids, and discrete phases are not rigorously fixed, since a continuous gradation of particle size is possible. Usually, however, colloids are defined as a separate class on the basis of size. When the particle size lies between about 1 and 1000 nm, the dispersion is called a *colloid*, a *colloidal suspension,* or a *colloidal solution.*

The size of a dispersed particle does not tell anything about what makes up the particle. It may consist of atoms, small molecules, or one giant molecule. For example, colloidal gold consists of various-sized particles each containing many thousands or more gold atoms. Colloidal sulfur can be made with particles containing a thousand or so S_8 molecules. An example of a huge molecule is hemoglobin, the protein responsible for the red color of blood. The molecular weight of this molecule is 66,800. It has a radius of some 3 nm.

Colloids are frequently classified on the basis of whether the component phases are solid, liquid, or gas, even though the separate phases are not visibly distinguishable once the colloid is formed. The more important classifications are *sols, emulsions, gels,* and *aerosols.*

In *sols,* a solid is dispersed through a liquid, so that the liquid forms the continuous phase and bits of solid form the discontinuous phase. Milk of magnesia is a sol consisting of solid particles of magnesium hydroxide dispersed in water. Sols can be made by breaking down large particles or building up small particles to colloidal dimensions. Colloidal gold can be made by striking an electric arc between two gold electrodes under water. It can also be made by the chemical reduction of a solution containing a gold compound. A gold sol made in 1857 by Michael Faraday is still in the London Museum and shows no detectable settling. Investigation of gold sol by X rays has shown that the particles of gold which are dispersed throughout the water are crystalline in nature.

Emulsions are colloids in which one liquid is dispersed through another liquid. A common example is milk, which consists of butterfat globules dispersed through an aqueous solution. In the creaming process the larger globules separate out. In the process of homogenization, milk is forced under pressure through small openings in a metal plate to break up the globules and thereby reduce creaming.

A *gel* is an unusual type of colloid in which a liquid contains a solid arranged in a fine network extending throughout the system. Both the solid and the liquid phases are continuous. Examples of gels are jellies, gelatin, agar, and slimy precipitates such as aluminum hydroxide. The so-called "canned heat," or "solid alcohol," is a gel made by mixing aqueous calcium acetate with alcohol. Napalm is a gel made from gasoline and a jellied substance such as polystyrene.

An *aerosol* is a colloid made by dispersing either a solid or a liquid in a gas. The former is called a smoke and the latter a fog. Cigarette smoke is an aerosol of solid ash dispersed in air. The fog spray produced from insecticide bombs is an aerosol of an insecticide solution dispersed in air.

8.6 Light Scattering by Colloids

A characteristic feature that distinguishes colloids from solutions is the *Tyndall effect*. When a beam of light is passed through a true solution, the path of the beam is not visible from the side. The dissolved particles are too small to scatter light, and therefore the light goes on through. In a colloid, the particles are big enough to scatter the light. Therefore, when a beam of light is turned on a colloid, an observer at one side can see the path of the beam. The situation is shown in Figure 8.5. The Tyndall effect can be produced readily by turning a flashlight on an aqueous solution of sodium thiosulfate ($Na_2S_2O_3$) and adding a few drops of dilute acid. A chemical reaction occurs which produces elemental sulfur. The light beam is invisible until the sulfur particles aggregate to colloidal dimensions.

By taking into consideration the wave nature of light, information can be obtained from the Tyndall effect about the size and shape of the scattering particles. In an ordinary solution, the particles of solute are much smaller than the wavelength of the light. Visible light has wavelengths ranging from 400 to 720 nm. Solute particles that are 0.5 nm or so in diameter are too small

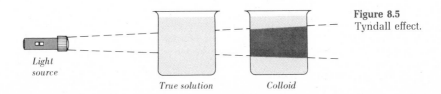

Light source

True solution   *Colloid*

Figure 8.5
Tyndall effect.

to affect a wave of such length. However, when solute particles are of the order of hundreds of nanometers in diameter, the light beam is scattered or diffracted and becomes visible. Careful studies of this scattering have been used to determine the size of particles in colloids such as suspensions of rubber.

The particles of a colloid are too small to be visible under a microscope. However, they are big enough to reflect light. When a microscope is focused on a Tyndall beam, light is reflected into the microscope. The details of the shape of the particle are not apparent, but the position of the particle may be fixed by noting the position at which the light pinpoint appears. When observed in this way, colloidal particles are seen to undergo Brownian motion, the rapid, random, zigzag motion previously mentioned in discussing gases. The smaller the colloidal particle, the more violent is its Brownian motion.

Under ordinary circumstances, it is observed that a colloid in an uninsulated container does not settle out. However, when the colloid is kept in a well-insulated container, some settling is observed, and after a time there will be a gradation in the concentration of colloidal particles from the top to the bottom of the sample. This gradation in concentration develops because there are two opposing effects: (1) the attraction due to gravity, which tends to pull heavier particles down, and (2) the dispersing effect due to Brownian motion. The more massive the particles, the more important is the first effect and the more pronounced is the concentration gradation. Why then is no appreciable concentration gradient observed for colloids in uninsulated containers? The main reason is that in an uninsulated container there are convection currents due to nonuniform temperature. These currents keep the colloidal suspension constantly stirred up.

8.7   Adsorption

In some colloids, the colloidal particles adsorb electric charges. For example, ferric oxide sol consists of positively charged aggregates of hydrous ferric oxide (rust). The positive charge enhances the stability of the colloid. Normally, when an uncharged particle in its Brownian motion hits another, they coagulate to form a larger particle. From collisions between large particles, a particle results which is so large that Brownian motion cannot keep it in suspension. However, ferric oxide has great adsorption power for H^+. By the proper method of preparation, it is possible to form ferric oxide particles with H^+ adsorbed on the surface, as shown on the left in Figure 8.6. Presumably the H^+ ions are stuck on oxygen atoms which protrude from the particles. A particle which has H^+ adsorbed on it has a net positive charge and thereby repels any similarly charged particle. The charged ferric oxide particles try to stay as far apart from each other, on an average, as possible.

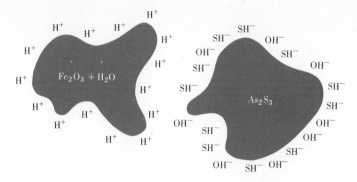

Figure 8.6
Adsorption of charges on colloidal aggregates.

There is little chance that they will come together to form a large mass which settles out. Arsenious sulfide (As_2S_3) forms a negative sol by adsorbing SH^- or OH^- ions, as shown on the right in Figure 8.6. Not surprisingly, mixing a positively charged ferric oxide sol with a negatively charged arsenious sulfide sol coagulates them both.

That some colloidal particles are electrically charged can be shown by studying *electrophoresis, the migration of colloidal particles in an electric field.* Figure 8.7 shows the experimental setup. A U tube is partly filled with the colloidal solution. Very carefully, so as not to disturb the colloid, the remainder of the U tube is filled with a conducting salt solution. Electrodes are inserted into the solution, with one electrode positively charged and the other negatively charged. After a time that may range from 30 min to 48 h, the boundaries between the colloid and the salt solution have shifted because of migration of the colloid. From the direction of migration, the sign of the charge of the colloidal particles can be determined. For example, when the electrophoresis of ferric oxide sol is observed in this cell, the boundary moves toward the negative electrode and away from the positive electrode, suggesting that the colloid is positively charged. By observing the rate at

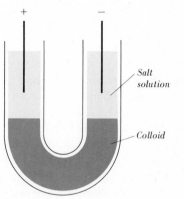

Figure 8.7
Electrophoresis.

which migration occurs, it is also possible to get information about the size and the shape of the colloidal particles.

Electrophoresis has been applied with great success to protein molecules, which in acid solution pick up hydrogen ions to become positively charged. Some proteins, like gamma globulin, seem to be spherical in shape. Others, such as fibrinogen, the blood-clotting agent, are shaped like cigars. Knowledge of the size and shape of protein molecules helps in understanding their behavior.

Some colloidal particles adsorb films of molecules, which shield them from other particles. An example of this adsorption is found in gelatin. Gelatin is a high-molecular-weight protein which has the property of tying to itself a sheath of water. This film of very tightly bound water prevents the gelatin particle from coagulating with another gelatin particle. If two gelatin particles collide, they do not coagulate, because the gelatin parts have not been able to get in contact with each other. This property of gelatin is used in stabilizing colloids of silver bromide in manufacturing photographic film. If finely divided silver bromide were stirred up with water, it settles out. However, the silver bromide is mixed with gelatin and the gelatin forms a film on the outside of the silver bromide. The gelatin in turn adsorbs a layer of water, so that essentially two protective films have formed on silver bromide to keep it in suspension.

There are two reasons for the high adsorption properties shown by colloids. One is the extremely large surface area. Subdivision of a solid, for example, may increase its surface area by many factors of 10. One cubic centimeter of sulfur in the form of a cube 1 cm on edge has a surface area of 6 cm^2, or about 1 in.^{2}. However, 1 cm^3 of sulfur ground up into cubes which are 10^{-5} cm on the edge has its surface area increased to 6×10^5 cm^2, or about 700 ft^2. The second reason for great adsorption is that surface atoms have special properties. The valence of a surface atom is usually not satisfied, since the atom normally bonds in three dimensions, which it cannot do at the surface. The greater the state of subdivision, the greater is the fraction of atoms that have unsatisfied valence forces.

Charcoal is a substance in which the surface atoms are an appreciable fraction of the total number. It consists of solid carbon with a fine network of tunnels extending through the specimen. The surface area has been determined to be of the order of 1000 ft^2/g. On all this large surface area are carbon atoms which have unsaturated valence. They can attract molecules, especially polar molecules, thus accounting for the high adsorption that is characteristic of charcoal. When a mixture of hydrogen sulfide, H_2S, and oxygen is passed over a charcoal surface, the H_2S is selectively adsorbed. Because H_2S is a polar molecule with the sulfur end more negative than the hydrogen end, it is more strongly adsorbed than the oxygen molecule, which is symmetrical and nonpolar. The charcoal gas mask makes use of this principle of selective adsorption. The charcoal selectively adsorbs poisonous gases, which are usually complicated polar molecules, and lets the oxygen through for respiration.

Surface adsorption by colloidal material has recently become an area of increased interest to scientists and engineers. The reason for this greater interest is the urgent necessity of conserving both our environment and our dwindling energy supplies. Surface adsorption provides the means for purifying both air and water. Surface adsorption is required also in most catalytic processes needed to increase the efficiency of energy use. Scientific interest in colloids represents an example of a field where scientific understanding outstripped technological application. Now that the world's problems have demanded increased technological applications, the backlog of basic understanding of colloids is about to be exhausted and vigorous research into the nature of colloidal materials is beginning to be pursued once again.

QUESTIONS

8.1 *Properties of Liquids* In terms of kinetics theory, account for observed differences between liquids and both gases and solids with respect to each of the following: (*a*) compressibility, (*b*) retention of shape, (*c*) rate of diffusion.

8.2 *Evaporation* On a molecular picture, account for the fact that a liquid placed in a well-insulated container decreases in volume with a noticeable drop in temperature.

8.3 *Evaporation* Consider a pure liquid in a closed container. As it is heated up (past its boiling point) the liquid level gradually drops. Eventually a temperature is reached where (no matter how much liquid you started with) all the liquid evaporates. Account for these observations and tell why nothing special (no bubbles) was observed at the normal boiling point but why there is a definite temperature for total evaporation.

8.4 *Equilibrium Vapor Pressure* Why is the equilibrium vapor pressure considered as a property of the liquid and not of the gas?

8.5 *Boiling* No matter how much heat you put under a pan of water on a stove, you can't heat it above the boiling point. Why not?

8.6 *Amorphous Solids* (*a*) Without use of X rays how might you determine whether a solid was amorphous or crystalline? (*b*) With X rays how might you distinguish the two possibilities?

8.7 *Properties of Liquids* Liquids A and B are at the same temperature. When a drop of each is placed on the back of your hand, A feels much colder than B. Account for this and tell what you conclude about molecular differences for the two liquids.

8.8 *Colloids* Distinguish clearly between each of the following pairs: (*a*) solution and a sol, (*b*) sol and emulsion, (*c*) emulsion and gel, (*d*) aerosol and sol.

8.9 *Tyndall Effect* (*a*) What relationship, if any, is there between Brownian motion and the Tyndall effect? (*b*) How would you go about observing each of these? (*c*) What relationship, if any, is there between Brownian motion and the nonsettling out of colloidal suspensions?

8.10 *Electrophoresis* A sol is observed in an electrophoresis study to move toward the positively charged electrode. (*a*) Account for this observation. (*b*) What might be expected to happen if this sol were mixed with one which showed the opposite behavior on electrophoresis?

8.11 *Adsorption* Which of the following would be most strongly adsorbed on charcoal: Cl_2, CO_2, $CHCl_3$?

8.12 *Vapor Pressure* How are the vapor pressures of liquids related to the attractive forces between the molecules?

8.13 *Amorphous Solids* (*a*) How do crystalline solids and amorphous solids differ with regard to internal structure? (*b*) Give three examples of amorphous solids.

8.14 *Vapor Pressure* (*a*) If 2.20 g of H_2 and 16.0 g of O_2 are placed in a 10.0-liter container at 25°C, what is the pressure of gas in the container? (*b*) If a spark is passed through the gas so as to give the reaction

$$2H_2(g) + O_2(g) \longrightarrow 2H_2O(l)$$

what is the final pressure of the gas after it has cooled to 25°C?

8.15 *Vapor Pressure* Describe and explain in terms of kinetic molecular theory the effect of each of the following on the vapor pressure of a liquid: (*a*) doubling the volume of the vapor above the liquid at constant temperature, (*b*) adding hydrogen gas to the system until its pressure is 15 atm.

8.16 *Model of a Liquid* The density of liquid mercury is 13.61 g/ml at 20°C. If the mercury atom is assumed to be a sphere of radius 1.44×10^{-8} cm, what percentage of liquid mercury is empty space?

8.17* *Energy Distribution* Suppose in a given liquid sample the distribution of kinetic energies at a given time is as follows: 25 percent of the molecules each have 0.30 unit of kinetic energy; 40 percent, 0.50; 15 percent, 1.00; 6 percent, 1.25; 9 percent, 1.50; 3 percent, 1.90; and 2 percent, 2.60. (*a*) Make a graph showing this distribution. (*b*) Calculate the average kinetic energy per molecule. (*c*) Suppose the molecules with 1.90 and 2.60 units of kinetic energy escape. What will be the average kinetic energy of those left behind?

8.18 *Colloidal Dimensions* Consider a sol consisting of 1.00 cm³ of sulfur suspended in 999 ml of water. (*a*) If the average sulfur particle is a cube of edge length 100 nm, how many of them are there in the suspension? (*b*) On the average (in nanometers), how far apart are two adjacent sol particles? (Assume a cubic array of sol particles in a cubic suspension.)

8.19* *Surface Adsorption* The great adsorption of charcoal for various molecules is attributed to its great surface area and the unsaturated valence forces of the C atoms at the surface. For a sample of charcoal having a surface area of 100 m^2/g, what fraction of the C atoms are at the surface? *Note:* The atomic diameter of C is 0.15 nm, and you may therefore assume that each atom occupies an average of 2.3 × 10^{-2} nm^2.

8.20 *Boiling* (a) Why does water boil at a lower temperature in the mountains than at the seashore? (b) How does a pressure cooker work?

8.21* *Equilibrium Vapor Pressure* What weight of water is needed to form the equilibrium vapor pressure (so-called 100% humidity) in a room 2.6 m high with a floor area of 20 m^2 at 25°C where the equilibrium vapor pressure is 0.031 atm?

8.22 *Boiling Point* How can you account for the fact that the normal boiling point of HF (19.9°) is higher than that of HCl (−85°), while that of F$_2$ (−188°) is lower than that of Cl$_2$ (−34°)?

8.23 *Evaporation* According to Figure 8.1, it is the molecules with higher than average kinetic energy which evaporate from a liquid. For the vapor in equilibrium with the liquid, what can you say about its temperature and the average kinetic energy of its molecules? Explain what happened.

8.24 *Equilibrium Vapor Pressure* Assume that in Figure 8.2 the liquid volume is initially $\frac{1}{30}$ of the gas volume. What fraction of the water must evaporate to give the equilibrium vapor pressure of 0.023 atm at 20°C?

8.25* *Colloidal Dimensions* Consider 1 mole of solid gold, density 19.3 g/cm^3. (a) If the mole of gold is in the form of a cube, what is its surface area? (b) If the gold is ground into tiny cubes 10 nm on an edge, what is its surface area? (c) How many gold atoms are there in each of the cubes in (b)?

8.26 *Structure* Experiments with very fine spheres containing surface charges like those in Figure 8.6 show that the volumes occupied by mixtures of oppositely charged spheres may exceed the sums of the volumes occupied by the two kinds of particles. (a) When separated, what kind of packing would you expect? (b) Account for the apparent increase in volume. (c) Why is it that this effect is large enough to observe only with very small particles? (d) How might this kind of effect be important in accounting for density changes on melting solids which have close-packed structures and also those which like ice do not?

8.27* *Avogadro Number* If we could measure colloidal sizes of large molecules sufficiently accurately, we could use such measurements to determine the Avogadro number from known densities. Assume the hemoglobin molecule has a radius of 3 nm (Section 8.5), a molecular weight of 66,800, and a density of 1 g/cm^3. Calculate a *rough* value for the Avogadro number.

Changes of State and Thermo-dynamics

9

The chemistry of matter is concerned not only with properties of substances but also with the changes they undergo. Similarly our study of the states of matter is concerned with changes from one state to another. Such changes are of both intrinsic and practical interest. Furthermore they are directly involved in many chemical reactions and play a role in the final outcome. For all these reasons we will now turn our attention to the changes which matter undergoes as when a solid melts or a liquid evaporates. In our study we will be particularly concerned with energy considerations accompanying changes of state. The systematization of energy changes and of the influence of temperature and pressure on them constitute an important part of what is called *thermodynamics*. Thermodynamics, in turn, is one of the most powerful tools scientists have in their study of nature.

9.1 Heating Curves

The temperature variations which accompany changes of states are represented in Figure 9.1. The curve shown is a *heating curve* corresponding to the uniform addition of heat to a substance. Since heat is added at a constant rate, distance on the time axis is also a measure of the amount of heat added. Let us start at time t_0 with the temperature at absolute zero and ask what will happen to the atoms or molecules of crystalline substance. As heat is

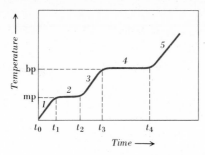

Figure 9.1
Heating curve.

added, each particle vibrates back and forth about a lattice point, which thus represents the center of this motion. As more heat is added, the vibration becomes greater. Although no change is visible, because the vibration is so small, the crystal progressively becomes slightly less ordered. The added heat increases the motion (kinetic energy) of the particles. Since temperature measures average kinetic energy, temperature rises along portion 1 of Figure 9.1. This continues until the melting point of the substance is reached.

At the melting point (abbreviated mp) the vibration of particles is so vigorous that any added heat serves to loosen binding forces between neighboring particles. Consequently, from time t_1 to t_2 (along portion 2 of the curve), added heat goes not to increase the average kinetic energy but to increase the potential energy of the particles. Potential energy is increased because work is done against attractive forces. During this period, there is no change in average kinetic energy, and so the substance stays at the same temperature. From t_1 to t_2 the amount of solid gradually decreases, and the amount of liquid increases. The *temperature at which the solid and liquid coexist* is defined as the *melting point* of the substance.

Eventually (at time t_2), sufficient heat has been added to break up all the crystal structure. Along portion 3 of the curve, added heat again increases the average kinetic energy of the particles of the liquid, and the temperature rises. This continues until the boiling point (bp) is reached. At the boiling point, added heat is used to overcome the attraction of one particle for its neighbors in the liquid. Along portion 4 of the curve, there is an increase in the potential energy of the particles but no change in their average kinetic energy. From time t_3 to t_4, liquid converts to gas. Finally, after all the liquid has been converted to gas, added heat raises the kinetic energy of the particles, and the temperature rises as shown by portion 5 of the curve.

9.2   Cooling Curves

The *cooling curve* results when heat is removed at a uniform rate from a substance. For a pure substance that is initially a gas, the temperature as a

Changes of State and Thermodynamics

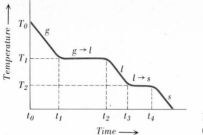

Figure 9.2
Cooling curve.

function of time is shown in Figure 9.2. As heat is removed from the gas, its temperature drops along the line marked "g" (gas). During this time, the average kinetic energy of the gas particles must decrease in order to compensate for the removal of energy to the outside. This slowing down proceeds until the particles are so sluggish that the attractive forces become dominant.

At time t_1 the particles start to coalesce to form a liquid. In the liquefaction process, particles leave the gas and enter the liquid state. Since it requires energy to take a particle from the liquid to the gas state, the reverse process, in which a particle goes from the gas to the liquid, releases energy. This decrease of potential energy on condensation supplies heat, which compensates for that being removed from the substance. Thus, as liquefaction proceeds, the temperature does not fall, and on average the particles do not slow down in their motion. As a result, the gas and the liquid are both at the same temperature, and the average kinetic energy of the particles in both phases is the same. From time t_1 to t_2 the temperature remains constant at T_1, the *condensation, or liquefaction, temperature.*

At time t_2 all the gas particles have condensed into the liquid state. Further removal of energy from the system causes the particles to slow down. As the average kinetic energy decreases, the temperature drops, as shown, along the line marked "l" (liquid). This drop continues until t_3, when the liquid begins to convert to solid.

In crystallization, the particles line up in a definite symmetry pattern, and as they go from the liquid state to the solid state, their *freedom* of motion is diminished. As each particle moves into position to form the crystal structure, the potential energy of the particle drops. The removal of heat energy to the outside is compensated for by the energy available from this decrease in potential energy. The average kinetic energy of the particle stays constant during the crystallization process. At T_2, the crystallization temperature, the motion of the particles is not slower in the solid than it is in the liquid, but it is a more restricted motion. From time t_3 to t_4 the temperature remains constant as the liquid converts to solid. When all the particles have formed the crystal structure, further removal of heat drops the temperature, as shown, along the final part of the curve marked "s" (solid).

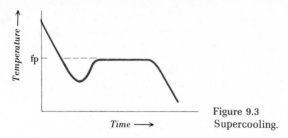

Figure 9.3
Supercooling.

The cooling curve is just the reverse of the heating curve. The temperature at which gas converts to liquid (liquefaction point) is the same as the temperature at which liquid converts to gas (boiling point). Similarly, the temperature at which liquid converts to solid (freezing point) is the same as the temperature at which solid converts to liquid (melting point).

Most cooling curves are not quite so simple as the one described above. The complication usually occurs in the portion corresponding to the transition from liquid to solid. Instead of following the dashed flat portion, as shown in Figure 9.3, the temperature follows the dip. The liquid does not crystallize at the freezing point but *supercools*. Supercooling arises in the following way: The particles of a liquid have little long-range pattern and move around in a disordered manner. At the freezing point, they should line up in characteristic crystalline arrangement, but only by chance do the particles start crystallizing correctly. Often they do not fall into the correct pattern immediately, and when heat continues to be removed from the substance in the absence of crystallization, the temperature falls below the freezing point. Particles continue moving, albeit more slowly, through various patterns until by accident they hit on the right one. Once this pattern has been built up to sufficient size, other particles rapidly crystallize on it. When a multitude of particles crystallize simultaneously, enough potential energy is converted suddenly to kinetic energy to heat up the whole system. The temperature actually increases until it coincides with the freezing-point temperature. From there on, the behavior is normal. Supercooling may be reduced by the introduction of a seed crystal, on which crystallization can occur, thereby initiating the proper structure. In rare instances even seeding does not induce crystallization. Further cooling then produces a rigid, supercooled liquid, or glass.

9.3 Heat Changes

As the heating curve of Figure 9.1 shows, the addition of heat to a substance is not necessarily accompanied by a rise in temperature. At the melting point, for example, the added heat goes not to raise the temperature but instead to increase what is called the *enthalpy,* or *heat content,* of the substance. Enthalpy is generally designated by the symbol H. Actually, it is the change in enthalpy during a process that is important, and the increase in enthalpy during that process is called ΔH (read "delta aitch"). As an example, when 1 mole of H_2O goes from the solid state to the liquid state, there is required 6.01 kilojoules of heat. In other words, the heat content of the H_2O increases by 6.01 kilojoules, and ΔH is equal to 6.01 kilojoules/mole. This is called the *heat of fusion* for ice.

The molar heat of fusion of a solid is characteristic of the substance and differs widely from one substance to another. For example, for NaCl the ΔH of fusion is 30.3 kilojoules/mole. In general, heats of fusion are greater for ionic solids than for molecular solids. The reason for this is that forces of attraction are generally stronger in ionic solids, so that more heat must be added to break up the structure and raise the potential energy of the particles. However, we need to be a bit cautious in making a direct relation between heat added to a substance and its energy increase. For example, when NaCl melts it expands and therefore work must be done to push aside the atmosphere to make room for the increased volume. Hence, the added heat goes not only to raising the internal energy of the NaCl but also to doing work against the atmosphere. If we represent the increase in internal energy by ΔE and the volume increase by ΔV, then we can write

$$\Delta H = \Delta E + P\,\Delta V$$

for the case where the only work done is against a constant atmospheric pressure P. For the process of fusion, the volume change is generally small, so that the $P\,\Delta V$ term is small compared with ΔE. Therefore, in these cases ΔH gives a good measure of ΔE; that is, the increase of heat content is a good measure of the increase of internal energy.

It should be noted that in the above discussion Δ stands for the change in some property. It represents the property in the final state minus that property in the initial state. Thus, if ΔH is positive, it means that H of the final state is greater than H of the initial state. Of course, the opposite can be true, as for the freezing of water

$$H_2O(l) \longrightarrow H_2O(s)$$

for which process ΔH is negative (-6.01 kilojoules). A negative ΔH means that H of the final state is less than the H of the initial state. In other words, heat would have to be liberated to the surroundings. It follows then that a negative ΔH corresponds to an exothermic change (heat liberated to the surroundings); a positive ΔH, to an endothermic change. We can keep the sign straight by noting that H tells us what is happening to the substance, which is the opposite of the effect it is having on the surroundings. In an exothermic change, heat is evolved and the heat content of the substance decreases.

Just as the heat content increases on melting, it increases on vaporization. The increase in heat content as 1 mole of substance passes from the liquid to the gaseous state is called the molar *heat of vaporization*. For H_2O, ΔH of vaporization is 40.7 kilojoules/mole at the normal boiling point. Thus we can write

$$H_2O(l) \longrightarrow H_2O(g \text{ at } 1 \text{ atm}) \qquad \Delta H = +40.7 \text{ kilojoules/mole}$$

or, for the reverse process of condensation,

$$H_2O(g \text{ at } 1 \text{ atm}) \longrightarrow H_2O(l) \qquad \Delta H = -40.7 \text{ kilojoules/mole}$$

Finally, it should be pointed out that once we know the ΔH's for two consecutive processes, e.g., melting and vaporization, we can calculate the ΔH for a combined process, e.g., sublimation. The increase in heat content on sublimation (passage from solid to gas) is simply the sum of the increase in heat content on melting (solid to liquid) plus the increase in heat content on vaporization (liquid to gas). If the melting and the vaporization data are not for the same temperature, a correction term must be added for the increase in heat content that accompanies warming up the substance from the melting temperature to the vaporization temperature. Once that has been done the increase in heat content (as is true for any thermodynamic property) depends only on the initial and final states of the system.

9.4 Vapor Pressure of Solids

As in a liquid, particles of a solid can escape into the vapor phase to establish vapor pressure. In a solid, not all the particles have the same energy. There is a distribution of energy in which most of the particles have the average energy, some have less energy than average, and others have more. Those particles which at any one time are of higher than average energy and are near the surface can overcome the attractive forces of their neighbors and escape into the vapor phase. If the solid is confined in a closed container, eventually there will be enough particles in the vapor phase that the rate of escape is equal to the rate of return. A dynamic equilibrium is set up, in which there is an equilibrium vapor pressure characteristic of the solid at a given temperature. Since the escaping tendency of particles depends on the magnitude of the intermolecular forces, the equilibrium vapor pressure differs from one substance to another. If the attractive forces in the solid are small, as in the case of a molecular crystal, such as camphor, the

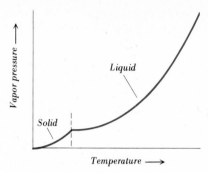

Figure 9.4
Temperature variation of vapor pressure.

escaping tendency is great, and vapor pressure is high. In an ionic crystal, such as sodium chloride, the binding forces are usually large, and the vapor pressure is low.

The vapor pressure of solids also depends on temperature. The higher the temperature, the more energetic the particles, and the more easily they can escape. The more that escape, the higher the vapor pressure. Quantitative measurements of the vapor pressure of solids can be made in the same way as for liquids. Figure 9.4 shows how the vapor pressure of a given substance changes with temperature. At absolute zero, the particles of a solid have no escaping tendency, and so the vapor pressure is zero. As the temperature is raised, the vapor pressure rises. It rarely gets to be very high before the solid melts. Above the melting point the vapor pressure is that of the liquid.

At any point along the portion of the curve marked "Solid," there is equilibrium between vapor and solid. The number of particles leaving the solid is equal to the number returning. When the temperature is raised, there are more particles breaking loose from the crystal than returning. This causes a net increase in the concentration of particles in the vapor phase. There also results an increased rate of condensation to the solid phase, which eventually becomes equal to the increased rate of evaporation. Equilibrium is reestablished at the new temperature.

The behavior of an equilibrium system when it is upset by the action of an external force is the subject of the famous *principle of Le Chatelier*, first published in 1884. Henry Le Chatelier stated that, *if a stress is applied to a system at equilibrium, then the system readjusts, if possible, to reduce the stress.* Raising the temperature of a solid-vapor equilibrium system amounts to applying to the system a stress in the form of added heat. Since the conversion from solid to gas is endothermic (requires heat),

Solid + heat $\rightleftharpoons$ gas

The stress of added heat can be absorbed by converting some of the solid to gas. On being cooled, the system adjusts itself to the lower temperature by producing heat from the conversion of gas to solid.

At the point where the vapor-pressure curve of a liquid intersects that for the solid (i.e., where vapor pressure of solid equals vapor pressure of

liquid), there is simultaneously an equilibrium between solid and gas, between liquid and gas, and between solid and liquid. This point of intersection
at which *solid, liquid, and gas coexist in equilibrium* is called the *triple point.*
Every substance has a characteristic triple point fixed by the nature of the
attractive forces between its particles. For water, the triple-point temperature
is 0.01°C, and the triple-point pressure is 0.006 atm. It should be noted that
the triple-point temperature of water is not quite the same as the normal
melting point, which is 0°C. This difference comes about because the normal
melting point is defined as the melting point at 1 atm pressure and because
the melting point changes as pressure is changed. At the triple point, the
only pressure exerted is the vapor pressure of the substance.

9.5 Phase Diagrams

The relation between the solid, liquid, and gaseous states of a given substance as a function of the temperature and pressure can be summarized
on a single graph known as a *phase diagram.* Each substance has its own
particular phase diagram, which can be worked out from experimental
observation. Figure 9.5 gives the phase diagram of H_2O. On this diagram,

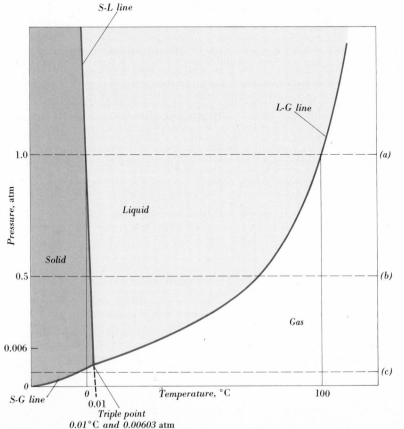

Figure 9.5
Phase diagram of H_2O (scale
of axes somewhat distorted).

various points represent the substance H_2O in the solid, liquid, or gaseous state, depending on its temperature and pressure. Each of the three regions corresponds to a one-phase system. For all values of pressure and temperature falling inside such a single-phase region, the substance is in the state specified. For example, at 0.5 atm pressure, H_2O at $-10°C$ is in the solid state, at $+10°C$ in the liquid state, and at $+100°C$ in the gas state. The lines which separate one region from another are equilibrium lines, representing an equilibrium between two phases. In the diagram the S-L line represents equilibrium between the solid and the liquid; the L-G line, equilibrium between liquid and gas; and the S-G line, equilibrium between solid and gas. The intersection of the three lines corresponds to the triple point, where all three phases are in equilibrium with each other.

The usefulness of a phase diagram can be illustrated by considering the behavior of H_2O when heat is added at a constant pressure. This corresponds to moving across the phase diagram from left to right. We distinguish three typical cases, indicated by a, b, and c in Figure 9.5:

a   **Pressure of the H_2O is kept at 1 atm**   The experiment is this: A chunk of ice is placed in a cylinder so as to fill the cylinder completely, with no empty space. A piston resting on the ice carries a weight which corresponds to 1 atm pressure. The H_2O starts as a solid. As heat is added, its temperature is raised. This corresponds to moving from left to right along the horizontal line a in Figure 9.5. When the S-L line is reached, the added heat melts the ice. Solid-liquid equilibrium persists at the normal melting point of 0°C until all the solid is converted to liquid. There is no gaseous H_2O thus far, because the vapor pressure of solid ice is much lower than the pressure required to push the piston out to make room for the vapor. As heating continues, liquid H_2O warms up from 0°C until the L-G line is reached, at a temperature which corresponds to 100°C. At this temperature, liquid-gas equilibrium is established. The system stays at 100°C as the liquid converts to gas. At 100°C the vapor pressure is great enough to exceed 1 atm, hence to move the piston and make room for the voluminous vapor phase. Since the external pressure is fixed at 1 atm, liquid continues to convert to vapor, and the piston moves out as heat is added until the liquid has converted completely to gas at the normal boiling temperature. From then on, the gas simply warms up.

b   **Pressure of the H_2O is kept at 0.5 atm**   Again the H_2O starts as the solid. The temperature is raised, moving to the right along horizontal line b. The H_2O stays as a solid until it reaches the temperature that corresponds to melting. Because of the tilt of the S-L line toward the left (the tilt has been somewhat exaggerated in Fig. 9.5), the temperature at which melting occurs is slightly higher at 0.5 atm than at 1 atm pressure. At 0.5 atm pressure, ice melts, not at 0°C, but slightly above zero, about 0.005°C. After all the ice has been converted to liquid at +0.005°C, further addition of heat warms up the liquid until boiling occurs at the L-G line. Boiling occurs when the vapor pressure of the water reaches 0.5 atm. The temperature at which this happens is 82°C, considerably lower that the normal boiling point of 100°C. Above 82°C only gaseous water exists at 0.5 atm pressure.

c **Pressure of the H_2O is kept at 0.001 atm** If the pressure exerted by the H_2O is kept at 0.001 atm, the H_2O can exist only along the horizontal line marked c. As the temperature is raised, solid H_2O warms up until it reaches the S-G line. Solid-gas equilibrium is established; solid converts to gas. When all solid has been converted, the temperature of the gas rises. There is no melting in this experiment, and no passage through the liquid state.

An interesting aspect of the H_2O phase diagram is that the S-L line, representing equilibrium between solid and liquid, tilts to the left with increasing pressure. This is unusual; for most substances the S-L line tilts to the right. The direction of the tilt is important, since it tells us whether the melting point rises or falls with increased pressure. In the case of H_2O, as pressure is increased (*up* on the phase diagram), the temperature at which solid and liquid coexist decreases (*left* on the phase diagram). The melting-point decrease is approximately 0.01°C/atm.

The lowering of the ice melting point by increased pressure is predicted by Le Chatelier's principle. The density of ice is 0.9 g/cm³; the density of water is 1.0 g/ml. One gram of H_2O in the solid state occupies a volume of 1.1 cm³; in the liquid state, 1 g of H_2O occupies 1.0 cm³. Thus, a given mass of H_2O occupies a larger volume as solid than as liquid. An equilibrium system consisting of water and ice at 1 atm is at the normal melting point of 0°C. If the pressure on the H_2O is increased, a stress exists which the system can relieve by shrinking in volume. It can shrink in volume by converting from ice to water; hence melting is favored. But melting is an endothermic process and requires heat. If the system is insulated, the only source of heat is the kinetic energy of the molecules. The molecules slow down, and the temperature drops. The result is that solid ice and liquid water under increased pressure coexist at a lower temperature.

Another phase diagram of interest is that of carbon dioxide (CO_2). In general appearance, as shown in Figure 9.6, it is similar to that of H_2O. However, the solid-liquid equilibrium line tilts to the right instead of to the left, since the melting point of CO_2 rises with increased pressure. The triple-point pressure of CO_2 is 5.2 atm; the triple-point temperature is −57°C. Since the triple-point pressure is considerably above normal atmospheric pressure, liquid carbon dioxide is not observed under usual conditions. In

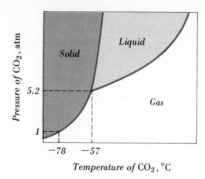

Figure 9.6
Phase diagram of CO_2.

order to get liquid carbon dioxide, the pressure must be higher than 5.2 atm. At 1 atm, only solid and gas can exist. When solid carbon dioxide in the form of dry ice is used as a refrigerant at 1 atm pressure, the conversion of solid to gas occurs at $-78°C$. There is no increase in temperature of the carbon dioxide until all the solid disappears.

9.6 Entropy, Free Energy, and Spontaneous Change

Having dealt with various aspects concerning changes of state including accompanying energy changes, we can now explore further the reasons why changes of state take place. As we will see, the thermodynamic principles underlying phase changes are broadly applicable to chemical changes.

In predicting whether a given change should occur in a system under particular conditions, it would seem intuitively clear that the answer depends on whether the change produces a state of lower energy. In the terms discussed earlier this means that our intuition would most likely say that systems should tend to lower their enthalpy, or heat content. However, intuition cannot always be trusted. For example, a chunk of ice at room temperature *spontaneously* melts, i.e., goes by itself to a state of higher energy (or heat content). Of course, the process is made possible by the flow of heat from the surroundings but the question is why does the chunk of ice use the added heat to melt rather than simply to warm up? By melting, it has spontaneously gone to a state of higher potential energy. Evidently, more than energy is involved in determining the direction of spontaneous changes.

The additional factor which must be considered is the tendency of a system to assume the most random molecular arrangement possible. In other words, systems tend to become disordered. The reason for this tendency is that there are many more ways of producing disordered arrangements than an ordered arrangement. The disorder is described quantitatively in terms of a property called the *entropy*, and we say that a more disordered state has a higher entropy than an ordered state. As a specific example, H_2O in the form of liquid water has a higher entropy (a random collection of various molecular arrangements) than does H_2O in the form of solid ice (a single ordered array). It is more probable that the higher entropy state will be formed rather than the lower entropy state.

Why is a state of high entropy (a so-called disordered state) more probable than one of low entropy (a so-called ordered state)? In ice there is a recognizable repeat pattern that is unique for ice. In liquid water there is not a unique arrangement of molecules, and hence there are a number of arrangements all of which together constitute the state of the liquid. Each one of these arrangements is just as probable as any other, and, in fact, each is as probable as the ice structure. However, there are many arrangements for the liquid compared to the one for ice. Hence, the sum of the molecular arrangements characteristic of liquid water is more probable than that for ice. As an analogy, consider a pair of dice. Any configuration of the two dice is equally probable, yet it is more likely that you roll a 7 than a 2. The reason is that there are six ways to roll a 7 ($6 + 1, 5 + 2, 4 + 3, 3 + 4, 2 + 5, 1 + 6$) but only a single way to roll a 2 ($1 + 1$). It follows that the state called "snake eyes" has a lower entropy than the state called "seven." It is for this reason that rolling a 7 is favored.

Just as changes are favored which result in a decrease of energy, those are also favored which result in an increase of entropy. In certain cases, as in melting ice, the two factors may oppose each other so that there is a question of which one wins out. Above the melting point the entropy increase is dominant, and so spontaneous melting occurs; below the melting point the energy decrease is dominant, and so spontaneous freezing occurs. The temperature itself is the critical factor and governs how important the entropy increase is, relative to the energy change. At the absolute zero of temperature, the entropy term contributes nothing in determining the direction of spontaneous change, hence the most stable state is that of lowest energy. As the temperature rises, molecular motion increases and the tendency to disorder becomes more important in determining the direction of change. At sufficiently high temperatures, the entropy factor becomes large enough to overcome even an unfavorable energy change.

Quantitatively, the interplay of the energy and entropy factors can be described by using a concept called the *free energy*. This is generally symbolized by G, in honor of Willard Gibbs, the Yale professor of mathematics who founded chemical thermodynamics. (F is sometimes used in place of G.) The definition of free energy is given by the relation

$$G = H - TS$$

where H is the heat content (or enthalpy) of a substance, S is its entropy, and T is the absolute temperature. During a change, the free-energy increase is

Changes of State and Thermodynamics

$$\Delta G = \Delta H - \Delta(TS)$$

That is, the free-energy change ΔG from the initial to the final states of a substance is equal to the change ΔH in heat content minus the change $\Delta(TS)$ in the temperature-entropy product. If G is identified with energy that is freely *available*, the above relation states that the available energy is less than the heat content by the amount TS. The temperature-entropy product TS is sometimes referred to as the "unavailable energy." We get a hint at the reason for this terminology by recalling that S measures the probability of a state of a substance and T gives the thermal disordering influence. The higher the temperature, the more the disorder and the less available the energy becomes.

For an *isothermal* change, that is, one that occurs at a *fixed temperature* T, the free-energy change is given by

$$\Delta G = \Delta H - T\,\Delta S$$

This is a most important equation because at constant temperature and pressure a *process can occur spontaneously only if there is a decrease in free energy*. In other words, for a spontaneous process, ΔG must be negative, corresponding to a free energy lower in the final state than in the initial.

A negative ΔG can result either from a decrease in heat content or an increase in entropy. The former (where $\Delta H < 0$) would correspond in general to an energetically favorable process, that is, one in which energy decreases; the latter ($\Delta S > 0$) corresponds to one in which disorder increases and so is favored by random thermal motion. For many processes, ΔH and ΔS have the same sign; for instance, in the melting of a solid both are positive (i.e., both the heat content and the entropy increase). For such cases, T will be all-important in deciding which term prevails. At sufficiently low T, ΔH will predominate. For the melting process, where ΔH is positive, ΔG will also be positive at sufficiently low T, and melting will not occur. At high temperature, the $T\,\Delta S$ term predominates over ΔH. Since ΔS for melting is positive, $-T\,\Delta S$ can cause ΔG to be negative. Hence, melting should occur. At one temperature, $T\,\Delta S$ just matches ΔH. ΔG at this temperature is equal to zero, and solid and liquid coexist in equilibrium. The temperature of this coexistence is, of course, the melting or freezing point.

We should note that the criterion of equilibrium between two states at constant temperature and pressure is that the free energy be equal in the two states, or that $\Delta G = 0$. The quantitative application of this criterion to a typical system is illustrated in the following example.

EXAMPLE 1   The heat required to melt sodium chloride is 30.3 kilojoules/mole. The entropy increase is 28.2 joules/mole · deg. Calculate the melting point from these data.

At the melting point,

$$\Delta G = 0 = \Delta H - T_{mp}\, \Delta S$$
$$\Delta H = T_{mp}\, \Delta S$$

$$T_{mp} = \frac{\Delta H}{\Delta S} = \frac{30,300 \text{ joules/mole}}{28.2 \text{ joules/mole} \cdot \text{deg}}$$

$$= 1070 \text{ K}$$

In summary and for future reference we should note that predictions concerning possible changes must take into consideration more than just the lowering of a system's energy. In the frequently encountered case of a substance in contact with its surroundings so that (1) pressure is maintained constant and (2) heat can flow in or out to maintain constant pressure, then the quantity of interest is the free energy.

1 At constant temperature and pressure, changes should occur which decrease the free energy, G.

2 At equilibrium, $\Delta G = 0$.

3 The free energy takes into account both the enthalpy and entropy of the system in question.

4 Enthalpy (H), or heat content, measures the internal energy of a system adjusted for any external work done (e.g., by expansion of volume).

5 Entropy (S) measures the probability that a system will assume a particular form which tends to maximize.

6 Spontaneous changes at constant temperature and pressure must be accompanied by a negative ΔG, where $\Delta G = \Delta H - T\, \Delta S$.

These points can be made clear by the following: A familiar example of a spontaneous change at constant temperature and pressure is the snapping back of a stretched rubber band. Why does it snap back? It is not the result of a lowering of energy or enthalpy. The volume is changed very little, and the heat change can be detected by the experiment of Figure 9.7, which you might try. If a rubber band is stretched tight while touching your lip, you will notice that heat is given off on stretching (feels warm) and heat is picked up on contracting (feels cool). This means that stretching the band lowers its enthalpy, or internal energy; contraction increases it. The spontaneous snapping back occurs despite the fact that it means increasing the energy of the band!

Because the snapping back will occur spontaneously at constant temperature and pressure, it must be accompanied by a decrease of free energy. This favorable free-energy change results from two contributions. The

Figure 9.7
Detection of heat changes on stretching a rubber band.

enthalpy contribution is unfavorable (it increases); therefore the entropy contribution must be favorable. A favorable entropy contribution (increase) means that the system (rubber band) has spontaneously reverted to a more probable (more random) state. Because the temperature at which the observation is made (room temperature) is sufficiently high, $-T \Delta S$ overcomes the unfavorable ΔH. At very low temperature, $-T \Delta S$ will be too small to compensate for ΔH and the rubber band will not snap back.

On a molecular level, the rubber-band experiment can by understood as follows. Rubber molecules are long chainlike hydrocarbons, $(C_5H_8)_n$, which can be either extended in more-or-less straight chains or coiled and kinked in many different disordered configurations. When rubber is stretched, the molecules are more extended, which turns out to be a lower energy configuration for the molecular collection. But it is a more ordered and less probable configuration because there are fewer possibilities for forming it. In the contracted state, many more configurations are possible for the molecules. Even though the energy is somewhat higher in this disordered array, the randomizing thermal motion causes it to be formed. Hence the temperature-entropy term wins out over the unfavorable enthalpy term, ΔG decreases, and the rubber band snaps back.

QUESTIONS

9.1 *Heating Curves* (a) Describe what experiment you would do to measure the heating curve of a pure solid. (b) What would you see during the two periods in your experiment when the temperature doesn't change?

9.2 *Cooling Curve* (a) Draw a cooling curve that starts with gas, ends with solid, and shows supercooling. (b) What accounts for supercooling? (c) Why do you not normally observe superheating?

9.3 *Enthalpy* (a) How is enthalpy related to internal energy? (b) Why is ΔH negative for exothermic changes?

9.4 *Vapor Pressure of Solids* (a) From your experience, quote one observation which shows that solids have vapor pressure. (b) Give a molecular explanation for the vapor pressure of solids.

9.5 *Le Chatelier* Consider a piston in a cylinder containing dry ice and CO_2 gas in equilibrium. The piston is pushed in by exerting pressure. (a) Use the Le Chatelier principle to predict what happens to the relative amounts of solid and gas. (b) What happens to the temperature? Explain.

9.6 *Heat of Fusion* The heat of fusion of NaCl and H_2O are 30.3 and 6.01 kilojoules/mole. (a) What would you conclude about the relative attractive forces for these two solids? (b) Is your answer to (a) affected by the fact that on melting NaCl expands and H_2O contracts? (c) How many kilojoules of energy are needed to melt 1.00 g of each of these solids at its normal melting point?

9.7 *Phase Diagram* (a) Draw a labeled phase diagram for a substance whose normal melting point is 20°C, normal boiling point is 90°C, and triple point is −5°C and 0.10 atm. (b) Compare the relative density of the gas, liquid, and solid for this substance at its triple point. Give reasons for your answer.

9.8 *Thermodynamics* State what is meant by (a) enthalpy, (b) entropy, (c) free energy.

9.9 *Spontaneous Change* Consider the process whereby an extended rubber band snaps back. (a) What is meant in this case by initial state and final state? (b) What is the sign of ΔG, ΔH, and ΔS for this process? (c) If the temperature is dropped far enough that the change is no longer favored, what are the signs of the three functions in (b)?

9.10 *Heating Curves and Phase Diagrams* Describe how you could use a series of heating curves to construct a phase diagram which was previously unknown.

9.11* *Heat Change* Consider a drop of water at 20°C weighing 0.0500 g. How many joules of heat must be added to it or subtracted from it (tell which for each case) in performing each of the following (recall that the heat capacity of H_2O is 4.18 joules/g · deg, the molar heat of fusion is 6.01 kilojoules, and the molar heat of vaporization is 40.7 kilojoules): (a) the water is heated to 100°C, (b) the drop of water at 100°C is boiled, (c) the vapor in (b) is converted completely to ice at 0°C?

9.12 *Phase Diagram* Using Figure 9.5, explain why the slope of *S-L* line for H_2O is negative (decreases going left to right).

9.13* *Enthalpy* The heat of fusion of ethyl alcohol, C_2H_5OH, at $-114°C$ is 104 joules/g. What is the change in enthalpy on fusion of 1 mole of ethyl alcohol?

9.14 *Phase Diagram* Using the phase diagram shown in Figure 9.6, decide in what state CO_2 would be under the following conditions: (a) 9 atm, $-57°C$; (b) 5 atm, $-70°C$; (c) 1 atm, $0°C$.

9.15* *Heat of Sublimation* If it requires 225 joules to evaporate 1 g of liquid carbon tetrachloride, CCl_4, and 21.3 joules to melt 1 g of solid CCl_4, what is the molar ΔH for the following process at a fixed temperature?

$$CCl_4(s) \longrightarrow CCl_4(g)$$

9.16 $\Delta H, \Delta S, T$ If the molar heat of fusion of benzene is 9930 joules/mole and ΔS is 35.7 joules/mole $\cdot$ deg, what is the melting point of benzene?

9.17* $\Delta G, \Delta H, \Delta S$ For the process $H_2O(s) \longrightarrow H_2O(l)$, ΔH is 6.02 kilojoules/mole and ΔS is 22.0 joules/mole $\cdot$ deg. (a) Calculate ΔG for the above process at 220 and 290 K. (b) At what temperature is the process spontaneous?

9.18 $\Delta G, \Delta H, \Delta S$ For the elements Al, K, and Cs, the respective heats of fusion per gram are 395.0, 61.0, and 22.2 joules. Given that the respective melting points are 659.0, 63.4, and 28.3°C, (a) calculate the ΔH of fusion per mole of each metal and (b) calculate the ΔS per mole of each metal at its melting point.

9.19 *Heat Changes* When iron metal melts, the process is endothermic with the absorption of 266 joules of heat per gram. What is the ΔH per mole for the reaction $Fe(l) \longrightarrow Fe(s)$?

9.20 *Sublimation* The heat of sublimation of carbon dioxide is 16.2 kilojoules/mole and ΔS of sublimation is 88.5 joules/mole $\cdot$ deg. What is the temperature, in °C, at which carbon dioxide sublimes?

9.21* *Enthalpy* 10.5 g of compound A, which has molecular weight of 84.0, undergoes a two-step process: ΔH for the first step is 56.0 kilojoules/mole and in the second step ΔH is 37.6 joules/g. Determine the overall enthalpy change.

9.22 *Entropy* Explain whether the entropy change is favorable or unfavorable in each of the following processes: (a) evaporation of alcohol, (b) decomposition of $H_2O(l)$ into $H_2(g)$ and $O_2(g)$.

9.23 $\Delta G, \Delta H, \Delta S$ Tell whether reaction should be spontaneous if: (a) $\Delta S < 0$ and $\Delta H > 0$, (b) $\Delta H < 0$ and $\Delta S > 0$, (c) $\Delta G > 0$, and (d) $\Delta H < T \Delta S$.

9.24 *Enthalpy* What is the value of ΔG for boiling H_2O at 100°C and 1.00 atm?

9.25 *Terms* What is meant by (a) triple point, (b) supercooling, and (c) heat of fusion?

9.26 *Phases* A beaker contains ice, water, and steam at the triple point. Compare these three phases with respect to (a) average kinetic energy, (b) ordering of the molecules, and (c) density.

9.27 *Le Chatelier Principle* A solid, which has a high vapor pressure, is in equilibrium with its vapor. Discuss whether there is a change in vapor pressure if (a) the temperature of the system is lowered, (b) the volume of the vapor is doubled, (c) a small chunk of solid is added.

9.28 *Changes of State* On a molecular picture account for the facts that (a) for most substances the heat of vaporization is greater than the heat of fusion, and (b) these energetically unfavorable processes do occur.

9.29 *Le Chatelier* Why is it that a pressure change affects the boiling point more than the freezing point?

9.30 *Critical Point* (a) Draw a phase diagram showing a critical point. (b) For the substance described by your phase diagram, describe a heating experiment conducted above the critical pressure by drawing the heating curve and telling what you would see at various times along it.

9.31 *Supercooling* Draw a cooling curve showing supercooling. At each point along the curve give the signs of ΔG, ΔH, and ΔS for the process of conversion from liquid to solid. If any of the values change sign during the time depicted, explain why.

9.32 *Thermodynamics* The element sodium has a molar heat of fusion of 3.04 kilojoules at its melting point of 97°C. What is the entropy change on melting? (Give sign, value, and units.)

9.33 *Thermodynamics* (a) Heat is added to a liquid. Part of the added heat goes to increase the internal energy of the liquid, and part to expand the liquid and thereby do work against constant atmospheric pressure. In terms, symbols, and equations used in this chapter, write an equation for the process and explain the meaning of its terms. (b) Of the heat added, part becomes unavailable energy and part available energy. Write an appropriate equation and identify the terms as available and unavailable. (c) Show how the fraction of available energy changes with temperature. (d) On a molecular level tell what is meant by "unavailable energy."

9.34 *Entropy* (a) Any bridge hand is equally probable, yet a perfect hand of spades is much rarer than a hand with four spades. Why? (b) What relationship, if any, does your reasoning in (a) have to the process of melting?

9.35 *Triple Point* (a) At the triple point, what relationship must exist between ΔH fusion, ΔS fusion, ΔH vaporization, ΔS vaporization? (b) Since this relationship is true only at one pressure, at least one of the quantities is very dependent on pressure. State which one you think this is and justify your answer.

9.36 *Direction of Time* For the isolated universe, the energy content is constant. Yet changes are constantly occurring. Overall, there can be no energy change driving these changes. (a) What factor is important in determining the changes which occur? (b) For the total universe, what happens with time to the total value for the factor mentioned in your answer to (a)? (c) Why is time unidirectional in that the universe differs today from any other time in its past history?

Solutions

In the development of chemical science, the existence of solutions could have been a serious obstacle to progress. Here is a class of homogeneous materials, representing all three states of matter, which do not conform to the law of definite composition or even the law of multiple proportions. Fortunately, solutions were early distinguished from pure compounds by a fundamental difference that appeared when substances were partially melted or evaporated. For a compound, a change of state of one part of a sample results in no change of composition. For a solution, it is generally found that two states in equilibrium will differ in composition with one state richer in component A, the other in component B.

Rather than proving detrimental, then, solutions have been most useful in chemistry. Liquid solutions especially are the most usual medium for chemical reaction. They allow an intimate mixture of potential reactants so that atomic rearrangements responsible for chemical change can occur. For this reason, vital biological fluids are aqueous solutions in which controlled chemical changes serve to regulate life processes. Plasma, the liquid half of blood, is about 90 percent water, containing dissolved proteins, other organic compounds, and about 1 percent salts. The fact that the body fluids of all animals are salty suggests a common origin of animal life in the sea.

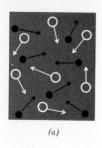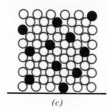

Figure 10.1
Models of solutions: (a) gaseous, (b) liquid, (c) solid.

10.1 Types of Solutions

Gaseous solutions are made by dissolving one gas in another. Since all gases mix in all proportions, any mixture of gases is homogeneous and is a solution. The kinetic picture of a gaseous solution is like that of a pure gas, except that the molecules are of different kinds. Figure 10.1a represents a gaseous solution of helium and oxygen, where the solid circles represent helium molecules and the open circles represent oxygen molecules. Ideally, the molecules move independently of each other.

Liquid solutions are made by dissolving a gas, liquid, or solid in a liquid. The kinetic picture of a sugar-water solution is represented in Figure 10.1b. The solid circles represent water molecules, and the open circles sugar molecules. The sugar molecules are distributed at random throughout the bulk of the solution. It is evident that on this molecular scale the term homogeneous has little significance. However, experiments cannot be performed with less than billions of molecules, so that for practical purposes the solution appears to be homogeneous.

Solid solutions are solids in which one component is randomly dispersed on an atomic or molecular scale throughout another component. An example of a solid solution is shown in Figure 10.1c, where the solid circles represent atoms of one component, say copper, and the open circles atoms of the other, say zinc. As in any crystal, the packing of atoms is orderly, even though there is no particular order as to which lattice points are occupied by which kind of atom. Solid solutions are of great practical importance, since they make up a large portion of the class of substances known as alloys.

An *alloy* may be defined as a combination of two or more elements which has metallic properties. Sterling silver, for example, is an alloy consisting of a solid solution of copper in silver. In brass, an alloy of copper and zinc, it is possible to have a solid solution in which some copper atoms of the face-centered cubic structure of pure copper have been replaced by zinc atoms. Some kinds of steel are alloys of iron and carbon and can be considered as solid solutions in which carbon atoms are located in some of the spaces between iron atoms. The iron atoms are arranged in the regular structure of pure iron. It must be pointed out, however, that not all alloys are solid solutions. Some alloys, such as bismuth-cadmium, are heterogeneous mixtures containing tiny crystals of the constituent elements. Others, such as $MgCu_2$, are intermetallic compounds which contain atoms of different metals combined in definite proportions.

Two words that are convenient in the discussion of solutions are the terms *solute* and *solvent*. Accepted procedure is to refer to the substance present in larger amount as the solvent and to the substance present in smaller amount as the solute. However, the terms can be interchanged whenever it is convenient. For example, in solutions of sulfuric acid and water, sulfuric acid is sometimes referred to as the solute and water referred to as the solvent, even when the water molecules are in the minority.

10.2 Concentration

Because solutions do not have fixed compositions they cannot be designated by fixed chemical formulas. Instead, we specify the relative amounts of solute and solvent by several alternate means which collectively are called *concentration*. It is the concentration of a solution which determines its properties, e.g., the color of a dye solution or the sweetness of a sugar solution. Note that it is not simply the amount of sugar which determines the sweetness but the relative amounts of sugar and water, or amount of sugar per given amount of water. This is the meaning of concentration in chemistry and it is specified in a number of ways.

The *percentage of solute* is a frequently used but ambiguous designation which may refer to percentage by weight or percentage by volume. If the former is meant, and this is usually the case, it is the percentage of the total solution weight contributed by the solute. Thus, 3% H_2O_2 by weight would be 3 g of H_2O_2 per 100 g of solution. Percentage by volume is the percentage of the final solution volume represented by the volume of solute taken to make the solution. For example, 12% alcohol by volume would represent a solution made from 12 ml of alcohol plus enough solvent to bring the total volume to 100 ml.

The *molarity* of a solute is the number of moles of solute per liter of solution and is usually designated by a capital M. A 6.0-molar solution of HCl is labeled 6.0 M. The label means that the solution has been made up in a ratio that corresponds to adding 6.0 moles of HCl to enough water to make a liter of solution.

The *molality* of a solute is the number of moles of solute per 1000 g of solvent. It is usually designated by a small m. The label 6.0 m HCl is read "6.0 molal" and represents a solution made by adding, to every 6.0 moles of HCl, 1000 g of water.

The *normality* of a solute is the number of equivalents (Section 5.8) of solute per liter of solution. It is usually designated by a capital N. The label 0.25 N $KMnO_4$ is read "0.25 normal" and represents a solution which contains 0.25 equiv of potassium permanganate per liter of solution.

10.3 Why Solution Occurs

When alcohol (C_2H_5OH) and water are mixed, they form a solution. However, when gasoline (approximately C_8H_{18}) and water are mixed, they do not dissolve in each other. Why is a solution formed in one case but not in the other? An obvious factor to consider is the change in energy as the molecules are mixed. To produce a solution of alcohol in water, the attractive forces between alcohol molecules must be overcome in separating them. Similarly, the attractions between water molecules must be overcome. Both processes require expenditure of energy. This energy comes from the attractions between water and alcohol in the resulting solution. In other words, the solute-solvent attractions compensate for loss of solute-solute plus solvent-solvent attractions. For gasoline-water the compensation is insufficient, because the gasoline-water attractions are very weak.

When alcohol dissolves in water, heat is released to the surroundings. In other words, the system is going to a state of lower heat content, or energy. A common misconception is to assume that dissolving can occur only if the system goes to a state of lower energy. However, many salts dissolve in water even though they go to states of *higher* energy. This is shown by KCl, for example, in that the dissolving process is endothermic; i.e., heat is absorbed from the surroundings and goes to raise the heat content of the system. How, then, can we explain the fact that KCl dissolves in water? As was noted in Section 9.6, there is another factor, the entropy, that needs to be considered in deciding whether a process should spontaneously occur.

In general, a solution represents a state of higher entropy (*ergo,* a state of higher probability) than the state in which the components are unmixed. The reason for this is that there are more molecular configurations possible in the mixed state than in the unmixed state. This can be seen by noting that interchange of two differently colored circles in Figure 10.1*b* produces a new distinguishable configuration for the solution, whereas interchange of one solid circle for another does not. In the unmixed phases, new configurations are not possible since each phase consists only of solid circles or only of open circles. In summary, the mixed state of solute plus solvent affords a greater number of distinguishable configurations than unmixed solute and solvent, and so mixing is a probable process. (Recall, as discussed in Section 9.6, that "seven" is more probable than "snake eyes.") However, even though mixing is always a probable process, it may not occur in cases where the energy consideration is too unfavorable.

In terms of the free-energy concept described in Section 9.6, the above discussion can be given as follows: For dissolving to occur, ΔG for the process must be negative (i.e., there must be a decrease of free energy). Because of the relation $\Delta G = \Delta H - T\,\Delta S$, a negative ΔG can arise from a negative ΔH or from a positive ΔS. As we have noted, ΔS for the dissolving process is always positive. Therefore, it follows that any dissolving process having negative ΔH will occur; any dissolving process having zero or nearly zero ΔH will also occur. If, however, ΔH is positive and too large, then, even though ΔS is positive, $-T\,\Delta S$ will not compensate, and ΔG will be positive. In other words, for dissolving processes that are energetically very unfavorable (highly endothermic), a solution will not be formed, except possibly at very high temperatures.

EXAMPLE 1   For dissolving a crystal of KCl in water, the ΔH is 8.4 kilojoules/mole and the ΔS is 96 joules/mole $\cdot$ deg. Show that the free-energy change is favorable for dissolving at 25°C.

$$\Delta G = \Delta H - T\,\Delta S$$
$$= (8400 \text{ joules/mole}) - (298 \text{ deg})(96 \text{ joules/mole} \cdot \text{deg})$$
$$= -20{,}000 \text{ joules/mole}$$

10.4   Properties of Solutions

How are the properties of a solvent affected by the addition of a solute? As a specific example, how are the properties of water affected by the addition of sugar? Suppose we consider the following experiment: Two beakers, one containing pure water (beaker I) and the other containing a sugar-water solution (beaker II), are set under a bell jar, as shown in Figure 10.2. As time passes, we observe that the level of pure water in beaker I drops, whereas the level of the solution in beaker II rises. There is transfer of water from pure solvent to solution through the vapor phase. This transfer occurs because the escaping tendency, or vapor pressure, of pure water is higher than the escaping tendency of H_2O from the sugar-water solution.

The transfer of a solvent to a solution where its escaping tendency is lowered is responsible for the biological process called *osmosis*. If the contents of two beakers in Figure 10.2 are placed in direct contact separated only by a membrane whose pores are large enough to allow passage of water but too small for sugar molecules, the results will be the same as those noted

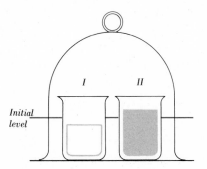

Figure 10.2
Change in levels due to lowered escaping tendency of water: (I) pure water, (II) sugar solution.

I   *II*

Initial level

above; i.e., water passes into the solution. Incidentally, if pressure is applied to the solution, the water molecules in it will be forced back to the pure-water side of the membrane. The *pressure needed to just balance the escaping tendency of the solvent into the solution* is called the *osmotic pressure* of the solution.

Another experimental observation which supports the idea that addition of a solute lowers the escaping tendency of solvent molecules is the lowering of the freezing point. For example, when sugar is added to water, it is observed that the solution needs to be cooled below 0°C in order to freeze out ice. The implication is that the tendency of H_2O to escape from the liquid phase (into solid) is decreased by the presence of solute.

The lowering of the freezing point and the reduction of the vapor pressure are found, at least in dilute solutions, to be directly proportional to the concentration of solute particles. This comes about because the escaping tendency of the *solvent* depends on its concentration, which in turn is lowered as the solute concentration increases. Consequently, since it is solvent concentration that is of concern, it matters little what the nature of the solute particles are that serve to decrease the solvent concentration.

Figure 10.3 shows by means of a phase diagram the effect on the solvent, water, of one particular concentration of solute. The solid lines represent the phase diagram of pure H_2O; the dashed lines, that of solution. The dashed line on the left corresponds to equilibrium between solid H_2O (pure ice) and the liquid solution. It represents the temperatures at which pure solid H_2O freezes out when the particular solution is cooled at different pressures. The dashed line on the right corresponds to equilibrium between gaseous

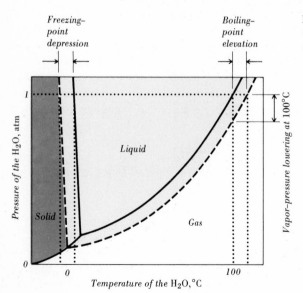

Figure 10.3

Comparison of phase diagrams of water and an aqueous solution. (Dashed lines refer to solution.)

H$_2$O and the liquid solution. It represents the temperatures at which pure gaseous H$_2$O boils off when the solution is heated at various pressures.

The most striking feature shown on the phase diagram is the extension, at all pressures, of the liquid range, both to higher temperatures and to lower temperatures. The liquid phase of water has been made more probable (i.e., more likely to occur) by the addition of solute. Associated with this is the fact that the vapor pressure of the water has been reduced. For example, as seen from Figure 10.3, at 100°C the vapor pressure of the water is not 1 atm, but less than that. Because the vapor pressure of water is lowered, the solution does not boil at 100°C, but at the temperature at which the vapor pressure becomes equal to 1 atm. If the solute is nonvolatile in that it contributes nothing to the vapor pressure (for example, sugar), the normal boiling point of the solution can be read directly from the phase diagram. Similarly, there is a depression of the normal freezing point.

In general, both the boiling-point elevation and the freezing-point depression depend on the nature of the solvent and on the concentration of solute particles. In water, the characteristic values for the freezing-point depression and the boiling-point elevation are 1.86 and 0.52°C, respectively, per Avogadro number of solute particles in a kilogram of water. These constants are referred to as the *molal freezing-point lowering* and the *molal boiling-point elevation* for water. They are called *molal* constants because the concentration of solute is expressed in molal units (Section 10.2). One thousand grams of H$_2$O in which there is dissolved the Avogadro number (1 mole) of particles of solute has a boiling point 0.52°C higher than that of pure water, Its freezing point is 1.86°C lower than that of pure water. In the same amount of water, twice the Avogadro number of particles causes twice the boiling-point elevation and twice the freezing-point depression. It does not matter much what the particles are. They may be neutral molecules or electrically charged ions. Since the main role of the solute particles is to increase the entropy by providing additional molecular arrangements, it is the concentration of the solute, not its nature, that determines these particular solution properties.

Freezing-point depression and boiling-point elevation can be used as a means of measuring the concentration of solute particles in a solution.† In principle, either property can be used. However, the freezing point is more easily measured accurately, and it is the preferred method of counting particles or moles in solution. From the number of particles in solution and the mass of solute, the molecular weight can be calculated. This can be done easily when the solute molecules remain intact in the solution.

---

†Once the freezing point of a compound is known, measurement of freezing point can be used to determine the purity of samples of that compound. For example, benzene has a freezing point of 5.5°C, and any sample of benzene showing a lower freezing point contains dissolved impurity.

EXAMPLE 2 When 45 g of glucose is dissolved in 500 g of water, the solution has a freezing point of $-0.93°$C. **(a)** What is the molecular weight of glucose? **(b)** If the simplest formula is CH_2O, what is the molecular formula of glucose?

a   The freezing point of H_2O is reduced $1.86°$C by 1 mole of particles per 1000 g of H_2O. If the freezing point of H_2O is reduced $0.93°$C, the solution must have:

0.93/1.86 mole of particles per 1000 g of H_2O, or
0.50 mole of particles per 1000 g of H_2O, or
0.25 mole of particles per 500 g of H_2O

Since there is 45 g of glucose per 500 g of H_2O,

$$45 \text{ g} = 0.25 \text{ mole}$$
$$1 \text{ mole} = 180 \text{ g}$$

Therefore, the molecular weight of glucose is 180 amu.

b   The molecular formula must be some multiple x of CH_2O, or $(CH_2O)_x$. The (simplest) formula weight is $12 + 2 + 16 = 30$ amu. The molecular weight is 180 amu, or x times 30, and so $x = 6$. The molecular formula is $(CH_2O)_6$, or $C_6H_{12}O_6$.

The above calculation assumes that the number of particles in the solution is the same as the number of molecules placed in solution. However, as discussed in the next section, the solution process itself may tear molecules apart.

10.5   Electrolytes

There are many cases in which the solution process is accompanied by dissociation, or breaking apart, of solute molecules. Because the dissociated fragments are usually electrically charged, it is possible by electrical measurements to show whether dissociation has occurred. Charged particles, or ions, moving in solution, constitute an electric current, so that a measurement of electric conductivity of the solution is all that is required. Figure 10.4 is a schematic diagram of an apparatus for determining whether a solute is dissociated into ions. The pair of electrodes is connected, in series with a meter, to a source of electricity. So long as the two electrodes are kept separated, no electric current can flow through the circuit, and the meter stays at zero. When the two electrodes are joined by an electric conductor, the circuit is complete and the meter moves. When the electrodes are dipped into a beaker of pure water, the meter does not move, indicating that water does not conduct electricity appreciably. When sugar is dissolved in the

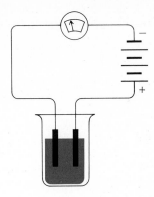

Figure 10.4
Experiment to determine conductivity of a solution.

water, the solution does not conduct, but when NaCl is dissolved in the water, the solution does conduct. By such experiments it is possible to classify substances as *electrolytes*, those which *produce conducting solutions,* and *nonelectrolytes* those which *produce nonconducting solutions.* Figure 10.5 gives the names and formulas of several examples.

Figure 10.5
CLASSIFICA-
TION OF
SOLUTES

Electrolytes		Nonelectrolytes	
HCl	Hydrochloric acid	$C_{12}H_{22}O_{11}$	Sucrose
H_2SO_4	Sulfuric acid	C_2H_5OH	Ethyl alcohol
$HC_2H_3O_2$	Acetic acid	N_2	Nitrogen
$NaOH$	Sodium hydroxide	O_2	Oxygen
$Ca(OH)_2$	Calcium hydroxide	CH_4	Methane
$NaCl$	Sodium chloride	CO	Carbon monoxide
Na_2SO_4	Sodium sulfate	CH_3COCH_3	Acetone

Electric conductivity requires the existence of charged particles. The greater the number of charges available for carrying electricity, the greater is the conductivity observed. By making full use of the ammeter in Figure 10.4 for quantitatively measuring the conductivity, it is possible to get information about the concentration of charges in the solution. When the conductivity of a solution labeled 1 m HCl (hydrochloric acid) is compared with the conductivity of 1 m $HC_2H_3O_2$ (acetic acid), it is found that the former conducts to a greater extent than the latter. Both solutions are made by dissolving 1 mole of solute in 1000 g of water. The inference is that HCl yields a higher concentration of charges than $HC_2H_3O_2$. From experimental observations of this kind, electrolytes may be subdivided into two more or less distinct groups: *strong* electrolytes, which give solutions that are good conductors of electricity, and *weak* electrolytes, which give slightly

Figure 10.6	Strong Electrolytes		Weak Electrolytes	
CLASSIFICA-TION OF ELECTRO-LYTES	HCl	Hydrochloric acid	$HC_2H_3O_2$	Acetic acid
	NaOH	Sodium hydroxide	TlOH	Thallous hydroxide
	NaCl	Sodium chloride	$HgCl_2$	Mercuric chloride

conducting solutions. Figure 10.6 lists representative compounds. Weak electrolytes differ from strong electrolytes in that weak electrolytes are only slightly dissociated into ions in solution, whereas strong electrolytes are essentially 100 percent dissociated into ions.

In a solution of a nonelectrolyte, molecules of solute retain their identity. For example, when sugar dissolves in water, as shown in Figure 10.7, the sugar molecules exist in the solution as solvated, or hydrated, molecules. The hydrated sugar molecule (consisting of a sugar molecule surrounded by a sheath of water molecules) is an uncharged, or neutral, species. When positive and negative electrodes are inserted in a solution containing hydrated sugar molecules, there is no reason for the particles to move one way or the other, since they are neutral. Hence there is no electric conductivity.

Electrolytes (before being dissolved) may be ionic or molecular substances. For ionic substances, it is not surprising that there are charged particles in solution, because the undissolved solid is already made up of charged particles. The solvent rips the lattice apart into its constituent pieces. Figure 10.8 shows what is thought to happen when the ionic solid NaCl dissolves in water. Since the chloride ion is negative, the positive ends of water molecules cluster about the chloride ion; i.e., the hydrogen atoms face the chloride ion. The chloride ion surrounded by its layer of water molecules moves off into the solution. It is now a hydrated chloride ion. The species is negatively charged because the chloride ion itself is negatively charged.

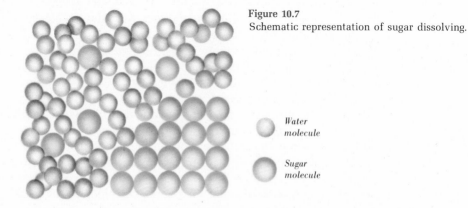

Figure 10.7
Schematic representation of sugar dissolving.

Water molecule

Sugar molecule

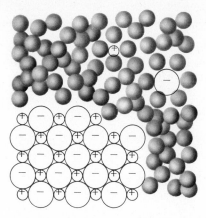

Figure 10.8
Schematic representation of NaCl dissolving.

At the same time the sodium ion undergoes similar hydration, with the difference that the negative or oxygen ends of the water molecules face the positive ion. Since the solution as a whole must be electrically neutral, equal numbers of hydrated sodium ions and of hydrated chloride ions are formed. When positive and negative electrodes are inserted into this solution, the positively charged hydrated sodium ions are attracted to the negative electrode, and the negatively charged hydrated chloride ions are attracted to the positive electrode. There is a net transport of electric charge as the positive charge moves in one direction and the negative charge moves in the opposite direction. (Electric conductivity is discussed in greater detail in Section 13.1.)

Ions may also be formed when certain *molecular substances* are dissolved in the proper solvent. For example, molecules of HCl are neutral, distinct species, which in the pure solid, liquid, or gaseous state do not conduct electricity because no ions are present. However, when HCl is placed in water, the resulting solution conducts electricity, indicating the formation of charged particles. As shown in the equation

$$H\!:\!\overset{\cdot\cdot}{\underset{H}{O}}\!: \;+\; H\!:\!\overset{\cdot\cdot}{\underset{\cdot\cdot}{Cl}}\!: \;\longrightarrow\; H\!:\!\overset{\cdot\cdot}{\underset{H}{O}}\!:\!H^{+} \;+\; :\!\overset{\cdot\cdot}{\underset{\cdot\cdot}{Cl}}\!:^{-}$$

HCl molecules interact with water molecules to form H_3O^+ and Cl^-. Positive and negative ions are formed, even though none is present in pure HCl. The positively charged H_3O^+ is referred to as a *hydronium ion*. The negative ion is the chloride ion. Both of these ions are hydrated; there are water molecules stuck on them just as with an ionic solute. The ionization of a solute by water can be considered to be a chemical reaction and can be described by a chemical equation, such as

$$HCl + H_2O \longrightarrow H_3O^+ + Cl^-$$

Since the hydronium ion, H_3O^+, can be considered a hydrated proton, and since water of hydration is often omitted from chemical equations, the above equation is written more simply as

$$HCl \longrightarrow H^+ + Cl^-$$

with the tacit understanding that all species are hydrated.[†]

10.6 Solubility

The term *solubility* is used in several senses. It describes the qualitative idea of the solution process. It also is used quantitatively to describe the composition of solutions. The solutions considered up to now represent *unsaturated solutions,* to which solute can be added a bit at a time to produce a whole series of solutions differing in concentration. For any solute and solvent, a large number of unsaturated solutions are possible. However, in most cases, the process of adding solute cannot go on indefinitely. Eventually, a stage is reached beyond which the addition of solute to a specified amount of solvent does not produce another solution of higher concentration. Instead, solute remains undissolved. In these cases there is a limit to the amount of solute which can be dissolved in a given amount of solvent. The solution which represents this limit is called a *saturated solution,* and the concentration of the saturated solution is called the *solubility* of the given solute in the particular solvent used.

The best way to ensure having a saturated solution is to have an excess of solute in contact with the solution. If the solution is unsaturated, solute dissolves until saturation is established. If the solution is saturated, the amount of excess solute remains unchanged, as does the concentration of the solution. The system is in a state of equilibrium. Apparently, it is a state of dynamic equilibrium, since, for example, an irregularly shaped crystal of solute dropped into the solution changes its shape although remaining constant in mass. In the equilibrium state, dissolving of solute still occurs but is compensated for by precipitation of solute out of solution. The number of solute particles going into solution per unit time is equal to the number of solute particles leaving the solution per unit time. The concentration of solute in the solution remains constant; the amount of solute in excess remains constant. The amount of excess solute present in contact with the

---

[†] Omitting the water of hydration is the most misleading in the case of the hydrogen ion. Whereas H^+ is essentially of zero size, H_3O^+ has a volume which is about 10^{15} times as great, comparable in size with other ions.

saturated solution does not affect the concentration of the saturated solution. In fact, it is possible to filter or separate the excess solute completely and still have a saturated solution. For convenience, a *saturated solution* is defined as one which *is or would be in equilibrium with excess solute.*

The concentration of a saturated solution, i.e., the solubility, depends on the nature of the solvent, the nature of the solute, the temperature, and the pressure.

1   **Nature of the solvent**   A useful generalization much quoted in chemistry is that like dissolves like. More specifically, high solubility occurs when the molecules of the solute are similar in structure and electrical properties to the molecules of the solvent. When there is a similarity of electrical properties, i.e., high dipole moment, between solute and solvent, then solute-solvent attractions are particularly strong. When there is dissimilarity, solute-solvent attractions are likely to be weak. For this reason a polar substance such as H_2O usually is a good solvent for a polar substance such as alcohol, but a poor solvent for a nonpolar substance such as gasoline. The nonpolar solvent benzene is a good solvent for gasoline.

In general, an ionic solid has a higher solubility in a polar solvent than in a nonpolar solvent. For example, at room temperature the solubility of NaCl in H_2O is 311 g/liter of solution, whereas the solubility of NaCl in gasoline is nil. Also, the more polar the solvent, the greater is the solubility of ionic solids. For example, at room temperature, the solubility of NaCl in ethyl alcohol is 0.51 g/liter of solution, compared with 311 g/liter of solution in water. The difference is ascribed to the lower polarity (lower dipole moment) of the ethyl alcohol molecule, with resulting lower attractions for the ions.

2   **Nature of the solute**   Changing the solute means changing the solute-solute and solute-solvent interaction. At room temperature the amount of sucrose that can be dissolved in water is 1311 g/liter of solution. This is more than four times as great as the solubility of NaCl. However, these numbers are rather misleading. The number of particles involved can better be seen by comparing the *molar* solubility. For NaCl, the saturated solution is 5.3 M, whereas for sugar the saturated solution is 3.8 M. On a molar basis, NaCl has a higher solubility in H_2O than does sugar. Since the attractions in solid NaCl are greater than those in sugar, the reason for the higher solubility of NaCl lies in the fact that the interactions of Na^+ and Cl^- with water molecules are greater than the interactions of sugar molecules with water molecules.

What effect does the presence of one solute in a solution have on the solubility of another solute in that same solution? As a crude approximation, unless the concentration of a substance is high, it has little effect on the solubility of other substances in that solution. For example, approximately the same concentration of NaCl can be dissolved in a 0.1 M sugar solution as in pure water. However, the solubility of NaCl is drastically lowered by a solute having an ion in common, such as KCl or $NaNO_3$.

3   **Temperature**   The solubility of *gases in water* usually decreases as the temperature of the solution increases. The tiny bubbles which form when water is heated are due to the fact that dissolved air becomes less soluble at higher temperatures. The flat taste characteristic of boiled water is largely due to the fact that dissolved air has been expelled. However, for *gases in solvents other than water* (and, in fact, even for water at higher temperatures), solubility of gases need not decrease with increasing temperature. Similarly, there is no general rule for the temperature change of solubility of *liquids* and *solids*. For example, with increasing temperature, lithium carbonate decreases in solubility in water, silver nitrate increases, and sodium chloride shows practically no change. Specific data are given in Figure 10.9 for the solubility of various substances in water.

Figure 10.9
CHANGE OF SOLUBILITY (In grams of solute per kilogram of H_2O)
WITH TEMPERATURE

Substance	0°	10°	20°	30°	40°	50°
$AgC_2H_3O_2$	7.2	8.8	10.4	12.1	14.1	16.4
$AgNO_3$	1220	1700	2220	3000	3760	4550
KCl	276	310	340	370	400	426
NaCl	357	358	360	363	366	370
Li_2CO_3	15.4	14.3	13.3	12.5	11.7	10.8
$CO_2(g)$ at 1 atm	3.3	2.3	1.7	1.3	0.97	0.76
$SO_2(g)$ at 1 atm	228	162	113	78	54	45
$O_2(g)$ at 1 atm	0.070	0.054	0.044	0.037	0.033	0.030

The change of solubility with temperature is closely related to the heat of solution of the substance. The *heat of solution* is the heat evolved when a solute dissolves to give the saturated solution and can be written as the heat that accompanies the following process:

Solute + solvent $\longrightarrow$ saturated solution + heat

The heat of solution as experimentally determined can be a positive quantity, in which case heat is given off to the surroundings, or it can be a negative quantity, in which case heat is absorbed from the surroundings. The heat of solution is related to the change in heat content of the system. It is equal to $-\Delta H$ for the solution process. The negative sign arises because, for example, as heat is liberated to the surroundings the heat content of the solution must decrease.

For lithium carbonate, the heat of solution is positive. Heat is evolved and usually appears as a rise in temperature of the solution. For silver nitrate, the heat of solution is negative. We can write for the latter process

$$AgNO_3(s) + H_2O \longrightarrow \text{solution} - \text{heat}$$

or

$$\text{Heat} + AgNO_3(s) + H_2O \longrightarrow \text{solution}$$

Heat must be supplied, since the process is endothermic. When a substance with negative heat of solution is dissolved, there is usually a drop in the temperature of the solution.

Whether the heat of solution is positive or negative depends on the nature of the solute and the solvent. Specifically, when solids are dissolved in water, the heat of solution depends on the relative magnitude of two energies: the energy required to break up the solid structure and the energy liberated when the particles are hydrated. If more heat is required to break up the lattice than is liberated on interaction of the solute and solvent particles, heat will be taken up (endothermic, negative heat of solution, positive ΔH). Conversely if less heat is required to break up the solute lattice than is liberated when the resulting solute particles interact with solvent, heat will be evolved (exothermic, positive heat of solution, negative ΔH).

How is the heat of solution related to the change of solubility with temperature? In a saturated solution in equilibrium with excess solute, the two processes, dissolving and precipitation, occur simultaneously. If the dissolving process is endothermic (ΔH positive), as with $AgNO_3$, the precipitation process is exothermic.

$$\text{Heat} + AgNO_3(s) + H_2O \rightleftharpoons \text{solution}$$

The upper arrow indicates the forward process of dissolving read from left to right, and the bottom arrow indicates the reverse process of precipitation read from right to left. In a beaker containing solid $AgNO_3$ in equilibrium with the saturated solution, the two processes occur equally. If heat is now added to the equilibrium system, Le Chatelier's principle predicts that the system will adjust to reduce the stress of added heat. The stress can be reduced in this case by favoring the dissolving process over the precipitation process. Until the stress is relieved, silver nitrate goes into solution faster than it comes out. The amount of solid in solution increases. The solubility of silver nitrate in water is therefore greater at the higher temperature. This behavior is typical of most solids. When placed in water, they dissolve by an endothermic process; hence raising the temperature increases solubility.

4   **Pressure**   The solubility of all *gases* is increased as the partial pressure of the gas above the solution is increased. Probably the most familiar example of this phenomenon is found in carbonated beverages. These are solutions of the gas CO_2 in a liquid solvent. Because of the way in which the beverages are bottled, the pressure of the carbon dioxide gas in the sealed bottle is rather high. The concentration of CO_2 dissolved in the solution is dependent directly on the partial pressure of CO_2 in the gas phase. When the bottle is opened, the pressure of carbon dioxide drops, its solubility is diminished, and bubbles of carbon dioxide form and escape from the beverage. So far as liquids and solids are concerned, there is practically no change in solubility with pressure.

In closing this section on solubility, we need to note that it is sometimes possible to prepare solutions which have a higher concentration of solute than that of the saturated solution. Such solutions are *supersaturated* and are *unstable with respect to the precipitation of excess solute*. A supersaturated solution of sodium acetate ($NaC_2H_3O_2$), for example, can be made as follows: A saturated solution of $NaC_2H_3O_2$ and H_2O in contact with excess solute is heated until the increase of solubility with temperature is sufficient to dissolve all the excess solute. At sufficiently high temperature an unsaturated solution results. This unsaturated solution is then cooled very carefully. The system ought to return to its original equilibrium state with the excess solute crystallized out. This, in fact, happens with most solids. However, for some, such as sodium acetate, cooling can be accomplished without crystallization. The resulting solution has a concentration of solute higher than would correspond to the saturated solution at the lower temperature. It is supersaturated. The situation is reminiscent of that observed for supercooling of a liquid below its freezing point.

Supersaturation can usually be destroyed in the same manner, i.e., by seeding. When a tiny seed crystal of sodium acetate is placed in a supersaturated solution of sodium acetate, excess solute crystallizes on it until the remaining solution is just saturated. Occasionally, a mechanical disturbance such as a sudden shock may suffice to break the supersaturation. Dust particles or even scratches on the inner surface of the container may act as centers on which crystallization can start.

10.7   Acids and Bases

In the preceding discussions, it has been assumed that water is a nonconductor of electricity in conductivity determinations. It is found, however, by precise measurement that even very highly purified water does conduct electric current to a slight extent. It must be that water itself is dissociated into positive and negative ions and should be classified as a weak electrolyte. The electrolytic dissociation of water can be represented most simply as the splitting of a water molecule to form a positive *hydrogen* ion and a negative *hydroxide* ion:

$$H_2O \longrightarrow H^+ + OH^-$$

Each of these ions is hydrated. An alternate way of describing the electrolytic dissociation of water is:

$$H:\overset{..}{\underset{H}{O}}: + H:\overset{..}{\underset{H}{O}}: \longrightarrow H:\overset{..}{\underset{H}{O}}:H^+ + :\overset{..}{\underset{..}{O}}:H^-$$

where a proton shifts from one oxygen atom to the other atom. However, we will use the simpler formulation and use "hydrogen ion" to denote the hydrated proton, H^+, rather than the hydronium ion, H_3O^+.

The degree of dissociation of water is very small. In pure water the concentration of hydrogen ion is $1.0 \times 10^{-7}\,M$. The concentration of hydroxide ion is, of course, the same, since each time a water molecule is split, one hydrogen ion and one hydroxide ion are formed. In a liter of water at room temperature there are approximately 1000 g, or 55 moles, of H_2O. Thus, the fraction of water dissociated is $(1.0 \times 10^{-7})/55$, or 0.0000002 percent. On an average, only 1 out of 500 million molecules of H_2O is dissociated. Although very small and seemingly trivial, this small percentage of dissociation is one of the most important properties of water. For example, many of the metabolic reactions in the body are critically dependent on the hydrogen-ion and the hydroxide-ion concentration in the system.

Substances added to water can upset the hydrogen-ion–hydroxide-ion equality. Those substances which *increase the hydrogen-ion concentration* are called *acids;* those substances which *increase the hydroxide-ion concentration* are called *bases.* For example, HCl and $HC_2H_3O_2$ are acids. HCl is a *strong acid,* one which is *highly dissociated* into ions; $HC_2H_3O_2$ is a *weak acid,* one which is *slightly dissociated.* As a class, acids have a set of characteristic properties which are referred to as *acid properties.* Acids have a noticeable sour taste. They affect dye materials in a specific way. For example, the purple dye litmus is turned pink in acid solution, and it serves as an *indicator* for the presence of acids. Litmus paper consists of absorbent paper impregnated with this indicator. Similarly, bases have characteristic *basic properties.* For example, the typical strong base NaOH, sodium hydroxide, has a bitter taste and a slippery feel, and it turns litmus blue.

It is not obvious from a formula whether a chemical compound is an acid or a base; it is not safe to generalize that any molecule containing the

OH group is a base. For example, the substance $SO_2(OH)_2$ contains two OH groups. When placed in water, the compound shows typical acid properties, indicating dissociation to form a hydrogen ion and a negative $SO_2(OH)O^-$ ion. The negative ion is usually written HSO_4^-. To emphasize that the compound is an acid, it is customary to write the formula as H_2SO_4 instead of $SO_2(OH)_2$. In general, the formulas of acids are written so that the available protons appear first in the formula. Thus, the formula of hypochlorous acid is usually given as HOCl instead of ClOH. On the other hand, it is not safe to assume that all formulas ending with OH denote bases. For example, C_2H_5OH is not a base, as defined above, but NaOH is. Only a knowledge of the experimental behavior can give an unequivocal decision as to whether a given compound is an acid, a base, or neither.

It might be useful to consider the full implications of the definition of acids and bases, as given in this section, that acids are substances which increase the hydrogen-ion concentration in water and bases are substances which increase the hydroxide-ion concentration in water. One obvious implication is that the definition is limited to aqueous systems. Another implication is that a hydrogen-ion increase can be brought about by substances which themselves do not contain hydrogen. For example, CO_2 is an acid because its addition to water increases the hydrogen-ion concentration so that the resulting solution has acid properties. This can best be represented by the net equation

$$CO_2 + H_2O \longrightarrow H^+ + HCO_3^-$$

where HCO_3^- represents the bicarbonate ion. Similarly, NH_3 is a base because its addition to water increases the hydroxide-ion concentration. This can be represented as

$$NH_3 + H_2O \longrightarrow NH_4^+ + OH^-$$

There are other, less evident reactions (often called hydrolysis reactions; Section 12.9) which change the H^+ and OH^- concentrations in water. For example, when certain aluminum compounds are added to water, the solution becomes acidic; when certain sulfides are added to water, the solution becomes basic. Thus, for aluminum chloride ($AlCl_3$) dissolved in water, the increase of hydrogen-ion concentration can be attributed to the net reaction

$$Al^{3+} + H_2O \longrightarrow H^+ + AlOH^{2+}$$

and for sodium sulfide (Na_2S) dissolved in water the increase of hydroxide-ion concentration can be attributed to the net reaction

$$S^{2-} + H_2O \longrightarrow OH^- + SH^-$$

In summary, we should note that the definition we are using for acid (increases H^+) and base (increases OH^-) is a useful one because it is simple and specific. These simple definitions, rather than more general ones are sufficient for our discussions in which acids and bases are limited to aqueous solutions.

10.8   Neutralization

The fact that water is only slightly dissociated into H^+ and OH^- indicates that the two ions have great affinity for each other in aqueous solution. Therefore, it is not surprising that, when H^+ and OH^- from separate sources are brought together, they readily combine to form water molecules. In this reaction, which is known as *neutralization,* the H^+ destroys the OH^- by forming H_2O. The process of neutralization is typical of acids and bases and is frequently made part of their definitions.

In the neutralization process there is produced a third class of substances known as *salts.* They consist of the positive ion of a base and the negative ion of an acid and can be defined as the neutralization products of an acid-base reaction. As a specific example, when a solution of H^+ and Cl^- is mixed with a solution of Na^+ and OH^-, the H^+ reacts with the OH^- to form H_2O, and Na^+ and Cl^- are left in the solution. When the solvent water is boiled off, the Na^+ ions and the Cl^- ions conglomerate to form the salt sodium chloride. The net equation for the neutralization reaction is

$$H^+ + OH^- \longrightarrow H_2O$$

The net equation for the subsequent formation of the salt on evaporation of the solvent is

$$Na^+ + Cl^- \longrightarrow NaCl(s)$$

The equation

$$H^+ + OH^- \longrightarrow H_2O$$

describes the neutralization of any *strong acid* by any *strong base.* In the reaction, H^+ and OH^- are used up, and H_2O molecules appear. Neither Cl^- nor Na^+ takes part in the neutralization reaction, and therefore neither appears in the net equation. The neutralization of HCl with NaOH is sometimes written as

$$HCl + NaOH \longrightarrow H_2O + NaCl$$

but since HCl, NaOH, and NaCl are all strong electrolytes, the species present in solution are ions. The equation is better written as

$$H^+ + Cl^- + Na^+ + OH^- \longrightarrow H_2O + Na^+ + Cl^-$$

The Na^+ and Cl^- are canceled because they appear on both sides of the equation. The net equation

$$H^+ + OH^- \longrightarrow H_2O$$

is therefore preferred. It does not concern itself with any species not pertinent to the reaction but tells only what disappears and what appears.

For the neutralization of a *weak acid* by a *strong base,* the net equation can be represented in general terms,

$$HA + OH^- \longrightarrow H_2O + A^-$$

where HA stands for any weak acid such as acetic acid, $HC_2H_3O_2$. Since weak acids are only slightly dissociated in aqueous solution, the original solution of the weak acid contains predominantly HA molecules. In the neutralization, it is the HA molecules that ultimately disappear, and this must be shown in the net equation. It may well be that the actual mechanism of the neutralization involves, first, dissociation of HA into $H^+ + A^-$, with subsequent union of $H^+ + OH^-$ to give H_2O. The net equation represents only the overall reaction.

10.9   Polyprotic Acids

The term *polyprotic acid* (sometimes called polybasic acid) is used to describe those acids which furnish more than one proton per molecule. Two examples of polyprotic acids are H_2SO_4, sulfuric acid, and H_3PO_4, phosphoric acid. In dissociation, polyprotic acids usually dissociate only one proton at a time. For example, when placed in water, H_2SO_4 gives H^+ and HSO_4^-. This reaction

$$H_2SO_4 \longrightarrow H^+ + HSO_4^-$$

is complete, and in this sense H_2SO_4 is called a strong electrolyte. When a solution containing 1 mole of sodium hydroxide is mixed with a solution containing 1 mole of sulfuric acid, 1 mole of H^+ reacts with 1 mole of OH^-. Evaporation of the resulting solution gives 1 mole of the salt $NaHSO_4$, sodium hydrogen sulfate. The ion HSO_4^- is an acid in its own right. Although fairly weak, it can give H^+ and SO_4^{2-}. This dissociation,

$$HSO_4^- \longrightarrow H^+ + SO_4^{2-}$$

can be considered as the second step in the dissociation of the diprotic acid H_2SO_4 and occurs only when there is a large demand for hydrogen ions. For example, when 1 mole of H_2SO_4 is mixed in solution with 2 moles of sodium hydroxide, the 2 moles of OH^- neutralize 2 moles of H^+. Evaporation of the solution produces the salt Na_2SO_4, sodium sulfate.

10.10 Equivalents of Acids and Bases

Acid-base neutralization, requires that an equal number of H^+ ions and OH^- ions be used up in the reaction. This can be expressed by writing a non-net equation. For example, in the complete neutralization of $Ca(OH)_2$, calcium hydroxide, by H_3PO_4, the non-net equation is

$$3Ca(OH)_2 + 2H_3PO_4 \longrightarrow Ca_3(PO_4)_2 + 6H_2O$$

Since each mole of $Ca(OH)_2$ furnishes 2 moles of OH^- and each mole of H_3PO_4 furnishes 3 moles of H^+, complete neutralization occurs if 3 moles of $Ca(OH)_2$ per 2 moles of H_3PO_4 are used. From such an equation, calculations can be made.

It is more convenient, however, to consider neutralization reactions by fixing attention only on the hydrogen and hydroxide. For this purpose, equivalents are convenient. *One equivalent of an acid is the quantity of acid required to furnish one mole of* H^+; *one equivalent of a base is the quantity of base required to furnish one mole of* OH^- *or accept one mole of* H^+. One equivalent of any acid just reacts with one equivalent of any base.

One of the simplest acids is HCl, hydrochloric acid, 1 mole of which weighs 36.5 g. Since 1 mole of HCl can furnish 1 mole of H^+, 36.5 g of HCl is 1 equiv. For HCl, and for all other monoprotic acids, 1 mole is the same as 1 equiv. This means that the molarity of such solutions numerically equals the normality (Section 10.2). For example, a solution labeled 0.59 M HCl requires 0.59 mole of HCl per liter of solution, or 0.59 equiv of HCl per liter, and so can also be labeled 0.59 N HCl.

For a diprotic acid such as H_2SO_4, 1 mole of acid can furnish on demand 2 moles of H^+. By definition, 2 moles of H^+ is the amount furnished by 2 equiv of acid. Therefore, for complete neutralization, 1 mole of H_2SO_4 is identical with 2 equiv. Since 1 mole = 98 g = 2 equiv, 1 equiv of H_2SO_4 weighs 49 g. A solution labeled 1 M H_2SO_4, indicating 1 mole of H_2SO_4 per liter of solution, can also be labeled as 2 N H_2SO_4. For complete neutralization of a triprotic acid such as H_3PO_4, 1 mole is equal to 3 equiv, and so the normality of any solution is three times the molarity.

The situation is similar for bases. For NaOH, 1 mole gives 1 mole of OH⁻. Therefore, 1 mole of NaOH is 1 equiv. For all solutions of NaOH the normality is equal to the molarity. For $Ca(OH)_2$, the normality of any solution is twice the molarity.

10.11   Solution Calculations

Labels on reagent bottles specify what the solution was made from, but not necessarily what the solution contains. For example, the label 0.5 M HCl appears on a solution made from 0.5 mole of HCl and sufficient water to give 1 liter of solution. Despite the label, there are no HCl molecules in the solution. HCl is a strong electrolyte and is 100 percent dissociated into H^+ and Cl^-. For most quantitative considerations, however, it is not necessary to know what species are actually in the solution. It is necessary to know only what is ultimately available. The label 0.5 M $HC_2H_3O_2$ also tells what the solution was made from, but in this case the solution actually contains $HC_2H_3O_2$ molecules, since acetic acid is a weak electrolyte and is very slightly dissociated. There is only a trace of H^+ and $C_2H_3O_2^-$ in the solution. However, if this solution is used for a neutralization reaction, not only the trace amount of H^+ but also the $HC_2H_3O_2$ are neutralized.

The use of solutions for chemical reactions requires a clear distinction between the *number of moles* of solute in a solution and its *concentration*. To illustrate, let us suppose 15.8 g of $KMnO_4$, potassium permanganate, is dissolved to make a 0.100 M $KMnO_4$ solution. The formula weight of $KMnO_4$ is 158 amu; hence 15.8 is equal to 0.100 mole. To make up the solution, the solute is placed in a graduated container, and water is added to it. Not necessarily 1 liter of water is added, only enough to bring the volume to a liter of solution. (Usually, the volume of solute plus the volume of the solvent is not exactly equal to the volume of the solution.) The solution can now be labeled 0.100 M $KMnO_4$, since it contains 0.100 mole of $KMnO_4$ in 1 liter of solution. The concentration does not depend on how much of this solution is taken. Whether one drop or 200 ml is considered, the solution is still 0.100 M $KMnO_4$. However, the number of moles of $KMnO_4$ taken does depend on the volume of solution. If the volume and the concentration of a sample are known, the number of moles of solute in the sample is the number of moles per liter multiplied by the volume of the sample in liters. In 200 ml of 0.100 M $KMnO_4$, there is (0.200 liter)(0.100 mole/liter), or 0.0200 mole of $KMnO_4$.

Solutions are extremely convenient because they permit measuring amounts of solute not by weighing the solute out but by measuring a volume of solution. For example, suppose a given chemical reaction requires 0.0100 mole of $KMnO_4$. This amount of $KMnO_4$ can be provided by 1.58 g of $KMnO_4$ or by 100 ml of 0.100 M $KMnO_4$ solution.

To summarize:

Liters of solution $\times$ molarity of solution = moles of solute in sample
Liters of solution $\times$ normality of solution = equivalents of solute in sample

---

EXAMPLE 3   To what volume must 50.0 ml of 3.50 M H_2SO_4 be diluted in order to make 2.00 M H_2SO_4?

Since 50.0 ml of 3.50 M H_2SO_4 contains

(0.050 liter)(3.50 moles/liter) = 0.175 mole H_2SO_4

Therefore, the question is how many liters of 2.00 M H_2SO_4 will contain this.

(x liters)(2.00 moles/liter) = 0.175 mole H_2SO_4
x = 0.0875 liter, or 87.5 ml

EXAMPLE 4   How many liters of 0.025 M H_3PO_4 are required to neutralize 25 ml of 0.030 M $Ca(OH)_2$?

Method 1: moles   In 25 ml of 0.030 M $Ca(OH)_2$, there is (25.0/1000 liters) $\times$ (0.030 mole/liter), or 0.00075 mole.
From the non-net equation

$$3Ca(OH)_2 + 2H_3PO_4 \longrightarrow Ca_3(PO_4)_2 + 6H_2O$$

3 moles of $Ca(OH)_2$ require 2 moles of H_3PO_4, or 0.00075 mole of $Ca(OH)_2$ requires 0.00050 mole of H_3PO_4.
0.025 M H_3PO_4 means 0.025 mole of H_3PO_4 per liter of solution.
To get 0.00050 mole of H_3PO_4 as required, we take 0.00050/0.025, or 0.020, liter of 0.025 M H_3PO_4.

Method 2: equivalents   0.030 M $Ca(OH)_2$ is 0.060 N $Ca(OH)_2$, and 0.025 M H_3PO_4 is 0.075 N H_3PO_4.
In 25 ml of 0.030 M $Ca(OH)_2$ there is (25.0/1000 liters)(0.060 equiv/liter), or 0.0015 equiv of base.
We need 0.0015 equiv of acid for neutralization.
The acidic solution has 0.075 equiv/liter, or 0.0015 equiv in 0.0015/0.075, or 0.020, liter of solution.

10.12   Oxidation-Reduction in Solution

In the quantitative consideration of oxidation-reduction reactions in aqueous solution, no new principles need to be introduced. In fact, electrolytic dissociation simplifies consideration of oxidation-reduction. Only the net reaction need be considered; other ions present in the solution can be ignored. As a specific case, consider the reaction of an acidified solution of $KMnO_4$ with a solution of ferrous sulfate ($FeSO_4$). Before reaction occurs, the mixture contains K^+, MnO_4^-, H^+, HSO_4^-, Fe^{2+}, and SO_4^{2-}. After the reaction is complete, the mixture contains K^+, HSO_4^-, Mn^{2+}, Fe^{3+}, and SO_4^{2-}. The K^+, HSO_4^-, and SO_4^{2-} are present both in the initial mixture and the final mixture and can be ignored. The net reaction shows the disappearance of MnO_4^-, Fe^{2+}, and H^+ and the appearance of Mn^{2+}, Fe^{3+}, and H_2O. It can be written

$$Fe^{2+} + MnO_4^- + H^+ \longrightarrow Fe^{3+} + Mn^{2+} + H_2O$$

This equation can be balanced by using the procedure outlined previously (Section 5.7). The first step is to balance the oxidation-reduction from the electron transfer. Since the manganese atom changes oxidation state from $7+$ to $2+$, it appears to pick up five electrons. Since the iron atom changes oxidation state from $2+$ to $3+$, it appears to release one electron. The electron gain and electron loss must compensate; hence for every MnO_4^-, five Fe^{2+} are used. The balancing of electron transfer is shown as follows:

$$5Fe^{2+} + 1MnO_4^- + H^+ \longrightarrow H_2O + Mn^{2+} + 5Fe^{3+}$$

$$2+ \qquad 7+ \qquad\qquad\qquad 2+ \qquad 3+$$

$$5 \times \boxed{1e^- = 1 \times \boxed{5e^-}}$$

Since five atoms of iron disappear on the left, five atoms of iron as Fe^{3+} must appear on the right. Since one manganese atom disappears on the left, one Mn^{2+} must appear on the right. The net charge on the right is $17+$ (one Mn^{2+} and five Fe^{3+}), and so to balance this on the left we place $8H^+$ to go with five Fe^{2+} and one MnO_4^-. Finally, to conserve mass, the four atoms of oxygen from the MnO_4^- must appear on the right as four H_2O molecules. The final *net ionic equation* reads

$$5Fe^{2+} + MnO_4^- + 8H^+ \longrightarrow 5Fe^{3+} + Mn^{2+} + 4H_2O$$

Actually, the balanced net ionic equation can be written if the only information given is that MnO_4^- is reduced in the reaction by Fe^{2+} to form Mn^{2+} and Fe^{3+} in acidic solution. The steps are as follows:

1 Balance electron transfer.

2 Balance the net charge by placing H^+ where required in order to maintain net charge balance.

3 Place H_2O where it is required to balance the oxygen.

Remember these three steps!

The great advantage of approaching the oxidation-reduction reaction in this way is that it is necessary to remember only the oxidizing and reducing agents and their products. The rest can be figured out. In the reaction between potassium dichromate ($K_2Cr_2O_7$) and sulfur dioxide (SO_2) in acidic solution, the oxidizing agent $Cr_2O_7^{2-}$ and the reducing agent SO_2 disappear to form Cr^{3+} and HSO_4^-. The steps used to arrive at the balanced net equation are as follows:

Given: In acidic solution, $Cr_2O_7^{2-} + SO_2 \longrightarrow HSO_4^- + Cr^{3+}$:

1 $1Cr_2O_7^{2-} + 3SO_2 \longrightarrow 3HSO_4^- + 2Cr^{3+}$

   $\quad\ 6+ \qquad\qquad 4+ \qquad\qquad 6+ \qquad\quad 3+$

   $1 \times 3e^- \times 2 = 3 \times 2e^-$

Note that, since $Cr_2O_7^{2-}$ contains two chromium atoms, the gain of $3e^-$ per Cr atom must first be doubled to give the electron gain per dichromate ion, which is then compared with the electron loss of the sulfur.

2 $Cr_2O_7^{2-} + 3SO_2 \longrightarrow 2Cr^{3+} + 3HSO_4^-$

   Net charge: *Net charge:*

   $(2-) + (3)(0) = 2- \qquad\quad (2)(3+) + (3)(1-) = 3+$

   To maintain the net charge, add $5H^+$ to the left.

3 $Cr_2O_7^{2-} + 3SO_2 + 5H^+ \longrightarrow 2Cr^{3+} + 3HSO_4^-$

   To balance the oxygen atoms, add $1H_2O$ to the right:

   $Cr_2O_7^{2-} + 3SO_2 + 5H^+ \longrightarrow 2Cr^{3+} + 3HSO_4^- + H_2O$

This method of balancing equations works for basic solutions as well, except that, in basic solution, hydrogen ions do not exist in any appreciable concentration. The balancing of the net charge is done by placing hydroxide ions where needed. As an example, consider the preceding oxidation-reduction as carried out in basic solution. In basic solution, the oxidizing agent exists in the form of CrO_4^{2-} and the reducing agent in the form of SO_3^{2-}. The products are CrO_2^- and SO_4^{2-}. The sequence of steps follows:

Given: In basic solution, $CrO_4^{2-} + SO_3^{2-} \longrightarrow CrO_2^- + SO_4^{2-}$:

1     $2CrO_4^{2-} + 3SO_3^{2-} \longrightarrow 3SO_4^{2-} + 2CrO_2^-$

       6+   4+   6+   3+

       $2 \times |3e^- = 3 \times |2e^-$

2     $2CrO_4^{2-} + 3SO_3^{2-} \longrightarrow 2CrO_2^- + 3SO_4^{2-}$

    Net charge: *Net charge:*

    $(2)(2-) + (3)(2-) = 10-$ $(2)(1-) + (3)(2-) = 8-$

To maintain the net charge, we need two plus charges on the left, or two minus charges on the right. To place $2H^+$ on the left is forbidden, since the given solution is basic and does not contain any appreciable concentration of hydrogen ions. The alternative is to place two hydroxide ions on the right.

3     $2CrO_4^{2-} + 3SO_3^{2-} \longrightarrow 2CrO_2^- + 3SO_4^{2-} + 2OH^-$

Place $1H_2O$ on the left to balance the oxygen atoms:

    $2CrO_4^{2-} + 3SO_3^{2-} + H_2O \longrightarrow 2CrO_2^- + 3SO_4^{2-} + 2OH^-$

Net equations show both the species involved in the reactions and the stoichiometry. The number of moles of reactants can be calculated from the equation. The volumes of solutions (of given molarities) necessary for complete reaction are thus specified. In general, the principal interest is focused on the oxidizing agent and the reducing agent, since the acidic or basic nature of the solution is usually provided by an excess of an acid or base.

EXAMPLE 5   How many milliliters of 0.20 M $KMnO_4$ are required to oxidize 25.0 ml of 0.40 M $FeSO_4$ in acidic solution? The reaction which occurs is the oxidation of Fe^{2+} by MnO_4^- to give Fe^{3+} and Mn^{2+}.

The balanced net equation obtained above is

$5Fe^{2+} + MnO_4^- + 8H^+ \longrightarrow 5Fe^{3+} + Mn^{2+} + 4H_2O$

25.0 ml of 0.40 M $FeSO_4$ supplies (0.0250 liter)(0.40 mole/liter), or 0.010 mole of Fe^{2+}.

From the equation, 5 moles of Fe^{2+} requires 1 mole of MnO_4^-.

Hence 0.010 mole of Fe^{2+} requires 0.0020 mole of MnO_4^-.

Therefore, the question is how many liters of 0.20 M of $KMnO_4$ will supply this.

(x liters)(0.20 mole/liter) = 0.0020 mole

              x = 0.010 liter, or 10 ml

QUESTIONS

10.1 *Types of Solutions* All gases are soluble in each other; relatively few solid solutions are found; liquid solutions are intermediate between these extremes in terms of frequency of occurrence. Why do you think this is so?

10.2 *Concentration* What is the molar concentration of the following solutions: (a) 9.2 g of C_2H_5OH per liter of solution, (b) 9.2 g of C_2H_5OH per 250 ml of solution, (c) 1.0×10^{-3} mole of C_2H_5OH per 2.0 ml of solution?

10.3 *Solution Process* How is it possible that a solute can be quite soluble despite the solution process being endothermic?

10.4 *Osmosis* For osmosis to occur what is required (a) of the cell membrane, (b) of the solutions on the two sides of the membrane?

10.5* *Properties of Solutions* Calculate the freezing point and the normal boiling point of a solution containing 0.100 mole of the sugar glucose per 250 ml of aqueous solution.

10.6 *Electrolytes* Describe experiments which would show whether UO_2SO_4 is a nonelectrolyte, a weak electrolyte, or a strong electrolyte.

10.7 *Solubility* Use the Le Chatelier principle to account for each of the following: (a) The solubility of $AgNO_3$ in water increases as the temperature rises. (b) The solubility of air in water increases as the pressure rises.

10.8 *Supersaturation* Why can you not prepare a supersaturated solution by mixing solute and solvent at constant temperature?

10.9 *Acids* Which will neutralize more 0.10 M NaOH: 100 ml of 0.10 M HCl, or 100 ml of 0.10 M $HC_2H_3O_2$? Explain.

10.10 *Bases* Which will neutralize more 0.10 M HCl: 100 ml of 0.10 M $Ca(OH)_2$, or 100 ml of 0.10 N $Ca(OH)_2$? Explain.

10.11* *Neutralization* How much 0.100 M NaOH will be just neutralized by each of the following: (a) 100 ml of 0.0500 M HCl, (b) 200 ml of 0.200 M H_2SO_4, (c) 20.0 ml of 0.200 N H_2SO_4.

10.12 *Polyprotic Acid* (a) Write net equations for three steps of dissociation for H_3PO_4. (b) Describe how you might prepare three different sodium salts of phosphoric acid and write their formulas.

10.13* *Equivalents* How many equivalents of acid are there in each of the following: (a) 0.200 mole of H_2SO_4, (b) 4.90 g of H_2SO_4, (c) 100 ml of 0.100 M H_2SO_4, (d) 100 ml of 0.100 N H_2SO_4?

10.14 *Solution Calculations* To what volume must 20.0 ml of 6.00 M HCl be diluted to prepare a solution that is 0.100 M?

10.15* *Solution Calculations* How much 0.100 M NaOH just neutralizes 25.0 ml of an HCl solution prepared by diluting 25.0 ml of 2.00 N HCl to 100 ml?

10.16 *Solution Calculations* How many molecules of glucose are there in 0.100 ml of a 0.0500 M glucose solution?

10.17 *Terms* What is meant by (*a*) an acid, (*b*) a base, (*c*) molarity, (*d*) molality, (*e*) normality, and (*f*) osmosis?

10.18 ΔG, ΔH, ΔS A compound AB, which has the molecular weight of 40.0 g, is dissolved in water. Given that the lattice energy of the compound is 3.75 kilojoules/g, hydration energy is 140 kilojoules/mole, and ΔS is 50.3 joules/mole · deg, calculate ΔG for the dissolving of 1.00 mole of AB at 25°C.

10.19* ΔG, ΔH, ΔS 10.0 g of compound C, whose molecular weight is 98.0, is dissolved in water. If the water absorbed 8.16 kilojoules of heat, determine: (*a*) ΔH of solution per mole of C, and (*b*) ΔS of solution per mole of C at 27°C, if ΔG of solution is −160 kilojoules/mole.

10.20 *Solubility* Solubility of KCl in water is 276 and 426 g/kg of H_2O at 0°C and 50°C, respectively. Calculate the molality of the saturated solution of each temperature.

10.21* *Percent by Weight* A solution is made up of 2.00 moles of $NaNO_3$, 3.50 moles of K_2SO_4, and 250 g of H_2O. What is the percentage by weight of each compound?

10.22 *Molarity* Calculate the molarity of each of the following solutions: (*a*) 2.00 g of KCl in enough water to make 100 ml of solution, (*b*) 2.00 g of $Ca(OH)_2$ in enough water to make 100 ml of solution, and (*c*) 2.00 g of LiBr in enough water to make 100 ml of solution.

10.23* *Molecular Weight* A beaker containing 250 ml of aqueous solution freezes at −2.21°C. The density of the solution is 1.37 g/ml. If the solution is 72.1% by weight water, what is the molecular weight of solute in the solution?

10.24 *Equivalents* The neutralization of 1.21 g of solid acid requires 45.0 ml of 0.210 N LiOH. What is the equivalent weight of the solid acid?

10.25* *Oxidation-Reduction* (*a*) Complete and balance the equation for the following change in acid solution:

$$Zn(s) + NO_3^- \longrightarrow Zn^{2+} + NH_4^+$$

(*b*) Calculate the volume of 0.250 M solution of HNO_3 required just to oxidize 5.00 g of zinc.

10.26 *Oxidation-Reduction* (*a*) Balance the equation for the change in acid solution:

$$H_2O_2 + Cr_2O_7^{2-} \longrightarrow Cr^{3+} + O_2(g)$$
Peroxide

(b) How many liters of oxygen will be produced at 1 atm and 25°C by 15.2 g of H_2O_2 and 60.0 g of $Li_2Cr_2O_7$?

10.27* *Normality* An excess of Mg reacts with 500 ml of HCl and 3.05 liters of H_2 gas is collected over water at 25°C and 753 mmHg. What is the normality of the HCl solution?

10.28 *Equivalents* Determine the number of equivalents in 1 liter of each of the following 0.01 M solutions: $Ca(OH)_2$, NaOH, H_2SO_4, $HC_2H_3O_2$, H_3PO_4.

10.29* *Equivalents* What is the weight of 1 equiv of (a) the acid H_3AsO_4, (b) the base $Th(OH)_4$.

10.30 *Acids and Bases* When compound YO is dissolved in water, the solution becomes acidic. Is Y a transition metal, an alkaline-earth metal, or a nonmetal? Explain.

10.31* *Concentration* Given that a solution that is 20.0% by weight acetic acid, CH_3COOH, has a density of 1.028 g/ml, calculate (a) the molarity and (b) the molality of the solution.

10.32 *Freezing-Point Depression* 100 ml of solution containing 0.500 mole of acid HY freezes at −13.9°C. What is the apparent percent dissociation of acid in the solution?

10.33* *Freezing-Point Depression* 23.0% of 3.00 m acid, HR, dissociates in water. What is the freezing point of the solution?

10.34 *Oxidation-Reduction* Write net balanced equations for each of the following changes in acid solution:

(a) $Sn^{2+} + H_3AsO_4 \longrightarrow Sn^{4+} + HAsO_2$
(b) $Fe^{2+} + MnO_2(s) \longrightarrow Fe^{3+} + Mn^{2+}$
(c) $IO_3^- + HAsO_2 \longrightarrow I^- + H_3AsO_4$
(d) $Cr_2O_7^{2-} + SO_2 \longrightarrow Cr^{3+} + HSO_4^-$

10.35 *Oxidation-Reduction* Write net balanced equations for each of the following changes in basic solution:

(a) $ClO_3^- + SO_3^{2-} \longrightarrow Cl^- + SO_4^{2-}$
(b) $Mo(s) + NO_3^- \longrightarrow MoO_4^{2-} + NO_2^-$
(c) $H_3IO_6^{2-} + Cr(OH)_3(s) \longrightarrow IO_3^- + CrO_4^{2-}$
(d) $HOsO_5^- + Cu_2O(s) \longrightarrow Cu(OH)_2(s) + Os(s)$

10.36 *Oxidation-Reduction* (a) Write a balanced equation for the reaction of MnO_4^- and SO_2 in acid solution to form $Mn^{2+} + HSO_4^-$. (b) How many moles of Mn^{2+} will be formed by this reaction in a solution which initially contains 0.100 mole of oxidizing agent and 0.200 mole of reducing agent? (c) How many milliliters of 0.100 M oxidizing agent are needed to react with 20.0 ml of 0.200 M reducing agent in this reaction?

10.37 *Normality* Given a solution labeled 0.200 M $KMnO_4$. What is the normality of this solution for each of the following changes: (a) In acid solution, it is reduced to Mn^{2+}. (b) In basic solution, it is reduced to MnO_4^{2-}. (c) In neutral solution, it is reduced to $MnO_2(s)$.

10.38 *Concentration* Given a solution of sulfuric acid in water which contains 10.0% H_2SO_4 by weight. The density of this solution is 1.07 g/ml. Calculate the solute's (a) mole fraction, (b) molarity, (c) molality, (d) normality.

10.39* *Concentration of Electrolytes* 10.0 ml of 0.20 M NaCl is mixed with 15.0 ml of 0.40 M $CaCl_2$ to give 25.0 ml of final solution. Assuming all electrolytes are completely dissociated and there are no chemical reactions, calculate the concentration of each ionic species in the final solution.

10.40 *Concentration* (a) At 0°C and one standard atmosphere, 1.00 liter of CO_2 gas is dissolved in 1.00 liter of H_2O. The volume change is negligible. Calculate the molarity of the solution. (b) Is the solution acid or base? Account for your answer with a net equation.

10.41* *Saturated Solutions* At 28°, the weight of KCl, 36.2 g, is exactly the same as the weight of NaCl which can dissolve in 100 ml of H_2O. (a) Calculate the molality of each saturated solution. (b) What additional information is needed to get the molarities?

10.42 *Solubility* In going from 0 to 40°C, the solubility of $AgNO_3$ triples. (a) What can you say about the relative enthalpy changes of Ag^+ and NO_3^- ions forming a lattice and forming an aqueous solution? (b) Why is $AgNO_3$ soluble in water?

10.43 *Acids and Bases* It is found that NaF solutions are basic and $CuCl_2$ solutions are acidic. Write net equations which can account for these observations.

10.44 *Solution Properties* Account for each of the following: (a) Reactions in aqueous solution are generally much faster than reactions between the same solid reactants at the same temperature. (b) As opposed to a compound, when a solution is partly frozen, the two phases have different compositions.

10.45* *Acid-Base* 20.0 ml of 0.200 M HCl is mixed with 30.0 ml of 0.300 M NaOH and the mixture is diluted with water to 400 ml. Calculate the molar concentration of each ion in the final solution.

10.46 *Oxidation-Reduction*   20.0 ml of solution containing 0.200 M $Cr_2O_7^{2-}$ is mixed with 30.0 ml of a solution containing 0.300 M Fe^{2+} and 1.00 M H^+. Reaction occurs to produce Cr^{3+} and Fe^{3+}. If the final volume is 50.0 ml, calculate the final molar concentrations of $Cr_2O_7^{2-}$, Fe^{2+}, H^+, Cr^{3+}, and Fe^{3+}.

10.47 *Freezing of Solutions*   In special cases, a solvent-solute solution shows a higher freezing point than that of the pure solvent. How can this be true?

Chemical Dynamics

We have now completed our study of the general properties which characterize matter in its various pure and mixed states. However, except for weight relationships accompanying chemical change, we have not discussed in any detail the nature of chemical reaction. Before our discussions of individual chemical processes, it is desirable to describe the fundamentals common to all chemical changes. In this and the two succeeding chapters we will discuss the dynamic, the equilibrium, and the electrical aspects of chemical change.

In the development of the science of chemistry it was necessary that studies be directed initially toward working out molecular structures. In recent years there has been a significant increase of emphasis on studies of the molecular changes that occur during the course of a chemical reaction. This increased interest in chemical dynamics has come about because powerful instrumental techniques have made possible the study of complicated molecular changes such as those occurring in biological systems. Life processes consist of a series of interlinked molecular changes; unraveling them will lead to new means for regulating the anomalous interactions responsible for body malfunctions. Although we are well aware of the dramatic effects produced by drugs on human well-being, there is very little knowledge of the dynamics of the pertinent reactions.

At the basis of chemical dynamics is the study of the rates of chemical reactions and the mechanisms by which they occur. This branch of chemistry is called *chemical kinetics.* For most reactions, a sequence of consecutive steps is involved in going from reactants to products. This series of steps constitutes the *reaction mechanism,* and it is the job of chemical kinetics to deduce the most probable mechanism from observations on chemical reaction rates.

What are the factors which influence the rate of chemical reaction? Experiments show that four important factors are (1) nature of reactants, (2) concentration of reactants, (3) temperature, and (4) catalysis.

11.1   Nature of Reactants

In a chemical reaction, bonds are formed, and bonds are broken. The rate should therefore depend on the specific bonds involved. Experimentally, the reaction velocity depends on the specific substances brought together in reaction. For example, the reduction of permanganate ion (MnO_4^-) in acidic solution by ferrous ion (Fe^{2+}) is practically instantaneous. MnO_4^- disappears as fast as ferrous solution is added; the limiting factor is the rate of mixing the solutions. On the other hand, the reduction of permanganate ion in acidic solution by oxalic acid ($H_2C_2O_4$) is not instantaneous. The violet color characteristic of MnO_4^- persists long after mixing the solutions. In these two reactions everything is identical except the nature of the reducing agent, but still the rates are quite different.

The rates observed for different reactants vary widely. There are reactions, such as occur in acid-base neutralization, which may be over in a fraction of a microsecond, so that the rate is difficult to measure. There are also very slow reactions, such as those occurring in geological processes, which may not reach completion in a million years. The changes in a lifetime may be too small to be detected. Most information has, of course, been accumulated about reactions that occur at rates intermediate between these extremes.

11.2   Concentration of Reactants

It is found by experiment that the rate of a homogeneous chemical reaction depends on the concentration of the reactants. A homogeneous reaction is one which occurs in only one phase. Heterogeneous reactions involve more than one phase. It is found for *heterogeneous* reactions that the rate of reaction is proportional to the area of contact between the phases. An example is the rusting of iron, a heterogeneous reaction involving a solid phase, iron, and a gas phase, oxygen. Rusting is slow when the surface of contact is small, as with a bar of iron. If the bar is ground into powder, rusting is more rapid because of greater area of contact.

For *homogeneous* reactions, the rate depends on the concentration (amount per unit volume) of the reactants in solution. The solution may be liquid or gaseous. The concentration of a reactant can be changed either by its addition or removal or by changing the volume of the system, as by expansion of a gas or by addition of solvent to a liquid system. The specific effect on the rate has to be determined by experiment. Thus, in the reaction of substance A with substance B, the addition of A may cause an increase, a decrease, or no change in rate, depending on the particular reaction. Quantitatively, the rate may double, triple, become half as great, etc. A priori, it is not possible to look at the net equation for a chemical reaction and tell how the rate is affected by a change of concentration of reactants. The quantitative influence of concentration on the rate can be found only by experiment.

The determination of how the rate of reaction changes with concentration of reactants is an experimental problem beset by many difficulties. The usual procedure is to keep everything constant except the concentration of one reactant. As the concentration of the one reactant is systematically changed, the reaction rate is measured. This may be done by noting the rate of disappearance of a reactant or the rate of formation of a product. Experimental difficulties usually come in determining the instantaneous concentration of a component as it is changing.

The reaction between hydrogen and nitric oxide,

$$2H_2(g) + 2NO(g) \longrightarrow 2H_2O(g) + N_2(g)$$

is a homogeneous reaction which can be investigated kinetically by following the change in pressure of the gaseous mixture as the reaction proceeds. The pressure drops because 4 moles of gas is converted to 3 moles of gas. Typical data for several experiments at 800°C are given in Figure 11.1. Since

Figure 11.1 RATE OF REACTION OF NO AND H_2	Initial Molar Concentration		
Experiment	NO	H_2	Initial Rate, atm/min
I	0.006	0.001	0.025
II	0.006	0.002	0.050
III	0.006	0.003	0.075
IV	0.001	0.009	0.0063
V	0.002	0.009	0.025
VI	0.003	0.009	0.056

reactants are being used up during the course of the reaction, their concentrations and their rate of reaction are constantly changing. The concentrations and rates listed are those at the very beginning of the reaction when little change has occurred. The first three experiments have the same initial concentration of NO but different initial concentrations of H_2. The last three experiments have the same initial concentration of H_2 but different initial concentrations of NO.

The data for experiments I and II show that, when the initial concentration of NO is constant, doubling the concentration of H_2 doubles the rate; I and III show that tripling the concentration of H_2 triples the rate. The rate of reaction is therefore found to be directly proportional to the concentration of H_2. The data for experiments IV and V show that, when the initial concentration of H_2 is constant, doubling the concentration of NO quadruples the rate; IV and VI show that tripling the concentration of NO increases the rate ninefold. The rate of reaction is therefore found to be proportional to the square of the concentration of NO. Quantitatively, the data can be summarized by stating that the reaction rate is proportional to (concentration of H_2) $\times$ (concentration of NO)^2. This can be written mathematically as

$$\text{Rate} = k[H_2][NO]^2$$

The equation is known as the *rate law* for the reaction and states that the rate is equal to a proportionality constant times the concentration of H_2 to the first power times the concentration of NO to the second power. Brackets represent concentration of a substance in moles per liter. The proportionality constant k is called the *specific rate constant* and is characteristic of a given reaction although it may vary with temperature.

The general form of any rate law is

$$\text{Rate} = k[A]^n[B]^m\cdots$$

where n is the appropriate power to which the concentration of A must be raised, and m is the appropriate power to which the concentration of B must be raised, in order to summarize the data. The dots represent other reactants which may be involved in the rate law. The important thing to note is that the rate law is determined by experiment. A common error is to assume that the coefficients in the balanced net equation are the exponents in the rate law. This, in general, is not true. For example, in the reaction between H_2 and NO the exponents in the rate law are 1 and 2, whereas the coefficients in the balanced equation are 2 and 2. The only way to determine unambiguously the exponents in the rate law is to do the experiments.

11.3 Temperature

How does the temperature of a reaction affect its rate? Observations on rate experiments like that described in the previous section indicate that a rise in temperature almost invariably increases the rate of any reaction. Furthermore, a decrease in temperature generally decreases the rate, no matter whether the reaction is exothermic or endothermic. The increase of rate with temperature is expressed by an increase in the specific rate constant k. As to the magnitude of the effect, no generalization can be made. The magnitude varies from one reaction to another and also from one temperature range to another. A rule, which must be used with caution, is that a 10°C rise in temperature approximately doubles or triples the reaction rate. For each specific reaction it is necessary to determine from experiment the quantitative effect of a rise in temperature.

11.4   Catalysis

It is found by experiment that some reactions can be speeded up by the presence of substances which themselves remain unchanged after the reaction has ended. Such substances are known as *catalysts,* and their effect as *catalysis.* Often only a trace of catalyst is sufficient to accelerate the reaction. However, there are many reactions in which the rate of reaction is proportional to some power of the concentration of catalyst. The actual dependence of rate on catalyst concentration must be determined by experiment. If the experiments show there is such a dependence, then the catalyst concentration to the appropriate power becomes part of the rate law in the same way as the reactants are a part of it.

There are numerous examples of catalysis. When $KClO_3$ is heated so that it decomposes into KCl and oxygen, it is observed that a pinch of manganese dioxide (MnO_2) considerably accelerates the reaction. At the end of the reaction the $KClO_3$ is gone, but all the MnO_2 remains. It appears as if the catalyst is not involved in the reaction, because the starting amount can be recovered. However, the catalyst must take some part in the reaction, or else it could not change the rate.

When hydrogen gas escapes from a cylinder into the air, no change is visible. However, if the escaping hydrogen is directed at finely divided platinum, it is observed that the platinum glows and soon ignites the hydrogen. In the absence of platinum the H_2–O_2 reaction is too slow to observe. In contact with platinum, hydrogen reacts with oxygen from the air to form water. As they react, they give off energy which heats the platinum. As the platinum gets hotter, it heats the hydrogen and oxygen, so that their rate of reaction increases, until eventually ignition occurs and the reaction of hydrogen with oxygen becomes self-sustaining. Platinum is a general catalyst for many other reactions involving hydrogen. These include the hydrogenation of oils to form synthetic fats such as oleomargarine and the rearranging of hydrocarbons in making gasoline. Because of the economic importance of these reactions most of the platinum in the United States is in the form of catalysts.

Enzymes are complex substances in biological systems which act as catalysts for biochemical processes. Pepsin is the gastric juice and ptyalin in the saliva are examples. Ptyalin is the catalyst which accelerates the conversion of starch to sugar. Although starch will react with water to form sugar, it takes weeks for the conversion to occur. A trace of ptyalin is enough to make the reaction proceed at a biologically useful rate. Catalase, an enzyme which prevents accumulation of hydrogen peroxide in biological systems by accelerating the decomposition reaction

$$2H_2O_2 \longrightarrow 2H_2O + O_2$$

is the most potent catalyst known for this otherwise extremely slow reaction. Catalase is a protein of molecular weight 240,000 and contains four iron atoms per molecule.

11.5   Collision Theory

Many of the observed facts of chemical kinetics have been interpreted in terms of the *collision theory*. This theory starts from the basic assumption that, for a chemical reaction to occur, particles must collide. For substance A to react with substance B, it is necessary that the particles A, be they molecules, ions, or atoms, collide with particles B. In the collision, atoms and electrons are rearranged. There is a reshuffling of chemical bonds leading to the production of other species.

According to the collision theory, the rate of any step in a reaction is directly proportional to (1) the *number of collisions per second* between the reacting particles involved in that step and (2) the *fraction of these collisions that are effective*. That the rate should depend on the number of collisions per second seems obvious. For instance, in a box that contains A molecules and B molecules there is a certain frequency of collision between A and B. If more A molecules are placed in the box, the collision frequency between A molecules and B molecules is increased. With more collisions between reacting molecules, the reaction between A and B should go faster. However, this cannot be the full story. Calculations of the number of collisions between particles indicate that the collision frequency is very high. In a mixture containing 1 mole of A molecules and 1 mole of B molecules as gases at STP, the number of collisions is more than 10^{30}/s. If every one of these collisions led to reaction, the reaction would be over in an instant, and all reactions would be very fast. By observation, this is not true. It must be that only some of the collisions lead to reaction.

Collisions between A molecules and B molecules may be so gentle that there is no change in the identity of the molecules after collision. The colliding particles separate to resume their original identity. However, if A or B or both A and B have much kinetic energy before collision, they may penetrate each other far enough that electron rearrangement ensues. One or more new species may be formed.

The *extra amount of energy required to produce chemical reaction* is known as the *energy of activation.* Its magnitude depends on the nature of the reactants. For some reactions, the energy of activation is large. Such reactions are slow, since only a small fraction of the reactant particles have enough kinetic energy to furnish the required high energy of activation. Other reactions have a small energy of activation. These reactions are fast, since more of the particles have sufficient kinetic energy to furnish the required energy of activation; hence, a greater fraction of the collisions are effective.

Qualitatively, the collision theory quite satisfactorily accounts for the four factors which influence reaction rates: (1) The rate of chemical reaction depends on the *nature of the chemical reactants,* because the energy of activation differs from one reaction to another. (2) The rate of reaction depends on the *concentration of reactants,* because the number of collisions increases as the concentration is increased. (3) The rate of reaction depends on the *temperature,* because an increase of temperature makes molecules move faster. They collide more frequently with other molecules, and, more important, the collisions are more violent and more likely to cause reaction. In any collection of molecules there is a distribution of energies. According to collision theory, only the highly energetic molecules have enough energy to react. As shown in Figure 11.2, a rise in temperature shifts the whole energy-distribution curve to higher energies, and a larger fraction of the molecules are energetic enough to produce reaction. More of the collisions are therefore effective at high temperatures than at low temperatures. (4) The rate of reaction depends on the presence of *catalysts,* because somehow, in catalysis, collisions are made more effective. This may be done by a preliminary step involving reaction between one or more of the reactants and the catalyst. New reactants may be produced which react more readily.

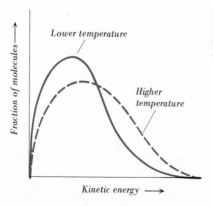

Figure 11.2
Energy distribution of molecules in gas at lower (solid curve) and higher temperatures (dashed curve).

One of the trickiest aspects of chemical kinetics is to account for the quantitative dependence of rate on concentration. For simplicity, let us consider one step of a reaction. Suppose that in this step one molecule of A reacts with one molecule of B to form a molecule AB. The balanced equation for *this step* is

$$A + B \longrightarrow AB$$

According to collision theory, the rate of formation of AB is proportional to the rate at which A and B collide. Let us imagine that we have a box that contains some B molecules and a single A molecule. The rate at which the A molecule collides with B molecules is directly proportional to the number of B molecules in the box. (If we should double the number of B molecules in the box, then we would have twice as many A–B collisions per second.) Suppose now we place a second A molecule in the box. We now have twice as many A molecules in the box, so that the total number of A–B collisions per second is doubled. In other words, the rate at which A and B molecules collide is directly proportional to the concentration of A and to the concentration of B. The rate of formation of AB should therefore be directly proportional to the concentration of A and to that of B. Thus, the rate law for this step is

$$Rate = k[A][B]$$

It should be noted that the exponents of [A] and of [B] in the rate law are unity, just as the two coefficients are unity in the balanced equation for the step.

What is the situation if the balanced equation for a step involves coefficients larger than 1? Consider the reaction

$$2A \longrightarrow A_2$$

In this step, an A molecule must collide with another A molecule to form A_2. The rate at which A_2 forms is thus proportional to the rate at which two A molecules collide. Again, we imagine a box, this time containing only molecules of type A. For *any one* A molecule, the rate at which it collides with any other A molecule is proportional to the number of other A molecules in the box. If we should double the number of other A molecules in the box, we would double the rate at which collisions occur with the one molecule under observation. Now suppose we extend our observation to all the molecules in the box. The total number of collisions per second is proportional to the number of collisions per second made by one A molecule times the total number of A molecules in the box. So if we double the total number of A molecules, *each one* makes twice as many collisions as before, and there are twice as many to take account of and so there are four times as many total collisions. Another way of saying the same thing is that the rate of collision is proportional to the number of molecules hitting multiplied by the number of molecules being hit. In any event, we can say that the rate at which two A molecules collide is proportional to the concentration of A times the concentration of A, or to the square of the concentration of A. Consequently, for the step

$$2A \longrightarrow A_2$$

we can write the rate law

$$Rate = k[A]^2$$

It might be noted that the exponent of the concentration of A in the rate law is 2, just as the coefficient of A in the balanced equation for the step is 2.

For the general case of a step for which the balanced equation shows the disappearance of n molecules of A and m molecules of B, we can write the rate law

$$Rate = k[A]^n[B]^m$$

indicating that the rate of that step is proportional to the concentration of A taken to the n power times the concentration of B taken to the m power.

We should note carefully the distinction between a single reaction step (for which we can deduce the rate law) and a net chemical change which may consist of a series of steps, and therefore without this knowledge we cannot deduce the rate law. Instead when we only know the overall equation, we must do rate experiments to determine the rate law. For example, in the reaction between NO and H_2, discussed in Section 11.2, the rate law obtained experimentally states that the reaction rate is proportional to the first power of the H_2 concentration times the second power of the NO concentration. Yet the balanced equation for the reaction is

$$2H_2(g) + 2NO(g) \longrightarrow N_2(g) + 2H_2O(g)$$

It appears from the above equation that reaction could occur by a collision between two H_2 molecules and the two NO molecules. The number of such collisions per second is proportional to the molar concentration of H_2 squared times the molar concentration of NO squared. This means that doubling the H_2 concentration ought to quadruple the number of collisions per second and therefore quadruple the rate. This does not agree with experiment. The collisions that determine the rate cannot be between the two H_2 molecules and two NO molecules.

To account for the observed rate law, collision theory assumes that this reaction, like many others, occurs in steps. In stepwise reactions, the slow step is the one which determines the rate. It is the bottleneck and determines the rate law for the reaction. The following example shows a two-step reaction:

$$A(g) + B(g) \longrightarrow [\text{intermediate}] \tag{1}$$
$$[\text{Intermediate}] + B(g) \longrightarrow C(g) \tag{2}$$
$$\text{Net reaction:} \quad A(g) + 2B(g) \longrightarrow C(g)$$

In step (1) a molecule of A collides with a molecule of B to form a short-lived intermediate. In step (2) the intermediate reacts with a molecule of B to form a molecule of C. If the first step is slow and if the second step is fast, the rate at which the product C forms depends only on the rate at which the intermediate forms. As soon as the intermediate appears, it is used up in the second reaction. The rate at which the intermediate is produced is determined by the collision of A and B. Thus, the rate of the slow step is proportional to the concentration of A times the concentration of B. Since the slow step is the rate-determining step in the overall change, the rate law for the net change is

Rate $= k[A][B]$

The net change is determined by adding steps (1) and (2):

$$A(g) + B(g) + B(g) \longrightarrow C(g)$$

The intermediate cancels out because it occurs on both sides of the equation. The coefficients which appear in the net equation are different from the exponents in the rate law.

For the specific reaction

$$2H_2(g) + 2NO(g) \longrightarrow N_2(g) + 2H_2O(g)$$

the rate law is determined by experiment to be

Rate $= k[H_2][NO]^2$

The reaction must occur in steps. One possible set of steps is

$$H_2(g) + NO(g) + NO(g) \longrightarrow N_2O(g) + H_2O(g) \tag{1}$$
$$H_2(g) + N_2O(g) \longrightarrow N_2(g) + H_2O(g) \tag{2}$$

The first step would have to be the slower and therefore would determine the rate law.

Chemists work hard to determine rate laws and, hence, reaction mechanisms, for two reasons. The first is to satisfy curiosity as to how chemical reactions occur. The second is to learn how to alter conditions in order to speed up desired reactions.

11.6   Transition-State Theory

The collision theory of chemical kinetics is the basis of the *transition-state theory,* which focuses attention on the path of the reaction during transition from reactants to products. To illustrate, we consider the one-step reaction in which one molecule of A collides with one molecule of B to form one molecule each of C and D. The collision is assumed to consist of the approach of an A molecule to a B molecule to form some kind of a transient complex particle. This complex particle, which is called the *activated complex,* can split apart to restore A and B molecules, or it can split in some other way to give the new particles C and D. Figure 11.3 shows how the potential energy of the system changes as A and B molecules come together to form the activated complex and then separate to give C and D molecules. On the vertical axis is plotted the potential energy of the system; on the horizontal axis is plotted a coordinate which tells how far the reaction has gone from the initial state toward the final state. In the initial state, A and B molecules are far enough apart not to affect each other. The potential energy is the sum of the potential energy of A by itself plus that of B by itself. As A and B come together, the forces of repulsion between the electron clouds become appreciable. Work must be done on the system to squash the molecules together. This means the potential energy must increase. It increases until it reaches a maximum that corresponds to the activated complex. The

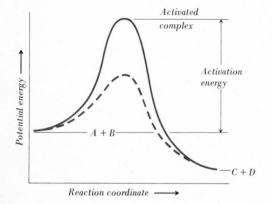

Figure 11.3

Potential-energy change during a reaction. (Broken line refers to catalyzed reaction.)

activated complex then splits into C and D molecules, and the potential energy drops as C and D move apart.

The difference (shown by the arrow) between the potential energy of the initial state A plus B and the potential energy of the activated complex is a measure of the energy which must be added to the particles in order for them to react. This is the activation energy of the reaction. It usually is supplied by converting some of the kinetic energy of the particles into potential energy. If A and B molecules do not have much kinetic energy, they are able, on collision, to go only part way up the side of the hump. All the kinetic energy may be converted into potential energy without getting the pair into the activated complex. In such case, A and B slide back down the hump and fly apart unchanged. The situation is similar to that of a ball rolled up the side of a hill. If the ball is rolled slowly (i.e., small kinetic energy), it goes part way up, stops, and rolls back again. If the ball is rolled rapidly (i.e., large kinetic energy), it goes completely to the top of the hill and down the other side. Similarly, if A and B molecules have high enough kinetic energy, they can attain the activated complex and get over the hump from A and B to C and D. In a reaction at higher temperatures more molecules get over the potential-energy hump per unit time, and the reaction occurs faster.

Two other aspects of Figure 11.3 are of interest. For the case represented, the final state C and D has lower potential energy than the initial state A and B. There is a net decrease in potential energy as the reaction proceeds. This energy usually shows up as heat, and so the particular change shown

$$A + B \longrightarrow C + D$$

is exothermic. The amount of energy required to distort A and B into the activated complex is more than made up for when the activated complex springs apart to form C and D. However, since the activation energy is large, the reaction is slow, even though the system finally goes to a lower potential-energy state. An example of a slow, exothermic reaction is

$$2H_2(g) + O_2(g) \longrightarrow 2H_2O(g)$$

When a reaction is catalyzed, there is a change of path or mechanism. Since the rate is now faster, the activation energy for the new path is lower than for the old path. The broken curve in Figure 11.3 shows what the potential-energy curve might look like for the new path. Since the barrier is lower, at any given temperature and concentration more particles per second can get over the hump; hence the reaction goes faster. As an example, the reaction between hydrogen and oxygen to form water is catalyzed by the presence of platinum. It has been suggested that the effect of the platinum is to react with H_2 molecules to produce H atoms. The oxygen molecules then collide with H atoms instead of with H_2 molecules. The new path has a lower activation energy in the rate-determining step.

11.7 Very Fast Reactions

Some of the most important chemical reactions, e.g., those in biological systems, take place in times shorter than a microsecond (10^{-6} s). This presents a real problem to the experimental investigator because it is not possible to mix together solutions containing the reactants in times much shorter than a millisecond (10^{-3} s). The fastest mixing times are produced by shooting together very fine streams of one solution containing one reactant and another containing the other reactant. Even so, the time required for the reactants to diffuse together is many times longer than the time required for the very fast reactions.

An example of a very fast reaction is that between the hydrated proton H^+ and hydroxide ion OH^-. In a solution having 0.1 molar concentration each of H^+ and OH^-, the reaction would be 99.9 percent complete in 10^{-8} s. Because this time is so much shorter than that required to mix together the solutions, special techniques have to be used to study it. One type of special technique which has been successfully employed, called *relaxation spectroscopy*, uses a sudden perturbation that brings about additional reaction in an already-reacted solution. For example, sudden compression, rapid temperature jump, or abrupt application of an electric field can disturb a system so as to bring about additional reaction. Measurements are then made, e.g., determining the light absorption by products or reactants, while the system adjusts to the perturbation. The time connected with the adjustment process is called the *relaxation time*.

From relaxation experiments it has been possible to determine the rates of a variety of very fast chemical reactions. As might be expected, reactions between oppositely charged simple ions go very fast. In fact, they generally occur at rates comparable to the rates at which collisions take place. In other words, the activation energy for the reaction is low compared with the average kinetic energy of the colliding particles. In the case of H^+ plus OH^-, the activation energy is sufficiently low that virtually every encounter of H^+ with OH^- leads to formation of H_2O.

Relaxation spectroscopy has become a major tool in unraveling the sequences of chemical reactions involved in biological processes. By comparison of measured rates and rate laws in biological reactions with those for simple processes, it is possible to decide which of these simple processes are important in living systems. One of the major findings is that proton transfer is extremely prevalent in a wide variety of reactions, such as those involving enzymes. Along with such transfer steps it is apparent that there are rearrangements of the giant enzyme molecule itself. Relaxation studies are invaluable in elucidating the nature of these protein rearrangements.

QUESTIONS

11.1 *Terms* State clearly what is meant by each of the following: (*a*) reaction mechanism, (*b*) chemical kinetics, (*c*) heterogeneous reaction, (*d*) activation energy.

11.2 *Nature of Reactants* Gas A reacts rapidly with gas B (molecular weight 20) but a million times slower with gas C (molecular weight 40). Discuss the difference in terms of rate of collisions and any other factors which might be important.

11.3 *Concentration of Reactants* Why is it that you cannot deduce the rate law for a reaction from the net balanced equation?

11.4 *Concentration of Reactants* Suppose that in Figure 11.1 the right-hand column read 0.025, 0.025, 0.025, 0.004, 0.008, 0.012. What would be the rate law?

11.5 *Temperature* For gases we can calculate the increase of rate of collisions with increase of temperature. However, we must measure the increase of reaction rate with increase of temperature. Why is this?

11.6 *Catalyst* (*a*) Criticize the statement: A catalyst controls the rate of a chemical reaction but does not take part in it. (*b*) By making as little change in the statement as possible, make it correct.

11.7 *Collision Theory* Consider a one-step reaction between two gaseous reactants, and decide whether each of the following changes the number of collisions per second, the effectiveness of the collisions, both, or neither: (*a*) addition of more reactants at constant volume, (*b*) increasing the volume, (*c*) adding an inert gas, (*d*) raising the temperature.

11.8 *Collision Theory* A reaction is being studied for which the net balanced equation shows 2 moles of B consumed for every 3 moles of A. How is it possible that the rate of this reaction depends only on the concentration of A squared so long as some B is present?

11.9 *Transition-State Theory* Two A molecules react to form A_2 in an endothermic change (ΔH positive). It is found experimentally that the activation energy is about twice the heat of reaction. (*a*) Draw a potential-energy curve for this reaction. (*b*) On the curve, show the activation energy for the reaction, the potential energy of reactants and product. (*c*) Describe the course of the reaction as it moves along the curve you have drawn.

11.10 *Fast Reactions* (*a*) How is it possible to measure a reaction rate that is much faster than the time required to mix the reactants? (*b*) For such a case, how could you experimentally determine the rate law?

11.11* *First-Order Reaction* A complex molecule undergoes a structural rearrangement accompanied by the loss of two of its atoms and the formation of a B molecule. For this process the rate of decomposition (moles per liter of A decomposed per second) is proportional to the molar concentration of A. (a) Write a rate law for this reaction. (b) If a 0.50 M solution of A becomes a 0.49 M solution after 10 s, what is the value of k (include units) in your rate expression?

11.12 *Rate Law* If the numbers in the right-hand column of Figure 11.1 were 0.036, 0.144, 0.324, 0.081, 0.324, and 0.729 for experiments I through VI, respectively, what would the rate law have been?

11.13 *General* Permanganate anion, which is a very strong oxidizing agent, oxidizes Sn^{2+} in acid solution according to the following equation:

$$16H^+ + 2MnO_4^- + 5Sn^{2+} \longrightarrow 5Sn^{4+} + 2Mn^{2+} + 8H_2O$$

If the rate law for the reaction is given by $k[MnO_4][Sn^{2+}]$, predict what effect each of the following changes would have on the rate of formation of Sn^{4+}: (a) addition of $KMnO_4$, (b) addition of acid, (c) addition of water.

11.14 *General* When gaseous SO is oxidized by O_2 it forms gaseous SO_2 at a rate given by $k[SO]^2[O_2]$ where k is 1.0×10^3 atm/M^3·s. We start with 0.50 mole of SO and 0.40 mole of O_2 in a 10-liter box at 25°C. (a) What is the initial rate of the reaction? (b) What are the concentrations of gases at the end of the reaction?

11.15 *Rate Law* Predict what the effect would be on the initial rate of SO_2 formation on making each of the following changes in the initial conditions given in problem 11.14: (a) Use of 1.00 mole of SO instead of 0.50, (b) use of 0.20 mole of O_2 instead of 0.40, (c) addition of 0.50 mole of helium, and (d) lowering the temperature.

11.16 *Activation Energy* Draw a labeled potential-energy diagram for a hypothetical reaction. By reference to the activation energy in your drawing, account for the fact that the reaction rate decreases with decrease in temperature.

11.17 *Reaction Mechanism* For the net change $3A + C \longrightarrow 2B + D$ the rate law is rate $= k[A]^2[C]$. Propose a stepwise mechanism consistent with this rate law.

11.18 *Terms* Differentiate clearly between rate and specific rate constant.

11.19* *Reaction Order* In the decomposition of PCl_5 to give PCl_3 and Cl_2 in benzene solution, the initial concentration of PCl_5 of 0.50 M gave a rate of 7.5×10^{-4} mole/liter·s. A concentration of 1.00 M gave a rate of 3.0×10^{-3} mole/liter·s. What is the order of this decomposition reaction?

11.20 *Rate Law* What would be the effect of each of the following on the value of the specific rate constant k: (a) decreasing the temperature, (b) adding a catalyst, (c) increasing the concentration of reactants, (d) adding an inert gas.

11.21 *Catalysts* Explain whether a catalyst can decrease a rate of reaction.

11.22 *Rate Constant* For the data listed in Figure 11.1, calculate the value, including units, of the rate constant, k.

11.23 *Half-Life* Consider a reaction which is "first order," i.e., rate proportional to the first power of a single reactant. (a) How will the initial rate in moles per liter per second change if the initial concentration is cut in half? (b) How will the time necessary for reaction of 1 percent of the material present compare in two experiments in one of which the initial concentration is twice the other? (c) How would you expect the time necessary for half the material present at any concentration to change? (d) Could the rate be expressed for this first-order case by telling the time necessary for half the material to react? (e) Do you think you could do the same for a reaction which is not first order?

11.24 *Reaction Mechanism* Suppose A and B react to form the molecule A_2B_2. (a) Write an equation for the overall change. (b) Suppose the rate of reaction is doubled when the concentration of either A or B is doubled. How will the rate change if A *and* B are both halved? (c) Write a rate law for this reaction. (d) Propose a mechanism consistent with your answers to (a) and (c).

11.25* *Rate Law* Consider the reaction between H_2 and NO discussed in Section 11.2. Calculate the initial rate if each of the following changes were made in experiment I of Figure 11.1: (a) 6×10^{-3} mole/liter of NO added, (b) 6×10^{-3} mole/liter of H_2 added, (c) concentration of both reactants halved, (d) volume of box halved, (e) 7×10^{-3} mole/liter of inert gas added.

11.26 *Initial Rate* Throughout our discussion we have said or implied that we are measuring initial reaction rate, i.e., rate at the first fraction of reaction. Explain why this is important by describing in general what you would expect to happen to the rate as the reaction proceeds. Give reasons for your answer.

11.27 *Catalyst* Suggest how a platinum catalyst might speed the reaction between hydrogen gas and an oil which is an unsaturated hydrocarbon that becomes saturated by the reaction, i.e., converts C=C double bonds to C—H bonds.

11.28 *Second-Order Reaction* Students are usually troubled by the fact that in a reaction involving collisions of like molecules, doubling the concentration quadruples the rate. Suppose you could ride on a molecule in such a reaction and count its collisions in a box containing a million molecules. (a) You measure the time required for your molecule to suffer a thousand collisions and call it Δt. During this time how many collisions would be suffered by each one of the other million molecules (forget about the difference between 1,000,000 and 999,999)? (b) Noting that a collision requires two molecules and we do not want to count one collision twice, how many collisions altogether occurred in the box during Δt? (c) Now we add another million molecules and you count collisions for the time Δt. What would you expect? (d) For the collection of 2 million molecules in the original box, how many collisions have occurred during Δt? (e) What is the relationship between collision rate for like molecules and their concentration?

11.29 *Reaction Mechanism* For the reaction of Fe^{2+} and H_2O_2 (hydrogen peroxide) to form Fe^{3+} and OH^-, the rate is found to double if either reagent is doubled in concentration and to quadruple if both are doubled. (a) Write a rate law for the reaction. (b) Write a balanced net equation for the reaction. (Do not use fractions.) (c) Propose a mechanism consistent with your answers to (a) and (b).

11.30 *Collision Theory* Consider a single gaseous molecule which could undergo reaction on striking another molecule. At two different time intervals the speed of this molecule differs by a factor of 2. Compare these two time intervals as quantitatively as you can in terms of the probability (a) of colliding with another molecule and (b) of a collision being effective in leading to chemical change.

11.31 *Enzymes* For biological reactions catalyzed by specific enzymes it is frequently the case that a particular reaction will have its rate decreased by the presence of a high concentration of product. (a) Show that this is helpful in controlling concentration of biological material. (b) Can you think of a way whereby so-called "product inhibition" comes about?

Chemical Equilibrium

12

It is found by experiment that, when reagents are brought together so as to undergo chemical reaction, the conversion of reactants to products is often incomplete, no matter how long the reaction is allowed to continue. In the initial state, the reactants are present at a definite concentration. As the reaction proceeds, the concentrations of reactants decrease. Sooner or later, however, they level off and become constant. A state is established in which the concentrations no longer change. This state is known as the state of *chemical equilibrium.*

The equilibrium state is of great interest in chemistry for several reasons. In the first place there are a number of vital chemical reactions which remain in a state of equilibrium and undergo only those changes which are permitted by the laws of chemical equilibrium. Reactions of this type include many biological reactions; maintenance of the acidity of body fluids is a good example. Other reactions which we might want to carry out (e.g., making gasoline from CO_2 and water) are impossible because of unfavorable equilibrium constraints. The sooner we understand this impossibility, the sooner we can focus our efforts in more profitable directions. Finally, a number of reactions can be performed efficiently only by understanding the principles of chemical equilibrium which will govern their outcome. These same principles allow predictions of the extent to which reactions can proceed. In no case, though, will equilibrium considerations tell us how *fast* a reaction will proceed. Instead they tell us what is possible.

In this chapter we will discuss the general principles governing the equilibrium state and learn to apply these principles to common examples, particularly those in aqueous solutions.

12.1   The Equilibrium State

As an example of the attainment of chemical equilibrium, we consider a gaseous reaction

$$CO(g) + Cl_2(g) \longrightarrow COCl_2(g)$$

in which one molecule of CO reacts with one molecule of Cl_2 to form one molecule of $COCl_2$, phosgene. At the start of the experiment, CO and Cl_2 are mixed in a container. The concentration of CO and the concentration of Cl_2 are measured as time passes. Results of the measurements are plotted in Figure 12.1, where concentration is the vertical axis and time is the horizontal axis. The initial concentration of CO is some definite number, depending on the number of moles of CO and the volume of the container. As time goes on, the concentration of CO diminishes, at first quite rapidly but later less rapidly. Eventually, the concentration of CO levels off and becomes constant. The concentration of Cl_2 changes in similar fashion, even though it does not start out at the same value as CO. The initial concentration of $COCl_2$ is zero. As time goes on, $COCl_2$ is produced. Its concentration increases quite rapidly at first but then levels off. Once this equilibrium state has been established, it persists indefinitely and, if undisturbed, will last forever.

The constant state that characterizes equilibrium vapor pressure (Section 8.2) is due to equality of opposing processes. Similarly, the constant state that characterizes chemical equilibrium is due to equality of opposing reactions. CO and Cl_2 molecules react to form $COCl_2$ molecules. So long as CO and Cl_2 are present, $COCl_2$ formation continues. Reaction does not

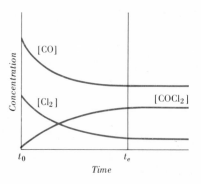

Figure 12.1
Approach to equilibrium.

stop. However, as soon as an appreciable number of $COCl_2$ molecules have formed, they can decompose to produce CO and Cl_2. At *equilibrium, the forward and reverse changes occur at the same rate.* The equality of opposing reactions is indicated by writing

$$CO(g) + Cl_2(g) \rightleftharpoons COCl_2(g)$$

or

$$CO(g) + Cl_2(g) = COCl_2(g)$$

An equilibrium of vital importance to life on earth is that between CO_2 of the atmosphere, H_2O, and limestone ($CaCO_3$).

$$CO_2(g) + H_2O + CaCO_3(s) \rightleftharpoons Ca^{2+} + 2HCO_3^-$$

It has served to maintain the CO_2 concentration of our atmosphere at a constant level appropriate for plant and animal life. It is, however, being gradually upset by man's rapid burning of fossil fuels, which has increased the CO_2 level faster than this geochemical equilibrium can reduce it.

12.2   Mass Action

It is found by experiment that every particular reaction has its own specific equilibrium state, in which there is a definite relation between the concentrations of the materials. To illustrate the relation, we again consider the reaction between CO and Cl_2 to produce $COCl_2$. In a series of experiments, all done at the same temperature but *differing in initial concentrations* of CO and Cl_2, the results shown in Figure 12.2 are obtained. The concentrations

Figure 12.2 EQUILIBRIUM CONCENTRATIONS REACHED FROM DIFFERENT STARTING CONDITIONS	Experiment	Concentrations, moles/liter		
		CO $+$	Cl_2 $\rightarrow$	$COCl_2$
	I	1.21	0.21	0.79
	II	1.0	1.0	3.1
	III	0.43	0.43	0.57
	IV	0.66	1.66	3.34

given, in moles per liter, are those in the equilibrium state and are called *equilibrium concentrations.* Although the equilibrium concentrations change from experiment to experiment, there is a single relationship which holds for all the experiments; viz., if the concentration of $COCl_2$ is divided by the concentration of CO times the concentration of Cl_2, the number 3.1 is obtained in every case. This number, the *equilibrium constant*, is characteristic of *this* specific reaction and varies only with changes in temperature. Whenever CO, Cl_2, and $COCl_2$ are present together in equilibrium, the concentrations must be such that they satisfy the condition

$$\frac{[COCl_2]}{[CO][Cl_2]} = 3.1$$

In the general case, we can describe the situation as follows: The balanced equation

$$nA(g) + mB(g) + \cdots \rightleftharpoons pC(g) + qD(g) + \cdots$$

can be read as n molecules of A plus m molecules of B react to form p molecules of C and q molecules of D. The dots represent other reactants, or products, so that this equation applies to any reaction. The letters n, m, p, and q represent numbers that are the coefficients in the balanced chemical equation. The letters A, B, C, and D represent the formulas of the various reactants and products. If the balanced equation is written in this way, the ratio that is constant at equilibrium is

$$\frac{[C]^p[D]^q \cdots}{[A]^n[B]^m \cdots}$$

This fraction is called the *mass-action expression*.† The brackets designate concentrations in moles per liter, and the exponents are the powers to which these concentrations must be raised. By convention, in writing the mass-action expression the concentrations of the materials on the right-hand side of the chemical equation are placed in the numerator and the concentrations of the materials on the left-hand side of the equation, in the denominator. At equilibrium the mass-action expression is numerically equal to the value of the equilibrium constant K for the particular reaction.

$$\frac{[C]^p[D]^q \cdots}{[A]^n[B]^m \cdots} = K$$

†The term *mass action* derives from the original work of Cato Maximilian Guldberg and Peter Waage, Norwegian chemists, who in 1864 proposed that the reaction A + B = C + D could be treated as follows: The "action force" between A and B is proportional to the "active mass" of A and that of B. This is called the law of mass action. Similarly, the "action force" between C and D is proportional to the "active masses" of C and D. At equilibrium, the "action force" between A and B equals the "action force" between C and D. Although Guldberg and Waage were not clear in what was meant by "action force" and "active mass," their work was a milestone in the development of a suitable description of chemical equilibrium.

This equilibrium condition is called the *law of chemical equilibrium*. The law states that, in a system at chemical equilibrium, the concentrations of the materials which participate in the reaction must satisfy the condition expressed by the constancy of the mass-action expression. There is no other restriction on what the individual concentrations must be.

The law of chemical equilibrium is an experimental fact. It can, however, be justified by using the principles of chemical kinetics and requiring at equilibrium the equality of the rates of forward and reverse reactions. For example, in the equilibrium

$$A(g) + B(g) \rightleftharpoons C(g) + D(g)$$

reaction may proceed in a single step or in a series of steps. Suppose the reaction proceeds through a single reversible step:

$$A(g) + B(g) \rightleftharpoons C(g) + D(g)$$

For the forward reaction,

$$Rate = k[A][B]$$

where k is the rate constant of the forward reaction. For the reverse reaction,

$$Rate = k'[C][D]$$

where k' is the rate constant of the reverse reaction. At equilibrium, the rate of the forward reaction is equal to the rate of the reverse reaction. Thus,

$$k[A][B] = k'[C][D]$$

or

$$\frac{[C][D]}{[A][B]} = \frac{k}{k'}$$

This proves that in this case the mass-action expression is equal to a constant.

If the reaction proceeds through more than one reversible step, the situation is somewhat more complicated. The equality of forward and reverse rates must be met for each individual step. Suppose, for example, that the above reaction occurs in a series of steps such as

$$A(g) + A(g) \rightleftharpoons C(g) + Q(g) \tag{1}$$
$$Q(g) + B(g) \rightleftharpoons A(g) + D(g) \tag{2}$$

where Q is some chemical intermediate. If k_1 and k_1' are the rate constants for the forward and reverse directions of reaction (1), and if k_2 and k_2' are the rate constants for the forward and reverse directions of reaction (2), then we can write

$$k_1[A][A] = k_1'[C][Q]$$

for the first step and

$$k_2[Q][B] = k_2'[A][D]$$

for the second step. These two algebraic equations must both be valid at equilibrium; they can be combined by solving each for [Q] and setting equal the two resulting expressions

$$\frac{k_1[A][A]}{k_1'[C]} = \frac{k_2'[A][D]}{k_2[B]}$$

Rearranging and simplifying gives

$$\frac{[C][D]}{[A][B]} = \frac{k_1 k_2}{k_1' k_2'}$$

Again, the mass-action expression is shown to be equal to a constant.

In summary, the mass-action expression is equal to a constant even though we may not know what the rates of the individual steps are. For example, for the phosgene equilibrium we know that the mass-action expression $[COCl_2]/[CO][Cl_2]$ equals 3.1 even though we may not know the steps that contribute to this constant.

The mass-action expression and the condition for equilibrium can often be simplified when the concentration of a substance cannot change. This happens, for instance, with heterogeneous equilibria, which involve two or more phases. For the equilibrium

$$2C(s) + O_2(g) \rightleftharpoons 2CO(g)$$

both solid and gas phases are involved. The solid phase consists of pure carbon and the gas phase of a mixture of oxygen and carbon monoxide. In mass-action expressions, the concentrations apply to the phase specified in the chemical equation. Hence, for the above equilibrium condition is

$$K' = \frac{[CO(g)]^2}{[C(s)]^2[O_2(g)]}$$

where [CO(g)] refers to CO in the gas phase, [C(s)] to solid carbon in the solid phase, and [O_2(g)] to oxygen in the gas phase. But the concentration of solid carbon in the solid phase cannot be changed, unlike CO and O_2 in the gas phase. If more solid carbon is added, its *concentration* is not changed because as the number of moles of carbon increases, the volume of carbon also increases. Given this, the constant concentration can be combined with the equilibrium constant K' to give a new constant which relates only the substances whose concentrations are variable. In other words we can write the equilibrium condition for the above reaction as

$$K = \frac{[CO(g)]^2}{[O_2(g)]}$$

where we have replaced $K'[C(s)]^2$ by the new constant K. *As a general rule, substances whose concentration is not variable are omitted from the mass-action expression.*

12.3   Equilibrium Constant

The numbers observed for equilibrium constants range from very large to extremely small, depending on the specific reaction. If the equilibrium constant is small ($K < 1$), the numerator of the mass-action expression is smaller than the denominator. This means that, in the equilibrium state, the concentration of at least one of the materials on the right of the chemical equation is small. Therefore, a small equilibrium constant implies that the reaction does not proceed far from left to right. For example, if for

$$N_2(g) + 3H_2(g) \rightleftharpoons 2NH_3(g) \qquad K = 2.37 \times 10^{-3}$$

then the reaction of N_2 and H_2 does not result in the production of much NH_3 at equilibrium. If the equilibrium constant is large ($K > 1$), the denominator of the mass-action expression is smaller than the numerator. This means that, in the equilibrium state, the concentration of at least one of the materials on the left of the chemical equation is small. Therefore, a very large equilibrium constant implies that the reaction proceeds from left to right essentially to completion. For example, if for

$$2NO(g) + O_2(g) \rightleftharpoons 2NO_2(g) \qquad K = 6.45 \times 10^5$$

then the reaction of NO with O_2 results in practically complete conversion to NO_2.

The value of the equilibrium constant is determined by experiment. For example, measurements have been made of the equilibrium involving hydrogen, iodine, and hydrogen iodide. The equilibrium is described by

$$H_2(g) + I_2(g) \rightleftharpoons 2HI(g)$$

In the equilibrium state all three components are present. The equilibrium condition is

$$\frac{[HI]^2}{[H_2][I_2]} = K$$

The number K can be determined by measuring all three concentrations in the equilibrium state. In an experiment at 490°C the following results might be obtained:

Concentration of H_2 = 0.0862 mole/liter
Concentration of I_2 = 0.263 mole/liter
Concentration of HI = 1.02 moles/liter

Since these concentrations are equilibrium concentrations, they satisfy the equilibrium condition:

$$\frac{[HI]^2}{[H_2][I_2]} = \frac{(1.02)^2}{0.0862 \times 0.263} = 45.9$$

For any equilibrium system of 490°C containing H_2, I_2, and HI, the mass-action expression must be equal to 45.9. If this condition is not satisfied, the system is not at equilibrium, and changes will occur until equilibrium is established.

12.4   Equilibrium Calculations

Once the value 45.9 has been determined, it can be used to describe any system containing H_2, I_2, and HI in chemical equilibrium at 490°C

EXAMPLE 1   If 1 mole of H_2 and 1 mole of I_2 are introduced into a 1-liter box at a temperature of 490°C, what will be the final concentrations in the box when equilibrium has been established?

Initially, there is no HI in the box. The system is not at equilibrium, since the mass-action expression is zero instead of 45.9. In order to establish equilibrium, changes must occur to produce HI. HI can come only from the reaction

$$H_2(g) + I_2(g) \rightleftharpoons 2HI(g)$$

This reaction proceeds to produce enough HI to satisfy the equilibrium condition.

Let n equal the number of moles of hydrogen that must disappear in order to establish equilibrium. Every time 1 mole of hydrogen disappears, 1 mole of iodine also disappears, and so n also represents the number of moles of iodine that disappear in order to establish equilibrium. According to the balanced equation, if 1 mole of hydrogen disappears, 2 moles of HI must be formed. If n moles of hydrogen disappear, $2n$ moles of HI must appear. Therefore, $2n$ is equal to the number of moles of HI formed in order to establish equilibrium. The situation is summarized as follows:

Initially	*At equilibrium*
$[H_2] = 1.000$ mole/liter	$[H_2] = (1.000 - n)$ mole/liter
$[I_2] = 1.000$ mole/liter	$[I_2] = (1.000 - n)$ mole/liter
$[HI] = 0$	$[HI] = 2n$ moles/liter

Since the volume of the box is 1 liter, the concentration of each component is identical to the number of moles of that component in the box. The equilibrium concentrations must satisfy the condition

$$\frac{[HI]^2}{[H_2][I_2]} = 45.9$$

Substitution gives

$$\frac{(2n)^2}{(1.000 - n)(1.000 - n)} = 45.9$$

for which

$$n = 0.772†$$

Therefore, at equilibrium

$[H_2] = 1.000 - n = 0.228$ mole/liter
$[I_2] = 1.000 - n = 0.228$ mole/liter
$[HI] = 2n = 1.544$ moles/liter

That these values represent equilibrium concentrations can be checked by calculating the value of the mass-action expression

$$\frac{[HI]^2}{[H_2][I_2]} = \frac{(1.544)^2}{0.228 \times 0.228} = 45.9$$

---

†This particular equation can be solved by taking the square root of both sides of the equation. For a more general case, we can use the ordinary algebraic methods for solving quadratic equations (see Appendix 2.5). Of the two roots necessarily obtained for the quadratic equation, one can be discarded as physically impossible. In this case, the root $n = 1.42$ would correspond to more than 100 percent reaction and, hence, must be discarded.

To emphasize the fact that it makes no difference from which side of the equation equilibrium is approached, we consider what happens when only HI is placed in the box at 490°C. Since initially there is no hydrogen or iodine in the system, decomposition of HI must occur in order to establish equilibrium.

EXAMPLE 2   If 2 moles of HI is injected into a box of 1-liter volume at 490°C, what will be the concentration of each species in the box at equilibrium?

The equilibrium is

$$H_2(g) + I_2(g) \rightleftharpoons 2HI(g)$$

Let x equal the number of moles of HI that must decompose in order to establish equilibrium. At equilibrium, the moles of HI will thus be $2 - x$. By the reverse reaction, for each 2 moles of HI that disappear, 1 mole of hydrogen and 1 mole of iodine are formed. If x moles of HI disappear, $x/2$ moles of hydrogen and $x/2$ moles of iodine appear. Hence, at equilibrium the moles of H_2 and I_2 will each be $x/2$. The initial and final concentrations are summarized:

Initially	*At equilibrium*
$[HI] = 2.000$ moles/liter	$[HI] = (2.000 - x)$ moles/liter
$[H_2] = 0$ mole/liter	$[H_2] = (x/2)$ moles/liter
$[I_2] = 0$ mole/liter	$[I_2] = (x/2)$ moles/liter

At equilibrium

$$\frac{[HI]^2}{[H_2][I_2]} = 45.9 = \frac{(2.000 - x)^2}{(x/2)(x/2)}$$

for which, solving the equation by taking the square root of both sides, we get

$$x = 0.456$$

Therefore, at equilibrium

$[H_2] = x/2 = 0.228$ mole/liter
$[I_2] = x/2 = 0.228$ mole/liter
$[HI] = 2.000 - x = 1.544$ moles/liter

The two examples show that it makes *no difference whether the equilibrium state is produced from the material on the left-hand side of the chemical equation or from the material on the right-hand side.* Change occurs so as to produce the material that is missing in sufficient concentration to establish equilibrium.

12.5   Equilibrium Changes

When a system at equilibrium is disturbed, chemical reaction occurs and equilibrium is reestablished. As an example, consider the equilibrium system consisting of H_2, I_2, and HI in a sealed box.

$$H_2(g) + I_2(g) \rightleftharpoons 2HI(g)$$

$$K = \frac{[HI]^2}{[H_2][I_2]}$$

At 490°C, K is 45.9. The concentrations of HI, H_2, and I_2 do not change until conditions are changed. Several kinds of changes are considered:

What is the effect of changing the concentration of one of the components by addition?   Suppose that more H_2 is added to a box which contains H_2, I_2, and HI in equilibrium at 490°C. What effect does this concentration increase have on the other components? Although this change and all those we will discuss can be treated by consideration of the mass-action expression, they can be handled qualitatively by the principle of Le Chatelier. According to the principle of Le Chatelier (Section 9.4), any equilibrium system subjected to a stress tends to change so as to relieve the stress. For a system in chemical equilibrium, changing the concentration of one of the components constitutes a stress. In the present case, if hydrogen is added to the box, the equilibrium system

$$H_2(g) + I_2(g) \rightleftharpoons 2HI(g)$$

adjusts itself so as to absorb the effect of the added hydrogen. The system can absorb the stress (i.e., added hydrogen) if some hydrogen molecules are used up in combining with iodine molecules to form HI. This means that the concentration of HI increases and the concentration of I_2 decreases.

What is the effect of decreasing the volume of the box?   Le Chatelier's principle predicts that when the volume of the box is reduced, the stress produced by crowding the molecules closer together can be relieved if the molecules are reduced in number. In the case

$$H_2(g) + I_2(g) \rightleftharpoons 2HI(g)$$

there is no device by which this can be accomplished. If one molecule of hydrogen and one molecule of iodine disappear, two molecules of HI are produced. There can be no change in the total number of molecules in the box. Neither the forward nor reverse reaction can absorb the stress of a decreased volume. There is no net change; the number of moles of H_2, I_2, and HI stays constant. Of course, since the volume is diminished, the *concentration* of each component is increased. But this increase applies equally to all components and there can therefore be no change in the equilibrium. However, in other equilibria there may be more pronounced change. An example is the equilibrium between nitrogen, hydrogen, and ammonia.

$$N_2(g) + 3H_2(g) \rightleftharpoons 2NH_3(g)$$

Here, the situation is different. When one molecule of N_2 reacts with three molecules of H_2, two molecules of NH_3 are formed. A decrease of the volume of the box can be compensated for by forming fewer molecules, i.e., by favoring the formation of ammonia. In the manufacture of ammonia, the yield is improved in this way. It is a general principle that, for reactions in which there is a change in the number of gas molecules, a decrease in the volume favors the reaction which produces fewer molecules.

What is the effect of raising the temperature?   The stress of added heat can be relieved by favoring the reaction direction which uses up heat. This, in turn, depends on the reaction in question. Since the equilibrium

$$H_2(g) + I_2(g) \rightleftharpoons 2HI(g) + 13 \text{ kilojoules}$$

is exothermic, when H_2 and I_2 form HI, heat is liberated. The reverse process absorbs heat. At equilibrium, the liberation of heat by the forward reaction is compensated for by the absorption of heat by the reverse reaction. If the temperature is increased, the system tries to relieve the stress by absorbing heat. Since the reverse reaction uses heat, it is favored. Favoring the reverse reaction causes a net decrease in the concentration of HI and a net increase in the concentration of I_2 and H_2. It is generally true, as the Le Chatelier principle predicts, that for exothermic reactions the reverse process is favored by raising the temperature. In other words, a temperature rise favors an endothermic change.

What effect does a catalyst have on an equilibrium system?   Because the catalyst does not appear in the net equation, its presence is ignored by both the mass-action expression and the Le Chatelier principle. That it is ignored is consistent with all observations since a catalyst cannot affect equilibrium concentrations. It will get the system to equilibrium faster, but the final state will be the same in its presence or absence.

If a catalyst could affect equilibrium concentrations, a perpetual-motion machine could be built, as follows: The cylinder shown in Figure 12.3 is fitted with a sliding piston and filled with N_2, H_2, and NH_3 in chemical equilibrium. The compartment with the trap door contains the solid catalyst, and a metal rod connects the trap door to the piston. The device operates in such a way that, when the piston is up, the catalyst is exposed. Suppose

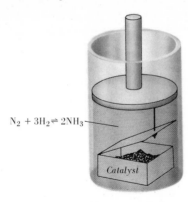

$N_2 + 3H_2 \rightleftharpoons 2NH_3$

Catalyst

Figure 12.3
Impossible perpetual-motion machine.

the catalyst could favor the net formation of NH_3. The total number of gas molecules would decrease; hence the pressure inside the cylinder would drop. As the piston moved in, the trap door would close, and the catalyst would no longer be exposed. In returning to the initial state, some of the ammonia would revert to nitrogen and hydrogen. The total number of gas molecules would increase, and the piston would move out. The trap door would open, and the catalyst would again be exposed. The cycle could repeat itself forever. Unfortunately, this, like all perpetual-motion devices, does not work. A catalyst cannot change equilibrium concentrations. All it does is change the opposing rates, and these it changes equally.

12.6   Water Equilibrium and pH

The discussion of the preceding sections was confined to gases, but the same principle holds in aqueous solutions; viz., the mass-action expression connecting the concentrations of species in aqueous solution must be equal to a constant if equilibrium is to exist. Let us consider the case of water. As already noted, H_2O is a weak electrolyte which dissociates according to the equation

$$H_2O \rightleftharpoons H^+ + OH^-$$

At equilibrium, the condition that must be satisfied is

$$K = \frac{[H^+][OH^-]}{[H_2O]}$$

In practice this equation can be simplified by taking advantage of the fact that the concentration of H_2O in pure water and also in dilute aqueous solutions is effectively constant (about 55 moles/liter). Therefore, we can write a new constant K_w in place of the constant product $K[H_2O]$ and arrive at the equilibrium condition

$$K_w = [H^+][OH^-]$$

K_w is called the *ion product,* or sometimes the dissociation constant, of water. It has the numerical value 1.0×10^{-14} at $25°C$.

In pure water, all the H^+ and the OH^- must come from the dissociation of water molecules. If x moles of H^+ are produced per liter, x moles of OH^- must be simultaneously produced. Therefore, we can reason as follows:

$$[H^+][OH^-] = 1.0 \times 10^{-14}$$
$$(x)(x) = 1.0 \times 10^{-14}$$
$$x^2 = 1.0 \times 10^{-14}$$
$$x = 1.0 \times 10^{-7}$$

Thus, in pure H_2O, the concentrations of H^+ and OH^- are each 1.0×10^{-7} M. This very small concentration is to be compared with an H_2O concentration of approximately 55 moles/liter. On an average, there is one H^+ ion and one OH^- ion for every 550 million H_2O molecules.

If an acid is added to water, the hydrogen-ion concentration increases above 1.0×10^{-7} M. The ion product must remain equal to 1.0×10^{-14}; consequently the hydroxide-ion concentration decreases below 1.0×10^{-7} M. Similarly, when a base is added to water, the concentration of OH^- increases above 1.0×10^{-7} M and the concentration of H^+ decreases below 1.0×10^{-7} M.

As a convenience for working with such small concentrations, the pH scale has been devised to express the concentration of H^+. The pH is defined as the negative logarithm of the hydrogen-ion concentration

$$pH = -\log[H^+] \quad \text{or} \quad [H^+] = 10^{-pH}$$

In pure water, where the concentration of H^+ is 1.0×10^{-7} M, the pH is 7. All neutral solutions have a pH of 7. Acid solutions have pH less than 7; basic solutions have pH greater than 7.

EXAMPLE 3   What is the pH of 0.10 M HCl?

In 0.10 M HCl, practically all the H^+ comes from the 100 percent dissociation of the strong electrolyte HCl.

$$[H^+] = 0.10 M = 1.0 \times 10^{-1} M$$
$$pH = -\log(1.0 \times 10^{-1})$$
$$= -\log 1.0 - \log 10^{-1} = 0 + 1 = 1$$

Note that the logarithm of a product is equal to the sum of the logs. The log of 1.0 is zero, and the log of 10^{-1} is -1.

EXAMPLE 4   What is the pH of 0.10 M NaOH?

NaOH is a strong electrolyte and accounts for essentially all the OH⁻ in the solution.

$$[OH^-] = 0.10 \, M$$

$$[H^+] = \frac{K_w}{[OH^-]} = \frac{1.0 \times 10^{-14}}{0.10} = 1.0 \times 10^{-13} \, M$$

$$pH = -\log(1.0 \times 10^{-13}) = 13$$

We should remember that an increase in pH means a decrease in acidity. This can be illustrated by the following approximate pH values for substances arranged in order of decreasing acidity: lime, pH 2; oranges, 3–4; carrots, 5; peas, 6; human saliva, which is nearly neutral, approximately 7; human blood, which is slightly basic, 7.3–7.5. About the only more alkaline material is egg white, which runs as high as pH 8.

So far, we have emphasized the dissociation of water to give ions. However, since equilibrium may be approached from the left or the right side of an equation, the same equilibrium constant that describes the dissociation of water also describes the association of H⁺ and OH⁻ to form water. Such association occurs in neutralization reactions, as discussed in Section 10.8, and is the basis of the process of *titration*, the progressive addition of an acid to a base, or vice versa. At each step in the titration, the expression $[H^+][OH^-] = 1.0 \times 10^{-14}$ must be satisfied in the solution. Figure 12.4 represents what happens to the pH as solid NaOH is added stepwise to 0.010 mole of HCl in 1 liter of water. As NaOH is progressively added, the original solution changes from acid (pH less than 7) to basic (pH greater than 7). The pH first rises very slowly, then rapidly through the neutral point, and finally very slowly as the solution gets more basic. Such a pH curve is typical of the titration of any strong acid with any strong base. The important thing to note is that, as the neutral point is approached, there is a sharp rise in

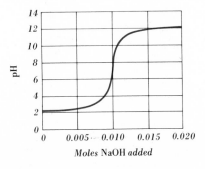

Figure 12.4
Titration curve of 1 liter of 0.01 M HCl.

pH. At this point even a trace of NaOH adds enough moles of base to increase the pH greatly. Thus, any method which locates the point at which the pH changes rapidly can be used to detect the *equivalence point* of a titration, i.e., the point at which equivalent amounts of base and acid have been mixed.

One method for determining the equivalence point makes use of the fact that many dyes have colors that are sensitive to hydrogen-ion concentration. Such dyes can be used as *indicators* to give information about the pH of a solution. Indicators are proton-containing organic molecules which change color when protons are transferred between them and the solvent. The pH at which proton transfer and hence color change occur is different for different indicators. This is shown in Figure 12.5. In Figure 12.4 the pH rises so sharply at the equivalence point that any one of the indicators of Figure 12.5, except possibly alizarin yellow, could be used to tell when enough NaOH has been added to neutralize 1 liter of 0.010 M HCl.

Figure 12.5 **INDICATOR** **COLORS** Indicator	pH at Which Color Changes	Color at Lower pH	Color at Higher pH
Methyl orange	4	Red	Yellow
Methyl red	5	Red	Yellow
Litmus	7	Red	Blue
Bromthymol blue	7	Yellow	Blue
Phenolphthalein	9	Colorless	Red
Alizarin yellow	11	Yellow	Red

12.7 Weak Acids

Unlike the dissociation of strong acids, which is complete, the dissociation of weak acids, for example, HF, is limited by the equilibrium

$$HF \rightleftharpoons H^+ + F^-$$

For handling such cases, the usual equilibrium considerations apply. Thus we can write

$$\frac{[H^+][F^-]}{[HF]} = K$$

The value of K, called the dissociation constant of HF, is 6.7×10^{-4} at $25°C$. Other representative constants are given in Figure 12.6.

Figure 12.6	Acid	Reaction	K (25°C)
DISSOCIATION OF WEAK ACIDS	Acetic	$HC_2H_3O_2 \rightleftharpoons H^+ + C_2H_3O_2^-$	1.8×10^{-5}
	Nitrous	$HNO_2 \rightleftharpoons H^+ + NO_2^-$	4.5×10^{-4}
	Hydrocyanic	$HCN \rightleftharpoons H^+ + CN^-$	4.0×10^{-10}
	Sulfurous	$H_2SO_3 \rightleftharpoons H^+ + HSO_3^-$	1.3×10^{-2}
		$HSO_3^- \rightleftharpoons H^+ + SO_3^{2-}$	5.6×10^{-8}

The smaller the value of K, the weaker is the acid. Thus, HCN is a weaker acid than HF, and much less dissociated for a given concentration. When K is 1 or greater, the acid is extensively dissociated, even in 1 M solution, and is classified as moderately strong. When K is 10 or greater, the acid is essentially 100 percent dissociated in all except very concentrated solutions. For example, perchloric acid ($HClO_4$) is one of the strongest acids and has K greater than 10. Similarly, HNO_3, HCl, and H_2SO_4 are common acids with high dissociation constants. For H_2SO_4, only one of the protons is completely dissociated. For the second, the dissociation constant is 1.3×10^{-2}. The equilibria can be written

$$H_2SO_4 \rightleftharpoons H^+ + HSO_4^- \qquad K_I > 10$$
$$HSO_4^- \rightleftharpoons H^+ + SO_4^{2-} \qquad K_{II} = 1.3 \times 10^{-2}$$

EXAMPLE 5

What is the concentration of all solute species in a solution labeled 1.00 M $HC_2H_3O_2$?

We have

$$HC_2H_3O_2 \rightleftharpoons H^+ + C_2H_3O_2^- \qquad K = 1.8 \times 10^{-5}$$

Let x equal the moles per liter of $HC_2H_3O_2$ that dissociate to establish equilibrium. According to the dissociation equation, each mole of $HC_2H_3O_2$ that dissociates produces 1 mole of H^+ and 1 mole of $C_2H_3O_2^-$. If x moles of $HC_2H_3O_2$ dissociates, then x moles of H^+ and x moles of $C_2H_3O_2^-$ must be formed. The initial and equilibrium concentrations are summarized as follows:

Initially	*At equilibrium*
$[HC_2H_3O_2] = 1.00$ mole/liter	$[HC_2H_3O_2] = (1.00 - x)$ mole/liter
$[H^+] = 0$ mole/liter	$[H^+] = x$ moles/liter
$[C_2H_3O_2^-] = 0$ mole/liter	$[C_2H_3O_2^-] = x$ moles/liter

At equilibrium

$$\frac{[H^+][C_2H_3O_2^-]}{[HC_2H_3O_2]} = 1.8 \times 10^{-5} = \frac{(x)(x)}{1.00 - x}$$

This is a quadratic equation and can be solved by the quadratic formula (Appendix 2.5). However, we can simplify by noting that only a small fraction of the weak acid will be dissociated and so x must be very small compared to 1.00, from which it is subtracted in the denominator. Thus we get the approximate equation

$$1.8 \times 10^{-5} \approx \frac{(x)(x)}{1.00}$$

obtained by assuming that $1.00 - x \approx 1.00$. From

$$1.8 \times 10^{-5} \approx x^2$$

we get

$$x = \sqrt{1.8 \times 10^{-5}} = \sqrt{18 \times 10^{-6}} = 4.2 \times 10^{-3}$$

Checking the approximation we find that $1.00 - x = 1.00 - (4.2 \times 10^{-3}) = 1.00$, as assumed, since the 1.00 is not known with accuracy beyond the second decimal place. Therefore, at equilibrium,

$$[HC_2H_3O_2] = 1.00 - x = 1.00 \ M$$
$$[H^+] = x = 0.0042 \ M$$
$$[C_2H_3O_2^-] = x = 0.0042 \ M$$

12.8   Buffer Solutions

In practically all biological processes as well as in many other chemical processes, it is important that the pH not deviate very much from a fixed value. For example, the proper functioning of human blood in carrying oxygen to the cells from the lungs is dependent on maintaining a pH very near to 7.4. In fact, for a particular individual, there is a difference of no more than 0.02 pH unit between venous and arterial blood in spite of numerous acid- and base-producing reactions in the cells.

The near constancy of pH in a system to which acid or base is added is due to the buffering action of an acid-base equilibrium. Let us consider, for example, a solution that contains acetic acid molecules and acetate ions (plus other ions, of course). The principal equilibrium in this solution can be written

$$HC_2H_3O_2 \rightleftharpoons H^+ + C_2H_3O_2^-$$

and it serves to buffer the solution as follows. If a strong acid is now added it serves to increase H^+ concentration, but the equilibrium adjusts by favoring the reverse reaction and so uses up much of the acid added. Similarly if base is added to the buffered solution, there is an initial drop in H^+

concentration. This favors the forward reaction which restores much (not all) of the H+ used. In other words, it "buffers" the effect of added base.

Quantitatively, we can see the effect of a buffer by writing the mass-action expression for the equilibrium

$$K = \frac{[H^+][C_2H_3O_2^-]}{[HC_2H_3O_2]}$$

If we solve for H+ we find

$$[H^+] = K \frac{[HC_2H_3O_2]}{[C_2H_3O_2^-]}$$

That is, the hydrogen-ion concentration is equal to the acid's dissociation constant times the concentration ratio of the two buffer components, $[HC_2H_3O]/[C_2H_3O_2^-]$. If these are made equal (that is, $[H^+] = K \times 1$) the solution will be buffered at a pH corresponding to $-\log K$. For acetic acid–acetate buffer, $-\log K = -\log(1.8 \times 10^{-5}) = 4.74$. Furthermore, if the concentration of buffer components is higher than any added acid or base, the buffer ratio will change little and so the pH will change little. For example, if $\frac{1}{10}$ mole of NaOH is added to a liter of buffer containing $HC_2H_3O_2$ and $C_2H_3O_2^-$ both at 0.50 M, the pH increases by only 0.02 unit to 4.76. By contrast, in an unbuffered system the pH would rise to 12. Buffering action can be understood qualitatively in terms of the Le Chatelier principle. Consider the dissociation equilibrium of a weak acid HA:

$$HA \rightleftharpoons H^+ + A^-$$

If to a solution containing sufficient A^-, strong acid is added, this system readjusts to use up the added H+ (back reaction) so that the pH is changed much less than if the buffering equilibrium were not present or if there were too little A^- to use with the added acid. Similarly, if strong base is added, the immediate effect is the reaction of OH− with H+, stressing the equilibrium. However, the stress is relieved by an adjustment which favors the forward reaction to replenish some of the H+ consumed by the base. Consequently, the pH is changed very little.

12.9 Hydrolysis

When the salt NaCl is placed in water, the resulting solution is observed to be neutral; i.e., the concentration of H+ and OH− are each $1 \times 10^{-7} M$ just as in pure water. However, when the salt NaF is dissolved in H_2O, the resulting solution is observed to be slightly basic. Other salts such as copper chloride or aluminum sulfate give slightly acid solutions. Such interactions between salts and water are called *hydrolysis*.

In order to understand what hydrolysis is, let us consider the solution of sodium fluoride. It is slightly basic because of the reaction

$$F^- + H_2O \rightleftharpoons HF + OH^-$$

which shows that fluoride ion, unlike chloride ion, has enough proton affinity to remove some H^+ from H_2O and tie it up as HF. However, the proton affinity of OH^- is considerably greater than that of F^-, so that in the competition for protons it is the OH^- that wins out most often. Another way of saying this is that the reverse reaction of the above equation has a greater tendency to occur than does the forward reaction. The net result is that when F^- is placed in H_2O, only a small fraction of the F^- is converted to HF. Stated another way, the equilibrium constant for the above reaction is small.

A fuller understanding of F^- hydrolysis can be obtained by noting that the extent of hydrolysis is governed by the relative proton affinities of F^- and OH^-:

$$F^- + H^+ \rightleftharpoons HF \tag{1}$$
$$OH^- + H^+ \rightleftharpoons H_2O \tag{2}$$

It should be noted that these two equations are the dissociation reactions of HF and H_2O written in reverse. Both HF and H_2O have small dissociation constants, and so the equilibrium constants for the reverse reactions are large. Note that reversing the chemical equation inverts the mass-action expression and hence the new K is the reciprocal of the old K:

$$K_{association} = \frac{1}{K_{dissociation}}$$

Specifically, the dissociation constant K_{HF} is 6.7×10^{-4} and therefore K for equation (1) is $1/6.7 \times 10^{-4}$ or 1.5×10^3. For equation (2), K is $1/K_w$ or $1.0 \times 10^{+14}$. Comparison shows that K for (2) is greater than K for (1); that is, the proton affinity of OH^- is enormously greater than the proton affinity of F^-.

For quantitative calculations, the equilibrium constant for hydrolysis, K_h, can be found by combining the dissociation constants of HF and H_2O:

$$F^- + H^+ \rightleftharpoons HF \qquad K = \frac{[HF]}{[F^-][H^+]} = \frac{1}{K_{HF}}$$

$$\underline{\qquad H_2O \rightleftharpoons H^+ + OH^- \qquad K = [H^+][OH^-] = K_w \qquad}$$

$$F^- + \cancel{H^+} + H_2O \rightleftharpoons HF + \cancel{H^+} + OH^- \qquad K_h = \frac{[HF][OH^-]}{[F^-]} = \frac{K_w}{K_{HF}}$$

The final equation is obtained (1) by writing the mass-action expression for the hydrolysis (omitting H_2O, since its concentration is constant); (2) substituting for $[HF]/[F^-]$ the value of this ratio found on the first line, that is, $[H^+]/K_{HF}$; and (3) for $[OH^-]$ its value found from the second line, that is, $K_w/[H^+]$. Cancellation of $[H^+]$ from the numerator and denominator gives $K_h = K_w/K_{HF}$.

Note that for any hydrolysis reaction, the equilibrium constant is equal to K_w divided by the dissociation constant of the weak electrolyte formed. It follows that the smaller the dissociation constant, the greater the extent of hydrolysis. However, unless the dissociation is so small that it approaches that of water (1.0×10^{-14}), hydrolysis will be slight.

What about the sodium ion that was present in the above NaF solution? Does it affect the water dissociation equilibrium? The answer is no. NaOH is a strong electrolyte, 100 percent dissociated into Na^+ and OH^- in aqueous solution. This means Na^+ when added to water will have no tendency to form undissociated NaOH. Hence, the equilibrium

$$H_2O \rightleftharpoons H^+ + OH^-$$

is not disturbed by Na^+. For this reason, Na^+ is said not to hydrolyze. Similarly, Cl^- does not hydrolyze, and the salt NaCl when added to H_2O produces neither acid nor basic reaction. However, addition of NaF to H_2O produces a basic solution, owing to the fact that F^- does hydrolyze.

Cu^{2+} or Al^{3+} ions, on the other hand, produce an acid solution when added to water. For example, some baking powders contain Al^{3+} to give an acid reaction with sodium bicarbonate which liberates CO_2. The acid reaction of Al^{3+} with water to produce H^+ can be viewed as the hydrolysis reaction:

$$Al^{3+} + H_2O \rightleftharpoons H^+ + Al(OH)^{2+}$$

As with F^- hydrolysis, there is a competition for one of the ions of water. In this case it is the OH^- ion, and Al^{3+} and H^+ are competing for it. The dissociation constant of $Al(OH)^{2+}$ is 1×10^{-9} (considerably larger than $K_w = 1.0 \times 10^{-14}$) so that the extent of hydrolysis is slight, and the resulting solution is only slightly acid.

The general equation for positive-ion hydrolysis can be written

$$M^+ + H_2O \rightleftharpoons MOH + H^+$$

whereas that for negative-ion hydrolysis can be written

$$X^- + H_2O \rightleftharpoons HX + OH^-$$

In the first case, the solution becomes acidic; in the second case, it becomes basic.

Although in principle the extent of hydrolysis can be calculated if the appropriate dissociation constant is known, it is convenient to keep in mind the qualitative differences between various ions. Figure 12.7 separates the common ions into three classes according to their extent of hydrolysis in 1 M solution. As the figure shows, for positive ions the extent of hydrolysis

Figure 12.7 EXTENT OF HYDROLYSIS	Not Hydrolyzed	Slightly Hydrolyzed	Extensively Hydrolyzed
	Na^+, K^+, Ca^{2+}, Ba^{2+}, Cl^-, NO_3^-, HSO_4^-	Zn^{2+}, Cu^{2+}, Al^{3+}, Fe^{3+}, Cr^{3+} F^-, $C_2H_3O_2^-$, CN^-, CO_3^{2-}	Sn^{4+}, Ce^{4+} S^{2-}, PO_4^{3-}

increases as the charge on the ion increases (owing to increased affinity for OH^-). The situation is not quite so obvious for negative ions although there is a general trend toward increased hydrolysis with increased negative charge.

We can summarize hydrolysis reactions by making use of the Le Chatelier principle. The equilibrium in question is the dissociation of water.

$$H_2O \rightleftharpoons H^+ + OH^-$$

Added salts may stress this equilibrium if they combine with either H^+ or OH^-. If an added M^+ associates with OH^-, then the concentration of OH^- will be diminished below its equilibrium value. In accord with the Le Chatelier principle, the system adjusts to the decreased OH^- concentration by favoring the forward reaction. In the process, more H^+ is formed, over and above that originally present; the solution becomes acid. Similarly, if added X^- associates with H^+ (thereby lowering the concentration of H^+), the forward reaction is again favored. The concentration of OH^- is increased over its original value and the solution becomes basic.

12.10 Solubility of Ionic Solids

When an ionic solid is placed in water, an equilibrium is established between the ions in the saturated solution and the excess solid phase. For example, with excess barium sulfate in contact with a saturated solution of barium sulfate, the equilibrium is

$$BaSO_4(s) \rightleftharpoons Ba^{2+} + SO_4^{2-}$$

The equilibrium condition is

$$\frac{[Ba^{2+}][SO_4^{2-}]}{[BaSO_4(s)]} = K$$

In this example two phases, solid and aqueous solution, are involved, and the equilibrium is called a *heterogeneous equilibrium*. Care must be taken that the concentrations of specified species only apply to the phase in which that species occurs. Thus, $[Ba^{2+}]$ and $[SO_4^{2-}]$ apply to molar concentrations in the aqueous phase and $[BaSO_4(s)]$ applies to the molar concentration of barium sulfate in the solid phase. Inasmuch as the concentration of $BaSO_4$ in solid $BaSO_4$ is constant, no matter how many moles are taken, the moles per liter are the same. Consequently, $[BaSO_4(s)]$ can be incorporated into the equilibrium constant:

$$[Ba^{2+}][SO_4^{2-}] = K[BaSO_4(s)] = K_{sp}$$

The constant K_{sp} is called the *solubility product* and the expression $[Ba^{2+}][SO_4^{2-}]$ the *ion product*. The equation states that the ion product must equal K_{sp} when the saturated solution is in equilibrium with excess solid. It should be noted that there is no separate restriction on what the concentration of Ba^{2+} and SO_4^{2-} must be. The concentration of Ba^{2+} can have any value, as long as the concentration of SO_4^{2-} is such that the product of Ba^{2+} concentration and SO_4^{2-} concentration is equal to K_{sp}.

The numerical value of K_{sp}, as of any equilibrium constant, must be determined by experiment. Once determined, it can be tabulated for future use. (Appendix 7 contains some typical values.) K_{sp} can be determined from a measurement of solubility in water. For $BaSO_4$ the solubility at 25°C is 3.9×10^{-5} mole/liter. Like practically all salts, $BaSO_4$ is a strong electrolyte and so is 100 percent dissociated into ions. Therefore, when 3.9×10^{-5} mole of $BaSO_4$ dissolves, it forms 3.9×10^{-5} mole of Ba^{2+} and 3.9×10^{-5} mole of SO_4^{2-}. In the saturated solution, the concentration of Ba^{2+} is 3.9×10^{-5} M, and the concentration of SO_4^{2-} is 3.9×10^{-5} M. Therefore, for the equilibrium

$$BaSO_4(s) \rightleftharpoons Ba^{2+} + SO_4^{2-}$$

we have the condition

$$K_{sp} = [Ba^{2+}][SO_4^{2-}] = (3.9 \times 10^{-5})(3.9 \times 10^{-5})$$
$$= 1.5 \times 10^{-9}$$

This means that, in any solution containing Ba^{2+} and SO_4^{2-} in equilibrium with solid $BaSO_4$, the product of the concentrations of Ba^{2+} and SO_4^{2-} is equal to 1.5×10^{-9}. If $[Ba^{2+}]$ multiplied by $[SO_4^{2-}]$ is less than 1.5×10^{-9}, the solution is unsaturated and $BaSO_4$ must dissolve to increase the concentrations of Ba^{2+} and SO_4^{2-}. If $[Ba^{2+}]$ times $[SO_4^{2-}]$ is greater than 1.5×10^{-9}, the system is not at equilibrium. $BaSO_4$ precipitates in order to decrease the concentrations of Ba^{2+} and SO_4^{2-}. Since K_{sp} is a very small number, $BaSO_4$ may be called an insoluble salt.

When $BaSO_4$ is placed in pure water, the concentrations of Ba^{2+} and

SO_4^{2-} must be equal. On the other hand, it is possible to prepare a solution in which unequal concentrations of Ba^{2+} and SO_4^{2-} are in equilibrium with solid $BaSO_4$. As an illustration, unequal amounts of barium chloride and sodium sulfate might be added to water. A precipitate of $BaSO_4$ forms if K_{sp} of $BaSO_4$ is exceeded. However, there is no requirement that $[Ba^{2+}] = [SO_4^{2-}]$, since the two ions come from different salts. Alternatively, barium sulfate solid might be added to an Na_2SO_4 solution. Some barium sulfate dissolves, but in the final solution the concentration of SO_4^{2-} is greater than the concentration of Ba^{2+}.

EXAMPLE 6   Given that the K_{sp} of radium sulfate ($RaSO_4$) is 4×10^{-11}. Calculate its solubility in (a) pure water and (b) 0.10 M Na_2SO_4.

    a   Let x = moles of $RaSO_4$ that dissolve per liter of water. Then, in the saturated solution,

$$[Ra^{2+}] = x \text{ moles/liter}$$
$$[SO_4^{2-}] = x \text{ moles/liter}$$
$$RaSO_4(s) \rightleftharpoons Ra^{2+} + SO_4^{2-}$$
$$[Ra^{2+}][SO_4^{2-}] = K_{sp} = 4 \times 10^{-11}$$
$$(x)(x) = 4 \times 10^{-11}$$
$$x = 6 \times 10^{-6} \text{ mole/liter}$$

Thus, the solubility of $RaSO_4$ is 6×10^{-6} mole/liter of water, giving a solution containing 6×10^{-6} M Ra^{2+} and 6×10^{-6} M SO_4^{2-}.

    b   Let y = moles of $RaSO_4$ that dissolve per liter of 0.10 M Na_2SO_4. This dissolving produces y moles of Ra^{2+} and y moles of SO_4^{2-}. The solution already contains 0.10 M SO_4^{2-}. Thus, in the final saturated solution,

$$[Ra^{2+}] = y \text{ moles/liter}$$
$$[SO_4^{2-}] = (y + 0.10) \text{ moles/liter}$$

where

$$[Ra^{2+}][SO_4^{2-}] = (y)(y + 0.10) = K_{sp} = 4 \times 10^{-11}$$

Since K_{sp} is very small, not much $RaSO_4$ dissolves and y is so small that it is negligible compared with 0.10.

$$y + 0.10 \approx 0.10$$
$$[Ra^{2+}][SO_4^{2-}] \approx y \times 0.10 \approx 4 \times 10^{-11}$$
$$y \approx \frac{4 \times 10^{-11}}{0.10} = 4 \times 10^{-10} \text{ mole/liter}$$

Thus, the solubility of $RaSO_4$ in 0.10 M Na_2SO_4 is 4×10^{-10} mole/liter, giving a solution in which the concentration of Ra^{2+} is 4×10^{-10} M and that of SO_4^{2-} is 0.10 M.

It is interesting to note that $RaSO_4$ is less soluble in a Na_2SO_4 solution than in pure water. This is an example of the *common-ion effect,* by which the solubility of an ionic salt is generally decreased by the presence of another solute that furnishes one of its ions. Thus, radium sulfate is less soluble in any solution containing either radium ion or sulfate ion that it is in water. The greater the concentration of the common ion, the less radium sulfate can dissolve. Of course, if the common ion is present in negligible concentration, it has no appreciable effect on the solubility. This is illustrated in the following example.

EXAMPLE 7   Given that magnesium hydroxide [$Mg(OH)_2$] is a strong electrolyte and has a solubility product of 8.9×10^{-12}, calculate the solubility of $Mg(OH)_2$ in water.

Let x = moles of $Mg(OH)_2$ that dissolve per liter. According to the equation

$$Mg(OH)_2(s) \rightleftharpoons Mg^{2+} + 2OH^-$$

x moles of $Mg(OH)_2$ dissolve to give x moles of Mg^{2+} and $2x$ moles of OH^-. Some hydroxide ion is also furnished by the dissociation of water. Since H_2O is a very weak electrolyte, it contributes only a negligible amount of OH^- compared with that furnished by the dissolving of $Mg(OH)_2$. Thus at equilibrium

$[Mg^{2+}] = x$ moles/liter
$[OH^-] \approx 2x$ moles/liter

For the saturated solution the equilibrium is

$$Mg(OH)_2(s) \rightleftharpoons Mg^{2+} + 2OH^-$$

and

$$K_{sp} = 8.9 \times 10^{-12} = [Mg^{2+}][OH^-]^2$$

Substituting, we get

$$(x)(2x)^2 = 8.9 \times 10^{-12}$$
$$4x^3 = 8.9 \times 10^{-12}$$
$$x = \sqrt[3]{2.2 \times 10^{-12}} = 1.3 \times 10^{-4} \text{ mole/liter}$$

Thus, 1.3×10^{-4} mole of $Mg(OH)_2$ dissolves per liter of water. The saturated solution contains 1.3×10^{-4} M Mg^{2+} and 2.6×10^{-4} M OH^-.

As noted in Section 12.2, the mass-action expression for a given reaction contains concentrations raised to powers that correspond to the coefficients in the chemical equation. Since the ion product is a mass-action expression, it must be formed by raising the concentrations of ions to powers that correspond to the coefficients in the solubility equation. An exponent applies to the concentration of the specified ion, no matter where that ion comes from. For example, in the following example, essentially all the OH^- comes from NaOH, but its concentration still must be squared.

EXAMPLE 8   Calculate the solubility of $Mg(OH)_2$ in 0.050 M NaOH.

Let x = moles of $Mg(OH)_2$ that dissolve

$$Mg(OH)_2(s) \longrightarrow Mg^{2+} + 2OH^-$$

per liter. This forms x moles of Mg^{2+} and 2x moles of OH^-. Since the solution already contains 0.050 mole of OH^-, equilibrium concentrations are

$$[Mg^{2+}] = x \text{ moles/liter}$$
$$[OH^-] = (2x + 0.050) \text{ moles/liter}$$
$$[Mg^{2+}][OH^-]^2 = (x)(2x + 0.050)^2 = K_{sp}$$
$$(x)(2x + 0.050)^2 = 8.9 \times 10^{-12}$$

Assuming that x is a very small number and that 2x can be neglected when added to 0.050, we have approximately

$$(x)(0.050)^2 \approx 8.9 \times 10^{-12}$$
$$x = 3.6 \times 10^{-9} \text{ mole/liter}$$

Since x is small compared with 0.050, the assumption is valid. The calculation indicates that 3.6×10^{-9} mole of $Mg(OH)_2$ can dissolve in 1 liter of 0.050 M NaOH to give a saturated solution containing 3.6×10^{-9} M Mg^{2+} and 0.050 M OH^-.

One of the most useful applications of the solubility product is to predict whether or not precipitation will occur when two solutions are mixed. In the saturated solution of a salt, the ion product equals K_{sp}. If two solutions containing the ions of a salt are mixed and if the ion product then exceeds K_{sp}, precipitation should occur.

EXAMPLE 9   Should precipitation occur when 50 ml of 5.0×10^{-4} M $Ca(NO_3)_2$ is mixed with 50 ml of 2.0×10^{-4} M NaF to give 100 ml of solution? The K_{sp} of CaF_2 is 1.7×10^{-10}.

In order to solve such a problem, it is convenient to calculate first the concentration of the ions in the mixture, assuming that no precipitation occurs. Thus, the Ca^{2+} from the 5.0×10^{-4} M $Ca(NO_3)_2$ solution is made 2.5×10^{-4} M in the final mixture because of the twofold dilution. Likewise, the F^- is diluted to $1.0 \times 10^{-4}\,M$ in the final mixture. Therefore, if no precipitation occurs, the final solution would have

$$[Ca^{2+}] = 2.5 \times 10^{-4}\,M \qquad \text{and} \qquad [F^-] = 1.0 \times 10^{-4}\,M$$

To determine whether precipitation should occur, it is necessary to see whether the ion product exceeds the solubility product. For a saturated solution of CaF_2 the equilibrium would be

$$CaF_2(s) \rightleftharpoons Ca^{2+} + 2F^-$$

for which the ion product is $[Ca^{2+}][F^-]^2$. In the present mixture the ion product has the numerical value

$$[Ca^{2+}][F^-]^2 = (2.5 \times 10^{-4})(1.0 \times 10^{-4})^2 = 2.5 \times 10^{-12}$$

Since this number does not exceed 1.7×10^{-10}, the K_{sp} of CaF_2, precipitation does not occur. The solution obtained as the final mixture is unsaturated with respect to precipitation of CaF_2.

In order to precipitate a salt, the ion product must be made to exceed the K_{sp} of that salt. Application of this principle for driving ions out of solution. For example, given a solution of $RaCl_2$, the Ra^{2+} can be made to precipitate as $RaSO_4$ by addition of Na_2SO_4. The more the concentration of SO_4^{2-} is increased in the solution, the lower the concentration of Ra^{2+} becomes. Essentially all the valuable Ra^{2+} can be removed from the solution in this way by adding a large excess of SO_4^{2-} ions.

In concluding this section on solubility, we consider how the solubility of an ionic solid can be modified by manipulating the hydrogen-ion concentration of the solution. This is a very useful analytical tool when trying to determine what positive ions are in a solution. The ions can be separated from each other by precipitating them as sulfides, the solubilities of which are strongly dependent on the hydrogen-ion concentration of the solution. The reason for this dependence is the fact that H_2S is a weak diprotic acid.

$$H_2S \rightleftharpoons H^+ + HS^- \qquad K_I = \frac{[H^+][HS^-]}{[H_2S]} = 1 \times 10^{-7}$$

$$HS^- \rightleftharpoons H^+ + S^{2-} \qquad K_{II} = \frac{[H^+][S^{2-}]}{[HS^-]} = 1 \times 10^{-14}$$

$$\overline{H_2S \rightleftharpoons 2H^+ + S^{2-} \qquad K = \frac{[H^+]^2[S^{2-}]}{[H_2S]} = K_I K_{II} = 1 \times 10^{-21}}$$

Note that the combined equilibrium, which is the sum of the two stepwise equilibria, has a mass-action expression and equilibrium constant which are products of those for the two steps. We should note also that $H_2S(g)$ is not particularly soluble (0.10 M), and so for any solution saturated with H_2S, we get the equilibrium expression

$$\frac{[H^+]^2[S^{2-}]}{[0.10]} = 1 \times 10^{-21}$$

or

$$[H^+]^2[S^{2-}] = 1 \times 10^{-22}$$

Now we can see why the concentration of S^{2-} in a saturated H_2S solution is so dependent on acidity.

$$[S^{2-}] = \frac{1 \times 10^{-22}}{[H^+]^2}$$

If the acidity is raised from $1 \times 10^{-3}[M]$ to 1.0 M, the $[S^{2-}]$ concentration decreases by 1×10^6, or a million fold! Thus, by controlling acidity we can control in a sensitive manner the concentration of S^{2-} such that it will selectively precipitate positive ions. This is the basis of a scheme of qualitative analysis.

EXAMPLE 10   A solution contains Zn^{2+} and Cu^{2+}, each at 0.02 M. The K_{sp} of ZnS is 1×10^{-22}; that of CuS, 8×10^{-37}. If the solution is made 1 M in H^+, and H_2S gas is bubbled in until the solution is saturated, should a precipitate form?

In a saturated H_2S solution, where $[H_2S] = 0.10$ M, K_I and K_{II} combine to give

$$[H^+]^2[S^{2-}] = 1 \times 10^{-22}$$

If $[H^+] = 1$ M,

$$[S^{2-}] = 1 \times 10^{-22} \text{ M}$$

For ZnS, the ion product is

$$[Zn^{2+}][S^{2-}] = (0.02)(1 \times 10^{-22}) \quad \text{or} \quad 2 \times 10^{-24}$$

For CuS, the ion product is

$$[Cu^{2+}][S^{2-}] = (0.02)(1 \times 10^{-22}) \quad \text{or} \quad 2 \times 10^{-24}$$

Since the ion product of ZnS does not exceed 1×10^{-22}, the K_{sp} of ZnS, ZnS does not precipitate. Since the ion product of CuS does exceed 8×10^{-37}, the K_{sp} of CuS, CuS does precipitate.

QUESTIONS

12.1 *Approach to Equilibrium* Consider the $COCl_2$ equilibrium described in Figure 12.1. (*a*) If the forward reaction depends simply on collisions between CO and Cl_2 molecules, how does the rate of the forward reaction change from the start of reaction until equilibrium is reached? (*b*) Discuss the reverse reaction rate in similar fashion, assuming it is first order. (*c*) What is special about the equilibrium state for these two reactions?

12.2 *Mass Action* Write the mass-action expressions for each of the following cases:
(*a*) $2NO_2(g) \rightleftharpoons N_2O_4(g)$
(*b*) $4NH_3(g) + 5O_2(g) \rightleftharpoons 2NO(g) + 6H_2O(g)$
(*c*) $S(s) + O_2(g) \rightleftharpoons SO_2(g)$
(*d*) $Fe_2O_3(g) + 3CO(g) \rightleftharpoons 2Fe(s) + 3CO_2(g)$

12.3* *Equilibrium Constant* Consider the equilibrium

$$A(g) + 2B(g) \rightleftharpoons 3C(g)$$

If at equilibrium the concentrations of A, B, and C are 0.0100, 0.0200, and 0.0300, what is the value of K?

12.4 *Equilibrium Calculations* For the gaseous equilibrium $CO + Cl_2 \rightleftharpoons COCl_2$, $K = 3.1$. What concentration of Cl_2 is in equilibrium with 0.10 M $COCl_2$ and 0.20 M CO?

12.5 *Equilibrium Changes* The reaction of CO and Cl_2 to form $COCl_2$ is exothermic. Describe how each of the following changes made on the equilibrium of the three gases will change the amount of $COCl_2$ present at equilibrium: (*a*) add CO, (*b*) remove Cl_2, (*c*) increase volume, (*d*) raise temperature, (*e*) add catalyst.

12.6 *pH* The pH's of three very different soil samples are found to be 4.0, 5.0, and 6.0. (*a*) What is the hydrogen-ion concentration of each? (*b*) If all three were desired to be 6.0, how many times as much lime per acre must be added to the first as to the second?

12.7* *Water Equilibrium* Samples of human gastric fluid range from pH 1.0 to 3.0. For each extreme calculate the OH^- concentration present.

12.8 *Weak Acid* For HCN the dissociation constant is 4.0×10^{-10}. Calculate the H^+ concentration in (*a*) 1.0 M HCN, (*b*) 0.040 M HCN.

12.9 *Buffer* (*a*) By consulting Figure 12.6, describe how you would prepare a buffer to maintain a solution at a definitely basic pH by stating what chemicals you would add to water. (*b*) To make this buffer as effective as possible you would want high and equal concentrations of the components. Explain why.

12.10 *Hydrolysis* By means of net equations show whether solutions of each of the following should be acid, base, or neutral: (a) NaCN, (b) NaCl, (c) $CuCl_2$.

12.11* *Solubility* The ionic solid $Sr^{2+}CrO_4{}^{2-}$ has a solubility product of 3.6×10^{-5}. (a) How many moles of $SrCrO_4$ should dissolve in a liter of water? (b) How many moles $SrCrO_4$ should dissolve in a liter of 0.50 M Na_2CrO_4?

12.12 *Solubility* The solubility product of MgF_2 is 8×10^{-8}. How many moles of MgF_2 should dissolve in (a) a liter of water, (b) a liter of 0.10 M BaF_2?

12.13* *Equilibrium Constant* Consider the reaction between two A molecules to form $A_2(g)$. For an initial concentration of A(g) equal to 0.10 M, and an equilibrium concentration of 0.020 M, calculate the numerical value of K.

12.14 *Solubility* A solution containing Cl^-, Br^-, and I^- all at 1.0×10^{-6} M is mixed with an equal volume of solution containing 1.0×10^{-6} M Ag^+. K_{sp} values for AgCl, AgBr, and AgI are 1.7×10^{-10}, 5.0×10^{-13}, and 8.5×10^{-17} What salt or salts should precipitate?

12.15 *Solubility* FeS is quite soluble in 1 M acid but will not dissolve in very slightly acid solution. Why is this?

12.16 *Gas Equilibrium* In Figure 12.2, all experiments are at the same temperature. (a) Compare experiments II and III by calculating how many times the total pressure of III is the total pressure of II. (b) Show that the relative conversion to $COCl_2$ in experiments II and III is consistent with the Le Chatelier principle.

12.17 *Mass Action* Consider the equilibrium between N_2, H_2, and NH_3 gases. If you wanted to increase the amount of NH_3 present, you could double the concentration of either N_2 or H_2. You find however that doubling H_2 has a much larger effect. Why is this?

12.18 *Equilibrium Changes* The equilibrium between N_2 and H_2 to form NH_3 is exothermic. (a) Will you get more or less conversion to NH_3 at equilibrium when the temperature is raised? (b) In the last chapter we saw that reaction rate increases as temperature increases. This applies to both forward and reverse reactions. Show how this can be true without affecting your answer to (a). (c) What must be different about the influence of temperature and a catalyst on forward and reverse rates?

12.19* *Equilibrium Constant* For the gaseous equilibrium between NO and O_2 to form NO_2, $K = 6.45 \times 10^5$. (a) At what O_2 concentration is the NO_2 concentration equal to the NO concentration? (b) At what O_2 concentration is the NO_2 concentration 100 times the NO concentration?

12.20 *Gaseous Equilibrium* Consider the equilibrium of A_2 gas decomposing to give two A molecules, also gaseous. (a) If at STP, only 0.01 percent of the A_2 is decomposed, what is K? (*Note:* Convert concentrations to moles per liter.) (b) What concentration of A is in equilibrium with 0.100 M A_2 at 0°C?

12.21* *pH* Calculate the pH of each of the following solutions: (a) 0.10 M HCl, (b) 0.010 M NaOH, (c) 100 ml of solution made by mixing 10 ml of (a) with 90 ml of (b).

12.22 *Hydrolysis* Should you expect M^{2+} or M^{4+} to hydrolyze to a greater extent? Justify your answer.

12.23* *Buffer Solution* Calculate the $[H^+]$ of a solution made by adding 3.0×10^{-2} mole of HCl to 100 ml of a buffer solution containing 1.0 M HF and 1.0 M NaF.

12.24 *pH* Calculate the pH of a solution that is 0.18 M HOBr. The K_{diss} for HOBr is 2.1×10^{-9}.

12.25* K_{diss} The pH of 0.25 M HCN solution is 5.0. Calculate K_{diss}.

12.26 K_{diss} The hydrogen-ion concentration of 1.5 M HNO_2 is 0.026 M. What is K_{diss} for HNO_2?

12.27 *pH* Given the equilibrium constants:
(a) $H_2SO_3 \rightleftharpoons H^+ + HSO_3^-$ $K = 1.3 \times 10^{-2}$
(b) $ZnOH^+ \rightleftharpoons Zn^{2+} + OH^-$ $K = 4 \times 10^{-5}$
(c) $HCN \rightleftharpoons H^+ + CN^-$ $K = 4.0 \times 10^{-10}$
(d) $NH_3 + H_2O \rightleftharpoons NH_4^+ + OH^-$ $K = 1.8 \times 10^{-5}$
(e) $H_2Se \rightleftharpoons H^+ + HSe^-$ $K = 1.9 \times 10^{-4}$

Arrange in order of increasing pH equimolar solutions of each species at the extreme left of each of the above equilibria.

12.28 *Percent Dissociation* Percent dissociation of 0.100 M HF solution is 7.86 percent. What is the equilibrium constant for HF $\rightleftharpoons H^+ + F^-$?

12.29* *Percent Dissociation* How does the percent dissociation change when 0.50 M $HC_2H_3O_2$ solution is diluted twentyfold?

12.30 *Change of Pressure* Given the equilibrium

$$H_2(g) + I_2(g) \rightleftharpoons 2HI(g) + 12.5 \text{ kilojoules}$$

What would be the effect on the equilibrium concentrations of a decrease of volume? Explain in terms of K, the principle of Le Chatelier, and kinetics.

12.31* *pH Titrations* Calculate the pH of the following solutions assuming volumes are additive: (a) 50.0 ml of 0.0100 M HCl and 10.0 ml of 0.200 M KOH,

(b) 120.0 ml of 0.500 M HCl and 20.0 ml of 0.200 M NaOH, (c) 30.0 ml of 0.250 M HCl and 15.0 ml of 0.0500 M NaOH.

12.32 *Complex Ions* Given a solution in which the equilibrium $Ag(CN)_2^- \rightleftharpoons Ag^+ + 2CN^-$ is established. How would the equilibrium concentration of $Ag(CN)_2^-$ be affected by addition of each of the following: (a) NaCN, (b) HNO_3, (c) H_2O, (d) NaCl to precipitate AgCl.

12.33 *Precipitation* The K_{sp} of $La(OH)_3$ is 1.0×10^{-19} and that of $BaSO_4$ is 1.5×10^{-9}. Show whether a precipitate would be expected on mixing 35.0 ml of 9.0×10^{-5} M $Ba(OH)_2$ with 75.0 ml of 2.5×10^{-5} M $La_2(SO_4)_3$.

12.34 *pH* Given a solution that is 0.60 M HNO_2. To what volume must 1 liter of this solution be diluted in order to (a) double the hydroxy-ion concentration, (b) double the pH.

12.35* *Hydrolysis* Given that $K = 2.5 \times 10^{-7}$ for $HSeO_3^- \rightleftharpoons H^+ + SeO_3^{2-}$, calculate the pH of 0.25 M K_2SeO_3.

12.36 *Hydrolysis* Given $K = 1.2 \times 10^{-2}$ for $HSO_4^- \rightleftharpoons H^+ + SO_4^{2-}$ and $K = 3 \times 10^{-13}$ for $HAsO_4^{2-} \rightleftharpoons + AsO_4^{3-}$. What are the hydrolysis constants for Na_2SO_4 and Na_3AsO_4?

12.37 *Mass Action* The mass-action expression for a reaction is given by

$$\frac{[Mn^{2+}][Fe^{3+}]^5}{[Fe^{2+}]^5[H^+]^8[MnO_4^-]}$$

(a) Write the balanced net equation for the reaction. (b) Explain the reason for any omissions from the mass-action expression.

12.38 *Amount vs. Concentration* (a) Students sometimes make a mistake in not distinguishing between moles and moles per liter. In spite of this mistake, they frequently obtain correct results. In which of the following gaseous equilibria will they get a right answer? Explain.
(a) $2NO + O_2 \longrightarrow 2NO_2$
(b) $N_2 + 3H_2 \longrightarrow 2NH_3$
(c) $2HI \longrightarrow H_2 + I_2$

When would K be unaffected by using pressure instead of molarity? Explain.

12.39 *Heterogeneous Equilibrium* Suppose the following species occur in equilibrium systems: $BaO(s)$, $Li(g)$, $H_2O(l)$, $HC_2H_3O_2(s)$, $I_2(s)$, $Mn^{+2}(s)$. Which can usually be omitted from the mass-action expression? Are there any exceptions? Explain.

12.40 *Heterogeneous Equilibrium* (a) Write the mass-action expression for the following reaction: $2HI(l) \longrightarrow I_2(g) + H_2(g)$. (b) Explain what happens if more HI(l) is added.

12.41* *Solubility*   The solubility product of $CaF_2(s) \rightleftharpoons Ca^{2+} + 2F^-$ is 1.7×10^{-10}. (a) How many moles of CaF_2 will dissolve completely in 2.0 liters of water? (Ignore hydrolysis.) (b) How many moles of CaF_2 should dissolve in 2.0 liters of 0.0050 M KF solution? (Ignore hydrolysis.)

12.42   *Equilibrium*   The equilibrium constant for $H_2(g) + I_2(g) \rightleftharpoons 2HI(g)$ is 44.5 at 1000°C. At $t = 0$, 4.00 moles of H_2, 2.00 moles of I_2, and 1.00 mole of HI are injected into a 2.00-liter box. What will the concentration of each species be at equilibrium at 1000°C?

12.43   *Precipitation*   100 ml of a solution containing 7.00×10^{-4} g of barium sulfate, $BaSO_4$, is mixed with 50 ml of solution that is 0.001 M Na_2SO_4. Should $BaSO_4$ precipitate?

12.44   *pH*   (a) What concentration of HF ($K = 6.7 \times 10^{-4}$) is needed to give a solution of pH = 2.0? (b) What concentration of NaF is needed to give a solution of pH = 8.0?

12.45* *Buffer*   (a) Tell what ratio of concentrations of $HC_2H_3O_2$ ($K = 1.8 \times 10^{-5}$) and of $C_2H_3O_2^-$ you would use to give a solution buffered at pH = 5.0. (b) Why isn't it practical to use these components to buffer at pH = 7.0?

12.46   *Weak Acid*   What is the H^+ concentration in a solution made by adding 0.027 g of HCN ($K = 4.0 \times 10^{-10}$) to 100 ml of water?

12.47* *Sulfuric Acid*   As noted in Section 12.7, sulfuric acid is a strong acid with K_I large and $K_{II} = 1.3 \times 10^{-2}$. Calculate the concentration of all three ions H^+, HSO_4^-, SO_4^{2-} in 1.0 M H_2SO_4 as follows: (a) Assuming the first step of dissociation is complete and the second is negligible, calculate the concentrations of H^+ and HSO_4^-. (b) Substitute your values from (a) into the mass-action expression for the second step and calculate SO_4^{2-}. (c) Check the assumption made in saying that all H^+ came from the first step by noting that the second step actually produces as much H^+ as SO_4^{2-}. (d) If the H_2SO_4 were much more dilute than 1.0 M, could you proceed this way?

12.48   *Percent Dissociation*   Show that the percent dissociation of a weak electrolyte increases on dilution by calculating the percent dissociation of 1.00 M $HC_2H_3O_2$ ($K = 1.8 \times 10^{-5}$) and 0.10 M $HC_2H_3O_2$.

12.49* *Solubility*   For NiS, K_{sp} is 3×10^{-21}. What is the minimum concentration of acid in a saturated H_2S solution which will permit a concentration of Ni^{2+} equal to 0.1 M without precipitation of NiS?

12.50   *Buffer*   (a) What is the H^+ concentration in a buffer which contains 0.5 M HF ($K = 6.1 \times 10^{-4}$) and 0.5 M NaF? (b) What is the H^+ concentration if 0.10 mole of HCl is added to a liter of the buffer in (a)? (c) What is the H^+ concentration if 0.10 mole of NaOH is added to 500 ml of the solution from (b)?

12.51* *Hydrolysis* K_{diss} for $Zn(OH)^+$ is 4×10^{-5} and that for $Cu(OH)^+$ is 1×10^{-8}. (*a*) Calculate hydrolysis constants for Zn^{2+} and Cu^{2+}. (*b*) Calculate the H^+ concentration and percent hydrolysis for a $0.1\ M\ Zn^{2+}$ solution and for another solution that contains $0.1\ M\ Cu^{2+}$.

12.52 *Limestone Caves* Water in contact with the atmosphere is slightly acidic due to dissolved CO_2 from the air which establishes the equilibrium $CO_2 + H_2O \rightleftharpoons H^+ + HCO_3^-$, for which $K = 4.2 \times 10^{-7}$. Limestone is the insoluble salt $CaCO_3$, for which $K_{sp} = 4.7 \times 10^{-9}$. The solubility of limestone is increased by acid solution since the acid ties up CO_3^{2-} by the equilibrium $HCO_3^- \rightleftharpoons H^+ + CO_3^{2-}$, for which $K = 4.8 \times 10^{-11}$. The net reaction for dissolving limestone in water made acid by dissolved CO_2 then becomes $CO_2 + CaCO_3(s) \rightleftharpoons Ca^{2+} + 2HCO_3^-$. (*a*) Calculate K for the combined reaction. (*b*) Assuming that water in contact with the atmosphere maintains a constant concentration of dissolved CO_2 at $1.0 \times 10^{-5}\ M$, calculate the number of liters of water necessary to dissolve 1.0 kg of limestone in cave formation.

Electro-
chemistry

13

Studies of the relation of electric energy and chemical change have proved fascinating and fruitful and they have been important in the development of modern ideas of the fundamental nature of atoms. For the future, such studies show promise of furthering our understanding of biological processes such as nerve conduction, brain mechanisms, and energy transport. Meanwhile, research in electrochemistry is of continuing practical utility in the direct interconversion of chemical energy and electric energy. In this chapter we will explore the relation by considering first the conduction of electricity through matter, then the conversion of electric energy into chemical energy by the process of electrolysis, and finally the reverse conversion of chemical energy into electric energy in devices such as simple cells, batteries, and fuel cells. All these topics belong to the field of electrochemistry.

13.1 Electric Conductivity

Electric energy may be transported through matter by the conduction of electric charge from one point to another in the form of an *electric current* (see Appendix 3.7 for a discussion of electrical terms). In order that the electric current flow, there must be charge carriers in the matter, and there must be a force that makes the carriers move. The charge carriers can be electrons, as in the case of metals, or they can be positive and negative ions,

281

as in the case of electrolytic solutions and molten salts. In the former case, conduction is said to be *metallic*; in the latter, *electrolytic*. The electric force that makes charges move is usually supplied by a battery or some similar source of electric energy. Any region of space in which there is an electric force is called an *electric field*.

As pointed out in Section 7.5, solid metals consist of ordered arrays of positive ions immersed in a sea of electrons. For example, silver consists of Ag^+ ions arranged in a face-centered cubic pattern with the entire structure permeated by a cloud of electrons equal in number to the number of Ag^+ ions in the crystal. The Ag^+ ions vibrate about fixed positions from which they do not move very far except under great stress. The electrons of the cloud, on the contrary, are free to roam throughout the entire crystal. When an electric field is impressed on the metal, the electrons migrate and thereby carry negative electric charge through the metal. In principle, it should be possible for an electric field to force all the loose electrons toward one end of a metal sample. In practice, it is extraordinarily difficult to separate positive and negative charges from each other without the expenditure of relatively enormous amounts of energy. The only way it is possible to keep a sustained flow of charge in a wire is to add electrons to one end of the wire and drain off electrons from the other end as fast as they accumulate. The metal conductor thus remains everywhere electrically neutral, since just as many electrons move into a region per unit time as move out.

Most of the electrons that make up the electron cloud in a metal are of very high kinetic energy. Metallic conductivity would therefore be extremely high were it not for a *resistance* effect. Electric resistance is believed to arise because lattice ions vibrate about their lattice points. By interfering with the migration of electrons, the ions keep down the conductivity. At higher temperatures, the thermal vibrations of the lattice increase, and therefore it is not surprising to find that, as the temperature of a metal is raised, its conductivity diminishes.

In solutions, the mechanism of conductivity is complicated by the fact that the positive ions are not fixed in position but are free to roam throughout the body of the solution. When an electric field is applied to such a solution, as shown in Figure 13.1, the positive ions experience a force in one direction, and the negative ions experience a force in the opposite direction. The simultaneous motion of positive and negative ions in opposite directions constitutes the *electrolytic current*. The current would stop if positive ions

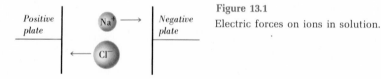

Figure 13.1

Electric forces on ions in solution.

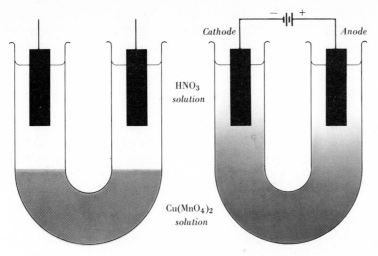

HNO$_3$
solution

Cathode   Anode

Cu(MnO$_4$)$_2$
solution

Figure 13.2
Experiment to show migration of
ions in electrolytic conductivity.

accumulated at the negative electrode and negative ions at the positive
electrode. In order that the electrolytic current continue, appropriate chemical reactions must occur at the electrodes to maintain electric neutrality.

That ions migrate when electrolytic solutions conduct electricity can
be seen from the experiment diagramed in Figure 13.2. The U tube is initially
half-filled with a deep purple aqueous solution of copper permanganate,
Cu(MnO$_4$)$_2$. The solution contains blue hydrated Cu^{2+} ions and purple
MnO$_4^-$ ions. A colorless aqueous solution of nitric acid, HNO$_3$, is floated
on top of the Cu(MnO$_4$)$_2$ solution. An electric field is maintained across the
solution by the two electrodes. After some time, it is observed that the blue
color characteristic of hydrated Cu^{2+} ions has moved into the region of the
cathode, suggesting a migration of Cu^{2+} toward the negative electrode. At
the same time, the purple color characteristic of MnO$_4^-$ has moved into the
region of the anode, indicating that negative ions move simultaneously
toward the positive electrode.

As in the case of metallic conduction, electric neutrality must be preserved in all regions of the solution at all times. Otherwise, the current would
soon cease. Figure 13.3 shows two of the possible ways by which electric

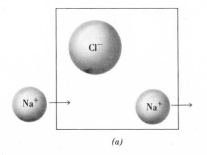

Cl$^-$

Na$^+$

Na$^+$

(a)

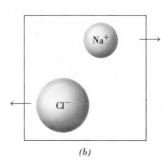

Na$^+$

Cl$^-$

(b)

Figure 13.3
Two ways in which migrating
ions could maintain electric
neutrality in a region of solution.

neutrality can be preserved for a given region of an NaCl solution. In (a), one Na⁺ ion enters the region defined by the box outline to compensate for the charge of the departing Na⁺ ion. In (b), as one Na⁺ ion leaves the region, one Cl⁻ ion departs in the opposite direction; hence the region shows no net change in charge. *Both* these effects occur simultaneously, their relative importance depending on the relative mobilities of the positive and negative ions.

In contrast to metallic conduction, electrolytic conduction generally increases when the temperature of the solution is raised.† This comes about because of two effects: (1) The average kinetic energy of the ions in the solution is raised by an increase of temperature, so that the ions, on an average, move faster. (2) The viscosity of the solvent is diminished by a rise in temperature. The second effect is the more important and is due to thermally increased motion of the solvent molecules to create spaces into which ions can move as they migrate from one electrode to the other.

13.2   Electrolysis

In order to maintain an electric current, it is necessary to have a complete circuit; i.e., there must be a closed loop whereby the electric charge can return to its starting point. If the complete circuit includes, as one component, an electrolytic conductor, chemical reaction must occur at the electrodes that dip into the electrolytic conductor. Electric energy is thus used to produce chemical change, and the process is called *electrolysis*.

A typical electrolysis circuit is shown in Figure 13.4. The vertical lines at the top of the diagram represent a battery in which the longer line is

†There are exceptions to this generalization. For example, with some weak electrolytes, the percentage dissociation may decrease with rising temperature. The resulting decrease in ion concentration may be big enough to cause a *decrease* in conductivity.

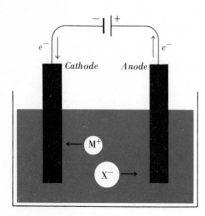

Figure 13.4
Electrolysis circuit.

the positive terminal, and the shorter line the negative one. The other lines are strips of wire, usually copper, that join the battery to the electrodes. The electrodes dip into the electrolytic conductor, containing the ions M^+ and X^- that are free to move. The battery creates an electric field which pushes the electrons in the wires in the directions shown by the arrows. Electrons are crowded onto the left-hand electrode and drained away from the right-hand electrode. The circuit is not complete unless there is some way by which electrons can be consumed at the left electrode and generated at the right electrode. Chemical changes must occur. At the left electrode, a *reduction* process must occur in which some ion or molecule accepts electrons and is thereby reduced. At the right-hand electrode, electrons must be released to the electrode in an *oxidation* process. The electrode at which reduction occurs is called a *cathode;* the electrode at which oxidation occurs is called an *anode.* In order for the reduction process to continue at the cathode, ions must keep moving toward it. These ions are the positive ions and are called *cations.* Simultaneously, negative ions move to the anode and are called *anions.*

Variations in the possible reactions for electrolysis are illustrated by the following examples.

13.3   Electrolysis of Molten NaCl

Molten NaCl contains Na^+ and Cl^- ions which are free to migrate. Figure 13.5 shows a schematic diagram of the electrolysis cell. Inert electrodes of carbon or platinum dip into the molten NaCl. As diagramed, a reduction process must occur at the left-hand electrode, which therefore is the cathode. Of the two ions, Na^+ and Cl^-, only Na^+ can be reduced. On electrolysis, Na^+ is reduced and forms metallic Na. The *cathode reaction* can be written

$$Na^+ + e^- \longrightarrow Na$$

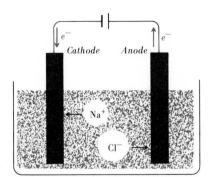

Figure 13.5

Electrolysis of molten NaCl.

indicating that at the cathode one Na^+ ion picks up one electron to form a neutral Na atom. During the change, mass and charge are conserved, and so the cathode reaction is in a sense a chemical reaction expressible by a balanced equation. Since the equation shows only a reduction process, it is referred to as a *half-reaction*.

At the anode, oxidation occurs. Of the two species in the cell, Na^+ and Cl^-, only the Cl^- can be oxidized. When oxidized, Cl^- releases an electron to the anode, and a neutral chlorine atom forms. Two chlorine atoms combine to produce a diatomic chlorine molecule, Cl_2. These Cl_2 molecules bubble off as a gas. The net *anode half-reaction* can be written

$$2Cl^- \longrightarrow Cl_2(g) + 2e^-$$

At the cathode, electric energy has been used to convert Na^+ into Na metal; at the anode, to convert Cl^- into Cl_2. By addition, the two electrode half-reactions can be combined into a single overall *cell reaction*. In order to keep electrons from piling up in the cell, as many must disappear at the cathode as appear at the anode. To ensure electron balance, the half-reactions are multiplied by appropriate coefficients so that, when the half-reactions are added, the electrons cancel out of the final equation. Thus, for the electrolysis of molten NaCl,

Cathode reaction: $\quad 2Na^+ + 2e^- \longrightarrow 2Na$

Anode reaction: $\qquad\quad 2Cl^- \longrightarrow Cl_2(g) + 2e^-$

Overall reaction: $\quad 2Na^+ + 2Cl^- \xrightarrow{\text{electrolysis}} 2Na + Cl_2(g)$

In order to emphasize that this reaction occurs by the consumption of electric energy, the word *electrolysis* is often written under the arrow.

13.4 Electrolysis of Aqueous NaCl

When aqueous NaCl solution is electrolyzed under appropriate conditions, it is observed that hydrogen gas is liberated at the cathode and chlorine gas is liberated at the anode. How can these observations be accounted for in terms of electrode reactions? Figure 13.6 shows the electrolysis cell, which now contains, besides Na^+ and Cl^- ions, H_2O molecules and traces of H^+ and OH^- from the dissociation of water. Molecules of H_2O can be either oxidized to O_2 and H^+ by removal of electrons or reduced to H_2 and OH^- by the addition of electrons. The H_2O must thus be considered as a possible reactant at each electrode.

At the cathode, reduction must occur. Three reactions are possible:

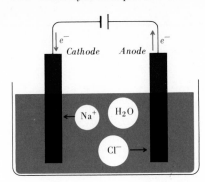

Figure 13.6
Electrolysis of aqueous NaCl.

$$Na^+ + e^- \longrightarrow Na(s) \tag{1}$$
$$2H_2O + 2e^- \longrightarrow H_2(g) + 2OH^- \tag{2}$$
$$2H^+ + 2e^- \longrightarrow H_2(g) \tag{3}$$

It is not easy to predict which of several possible reactions will occur at a cathode. It is necessary to consider which reactant is reduced most *easily* and which reactant is reduced most *rapidly*. The strongest oxidizing agent is not necessarily the fastest. Further complications appear when currents are very large and when concentrations of reactants are very small. The fact that hydrogen gas and not metallic sodium is formed in the electrolysis of aqueous NaCl indicates that reaction (2) or (3) occurs.† In NaCl solution, the concentration of H^+ is not large enough to make reaction (3) reasonable as a *net change*. Therefore, in the electrolysis of aqueous NaCl, reaction (2) is the one usually written for the cathode reaction.

In the electrolysis, OH^- forms in the region around the cathode, and positive ions (Na^+) move into this region to preserve electric neutrality. In addition, OH^- migrates away from the cathode. Both migrations are consistent with the requirement that cations migrate toward the cathode and anions toward the anode.

At the anode, oxidation must occur. Two reactions are possible:

$$2Cl^- \longrightarrow Cl_2(g) + 2e^- \tag{4}$$
$$2H_2O \longrightarrow O_2(g) + 4H^- + 4e^- \tag{5}$$

The observed fact that chlorine is evolved indicates that reaction (4) is the one which occurs. As the chloride-ion concentration around the anode is depleted, fresh Cl^- moves into the region and Na^+ moves out.

In summary, the equations for the electrolysis of aqueous NaCl are

† Years ago it was thought that the metal Na was first formed by reaction (1) and then subsequently reacted with water to liberate H_2. However, there is no evidence that any intermediate Na is ever formed in this electrolysis.

Cathode reaction: $\quad 2e^- + 2H_2O \longrightarrow H_2(g) + 2OH^-$

Anode reaction: $\qquad\qquad 2Cl^- \longrightarrow Cl_2(g) + 2e^-$

Overall reaction: $\quad 2Cl^- + 2H_2O \xrightarrow{\text{electrolysis}} H_2(g) + Cl_2(g) + 2OH^-$

As expressed by the overall reaction, during the electrolysis H_2 gas and Cl_2 gas are formed, the concentration of Cl^- diminishes, and the concentration of OH^- increases. Since there is always Na^+ in the solution, the solution is very gradually converted from aqueous NaCl to aqueous NaOH. In fact, in the commercial production of chlorine by the electrolysis of aqueous NaCl, solid NaOH is obtained as a by-product by evaporating H_2O from the residual solution left after electrolysis.

13.5 Electrolysis of Aqueous Na_2SO_4

When aqueous Na_2SO_4 is electrolyzed, it is observed that H_2 gas is formed at the cathode, and O_2 is formed at the anode. The changes at the electrodes can be demonstrated by running the electrolysis in the two-compartment cell shown in Figure 13.7. A few drops of litmus, initially added to the solution, take on a violet coloration, since the solution contains Na^+, SO_4^{2-}, and H_2O before electrolysis and is essentially neutral. After electrolysis has proceeded for a while, the litmus in the cathode compartment becomes blue, indicating the solution in that region to be basic; the litmus in the anode compartment becomes red, indicating the solution there to be acidic. Consistent with these observations are the following electrode reactions:

Cathode: $\quad 2e^- + 2H_2O \longrightarrow H_2(g) + 2OH^-$

Anode: $\qquad\quad 2H_2O \longrightarrow O_2(g) + 4H^+ + 4e^-$

The OH^- from the cathode reaction turns litmus blue; the H^+ from the anode reaction turns litmus red. The overall cell reaction is obtained by doubling the cathode reaction and adding it to the anode reaction. The four electrons

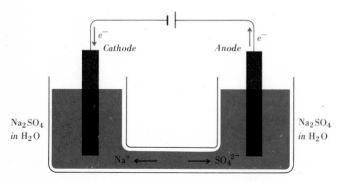

Figure 13.7
Two-compartment electrolysis cell.

cancel, and the result is

$$6H_2O \xrightarrow[\text{electrolysis}]{} 2H_2(g) + O_2(g) + 4H^+ + 4OH^-$$

In this equation, both H^+ and OH^- appear as products. The only reason neutralization does not occur is that the H^+ is formed in the anode compartment and OH^- in the cathode compartment. If the solution is now poured from the cell into a beaker in order that full mixing may take place, neutralization occurs and the litmus is restored to its original violet color. Addition of the neutralization reaction

$$4H^+ + 4OH^- \longrightarrow 4H_2O$$

to the above overall cell reaction gives for the net reaction

$$2H_2O \xrightarrow[\text{electrolysis}]{} 2H_2(g) + O_2(g)$$

In this electrolysis, only water disappears. The Na^+ and SO_4^{2-} initially present are also present at the conclusion of the electrolysis. Is the Na_2SO_4 necessary? Because of the requirements of electric neutrality, some kind of electrolytic solute must be present. Positive ions must be available to move into the cathode region to counterbalance the charge of the OH^- produced by the cathode reaction. Negative ions must be available to move to the anode to counterbalance the H^+ produced by the anode reaction.

Almost any ionic solute makes possible the electrolysis of water as described by the above equation. The only requirement is that the ions of the solute not be oxidized or reduced, as would happen, for example, when aqueous $CuSO_4$ is electrolyzed. Cu^{2+} is more easily and rapidly reduced than H_2O, and so during electrolysis copper plating would form on the cathode.

In some cases, the electrodes themselves may take part in the electrode reactions. In each of the above cells, the electrodes were assumed to be inert. This would almost always be the case if the electrodes were made of the inert metal platinum. If, however, the electrode material is reactive, it must be considered as a possible reactant. For example, copper anodes are frequently themselves oxidized during electrolysis when no other species present is more readily oxidized.

13.6 Quantitative Aspects of Electrolysis

By experimentation, Michael Faraday, the great English chemist and physicist, established early in the nineteenth century the laws of electrolysis that bear his name. The Faraday laws state that the weight of product formed at an electrode is proportional (1) to the amount of electricity transferred at the electrode and (2) to the equivalent weight of the substance produced. The laws can be accounted for by considering the electrode reactions. For example, in the electrolysis of molten NaCl, the cathode reaction

$$Na^+ + e^- \longrightarrow Na$$

tells us that one sodium atom is produced at the electrode when one sodium ion disappears and one electron is transferred. When the Avogadro number of electrons is transferred, 1 mole of Na^+ disappears and 1 mole of Na is formed. For this reaction, 1 equiv of Na is 23.0 g; hence transfer of the Avogadro number of electrons liberates 23.0 g of Na. Doubling the amount of electricity transferred doubles the weight of sodium produced.

The Avogadro number of electrons is such a convenient measure of the amount of electricity that it is designated by a special name, the *faraday*. In electrical units 1 faraday is equal to 96,500 coulombs of charge. As described in Appendix 3.5, a *coulomb* of charge is the amount of electricity that is transferred when a current of one ampere flows for one second. Thus the current in amperes, I, multiplied by the time in seconds, t, is equal to the number of coulombs, Q.

$$Q \text{ (coulombs)} = I \text{ (amperes)} \times t \text{ (s)}$$

The electric charge in coulombs divided by 96,500 is equal to the number of faradays.

Electrode half-reactions expressed in ions, electrons, and atoms can be read in terms of moles if the electricity is expressed in faradays. Thus,

$$Na^+ + e^- \longrightarrow Na$$

can be read either "One sodium ion reacts with one electron to form one sodium atom" or "1 *mole* of sodium ions reacts with 1 faraday of electricity to form 1 *mole* of sodium atoms."

EXAMPLE 1　How many grams of chlorine can be produced by the electrolysis of molten NaCl at a current of 10.0 amperes for 5.00 min?

We have

$$(10.0 \text{ amperes})(5.00 \text{ min})(60 \text{ s/min}) = 3000 \text{ coulombs}$$

$$\frac{3000 \text{ coulombs}}{96,500 \text{ coulombs/faraday}} = 0.0311 \text{ faraday}$$

The half-reaction is

$$2Cl^- \longrightarrow Cl_2(g) + 2e^-$$

Keeping in mind that the formation of 1 mole of Cl_2 requires 2 faradays, 0.0311 faraday produces

$$0.0311 \text{ faraday} \times \frac{1 \text{ mole } Cl_2}{2 \text{ faradays}} = 0.0156 \text{ mole of } Cl_2$$

0.0156 mole of Cl_2 weighs

$$0.0156 \text{ mole} \times \frac{70.9 \text{ g}}{1 \text{ mole } Cl_2} = 1.11 \text{ g}$$

13.7   Cells and Batteries

In the cells discussed above, electric energy in the form of a current was used to bring about oxidation-reduction reactions. It is also possible to do the reverse, i.e., use an oxidation-reduction reaction to produce electric current and thus convert chemical energy to electric energy. The main requirement is that the oxidizing and reducing agents be separated from each other so that electron transfer must occur through a wire. Any device which accomplishes this is called a *galvanic,* or *voltaic,* cell, after Luigi Galvani (1780) and Alessandro Volta (1800) who made the basic discoveries. A *battery* is a collection of two or more cells generally connected in series so that the anode of one is joined to the cathode of the preceding cell.

When a bar of zinc is dipped into a solution of copper sulfate, $CuSO_4$, copper plating is obtained. The net reaction is

$$Zn(s) + Cu^{2+} \longrightarrow Zn^{2+} + Cu(s)$$

In this change, Zn is oxidized and Cu^{2+} is reduced by the transfer of electrons from zinc atoms to copper ions. To emphasize this transfer of electrons, the net reaction can be split into two half-reactions:

$$Zn(s) \longrightarrow Zn^{2+} + 2e^-$$
$$Cu^{2+} + 2e^- \longrightarrow Cu(s)$$

The galvanic cell operates on the principle that two separated half-reactions can be made to take place simultaneously, with the electron transfer occurring through a wire. The typical galvanic cell shown in Figure 13.8 uses the reaction

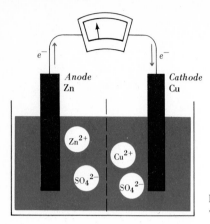

Figure 13.8

Typical galvanic cell.

$$Zn(s) + Cu^{2+} \longrightarrow Zn^{2+} + Cu(s)$$

A galvanic cell that uses this specific reaction is called a *Daniell cell*. The broken line represents a porous partition which separates the container into two compartments but still permits diffusion of ions between them. In the left-hand compartment is a solution of zinc sulfate, into which a zinc bar is dipped; in the right-hand compartment is a copper bar dipping into a solution of copper sulfate. When the two electrodes are connected by a wire, electric current flows, as shown by a meter in the circuit. In time, the zinc bar is eaten away, and copper is plated out on the copper bar.

The cell operates as follows: At the zinc bar, oxidation occurs, making Zn the anode. The half-reaction

$$Zn(s) \longrightarrow Zn^{2+} + 2e^-$$

produces Zn^{2+} ions and electrons. The zinc ions migrate away from the anode into the solution, and the electrons move through the wire, as indicated in the figure. At the copper bar, reduction occurs, making Cu the cathode. The electrons come through the wire and move onto the cathode, where they are picked up and used in the reaction

$$Cu^{2+} + 2e^- \longrightarrow Cu(s)$$

Copper ions in the solution are depleted, and new copper ions move into the vicinity of the cathode. The circuit is complete. Consistent with previous notation, cations (Zn^{2+} and Cu^{2+}) in the solution move toward the cathode (the copper bar), and anions (SO_4^{2-}) move toward the anode (the zinc bar). Electrons flow through the wire, and a current is obtained from an oxidation-reduction reaction. The cell runs until either the Zn or Cu^{2+} is depleted.

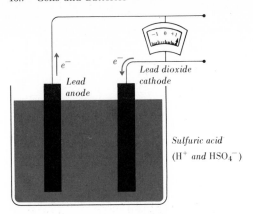

Sulfuric acid
(H^+ and HSO_4^-)

Figure 13.9
Cell of a lead storage battery during discharge.

Actually, to get a current from a Daniell cell, the Zn^{2+} ions and the Cu bar need not be initially present. Any metal support for the plating of Cu will serve in place of the Cu bar. Any positive ion that does not react with Zn metal will serve in place of Zn^{2+}. However, as the cell reaction proceeds, Zn^{2+} is necessarily produced at the anode. Furthermore, the porous partition serves mainly to keep Cu^{2+} from getting over to the Zn metal, where direct electron transfer would short-circuit the cell. The partition must be porous in order to allow diffusion of positive and negative ions from one compartment to the other. Otherwise, the solution would soon become positively charged in the anode compartment (because of accumulation of Zn^{2+}) and negatively charged in the cathode compartment (because of depletion of Cu^{2+}), causing the current to cease.

In principle, any oxidation-reduction reaction is separable into two half-reactions and can be made a source of electric current as a galvanic cell. Probably the most famous example is the *lead storage battery*, or *accumulator*. As shown in Figure 13.9, the basic features are electrodes of lead (Pb) and lead dioxide (PbO_2), dipping into aqueous H_2SO_4. When the cell operates, the reactions are

Anode: $$Pb(s) + HSO_4^- \longrightarrow PbSO_4(s) + 2e^- + H^+$$
Cathode: $$PbO_2(s) + HSO_4^- + 3H^+ + 2e^- \longrightarrow PbSO_4(s) + 2H_2O$$

Overall reaction:
$$Pb(s) + 2HSO_4^- + 2H^+ + PbO_2(s) \longrightarrow 2PbSO_4(s) + 2H_2O$$

The insoluble lead sulfate, $PbSO_4$, that is formed at each electrode adheres to the electrode. During the *charging* of a battery, the electrode reactions are reversed so as to restore the cell to its original condition. In *discharge*, as shown by the overall cell reaction, Pb and PbO_2 are depleted, and the concentration of H_2SO_4 is diminished. Since the density of the aqueous solution is chiefly dependent on the concentration of H_2SO_4, measurement of the density tells how far the cell is discharged.†

Another common galvanic cell is the Leclanché dry cell used in flashlights. The cell usually consists of a zinc can containing a centered graphite rod surrounded by a moist paste of manganese dioxide, MnO_2; zinc chloride, $ZnCl_2$; and ammonium chloride, NH_4Cl. The zinc can is the anode, and the graphite rod is the cathode. At the anode, Zn is oxidized; at the cathode, MnO_2 is reduced. The electrode reactions are extremely complex and vary, depending on how much current is drawn from the cell. For the delivery of very small currents, the following reactions are probable:

Anode:
$$Zn(s) \longrightarrow Zn^{2+} + 2e^-$$

Cathode:
$$2MnO_2(s) + Zn^{2+} + 2e^- \longrightarrow ZnMn_2O_4(s)$$

Overall reaction:
$$Zn(s) + 2MnO_2(s) \longrightarrow ZnMn_2O_4(s)$$

An increasingly important kind of galvanic cell is the *fuel cell,* in which a fuel such as a hydrocarbon or hydrogen is oxidized by oxygen. The hydrogen-oxygen fuel cell was primarily developed for use in the lunar exploration program, but it has promise for more general use in places where there is a requirement for highly efficient conversion of chemical into electric energy. Whereas the burning of a fuel and use of the heat liberated to drive a steam turbine are only about 40 percent efficient, the direct conversion from chemical reaction to electric current can be made up to 75 percent efficient. In the hydrogen-oxygen fuel cell, the overall reaction is

$$2H_2(g) + O_2(g) \longrightarrow 2H_2O$$

for which the half-reactions in basic solution are

Anode:
$$H_2(g) + 2OH^- \longrightarrow 2H_2O + 2e^-$$

Cathode:
$$O_2(g) + 2H_2O + 4e^- \longrightarrow 4OH^-$$

Use of basic solutions instead of acid reduces the problems of corrosion and improves the rate of reaction at the electrodes. Figure 13.10 shows one design which has been used successfully. Two chambers constructed of porous carbon dip into an aqueous solution of KOH. Hydrogen gas is pumped into one chamber where it is oxidized in the anode reaction. Simultaneously oxygen gas is passed into the other chamber where it is reduced in the cathode reaction. Because H_2 and O_2 each react very slowly at room temper-

---

† It would seem natural to say that, when the battery is discharging, the anion HSO_4^- moves to the anode. However, as is obvious from the cathode half-reaction, some of the HSO_4^- must also move to the cathode in order to form $PbSO_4$. Thus we have the unusual but not unique situation in which the anion moves toward both electrodes.

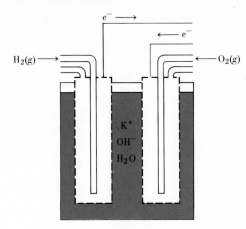

Figure 13.10
Hydrogen-oxygen fuel cell.

ature, catalysts are mixed in and pressed with the carbon. Much research is now going on to find improved catalysts; suitable catalysts at the anode are finely divided platinum or palladium and at the cathode, cobalt oxide (CoO), platinum, or silver. In orbiting spacecraft, special conditions (e.g., zero gravity) necessitated suitable design changes, for example, replacement of the liquid electrolyte solution by an ion-exchange resin (Section 15.4).

Although the hydrogen-oxygen fuel cell is appropriate for specialized uses, a more exciting prospect is the development of a fuel cell in which normal hydrocarbon fuels would be used as reducing agents. One possibility is to use "natural gas" in the reaction

$$CH_4(g) + 2O_2(g) \longrightarrow CO_2(g) + 2H_2O$$

where the half-reactions in acid would be

Anode: $CH_4(g) + 2H_2O \longrightarrow CO_2(g) + 8H^+ + 8e^-$
Cathode: $O_2(g) + 4H^+ + 4e^- \longrightarrow 2H_2O$

or in base, allowing for the fact CO_2 in base exists as CO_3^{2-} ion,

Anode: $CH_4(g) + 10OH^- \longrightarrow CO_3^{2-} + 7H_2O + 8e^-$
Cathode: $O_2(g) + 2H_2O + 4e^- \longrightarrow 4OH^-$

Cells using these reactions have been made to work but difficult technological problems still remain.

Modern society puts a great emphasis on the mobility of people and machines, and so there has been increasing demand for efficient portable power packs such as the above cells. The space-exploration programs, in particular, have given a decided impetus to the search for power sources to substitute for the traditional internal-combustion engine. Recently, it appeared that the decisive argument might be made by the pollution problem. Society finally appears fed up with the acoustic assault on our ears and the pollution of our atmospheric environment by the noxious emissions of poorly regulated internal-combustion engines. Given the reluctance of the citizenry to pay for cleaner and quieter surroundings, it is not surprising that the old-fashioned gasoline engine still thunders and spews away. Nonetheless, in the past few years both government and industry have begun to subsidize the search for suitable electrochemical replacements. Today, given the stark fact that the world's oil resources are decreasing in quantity and rising in price, there is little doubt that this search will continue at an accelerated pace.

13.8 Reduction Potentials

A voltmeter connected between the two electrodes of a galvanic cell shows a characteristic voltage whose magnitude depends on what reactants take part in the electrode reactions and on their concentrations. For example, in the Daniell cell, if Zn^{2+} and Cu^{2+} are at $1\,m$ concentration, and the temperature is $25\,°C$, the voltage measured between the Zn electrode and the Cu electrode is 1.10 volts, no matter how big the cell or how big the electrodes. This voltage is characteristic of the reaction

$$Zn(s) + Cu^{2+} \longrightarrow Zn^{2+} + Cu(s)$$

The voltage measures the force with which electrons would be moved around the circuit and therefore measures the tendency of this reaction to take place. Thus, galvanic cells give a quantitative measure of the relative tendency of various oxidation-reduction reactions to occur.

Figure 13.11 shows a galvanic cell set up to study the reaction

Figure 13.11

Voltage measurement of zinc-hydrogen cell.

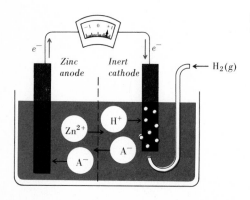

$$Zn(s) + 2H^+ \longrightarrow H_2(g) + Zn^{2+}$$

In the anode compartment, a zinc bar is suspended in a solution of a zinc salt. In the cathode compartment, H_2 gas is led in through a tube so as to bubble over an inert electrode, made, for example, of Pt, dipped in an acidic solution. The anode reaction is

$$Zn(s) \longrightarrow Zn^{2+} + 2e^-$$

The cathode reaction is

$$2H^+ + 2e^- \longrightarrow H_2(g)$$

When, at $25°C$, the concentrations of H^+ and of Zn^{2+} are 1 m and the pressure of the H_2 gas is 1 atm, the voltmeter reads 0.76 volt. The deflection is in such direction as to indicate that Zn has a greater tendency to give off electrons than has H_2. In other words, the half-reaction $Zn(s) \longrightarrow Zn^{2+} + 2e^-$ has a greater tendency to occur than $H_2(g) \longrightarrow 2H^+ + 2e^-$ and the tendency of the first exceeds that of the second by 0.76 volt.

The galvanic cell in Figure 13.12 makes use of the reaction

$$H_2(g) + Cu^{2+} \longrightarrow 2H^+ + Cu(s)$$

The anode reaction is

$$H_2(g) \longrightarrow 2H^+ + 2e^-$$

and the cathode reaction is

$$Cu^{2+} + 2e^- \longrightarrow Cu(s)$$

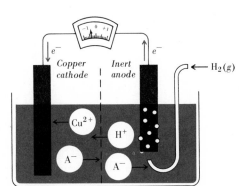

Figure 13.12
Voltage measurement of copper-hydrogen cell.

When, at 25°C, the concentrations of H^+ and Cu^{2+} are 1 m and the pressure of H_2 is 1 atm, the voltmeter reads 0.34 volt. The meter-deflection direction indicates that electrons tend to flow from H_2 to Cu. In other words, the half-reaction $Cu^{2+} + 2e^- \longrightarrow Cu(s)$ has a greater tendency to occur than $2H^+ + 2e^- \longrightarrow H_2(g)$ by 0.34 volt.

In all cells, the voltage observed arises from two sources: an electric potential at the anode and an electric potential at the cathode. If either of these electric potentials were known, the other could be obtained by subtraction. However, it is impossible to measure the voltage of an individual electrode, since any complete circuit necessarily contains two electrodes. We are forced to assign a completely arbitrary voltage contribution to one electrode. The voltage of the other electrode thereby becomes fixed. By international agreement, the voltage assigned to the standard hydrogen electrode (at 25°C, 1 atm H_2 pressure, and 1 m H^+ concentration) is set equal to zero. Consequently, in any cell which contains the standard hydrogen electrode, the entire measured voltage is attributed to the other electrode. Voltages thus assigned are called *oxidation-reduction potentials* or *redox potentials*. If the half-reaction is written with the electrons on the left, the associated voltage is called a *reduction potential;* if the half-reaction is written with the electrons on the right, the associated voltage is called an *oxidation potential.* For years both have been in use, and there has been confusion as to which was meant when electrode potential was specified. There is now an internationally accepted convention that electrode potentials are reduction potentials. We will follow this convention.

Figure 13.13 lists various half-reactions with their reduction potentials. A more extensive listing is given in Appendix 6. The voltage given applies

Figure 13.13 SOME HALF-REACTIONS AND THEIR REDUCTION POTENTIALS	Half-Reaction	Potential, volts
	$F_2(g) + 2e^- \longrightarrow 2F^-$	$+2.87$
	$MnO_4^- + 8H^+ + 5e^- \longrightarrow Mn^{2+} + 4H_2O$	$+1.51$
	$Cl_2(g) + 2e^- \longrightarrow 2Cl^-$	$+1.36$
	$O_2(g) + 4H^+ + 4e^- \longrightarrow 2H_2O$	$+1.23$
	$Br_2 + 2e^- \longrightarrow 2Br^-$	$+1.09$
	$Hg^{2+} + 2e^- \longrightarrow Hg(l)$	$+0.85$
	$Ag^+ + e^- \longrightarrow Ag(s)$	$+0.80$
	$I_2 + 2e^- \longrightarrow 2I^-$	$+0.54$
	$Cu^{2+} + 2e^- \longrightarrow Cu(s)$	$+0.34$
	$2H^+ + 2e^- \longrightarrow H_2(g)$	0
	$Fe^{2+} + 2e^- \longrightarrow Fe(s)$	-0.44
	$Zn^{2+} + 2e^- \longrightarrow Zn(s)$	-0.76
	$Al^{3+} + 3e^- \longrightarrow Al(s)$	-1.66
	$Mg^{2+} + 2e^- \longrightarrow Mg(s)$	-2.37
	$Na^+ + e^- \longrightarrow Na(s)$	-2.71
	$Li^+ + e^- \longrightarrow Li(s)$	-3.05

when the half-reaction proceeds in the forward direction. For the reverse direction, the sign of the voltage must be changed.

The forward reaction is a reduction in which an oxidizing agent, shown on the left, is reduced. The table is so arranged that the oxidizing agents are listed in order of decreasing strength. In other words, the forward half-reaction shows decreasing tendency to occur from the top of the table to the bottom. For example, in the list given, fluorine (F_2) is the best oxidizing agent (i.e., the most easily reduced), since it has the highest tendency to pick up electrons. Lithium ion (Li^+) is the poorest oxidizing agent (i.e., the most difficult to reduce, since it has the least tendency to pick up electrons.

The numerical values of the reduction potentials given in Figure 13.13 apply to aqueous solutions at $25°C$ in which the concentration of dissolved species is $1\ m$. A positive value of the reduction potential indicates that the oxidizing agent is stronger, or more easily reduced than H^+; a negative value indicates that the oxidizing agent is weaker, or less easily reduced than H^+. The magnitude of the potential is a quantitative measure of the relative tendency of the half-reaction to occur from left to right. Nothing is implied, however, about whether the reaction is actually *fast* enough to be observed.

Each oxidizing agent in Figure 13.13 is coupled in its half-reaction with its reduced form. For example, Cu^{2+} is coupled with Cu. The reduced form is capable of acting as a reducing agent when the half-reaction is reversed by some means. Thus, the reduction potentials in Figure 13.13 also give information about the relative tendency of reducing agents to give off electrons. If a half-reaction, such as the one at the top of the table, has a strong tendency to go to the right, it has little tendency to go to the left; i.e., it is hard to reverse, and the reducing agent (F^-) is a poor one. Of the reducing agents shown just to the right of the arrows in Figure 13.13, F^- is the poorest and Li(s) the strongest. The half-reaction

$$Li(s) \longrightarrow Li^+ + e^- \qquad +3.05 \text{ volts}$$

has greater tendency to occur than

$$2F^- \longrightarrow F_2(g) + 2e^- \qquad -2.87 \text{ volts}$$

The voltage of a half-reaction is a measure of the tendency of the half-reaction to occur. This voltage is independent of the other half of the complete oxidation-reduction reaction. The voltage of a complete reaction can be obtained by adding the voltages of its two half-reactions. The voltage so obtained gives the tendency of the complete reaction to occur and is the voltage measured for a galvanic cell which uses the reaction. For example, in the Daniell cell:

Anode:	$Zn(s) \longrightarrow Zn^{2+} + 2e^-$	$+0.76$ volt
Cathode:	$2e^- + Cu^{2+} \longrightarrow Cu(s)$	$+0.34$ volt
Complete cell:	$Zn(s) + Cu^{2+} \longrightarrow Zn^{2+} + Cu(s)$	$+1.10$ volts

The voltage so calculated, $+1.10$, is positive, which indicates that the reaction tends to go spontaneously as written. It should be noted that the value 1.10 volts applies when the concentrations of the ions are 1 m, since oxidation potentials are defined for concentrations of 1 m. If the concentrations are changed, the cell voltage changes, as can be predicted by the Le Chatelier principle. For example, an increase in Cu^{2+} concentration means that the reaction has greater tendency to go to the right, and the voltage is increased. Likewise, an increase in Zn^{2+} concentration decreases the voltage.

Any oxidation-reduction reaction for which the voltage is positive has the tendency to take place as written. Whether a given reaction should take place spontaneously can be determined from the relative positions of its two half-reactions in a table of reduction potentials. In Figure 13.13, for example, any oxidizing agent reacts with any reducing agent below it. I_2 oxidizes Cu, H_2, Fe, etc., but does not oxidize Br^-, H_2O, Cl^-, etc. Similarly, any reducing agent reacts with any oxidizing agent above it. Zn reduces Fe^{2+}, H^+, Cu^{2+}, but does not reduce Al^{3+}, Mg^{2+}, Na^+, etc.

In summary, reduction potentials are written as shown by the following:

$$A \quad + e^- \longrightarrow \quad B \qquad E$$

$$\underset{\text{agent}}{\text{Oxidizing}} \qquad\qquad \underset{\text{agent}}{\text{Reducing}}$$

A will oxidize any reducing agent of lesser (algebraically) E (voltage). B will reduce any oxidizing agent of greater E.

It might be noted that this relationship is true because those oxidation-reduction reactions which spontaneously occur result in a decrease of free energy for the materials involved. Thus there is a direct relationship between the free-energy change and the cell voltage: $\Delta G = -n\mathcal{F}E$. The negative sign indicates that a favorable voltage (positive E) results from a negative free-energy change, n is the number of electrons transferred in the half-reactions, and $\mathcal{F}$ is the faraday (96,500 coulombs) which converts volts to joules.

13.9 Balancing Equations by Half-Reactions

An oxidation half-reaction must always be paired with a reduction half-reaction, in order that the electron balance of the world remain undisturbed. This requirement of electron balance makes possible a method of balancing equations which differs from the oxidation-number method discussed in Section 5.7 in that the artificially devised oxidation number is not necessary.

The balanced equation for the change

$$Zn(s) + Ag^+ \longrightarrow Zn^{2+} + Ag(s)$$

can be written by noting that the Zn must release two electrons to form Zn^{2+} and that these two electrons must be picked up by two Ag^+ ions to form two Ag atoms. The principle of the method is to write the half-reactions and then match electron loss and gain. Thus,

$$Zn(s) \longrightarrow Zn^{2+} + 2e^-$$
$$\underline{2[e^- + Ag^+ \longrightarrow Ag(s)]}$$
$$Zn(s) + 2e^- + 2Ag^+ \longrightarrow Zn^{2+} + 2e^- + 2Ag(s)$$

or

$$Zn(s) + 2Ag^+ \longrightarrow Zn^{2+} + 2Ag(s)$$

The balanced equation for the change

$$Fe^{2+} + MnO_4^- \longrightarrow Fe^{3+} + Mn^{2+}$$

in acidic solution can be written from the two half-reactions

$$Fe^{2+} \longrightarrow Fe^{3+} + e^-$$
$$5e^- + 8H^+ + MnO_4^- \longrightarrow Mn^{2+} + 4H_2O$$

Multiplying the first by 5 and adding to the second gives

$$5Fe^{2+} + 8H^+ + MnO_4^- \longrightarrow 5Fe^{3+} + Mn^{2+} + 4H_2O$$

When given an equation to balance *in acidic solution*, the detailed steps to follow are:

1   Separate the change into half-reactions.

2   Balance each half-reaction separately:
    a   Change coefficients to account for all atoms except H and O.
    b   Add H_2O to the side deficient in O.
    c   Add H^+ to the side deficient in H.
    d   Add e^- to the side deficient in negative charge.

3   Multiply half-reactions by appropriate numbers needed to balance electrons, and add.

4   Subtract any duplications on left and right.

The following example shows the stepwise procedure used to write a balanced equation for the change

$$NO_2^- + Cr_2O_7^{2-} \longrightarrow NO_3^- + Cr^{3+}$$

in acidic solution.

Step 1:

$$NO_2^- \longrightarrow NO_3^-$$

$$Cr_2O_7^{2-} \longrightarrow Cr^{3+}$$

Step 2a:

$$NO_2^- \longrightarrow NO_3^-$$

$$Cr_2O_7^{2-} \longrightarrow 2Cr^{3+}$$

Step 2b:

$$H_2O + NO_2^- \longrightarrow NO_3^-$$

$$Cr_2O_7^{2-} \longrightarrow 2Cr^{3+} + 7H_2O$$

Step 2c:

$$H_2O + NO_2^- \longrightarrow NO_3^- + 2H^+$$

$$Cr_2O^{2-} + 14H^+ \longrightarrow 2Cr^{3+} + 7H_2O$$

Step 2d:

$$H_2O + NO_2^- \longrightarrow$$
$$NO_3^- + 2H^+ + 2e^-$$

$$Cr_2O_7^{2-} + 14H^+ + 6e^+ \longrightarrow$$
$$2Cr^{3+} + 7H_2O$$

(Two electrons have been added to the right side, since in step 2c the left side has a net charge of $1-$ and the right side has a net charge of $1+$. The right side was deficient in negative charge by 2 units.)

(Six electrons have been added to the left, since in step 2c the left side is $12+$ while the right side is $6+$.)

Step 3:

$$3(H_2O + NO_2^- \longrightarrow NO_3^- + 2H^+ + 2e^-)$$
$$\underline{Cr_2O_7^{2-} + 14H^+ + 6e^- \longrightarrow 2Cr^{3+} + 7H_2O}$$
$$3H_2O + 3NO_2^- + Cr_2O_7^{2-} + 14H^+ + 6e^- \longrightarrow$$
$$3NO_3^- + 2Cr^{3+} + 7H_2O + 6H^+ + 6e^-$$

where $3H_2O$, $6H^+$, and $6e^-$ are duplicated on left and right.

Step 4:

$$3NO_2^- + Cr_2O_7^{2-} + 8H^+ \longrightarrow 3NO_3^- + 2Cr^{3+} + 4H_2O$$

If the reaction occurs *in basic solution*, the equation must not contain H^+. In order to add H atoms in step 2c, add H_2O molecules equal in number to the deficiency of H atoms and an equal number of OH^- ions to the opposite side. The rest of the method is the same. An example of a reaction in basic solution is the change

$$Cr(OH)_3(s) + IO_3^- \longrightarrow I^- + CrO_4^{2-}$$

The half-reactions are

$$Cr(OH)_3(s) + 5OH^- \longrightarrow CrO_4{}^{2-} + 4H_2O + 3e^-$$
$$IO_3{}^- + 3H_2O + 6e^- \longrightarrow I^- + 6OH^-$$

and the final net equation is

$$2CR(OH)_3(s) + IO_3{}^- + 4OH^- \longrightarrow 2CrO_4{}^{2-} + I^- + 5H_2O$$

Whether or not you adopt this method in preference to the oxidation-number method for balancing equations, this method is needed for balancing half-reactions.

QUESTIONS

13.1 *Electric Conductivity* Contrast electric conduction through a metal wire and through an electrolyte solution in terms (for each case) of the nature of the charge carriers, preservation of electric neutrality by region, the resistance to conduction, and the reasons for the change of conductivity with temperature.

13.2 *Electrolysis* Diagram a typical electrolysis experiment, showing direction of flow of electrons, cations, and anions, label electrodes, and tell at which oxidation occurs and at which reduction occurs.

13.3 *Electrolysis of Molten* NaCl Account for each of the following for electrolysis of molten NaCl: (*a*) For any given time interval the weight of Cl_2 produced is 1.54 times that of Na produced. (*b*) At any given time, the left half of the molten NaCl and the right half are each electrically neutral (excluding the direct vicinity of the electrodes). (*c*) If the external direct-current source is replaced by an alternating-current source, no products are formed even though alternating current continues to flow.

13.4 *Electrolysis of Aqueous* NaCl As the electrolysis of aqueous NaCl proceeds would you expect (give reasons) (*a*) the composition of the solution to differ near the two electrodes, (*b*) the conductivity to increase or decrease, (*c*) the electrode reactions to change eventually?

13.5* *Electrolysis of Aqueous* Na_2SO_4 (*a*) How does the Na_2SO_4 serve to prevent the accumulation of ions of one polarity from accumulating in each electrode compartment of the cell in Figure 13.7? (*b*) If after a period of time 10 ml of H_2 has formed, how much O_2 has formed? (*c*) If electrolysis continued until a liter of water were used up, how many liters of dry O_2 at STP would be formed?

13.6 *Quantitative Electrolysis* How long must a current of 10.0 amperes be passed through molten NaCl in order to produce 4.60 g of Na?

13.7* *Galvanic Cell* From a cell such as that diagramed in Figure 13.8, how many grams of Zn electrode would dissolve in order to deliver 0.50 ampere for 1.0 h?

13.8 *Galvanic Cell* Set up a cell that uses the reaction

$$Mg(s) + Cl_2(g) \longrightarrow Mg^{2+} + 2Cl^-$$

(a) Write the half-reactions for the two electrodes. (b) Calculate the voltage that would be observed at 25°C if all concentrations were 1 M and the gas pressure were 1 atm.

13.9* *Electrolysis* Diagram a cell that can be used for the electrolysis of molten KCl. Label the electrodes as anode and cathode. Show by arrows the directions of electron motion in the circuit. Write the anode and cathode half-reactions, overall cell reaction, and the least amount of voltage that must be applied to the system.

13.10 *Electrolysis Calculations* Suppose you are electrolyzing aqueous K_2SO_4 with inert electrodes. By the time 250 ml of H_2 (STP) has formed at the cathode, how many milliliters of O_2 (STP) should have formed at the anode?

13.11 *Galvanic Cell* Sketch a cell that could use the reaction

$$Sn(s) + 2Ag^+ \longrightarrow Sn^{2+} + 2Ag(s)$$

Label the anode and cathode, indicate the direction of motion of ions and electrons in the circuit, and write half-reactions for the electrodes. Does the reaction proceed spontaneously?

13.12 *Electrolysis* How many amperes do you need to pass for 10.0 h through 200 ml of 2.2 M KCl to collect 10.0 g of Cl_2?

13.13 *Terms* What is meant by (a) anode, (b) cathode, (c) metallic conduction, (d) electrolytic conduction?

13.14 *Electrical Terms* Distinguish between faraday, coulomb, and ampere.

13.15* *Faraday's Law* Calculate the (a) moles, (b) grams, and (c) liters at STP of chlorine gas that can be produced by the electrolysis of aqueous LiCl with a current of 2.15 amperes for 15.5 min.

13.16 *Faraday's Law* How many (a) faradays, (b) coulombs, (c) amperes for 1.00 min are required to deposit 1.00 g of Fe from Fe^{2+}?

13.17* *Electrolysis* Often, the amount of charge passed is determined by measuring the weight of Cu(s) deposited by electrolysis of Cu^{2+} solution. If a cathode increases in weight by 0.358 g, how many coulombs have been transferred?

13.18 *Electrolysis* In a given electrolysis of Ag_2SO_4, the cathode shows a weight increase of 0.500 g in 15.0 min due to deposition of silver. What is the average current through the solution?

13.19 *Electrolysis* Predict what is oxidized and what is reduced on electrolysis of the following aqueous solutions: (*a*) LiCl, (*b*) K_2SO_4, (*c*) CuI_2.

13.20 *Lead Storage Cell* (*a*) Write the net equation for charging a lead storage cell. (*b*) Show why the density of the liquid increases on charging. (*c*) Why cannot the cell be completely recharged by adding concentrated sulfuric acid?

13.21 *Dry Cell* Why is it impractical to recharge a dry cell in the same way a lead storage cell is recharged?

13.22 *Fuel Cell* Why are the technical problems in making a fuel cell from the $CH_4 + O_2$ reaction greater than those for other galvanic cells?

13.23 *Reduction Potentials* Consider Figure 13.13. (*a*) Why is it that this table, constructed to order oxidizing agents, also orders reducing agents? (*b*) Pick the fourth best oxidizing agent and the fourth best reducing agent listed in the table. (*c*) Pick the best oxidizing agent and the best reducing agent from the following list by using the table: I^-, Br^-, Fe^{2+}, Mg^{2+}, Mn^{2+}, MnO_4^-. (*d*) Using oxidizing and reducing agents listed in (*c*) and Figure 13.13, write balanced equations for two complete reactions which should take place.

13.24 *Cell Voltage* Using Figure 13.13, give the expected cell voltage for the following pairs of possible reactants, write complete equations, and tell which should go spontaneously. (*a*) $Cl_2(g) + Al$, (*b*) $Zn(s) + Mg^{2+}$, (*c*) $I^- + Cu^{2+}$, (*d*) $Br_2 + I^-$, (*e*) $H_2(g) + MnO_4^-$.

13.25 *Half-Reactions* Write balanced half-reactions and net cell reactions for each of the following changes in acid solution.
(*a*) $Zr(s) + H_3AsO_4 \longrightarrow Zr^{4+} + HAsO_2$
(*b*) $H_3PO_3 + NO_3^- \longrightarrow NO(g) + H_3PO_4$
(*c*) $PH_3(g) + IO_3^- \longrightarrow P(s) + I^-$
(*d*) $H_5IO_6 + I^- \longrightarrow I_2$
(*e*) $Cr_2O_7^{2-} + V^{2+} \longrightarrow Cr^{3+} + V(OH)_4^+$

13.26 *Half-Reactions* Write balanced half-reactions and net cell reactions for each of the following changes in basic solution:
(*a*) $MnO_4^{2-} + H_5IO_6 \longrightarrow MnO_4^- + IO_3^-$
(*b*) $NH_3 + ClO_2^- \longrightarrow NH_2OH + Cl^-$
(*c*) $IO^- + MnO_4^- \longrightarrow IO_3^- + MnO_2(s)$
(*d*) $Cu_2O(s) + NiO_2(s) \longrightarrow Cu(OH)_2(s) + Ni(OH)_2(s)$
(*e*) $Ag_2O(s) + Ag_2O_3(s) \longrightarrow AgO(s)$

13.27* *Electrolysis* Suppose you are in business to make hydrogen and chlorine gas by electrolysis of saltwater. Assume the only significant cost you have is buying electricity which costs you 2 cents/kilowatthour (i.e., volts times amperes times 1000 times hours). You electrolyze at 10 volts. (*a*) How many grams of each can you produce for $1.00 worth of electricity? (*b*) If you divide the electric cost equally between the H_2 and corresponding weight of Cl_2, what is your cost per pound (454 g/lb)?

13.28 *Galvanic Cell* The cell pictured in Figure 13.8 is constructed such that each electrode weighs 60.0 g, each half of the liquid compartment contains 400 ml, and $ZnSO_4$ and $CuSO_4$ are each 2.00 M. The cell is run until it stops because one substance has been used up. (*a*) Which one is it that was used up? (*b*) If during the discharge the average current was 0.200 ampere, how long did the cell run?

13.29 *Galvanic Cell* For the cell shown in Figure 13.8 tell (*a*) why there is a voltage developed, (*b*) why each ion moving as indicated in the figure maintains essential electric neutrality, (*c*) whether the anode appears positive or negative as viewed both from solution and from the wire.

13.30 *Conductivity* Consider a current of 0.100 ampere flowing in a wire. How many electrons pass a point on this wire per second?

13.31 *Hydrogen Electrode* Criticize the following misstatements. (*a*) "Because the reduction potential for the hydrogen electrode is zero, it has equal tendency to go in either direction." (*b*) "Because the hydrogen electrode is zero, its voltage contribution cannot be changed by altering hydrogen pressure." (*c*) "The hydrogen reduction potential applies to any aqueous solution which contains some H^+, no matter how little."

13.32 *Automobile Battery* In starting a car, the battery delivers roughly 50 amperes. (*a*) During the 5 s that it might take to start a car, how many grams total of Pb and PbO_2 are consumed in the battery? (*b*) If the car were run strictly from batteries, how many grams total of Pb and PbO_2 would be consumed per mile if 50 amperes would make it go at 5 mph?

13.33* *Electrolysis* You start a clock at 1:00:00 and an electrolysis at 1.00 ampere at the same time. The solution electrolyzed is 100.0 ml of 0.1000 M HCl and 0.1000 M NaCl and is stirred. At what time on the clock is the pH of the solution (*a*) 2.0, (*b*) 3.0, (*c*) 4.0, (*d*) 10.0, (*e*) 11.0, (*f*) 12.0?

13.34 *Reduction Potentials* A table of reduction potentials summarizes a very large number of complete oxidation-reduction reactions and tells which should "go" and which should not. Conversely, by measuring electrode potentials for a new oxidizing-reducing pair a chemist makes but a single measurement and receives a large amount of information. Consider a table such as that in Appendix 6 containing 100 half-reactions. How many conceivable complete reactions does it represent? How many of these should take place spontaneously?

Hydrogen, Oxygen, and Water

14

We now begin a discussion of the specific behavior of the more important chemical substances. The discussion will make full use of the general principles which we have developed. These principles may allow us to predict observed behavior, but more importantly they provide a background and framework for bringing together key observations and systematic knowledge of the behavior of chemical substances. Such empirical knowledge is not merely collected examples of chemical principles. It is chemistry, and the principles are the tools utilized to systematize knowledge. An understanding of the physical world demands both a knowledge of the facts and of the unifying principles. If either must take precedence it is the empirical knowledge which comprises descriptive chemistry. However, even partial mastery of the vast body of descriptive chemistry requires full use of all possible unifying principles. This will be our approach in discussing some of the more important and fundamental chemistry of the elements.

The first element of the periodic table, hydrogen, has but one proton in its nucleus and one orbital electron. In its lowest energy state, the H atom has the electron in a 1s energy level. Because the 1s level can contain another electron of opposite spin, H atoms can normally attain a lower energy state by pairing up to form H_2 molecules. H_2 molecules react with many other elements to form a large variety of compounds.

The element oxygen, atomic number 8, has two 1s, two 2s, and four 2p electrons. Except for fluorine, it is more electronegative than any other element and forms compounds with all elements except some of the noble gases. The study of oxygen compounds has been important in unraveling the chemistry of other elements. One of these compounds, water, is the most important reaction medium in chemistry.

14.1   Occurrence of Hydrogen

Hydrogen appears to be the most abundant element in the universe. Analysis of light emitted by stars indicates that most stars are predominantly hydrogen. For example, approximately 90 percent of the sun's mass is hydrogen. On the earth, hydrogen is much less abundant. The earth's gravitational attraction, being much less than that of stars and larger planets, is too small to hold very light molecules. Considering only the earth's "crust" (atmosphere, oceans, and 10 miles of solid material), hydrogen is third in abundance by number of atoms. Of each 1000 atoms of crust, 530 are oxygen, 160 are silicon, and 150 are hydrogen. On a mass basis, hydrogen is ninth in order and contributes only 0.88 percent of the mass of the earth's crust.

On the earth, free, or uncombined, hydrogen is rare. It is found occasionally in volcanic gases. Also, as shown by study of the aurora borealis, it is found in traces in the upper atmosphere. On the other hand, combined hydrogen is quite common. In water, hydrogen is combined with oxygen and makes up 11.2 percent of its total mass. The human body, two-thirds of which is water, is approximately 10% H by weight. In coal and petroleum, hydrogen is combined with carbon as hydrocarbons. Clay and a few other minerals contain appreciable amounts of hydrogen, usually combined with oxygen. Finally, all plant and animal matter is composed of compounds of hydrogen with oxygen, carbon, nitrogen, sulfur, etc.

14.2   Preparation of Hydrogen

In producing an element for commercial use, the primary consideration is usually cost. For laboratory use, the important consideration is convenience. For *commercial* hydrogen, the primary sources are water and hydrocarbons.

Hydrogen can be made inexpensively by passing steam over hot carbon:

$$C(s) + H_2O(g) \xrightarrow[1000°C]{} CO(g) + H_2(g)$$

It is hard to get pure hydrogen from this source, because carbon monoxide, CO, is difficult to separate completely from hydrogen. The mixture of H_2 and CO is an industrial fuel, *water gas.*

Purer and still relatively inexpensive hydrogen can be made by passing steam over hot iron:

$$3Fe(s) + 4H_2O(g) \longrightarrow Fe_3O_4(s) + 4H_2(g)$$

The iron can be recovered by reducing Fe_3O_4 with water gas.

The purest (99.9 percent) but most expensive hydrogen available commercially is *electrolytic hydrogen*, made from the electrolysis of water:

$$2H_2O \xrightarrow{\text{electrolysis}} 2H_2(g) + O_2(g) \qquad \Delta H = +565 \text{ kilojoules}$$

The reaction is endothermic and requires energy, which must be supplied by the electric current. It is the power consumption, not the raw material, that makes electrolytic hydrogen expensive. In practice, alkaline (basic) solutions are electrolyzed with cells designed to keep anode and cathode products separate. The electrode reactions are

Anode: $\qquad 4OH^- \longrightarrow O_2(g) + 2H_2O + 4e^-$

Cathode: $\quad 2e^- + 2H_2O \longrightarrow H_2(g) + 2OH^-$

Net: $\qquad\qquad 2H_2O \longrightarrow 2H_2(g) + O_2(g)$

Considerable hydrogen is also formed as a by-product of the commercial preparation of Cl_2 and NaOH by the electrolysis of aqueous NaCl (Section 13.4).

In petroleum refineries, where gasoline is made by the catalytic cracking of hydrocarbons, hydrogen is a by-product. When gaseous hydrocarbons are passed over a hot catalyst, decomposition occurs to form hydrogen along with product hydrocarbons.

In the *laboratory*, pure hydrogen is usually made by the reaction of zinc metal with acid:

$$Zn(s) + 2H^+ \longrightarrow Zn^{2+} + H_2(g)$$

In principle, reaction should occur with any metal having a negative reduction potential (Section 13.8). For some metals, such as iron, the reaction is quite slow, even though the voltage is favorable. In water, where the concentration of H^+ is only 1.0×10^{-7} M, the reduction by metals is more difficult. The voltage for the half-reaction

$$2H^+(1.0 \times 10^{+7} M) + 2e^- \longrightarrow H_2(g)$$

is -0.41 volt. In order to liberate H_2 from water, a metal must have a reduction potential below -0.41 volt. Thus, the element sodium reacts with water to liberate H_2 by the reaction

$$2Na(s) + 2H_2O \longrightarrow H_2(g) + 2Na^+ + 2OH^-$$

In principle, zinc should liberate H_2 from H_2O by a similar reaction, but the reaction is too slow to be useful at room temperature.

Laboratory hydrogen can also be made conveniently from the reaction of aluminum metal with base or from the reaction of CaH_2 with water.

14.3   Properties and Uses of Hydrogen

Hydrogen at room temperature is a colorless, odorless, tasteless gas. The quantitative properties are summarized in Figure 14.1. The gas is diatomic

Figure 14.1 **PROPERTIES** **OF HYDROGEN**	Molecular weight	2.016 amu
	Bond length	0.0749 nm
	Bond energy	431 kilojoules/mole
	Approximate molecular diameter	0.2 nm
	Normal melting point	14.1 K
	Normal boiling point	20.4 K
	Critical temperature	33.2 K
	Density at STP	0.0899 g/liter
	Density of liquid (20 K)	0.07 g/ml

and consists of nonpolar molecules containing two hydrogen atoms held together by a covalent bond. In order to rupture the bonds in 1 mole of H_2 to form H atoms, 431 kilojoules of heat must be supplied. Because the dissociation is endothermic, it increases with temperature. At 4000 K and 1 atm pressure, H_2 is about 60 percent dissociated. When H_2 reacts, one of the steps is usually the breaking of the H—H bond. Because of the high energy required for this step, the activation energy is high, and H_2 reactions are slow. Most hydrogen compounds contain H covalently bound, since neither H^+ nor H^- is readily formed.

Molecular hydrogen is the lightest of all gases. It is one-fourteenth as heavy as air. A balloon filled with hydrogen rises in accordance with Archimedes' principle that the buoyant force on an object immersed in a fluid (such as air) is equal to the weight of fluid displaced by the object. In spite of its combustibility, meteorologists still frequently send aloft weather balloons inflated with hydrogen.

The very low melting and boiling points of hydrogen indicate that the intermolecular attractions are quite small. Because of the low boiling point, liquid hydrogen is used in the laboratory to produce low temperatures, but it can be kept for only a few hours, even in a Dewar flask immersed in liquid air.

Chemically, H_2 is able, under appropriate conditions, to combine directly with most elements. With oxygen, H_2 reacts to release large amounts of energy by the change

$$2H_2(g) + O_2(g) \longrightarrow 2H_2O(g) \qquad \Delta H = -485 \text{ kilojoules}$$

which occurs at an appreciable rate only at high temperatures or in the presence of a catalyst. In the oxyhydrogen torch the above reaction occurs to produce temperatures of about 2600°C, and the reaction is self-sustaining. Mixtures of H_2 and O_2 are explosive, and especially violently so when the ratio of H_2 to O_2 is approximately 2:1. With F_2, the reaction

$$H_2(g) + F_2(g) \longrightarrow 2HF(g) \qquad \Delta H = -544 \text{ kilojoules}$$

is explosive even at liquid-hydrogen temperatures.

With metals, the reaction of H_2 is not nearly so violent and often requires elevated temperatures. For example, sodium hydride (NaH) is formed by bubbling H_2 through molten sodium at about 360°C. Hydrides of group II elements are just as difficult to form.

Hydrogen also reacts with certain compounds. In some cases, it simply adds on the other molecules, as, for instance, in forming methyl alcohol (CH_3OH) from CO,

$$CO(g) + 2H_2(g) \xrightarrow[\text{catalyst}]{} CH_3OH(g)$$

Such addition reactions are called *hydrogenation* reactions and account for much of the industrial consumption of hydrogen. In other cases, hydrogen removes atoms from other molecules, as in the reduction of tungsten trioxide (WO_3) to W.

14.4 Compounds of Hydrogen

In its compounds, hydrogen is found in the three oxidation states 1+, 1−, and 0. In the first two cases, H forms compounds by losing a share of its lone electron or gaining a share of another electron, respectively. According to the rules for assigning oxidation numbers (Section 5.5), the relative electronegativity of H and the atom to which it is joined must be considered. In the general compound H_nX, the oxidation number of H is 1+ if X is the more electronegative atom and 1− if X is the less electronegative atom. The oxidation state 0 for hydrogen in compounds represents a rather special case.

Oxidation state 1+   This is the most important state since it includes most of the hydrogen compounds. In these compounds, H is combined with a more electronegative element such as any element from the right-hand side of the periodic table. In period 2, for example, the elements more electronegative than H are C, N, O, and F. With these elements, H forms compounds such as methane (CH_4), ammonia (NH_3), water (H_2O), and hydrogen fluoride (HF). It might be noted that, even though H is thought to be the more positive element in these compounds, the H is not always written first in the formula, as expected. In all these compounds the binding of hydrogen is covalent, and none of the compounds contains simple H^+ ion.

These compounds can be formed by direct union of the elements. The reactions are often slow, sometimes requiring a large activation energy, and so catalysts and high temperatures may be required. For example, the reaction between N_2 and H_2 to form NH_3 is usually carried out under pressure at about $500°C$ in the presence of a suitable catalyst, in this case, iron.

In compounds containing more than two elements, the H is usually in a positive oxidation state. In most such compounds (for example, $NaHSO_4$) the H is bonded to an atom more electronegative than itself.

Oxidation state 1−   When hydrogen is combined with an atom less electronegative than itself, the compound is said to be a *hydride*. These hydrides may be predominantly ionic, as with the elements of groups I and II, or covalent, as with the lighter elements of group III.

In the hydrides of elements of groups I and II, the H occurs as the negative hydride ion, H^-. The compounds at room temperature are ionic solids forming cubic or hexagonal crystals. When melted, they conduct electric current and on electrolysis form H_2 *at the anode* by the reaction

$$2H^- \longrightarrow H_2(g) + 2e^-$$

The hydride ion is unstable in water solution and is oxidized to H_2. Thus, for example, calcium hydride (CaH_2) in H_2O reacts as follows:

$$CaH_2(s) + 2H_2O \longrightarrow Ca^{2+} + 2OH^- + 2H_2(g)$$

The covalent hydrides such as silane (SiH_4) and arsine (AsH_3) are generally volatile liquids or gases. They are nonconductors and apparently contain no H^- ion. They are relatively mild reducing agents.

The term hydride is also applied to compounds in which H is joined to a less electronegative atom in a complex ion. Thus, for example, in the compound lithium aluminum hydride ($LiAlH_4$) the cation is Li^+, and the anion is the complex AlH_4^-. These complex hydrides are generally solids, react with water to liberate H_2, and are of great use as reducing agents.

Oxidation state 0   Hydrogen reacts with some metals such as uranium, copper, and palladium to form hard, brittle substances that conduct electricity and have typical metallic luster. In some cases, as with uranium (UH_3), the number of H atoms per metal atom is fixed and is a whole number. In other cases, as with palladium (PdH_n), the number of H atoms per metal atom is variable and can even be less than 1. It is believed that, in these metallic materials, the hydrogen is dissolved as elementary hydrogen. Consequently, H is assigned a zero oxidation state. It may be that in these substances hydrogen exists as H atoms, which might even be dissociated into protons and electrons.

The dissolution of hydrogen in metals is important, because metals which dissolve hydrogen are catalysts for hydrogenation reactions. The catalyst is thought to act by dissolving the hydrogen as H atoms, which react more rapidly than H_2 molecules. The catalysis, by finely divided nickel, of the hydrogenation of oils to give fats is explainable in this way.

When hydrogen dissolves in a metal, the H atom may go into the lattice as a lattice defect (Section 7.6) and simply expand the lattice of the metal, or it may completely alter the type of lattice. In either case, the change may be significant enough to make the metal lose some of its desirable properties. This phenomenon, called *hydrogen embrittlement*, occurs even with small amounts of dissolved hydrogen, amounts that may be unavoidable in the preparation of pure metals. Thus, the large-scale industrial use of the very valuable metal titanium was made possible only after the development of preparation methods that prevented hydrogen entrapment.

14.5 Hydrogen Bond

In some compounds a hydrogen atom is apparently bonded simultaneously to two other atoms. For example, in the compound potassium hydrogen fluoride (KHF$_2$) the anion HF_2^- is believed to have the structure (FHF)$^-$, in which the hydrogen acts as a bridge between the two fluorine atoms. The hydrogen bridge consists of a proton shared between two atoms and is called a *hydrogen bond*. Hydrogen bonds seem to be formed only between small electronegative atoms life F, O, and N.

Evidence in support of the existence of hydrogen bonds comes from comparing properties of hydrogen-containing substances. For example, Figure 14.2 shows the normal boiling points for the hydrogen halides (lower curve) and for the hydrogen compounds of group VI elements (upper curve). It is evident that the boiling points of HF and H_2O are abnormally high

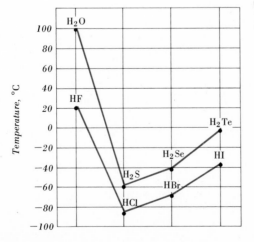

Figure 14.2
Boiling points of some hydrogen compounds.

compared with other members of each series. In the series HF, HCl, HBr, and HI, there is an increasing number of electrons per molecule, and, therefore, rising boiling points would be expected because of increased van der Waals attractions (Section 6.14). The unexpectedly high boiling point of HF is attributed to hydrogen bonds between fluorine atoms. The hydrogen bonding makes it more difficult to detach HF from the liquid. Independent evidence for hydrogen bonding in HF comes from studies of the vapor phase, which is found to contain aggregates such as $(HF)_6$, presumed to be held together by hydrogen bonds. The unexpectedly high boiling point of H_2O in the series H_2O, H_2S, H_2Se, and H_2Te is similarly attributed to hydrogen bonding. In Section 14.11, the importance of hydrogen bonding in the structure of water is discussed.

What is responsible for the hydrogen bond? The simplest view is that the positively charged proton is attracted by the negative electrons of two different atoms. When a hydrogen atom is bound to a very electronegative atom, the hydrogen has such a small share of the electron pair that it is almost like a bare proton. As such, it can be attracted to another electronegative atom. Because of its very tiny size, a given proton has room for only two atoms around it. This picture is consistent with the observations that hydrogen bonds are limited to compounds containing hydrogen that is bonded to very electronegative atoms and that one H can bridge between only two atoms.

Hydrogen bonds are important in biological systems. Proteins, for example, contain both >CO and >NH groups, and hydrogen bonds can be formed to bridge the space between the N and the O. The helical-coil structure so characteristic of many proteins depends on the existence of hydrogen bonds (see Section 22.4).

14.6   Isotopes of Hydrogen

Natural hydrogen consists of three isotopes: protium ($_1^1H$); deuterium, or heavy hydrogen ($_1^2H$, or D); and tritium ($_1^3H$, or T). The protium nucleus consists of a lone proton; the deuterium nucleus, of a proton and a neutron; and the tritium nucleus, of a proton and two neutrons. The protium nucleus is by far the most abundant of the three. In nature, there are 7000 times as many protium atoms as deuterium atoms and only 0.00000000000007 times as many tritium atoms. The scarcity of tritium atoms is due to the instability and consequent radioactivity of its nucleus.

In general, the properties of isotopes are *qualitatively* very similar. However, there may be *quantitative* differences, especially when the percentage difference in mass is appreciable. Figure 14.3 shows some of the properties of protium and deuterium.

In chemical reaction, protium and deuterium show a quantitative differ-

Figure 14.3 PROPERTIES OF HYDROGEN ISOTOPES	Property	Protium	Deuterium
	Mass of atom (H), amu	1.0078	2.0141
	Freezing point (H_2), K	14.0	18.7
	Boiling point (H_2), K	20.4	23.5
	Freezing point (H_2O), °C	0	3.8
	Boiling point (H_2O), °C	100	101.4
	Density at 20°C (H_2O), g/ml	0.998	1.106

ence both in their equilibrium and in their rate properties. Property differences arising from differences in mass are called *isotope effects*. For example, the dissociation constant of ordinary water in the equilibrium

$$H_2O \rightleftharpoons H^+ + OH^-$$

is 1.0×10^{-14} at room temperature. For the corresponding dissociation of heavy water,

$$D_2O \rightleftharpoons D^+ + OD^-$$

the constant is 0.2×10^{-14}, which is significantly smaller. The isotope effect on the rates of reactions is even more marked. A bond to a protium atom can be broken as much as 18 times faster than the bond to a deuterium atom. As an example, H_2 reacts with Cl_2 13.4 times as fast as D_2 does.

For elements heavier than hydrogen the isotope effect is much smaller. For example, $^{127}_{53}I$ reacts at most only 1.02 times faster than $^{129}_{53}I$, and the equilibrium properties show even greater similarity. The isotope effect becomes negligible for the heavier elements, where the percentage difference in mass between the isotopes is small.

The isotope effect in hydrogen is used as a basis for the separation of protium and deuterium. Since protium bonds are broken faster than deuterium bonds, electrolysis of water releases the light isotope faster than the heavy isotope. There is an enrichment of the heavy hydrogen in the residual water. By continuing the electrolysis until the residual volume is very small, practically pure deuterium oxide can be obtained. In a typical experiment, 2400 liters of ordinary water produces 83 ml of D_2O that is 99 percent pure.

14.7 Occurrence of Oxygen

Oxygen is by far the most abundant element in the earth's crust, either as mass or as number of atoms. Of the mass of the earth's crust, 49.5 percent is due to oxygen atoms. Silicon, the next most abundant, is only half as plentiful. In terms of numbers, oxygen atoms are more numerous than all other kinds of atoms combined.

In the free state, oxygen occurs in the atmosphere as O_2 molecules. Dry air is 21% oxygen by volume; i.e., for every 100 molecules of air, approximately 21 are oxygen. On a mass basis, dry air is 23% oxygen; for every 100 g of air approximately 23 g is oxygen. Since the oxygen in the air is largely the product of plant photosynthesis processes, it has been suggested that large-scale human pollution of the environment could lead to depletion of the free-oxygen supply in the atmosphere. However, measurements of clean air over the oceans from 1910 to 1970 have disclosed no change in oxygen content.

In the combined state, oxygen occurs naturally in many rocks, plants and animals, and water. Of the oxygen-containing rocks, the most abundant are those that contain silicon. The simplest of these is silica (SiO_2), the main constituent of sand found on beaches. The most abundant rock that does not contain silicon is limestone ($CaCO_3$). In plant and animal material, oxygen is combined with carbon, sulfur, nitrogen, or hydrogen.

14.8   Preparation of Oxygen

The industrial sources of oxygen are air and water. From air, oxygen is made by liquefaction and fractional distillation. Air, consisting by volume of 21% oxygen, 78% nitrogen, and 1% total of argon, neon, carbon dioxide, and water, is first freed of carbon dioxide and water, compressed, cooled, and expanded until liquefaction results to give liquid air. On partial evaporation, the N_2, being more volatile, boils away first, leaving the residue richer in O_2. Repeated cycles of this kind give oxygen that is 99.5 percent pure.

From water, very pure oxygen can be made by electrolysis as a by-product of hydrogen manufacture. Power consumption makes electrolytic oxygen more expensive than that obtained from air.

In the laboratory, oxygen is usually made by the thermal decomposition of potassium chlorate, $KClO_3$. The reaction

$$2KClO_3(s) \longrightarrow 2KCl(s) + 3O_2(g)$$

is catalyzed by the presence of various solids such as manganese dioxide (MnO_2), ferric oxide (Fe_2O_3), fine silica sand, or powdered glass. It is thought that the main function of the catalyst is to provide a surface on which evolution of oxygen gas can occur.

14.9   Properties and Uses of Oxygen

At room temperature, oxygen is a colorless, odorless gas. The molecule is diatomic and as described in Section 4.7 cannot be easily described by simple dot formulas. The problem is that the O_2 molecule contains two unpaired electrons. These occur because all three bonding molecular orbitals are filled and two electrons remain. They serve both to weaken the bonding strength to less than that expected for a triple bond and, more important, to impart magnetic properties to the molecule, owing to the fact that their spins are unpaired. As a consequence, both gaseous and liquid O_2 are paramagnetic (weakly attracted by magnetic fields).

Oxygen exhibits *allotropy*; i.e., it can exist as the element in more than one form. When energy is added to diatomic oxygen, the triatomic molecule ozone, O_3, is formed by the reaction

$$3O_2(g) \longrightarrow 2O_3(g) \qquad \Delta H = +285 \text{ kilojoules}$$

At room temperature, the equilibrium constant for this reaction is exceedingly small and is calculated to be only 10^{-54}. Thus, even though it increases with temperature, the equilibrium concentration of O_3 does not become appreciable at any temperature, and so not much O_2 can be converted to O_3 by the simple addition of heat. However, when energy is added in other forms, such as electric energy or high-energy radiation, significant amounts of O_3 can be produced because, once O_3 is obtained, it only slowly reverts to O_2. In the laboratory, ozone is easily made by passing air or oxygen between tinfoil conductors that are connected to the terminals of an electric-induction coil. Under the influence of the electric discharge, about 5 percent of the oxygen is converted to ozone. Ozone is also formed in appreciable amounts by lightning bolts, ultraviolet light, and by sparking electric motors. Trace amounts in the lower levels of the atmosphere can increase more than tenfold under the right combination of sunlight with industrial and automotive pollutants, as occurs under smog conditions. The minute amounts (0.02 to 0.03 part per million) that occur in unpolluted areas such as the mountains and the seashore are harmless, but it is not safe to breathe air containing 0.1 ppm ozone for long periods of time. Some of our large cities frequently reach the danger level. The ozone layer in the upper atmosphere is quite rich; its presence makes difficult astrophysical observations of light emitted by stars, because ozone absorbs some of the light, especially those wavelengths needed to identify nonmetallic elements. However, it is the ozone layer that protects life from destruction by this short-wavelength radiation. Maintenance of the protective ozone layer against destruction by reducing agents spewed into the atmosphere is of high priority to environmentalists and other thinking human beings.

The ozone molecule is not magnetic, and so all its electrons must be paired. The three oxygen atoms are arranged in the form of an isosceles triangle in which two of the atoms are not directly bound to each other. Ozone gas has a sharp, penetrating odor. Its solubility in water, in moles per liter, is about 50 percent higher than that of oxygen, probably because O_3 is a polar molecule whereas O_2 is not. When cooled to $-111.5°C$, ozone forms a deep blue liquid that is explosive because of the tendency of O_3 to decompose to O_2. The decomposition is normally slow but increases rapidly as the temperature is increased or a catalyst is added.

Both O_2 and O_3 are good oxidizing agents. Of the common oxidizing agents, ozone is second only to fluorine in oxidizing strength. In most reactions, at least at room temperature, O_2 is a slow oxidizing agent, whereas O_3 is more rapid.

Because of its cheapness and ready availability, oxygen is one of the most widely used industrial oxidizing agents. For example, in the manufacture of steel, oxygen is used to burn off impurities, such as carbon, phosphorus, and sulfur, which may give undesirable properties to steels. In the oxyacetylene torch, used for cutting and welding metals, temperatures in excess of 3000°C can be obtained by the highly exothermic reaction

$$2C_2H_2(g) + 5O_2(g) \longrightarrow 4CO_2(g) + 2H_2O(g)$$

Liquid oxygen is mixed with alcohol, charcoal, gasoline, powdered aluminum, etc., to give powerful explosives. The big rockets used to boost large payloads into space commonly are fueled with liquid oxygen (LOX) for oxidizing the rocket fuels.

The use of oxygen in respiration of plants and animals is well known. In human beings, oxygen, inhaled from the atmosphere, is picked up in the lungs by the hemoglobin in the blood and distributed to the various cells, which use it for tissue respiration. In tissue respiration, carbohydrates are oxidized to provide energy required for cellular activities. Since oxygen is a slow oxidizing agent, catalysts (enzymes) must be present in order that reaction may proceed at body temperature. In the treatment of heart trouble, pneumonia, and shock, the normal amount of oxygen in the air is supplemented with additional oxygen.

The uses of ozone depend on its strong oxidizing properties. For example, its germicidal use depends on its oxidation of bacteria. Inasmuch as oxidation of colored compounds often results in colorless ones, ozone is a bleaching agent for wax, starch, fats, and varnishes. When added to the air in small amounts, ozone destroys odors, but it can be used safely only in low concentration because it irritates the lungs.

14.10 Compounds of Oxygen

Except for the oxygen fluorides, O_2F_2 and OF_2, which are rarely encountered, the oxidation state of oxygen in compounds is negative. The oxidation numbers $1-$ and $2-$ are the ones most commonly observed.

Oxidation state 1− Compounds which contain oxygen with oxidation number $1-$ are called *peroxides*. They are characterized by a direct oxygen-oxygen bond, which usually breaks at high temperatures. Metals such as Na, Sr, and Ba form solid peroxides which contain peroxide ion, O_2^{2-}. When solid peroxides are added to acidic solutions, hydrogen peroxide (H_2O_2) is formed. For example,

$$BaO_2(s) + 2H^+ \longrightarrow Ba^{2+} + H_2O_2$$

If sulfuric acid is used, the barium ion precipitates as insoluble barium sulfate ($BaSO_4$), leaving a dilute solute solution of pure H_2O_2. Commercially, most H_2O_2 is prepared by the electrolysis of cold H_2SO_4 or NH_4HSO_4 solutions followed by distillation under reduced pressure. Because H_2O_2 is unstable, owing to the reaction

$$2H_2O_2 \longrightarrow 2H_2O + O_2(g)$$

it is difficult to keep. The decomposition is slow but is catalyzed by impurities such as dust and dissolved compounds. It is also accelerated in the presence of light. For these reasons, solutions of H_2O_2 are stored in dark bottles with various chemicals added which destroy catalysts.

Pure anhydrous H_2O_2, which can be obtained by distillation under reduced pressure, is a colorless liquid having a freezing point of $-0.9°C$ and an estimated boiling point of $151.4°C$. The structure corresponds to the formula H—O—O—H, where the bond angle H—O—O is $103°$. The four atoms of the molecule do not all lie in the same plane; rather, as shown in Figure 14.4, one H sticks out from the plane of the other three atoms.

In aqueous solution H_2O_2 is a weak acid dissociating

$$H_2O_2 \rightleftharpoons H^+ + HO_2^-$$

with a dissociation constant of about 10^{-12}. Because oxygen also shows oxidation states of 0 and $2-$, compounds containing peroxide oxygen ($1-$) can gain or lose electrons; hence they can act both as oxidizing agents and as reducing agents. In fact, in the decomposition

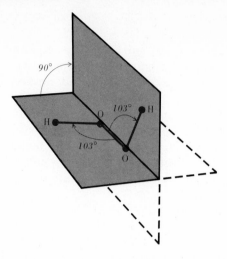

Figure 14.4
Structure of hydrogen peroxide molecule.

$$2H_2O_2 \longrightarrow 2H_2O + O_2(g)$$

hydrogen peroxide oxidizes and reduces itself. In the reaction

$$5H_2O_2 + 2MnO_4^- + 6H^+ \longrightarrow 5O_2(g) + 2Mn^{2+} + 8H_2O$$

hydrogen peroxide is a reducing agent (goes to O_2). In the reaction

$$H_2O_2 + 2I^- + 2H^+ \longrightarrow I_2 + 2H_2O$$

H_2O_2 is an oxidizing agent (goes to H_2O).

Oxidation state 2− The most common oxidation state of oxygen in compounds is 2−. These compounds include the *oxides,* such as BaO, and the *oxy compounds,* such as $BaSO_4$. In none of these is there an oxygen-oxygen bond. Instead, the oxygen atoms have gained a major share of two electrons from atoms other than oxygen.

All the elements except the lighter noble gases form oxides. Some of these oxides are ionic; others are covalent. In general, the more ionic ones are formed with the elements on the left of the periodic table. Thus, BaO contains Ba^{2+} and O^{2-} ions and, like all ionic substances, is a solid at room temperature. It can be heated to 2000°C without decomposition. When placed in water, the O^{2-} ion reacts to give basic solutions:

$$O^{2-} + H_2O \longrightarrow 2OH^-$$

The ionic oxides are therefore called *basic oxides*, or *basic anhydrides*, or, most simply, *bases*. They have the ability to neutralize acids. Thus, for example, when CaO is placed in acidic solution, neutralization occurs according to the equation

$$CaO(s) + 2H^+ \longrightarrow Ca^{2+} + H_2O$$

Elements on the right of the periodic table do not form simple ionic oxides but share electrons with oxygen atoms. Many of these molecular oxides, such as sulfur dioxide (SO_2), are gases at room temperature. They dissolve in water to give acidic solutions. For example,

$$SO_2(g) + H_2O \rightleftharpoons H^+ + HSO_3^-$$

The molecular oxides are therefore called *acidic oxides*, or *acidic anhydrides*, or, most simply, *acids*. They have the ability to neutralize bases. As an example, when SO_2 is bubbled through a basic solution, neutralization occurs as follows:

$$SO_2(g) + OH^- \longrightarrow HSO_3^-$$

It is not possible to classify all oxides sharply as either acidic or basic. Some oxides, especially those formed by elements toward the center of the periodic table, are able to *neutralize both acids and bases*. Such oxides are said to be *amphoteric*. An example of an amphoteric oxide is ZnO, which undergoes both the following reactions:

$$ZnO(s) + 2H^+ \longrightarrow Zn^{2+} + H_2O$$
$$ZnO(s) + 2OH^- + H_2O \longrightarrow Zn(OH)_4^{2-}$$

When any oxide reacts with water, the resulting compound contains OH, or *hydroxyl*, groups. If the hydroxyl group exists in the compound as the OH$^-$ ion, the compound is called a *hydroxide*. Hydroxides are formed by the reaction of ionic oxides with water, e.g.,

$$BaO(s) + H_2O \longrightarrow Ba(OH)_2(s)$$

Barium hydroxide [$Ba(OH)_2$] is a solid which contains Ba^{2+} and OH$^-$ ions in its structure. It, like all hydroxides except those of group I elements, reverts to the oxide when heated. Many hydroxides, e.g., aluminum hydroxide [$Al(OH)_3$], are insoluble in water. The soluble ones give basic solutions.

As mentioned in Section 10.7, some compounds contain the OH group not as an ion but covalently bound to another atom. For example, in H_2SO_4 there are two OH groups and two O joined to a central S atom. When placed in water, such compounds give acid solutions by rupture of the O—H bond. For this reason, they are called *oxy acids*. Most oxy acids can be dehydrated by heat to give oxides. They can also be neutralized to give *oxy salts* such as sodium sulfate (Na_2SO_4).

14.11 Water

The most important of all oxides, and in some respects the most important of all compounds, is H_2O. The water molecule is nonlinear, with the H—O—H angle equal to 104.5°. Each bond is polar covalent, with the H end of the bond positive with respect to the O end. The attraction between the positive end of one molecule and the negative end of another leads to the association of water molecules in both the liquid and solid states. A two-dimensional representation of the association is given in Figure 14.5. The cluster of water molecules is held together by hydrogen bonds (Section 14.5). The H atom, placed between two O atoms, may be considered bonded

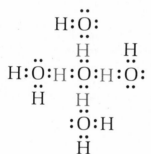

Figure 14.5
Clustering of water molecules due to hydrogen bonding. (Each hydrogen location shown is the average of two possible positions.)

equally to both.† The result of hydrogen bonding is to form a giant molecule in which each O atom is surrounded by four H atoms. (The simplest formula is still H_2O, because, of the four H atoms about a given O atom, only half of each H belongs to that O.) That there are four H atoms about each O is known from X-ray studies of ice. These studies do not detect the H atoms but do show that there are four oxygen atoms symmetrically placed about each oxygen. If the O atoms are joined to each other by H bonds, there must be four H atoms about each oxygen. This can be seen by considering the central atom in Figure 14.5.

† In Figure 14.5, the H atoms in color are shown midway between adjacent O atoms. Actually, a given H atom can jump back and forth from a position nearer one O atom to a position nearer the other. Thus, only the "average" position is shown.

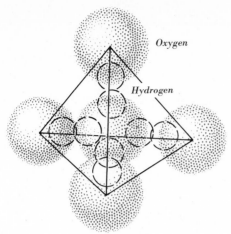

Oxygen

Hydrogen

Figure 14.6
Tetrahedral arrangement of oxygen atoms in ice structure. (Broken circles show the two possible positions for each hydrogen atom.)

The X-ray studies indicate also that the O atoms (of neighboring H_2O molecules) about a given O are located at the corners of a regular tetrahedron, as shown in Figure 14.6. Because of the tetrahedral arrangement, the ice structure extends in three dimensions and is not the flat, two-dimensional representation of Figure 14.5. Figure 14.7 is a better picture of the ice structure. It shows part of the crystal lattice, which extends in three dimensions. The large circles represent oxygen atoms, each of which is tetrahedrally surrounded by four H atoms, represented by the small circles. Every other oxygen atom (lighter color) has its fourth H hidden beneath it. This hidden H joins to another oxygen below, and so the structure continues in three dimensions. A notable feature of the structure is that it is honeycombed with hexagonal channels. Because of these holes, ice has a relatively small density.

When ice melts, the structure becomes less orderly but is not completely destroyed. In liquid water near the freezing point it is thought that the O atoms are still tetrahedrally surrounded by four H atoms as in ice. However,

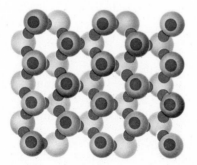

Figure 14.7
Ice structure.

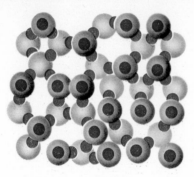

Figure 14.8
Water structure.

the overall arrangement of tetrahedra is more random and is constantly changing. An instantaneous view might be like that shown in Figure 14.8, where some of the hexagonal channels have collapsed to give a denser structure. Liquid water is denser than ice, as the data in Figure 14.9 indicate.

The fact that H_2O decreases in density when it freezes has important consequences for making earth the kind of planet it is. Imagine what it would be like if solid H_2O were more dense than liquid H_2O. On freezing, the H_2O in ponds and lakes would sink to the bottom, bringing fresh liquid up to the cold air where it would freeze also. There would be no formation of a floating insulator between the cold air and the lower body of liquid water, and so ponds and lakes would freeze solid in the wintertime. Not only would this drastically affect the ecology (fishes and frogs probably would not survive the rigors of winter) but also there would not be the moderating influence on climate that large under-the-ice bodies of unfrozen H_2O now exert. Oscillations in atmospheric temperature could then be so great that what is now called the geographic temperate zone could become quite inhospitable.

The data in Figure 14.9 also show that H_2O has a maximum density at 3.98°C. The maximum can be interpreted as follows: When ice is melted, the collapse of the structure leads to an increase in density. As the temperature of the liquid is raised, the collapse should continue further. However,

Figure 14.9 DENSITY OF H₂O AT VARIOUS TEMPERATURES	Temperature, °C	State	Density, g/ml
	0	Solid	0.917
	0	Liquid	0.9998
	3.98	Liquid	1.0000
	10	Liquid	0.9997
	25	Liquid	0.9971
	100	Liquid	0.9584

there is an opposing effect. The higher the temperature, the greater the kinetic motion of the molecules. Hydrogen bonds are broken, and the H_2O molecules move farther apart, on average. This effect becomes dominant only at temperatures above $3.98°C$. Below this temperature, collapse of structure is more important.

14.12   Water as a Solvent

Water is the commonest solvent both in nature and in the laboratory. However, it is far from being a universal solvent, since many substances are almost completely insoluble in water. The factors influencing solubility are many (Section 10.6), and therefore predictions concerning solubility are difficult. The situation for water is especially complex, because there is strong association of H_2O molecules. For solution to occur, considerable energy is required to tear water molecules from their neighbors in order to make room for solute particles.

In general, water is a rather poor solvent for substances which exist in solution as molecules. Thus, gasoline, oxygen, and methane are nearly insoluble in water. In these cases, water interacts so weakly with the molecular solute that not nearly enough energy is liberated to break down the water structure. There are, however, some molecular solutes which are highly soluble in water. Examples are ammonia (NH_3) and ethyl alcohol (C_2H_5OH). These substances interact strongly enough with the water to break its structure. In the case of NH_3, hydrogen bonds are established between the N of NH_3 and the O of H_2O. These hydrogen bonds can be used to justify the occasional practice of writing NH_4OH as the formula of dissolved NH_3. In aqueous solutions of ethyl alcohol, hydrogen bonds are formed between the O of C_2H_5OH and the O of H_2O. Sucrose ($C_{12}H_{22}O_{11}$) owes its appreciable solubility in large measure to hydrogen bonding, since it, like C_2H_5OH, has OH groups.

Although water is the best solvent known for substances which exist in solution as ions, many ionic solutes are practically insoluble in water. In general, the attractions between ions and polar H_2O molecules (called *hydration energies*) are great enough to break the water structure. However, there are strong attractions between the oppositely charged ions in the solid (*lattice energies*), which must be overcome in order that solution may occur. Both of these attractions must be considered in explaining solubility. In sodium chloride, ion-water attractions are great enough, and NaCl is quite soluble. In $BaSO_4$, ion-water attractions are not great enough, and $BaSO_4$ is quite insoluble, despite the fact that the ion-water attraction involved is greater than that for NaCl.

In comparing NaCl with $BaSO_4$, one important factor that needs to be considered is that in NaCl the ions are singly charged, whereas in $BaSO_4$

they are doubly charged. The higher the charge of an ion, the more strongly it will attract one end of a polar H_2O molecule. But the higher charge also causes a greater attraction between the ions in the solid. Thus increase of ionic charge seems to favor both solubility and insolubility. The problem is obviously complex, and there is no simple theory. It seems to be generally true that, if the charge on both anion and cation is increased, insolubility is favored. Thus, for example, $BaSO_4$ (both ions doubly charged) and $AlPO_4$ (both ions triply charged) are much less soluble than NaCl (both ions singly charged). On the other hand, if the charge of only one ion is increased, the solubility is not much changed. As an example, NaCl, $BaCl_2$, and $AlCl_3$ are all appreciably soluble. Similarly, NaCl, Na_2SO_4, and Na_3PO_4 are also soluble.

In addition to charge, there are other factors which affect solubility. One of these is size. In general, the smaller an ion, the more strongly it attracts other ions and water molecules. Another factor we have ignored is that there may be specific interactions either in the solid or in the solution. An example of specific interactions occurs in solid silver chloride (AgCl), where there are stronger van der Waals attractions between Ag^+ and Cl^- than between the ions of NaCl, thus favoring lower solubility for AgCl than for NaCl. In barium sulfide (BaS), there is a specific interaction with water. Here, the hydrolysis reaction of the sulfide ion, S^{2-}, with water occurs to help make BaS more soluble than $BaSO_4$. Thus, predictions as to solubility must be made with caution since it depends on a number of factors.

14.13   Hydrates

Analysis shows that many solids contain water molecules. These solids are called *hydrates* and are represented by formulas like that for nickel sulfate heptahydrate ($NiSO_4 \cdot 7H_2O$). The formula states that there are 7 moles of water per mole of $NiSO_4$ but does not specify how the H_2O is bound in the crystal. For example, in $NiSO_4 \cdot 7H_2O$ all seven H_2O molecules are not equivalent. Six are bound to the Ni^{2+} ion to give $Ni(H_2O)_6^{2+}$, and the seventh is shared between $Ni(H_2O)_6^{2+}$ and SO_4^{2-}. The solid is better given by the formula $Ni(H_2O)_6SO_4 \cdot H_2O$. In other hydrates, such as sodium carbonate decahydrate ($Na_2CO_3 \cdot 10H_2O$), water molecules are not bound directly to the ions, but their principal function seems to be to improve the packing of the ions in the crystal. Water of hydration can be driven off by heating to give *anhydrous* material. The loss of water is usually accompanied by a change in crystal structure. However, some substances, such as the silicate minerals called zeolites and proteins, lose water on heating without change in crystal structure. On reexposure to water, like sponges they take up water and swell. Apparently, water taken up in this way occupies semirigid tunnels within the solid.

Actually, water of hydration is more common than not in the usual salts of the chemistry laboratory. Blue copper sulfate, for example, is $CuSO_4 \cdot 5H_2O$ or, better, $Cu(H_2O)_4SO_4 \cdot H_2O$. Even acids and bases can exist as hydrates in the solid form. Examples are barium hydroxide $[Ba(OH)_2 \cdot 8H_2O]$ and oxalic acid $(H_2C_2O_4 \cdot 2H_2O)$.

Frequently, oxygen- and hydrogen-containing compounds are encountered whose composition may be known but whose structure is in doubt. Such a substance is obtained, for example, from the reaction of a base with a solution of aluminum salt. Under certain conditions, the product might have the composition AlO_3H_3. The most obvious conclusion is that the compound is the hydroxide, $Al(OH)_3$. However, it could just as well be the hydrated oxide, $Al_2O_3 \cdot 3H_2O$, for which the simplest formula also is AlO_3H_3. In order to distinguish the two possibilities, structure studies are needed, but in many cases these have not been done.

QUESTIONS

14.1* *Occurrence of* H   Taking 199×10^{31} g as the mass of the sun, calculate (*a*) the approximate number of moles of H atoms in the sun, (*b*) the number of H atoms.

14.2 *Preparation of* H   Tell how you could prepare hydrogen in each of the following: (*a*) from water without using a reducing agent, (*b*) from water by using a reducing agent, (*c*) from water by using a hydrogen compound as reducing agent, (*d*) from a substance other than water as the source of hydrogen.

14.3 *Properties of* H   (*a*) Why does H_2 have such a low boiling point? (*b*) Why is H_2 such an ideal rocket fuel? (*c*) Why are H_2 reactions at room temperature usually slow?

14.4 H *Compounds*   Write balanced equations for each of the following: (*a*) the formation of a hydrogen compound in each of its known oxidation states, (*b*) the oxidation-reduction reaction that you would expect to occur between two of the compounds of H formed in (*a*).

14.5 *Hydrogen Bond*   Tell clearly how hydrogen bonding accounts for each of the following: (*a*) the existence of the compound KHF_2, (*b*) the fact that HF has a higher boiling point than HCl, (*c*) the fact that HF gas has a molar volume some 20 percent that of an ideal gas, (*d*) ice is very far from a close-packed structure, (*e*) that proteins form helix structures as in Figure 22.4.

14.6 *Isotopes*   In a neutral solution of D_2O, what is the D^+ concentration?

14.7* *Occurrence of O* Calculate the percent by weight of oxygen in each of the following: (a) SiO_2, (b) $CaCO_3$, (c) partially liquefied air containing 40.0 moles of O_2 per 60.0 moles N_2, (d) the human body, ignoring all sources of O except that in water, which accounts for two-thirds of body weight.

14.8 *Preparation of O* Consider the preparation of O_2 from liquid air. (a) Why is it that in each cycle only a part of the air sample is allowed to liquefy or condense? (b) Which has the stronger attractive forces between molecules, N_2 or O_2? (c) What is the reason for the relative attractive forces being as you noted in (b)?

14.9 *Properties of O* Describe in your own words the bond in O_2 and explain why it is weaker than that in N_2?

14.10 *O Compounds* Write balanced, net equations for each of the following changes in acid solution: (a) H_2O_2 oxidizes Cu^+ to Cu^{2+}, (b) H_2O_2 reduces $Cr_2O_7^{2-}$ to Cr^{3+}, (c) dissolving of a basic oxide in acid, (d) dissolving of an acidic oxide in base, (e) dissolving of an amphoteric oxide in base, (f) dissolving of an amphoteric oxide in acid.

14.11 *Water* State whether each of the following statements is true or false and for those that are true whether each by itself would cause the density to increase or decrease as ice melts. (a) All the hydrogen bonds break. (b) The average distance between neighboring O atoms increases slightly. (c) The channels in ice structures collapse.

14.12 *Water as a Solvent* Account for each of the following differences of solubility in water: (a) NaCl is much more soluble than O_2. (b) NH_3 is much more soluble than CH_4. (c) KCl is much more soluble than $CaCO_3$. (d) CaS is much more soluble than $CaCO_3$.

14.13* *Hydrates* At 33°, $Na_2CO_3 \cdot 10H_2O$ melts (i.e., dissolves in its own water of hydration). Calculate the molality of the solution formed by melting this hydrate.

14.14 *Hydrogen-Oxygen Comparison* Account for each of the following differences: (a) H_2 can be oxidized or reduced, but O_2 can usually only be reduced. (b) O_2 is a major component of our atmosphere, H_2 is not. (c) O_2 converts somewhat to O_3, H_2 does not convert to H_3. (d) O_2 is paramagnetic, H_2 is not.

14.15* *Gas Volumes* Suppose that 1.500 liters of H_2O at 0.900 atm and 520 K is passed over hot carbon to give complete conversion to CO(g) and H_2(g). Subsequent passage over a catalyst gives formation of CH_3OH(g). What will be the volume of CH_3OH(g) produced at 0.900 atm and 520 K?

14.16 *Hydrogen Compounds* What is the oxidation state of hydrogen in each of the following compounds: H_2S, $NaBH_4$, LiH, $Pd_{0.7}H$, $NaHSO_4$, KOH, $NaHO_2$?

14.17* *Isotopes of Hydrogen* Given that the protium-deuterium abundance ratio is 7000:1, estimate how many HDO molecules there would be in a 0.030-g drop of water.

14.18 *Boiling Point* Given that the boiling points of H_2O, H_2S, H_2Se, and H_2Te are $100°C$, $-60.7°C$, $-41.5°C$, and $-2.2°C$, respectively, explain (a) why the boiling point of water is so high, and (b) why the others follow the observed trend.

14.19* *Preparation* The world's average annual production of H_2 is about 2.00×10^9 kg. (a) How many coulombs would be required to produce this amount by the electrolysis of water? (b) How many kilograms of iron would be required if hydrogen was produced according to the equation $3Fe(s) + 4H_2O(g) \longrightarrow Fe_3O_4 + 4H_2(g)$?

14.20 *Electrolysis* What volume of hydrogen at $30.0°C$ and 750 mmHg pressure can be produced by the electrolysis of 2.50 liters of H_2O?

14.21* *pH* What is the pH of a solution made by dissolving 1.53 g of BaO in 100 ml of water?

14.22 *Gas Stoichiometry* A mixture is made of equal volumes of $C_2H_2(g)$ and $O_2(g)$. A spark is passed through it so that all the O_2 is converted to CO_2 and H_2O by the reaction $2C_2H_2(g) + 5O_2(g) \longrightarrow 4CO_2(g) + 2H_2O(g)$. What is the fractional decrease in the volume of the system assuming that temperature and pressure remain constant?

14.23 *Equations* Write a balanced equation for each of the following: (a) $ZnO(s)$ dissolving in strong aqueous base, (b) oxidation of hydrogen peroxide in acid with MnO_4^- which forms Mn^{2+}, (c) $BaO_2(s)$ added to a dilute HNO_3 solution.

14.24 *Hydrate* In order to prepare 0.700 liter of 0.150 M Na_2CO_3, how many grams of each of the following must be used: (a) Na_2CO_3, (b) $Na_2CO_3 \cdot 10H_2O$?

14.25 *Solubility* Predict the solubility in water of each of the following: (a) C_3H_7OH, (b) NH_3, (c) $BaSO_4$, (d) NaI, (e) $AlPO_4$.

14.26 *Amphoterism* (a) What is meant by an "amphoteric compound?" (b) Give an example of such a compound. (c) Write balanced equations for reactions of the compound with dilute acid and dilute base.

14.27 *Anhydrides* Write a simplest formula for the anhydride of the following compounds: HNO_3, $HClO_4$, $LiOH$, H_2SO_3, H_2SO_4.

14.28 *Hydrate* Diprotic oxalic acid absorbs water if it is exposed to moist air to form $H_2C_2O_4 \cdot xH_2O$. 1.89 g of the hydrate neutralizes 200 ml of 0.150 M NaOH. What is the formula of oxalic acid hydrate?

14.29* *Hydrogen Balloon* What volume of a hydrogen-filled balloon would be needed to lift a 60-kg man? Assume the weight of the balloon is negligible and the average molecular weight of air is 29.0 under the lifting conditions (that is, STP).

14.30 *Hydrogen Peroxide* Assume K_{diss} of H_2O_2 to be 1×10^{-12}. (a) What is the pH of 1 M H_2O_2? (b) At what pH is H_2O_2 half dissociated to HO_2^-? (c) What is the H^+ concentration in a solution made by dissolving 0.050 mole of Na_2O_2 in 100 ml of 2.0 M HCl?

14.31* *Electrolysis* At an average current of 1.00 ampere, how long must you electrolyze an aqueous solution in order to produce 1.00 liter of H_2 and O_2 combined at STP?

14.32 *Atmosphere* What factors must be considered in accounting for the fact that whereas a majority of the atoms in the earth's crust are O atoms, a majority of the atoms in the atmosphere are N atoms?

14.33 *Rocket Fuels* Ignoring engineering difficulties, there are mentioned in this chapter two substitutes for O_2 in combining with H_2 for fueling rockets which might be superior to O_2. What are they?

14.34 *Weight Relations* As noted in this chapter, CH_3OH can be made by the catalytic combination of CO(g) and H_2(g). Water gas is an equimolar mixture of CO and H_2. What weight of water gas is needed to produce 100 g of CH_3OH?

14.35 *Boiling* As pure water boils, very careful measurements will show that its density slightly increases. Why is this?

14.36 *Bonding* How is it possible that O_3 is polar?

The Nontransition Metals

15

One classification of the chemical elements groups them into metals and nonmetals. A substance is said to be a metal if it has a shiny luster, is a good reflector of light, and is a good conductor of heat and electricity. As we shall see the division is not sharp, since there are elements called semi-metals (included in Chapter 17), and not equal, since metals outnumber nonmetals about 4 to 1. However, the metals may be further subdivided into transition metals and nontransition metals. The transition metals are more complicated because inner orbital electrons contribute significantly to their properties. Transition metals will be discussed in the next chapter following our discussion here of the generally simpler metals. It might be noted that as a group the nontransition metals are relatively soft compared to transition metals such as iron or tungsten. In this chapter, we will cover the general metallic properties, elements of the alkali-metal and alkaline-earth groups plus aluminum, tin, and lead.

15.1 Metallic Properties and Structures

When presented as a smooth surface, metals such as aluminum generally appear shiny; when finely divided, they may appear dull grey or even black. The shininess is rather special in that it is observed even at all viewing angles. This is in contrast to nonmetals, such as sulfur, which may also appear shiny but only when viewed at very low angles. The extraordinary ability of metals to reflect light at all angles, hence to be used as mirrors, is due to the numerous free electrons that permeate the metallic structure. These same free electrons account for the other metallic properties as well.

The simplest model of a metal is a three-dimensional assembly of positive ions suffused throughout by a cloud of electrons made up of the valence electrons of the atoms. In the case of magnesium, which has electron configuration $1s^2 2s^2 2p^6 3s^2$, the ions appear to be the dipositive Mg^{2+} cations (electron configuration $1s^2 2s^2 2p^6$ around a $12+$ nucleus), and the two valence electrons $3s^2$ have been contributed to the "free-electron cloud" that belongs to the whole sample. The Mg^{2+} cations are arranged as a hexagonal close-packed array of spheres (see Section 7.4), and, although the ions can vibrate somewhat from their equilibrium positions, the positions are more or less fixed in space. When an electric current is passed through solid magnesium metal, the Mg^{2+} cations do not move very far from the equilibrium positions. Not so with the electron cloud. Each atom of magnesium has contributed two electrons to the cloud, and every electron in the cloud can move freely through the whole crystal. The electrons are spoken of as being *delocalized*, in the sense that there is no one local site to which the electron is confined. In fact, it is accurate to think of each metallic free electron as an electronic wave that is spread out over the whole sample. In other words, the probability of finding *that* electron on a given atom is exactly equal for all the atoms of the sample.

The situation is somewhat like that shown in Figure 15.1. The large circles represent the Mg^{2+} cations in ordered array; the dashes, free electrons that belong to the electron cloud. When a voltage is applied, the whole body of free electrons tends to be displaced in response to the force. However, the cloud cannot move very far because it is strongly held by the general attraction of all the $2+$ ions for all the negatively charged electrons of the cloud. To get a continuing current fresh electrons have to be added to the metal by a connecting wire at one end of the sample and bled off the metal by a connecting wire at the other end of the sample. One way of viewing electric current is to think of the electrons coming in by one wire as repelling the electrons already in the metal so that they tend to move off the sample

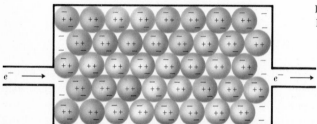

Figure 15.1
Model of electron current in magnesium metal.

via the other wire. In actual fact, not all the electrons move in the direction of the current. There is a distribution of speeds and directions of motion in space; it is only the net movement that constitutes the useful current.

Given the idea of electron migration as the mechanism of electric conductivity, it is easy to see why the conductivity should *decrease* when the temperature of the sample is raised. The higher the temperature, the more vigorous is the thermal vibration of the atomic cores that bear the positive charges. As with the motion of any wave, the electron wave moves through the structure most easily when the atom positions are most regularly spaced. The thermal vibrations introduce irregularities in the atomic positions so that the electron wave feels tugs in other directions that interfere with its regular motion. A crude analog for people is the irregular spacing of ties along a railroad track. To discourage track-walking, the railroad ties are not spaced evenly so that an even stride can be developed. Rather, uneven spaces are irregularly introduced to require a shorter or longer step and hence break one's stride. In a similar way, we can imagine the electron experiencing obstruction to its regular motion each time it meets an irregularity in the lattice. This increasing irregularity at higher temperatures is referred to as *scattering* of the electron motion. Impurity atoms in a metal also act as scattering centers, and so the best conductivity is observed on the very purest specimens. Copper and aluminum wires, two of the most important commercial electric conductors, need to be 99.999 percent pure to be economically satisfactory.

A good conductor of electricity is generally a good conductor of heat also. Witness the great use of copper and aluminum utensils for cooking purposes. The conductivity of heat is a much more complicated process, however, since the transmission of heat energy can proceed not only through the motion of free-electron carriers from one part of a sample to another but also by cooperative vibrations of the positive ions. As one ion moves away from its equilibrium position, it approaches a neighbor and repels it so that it also tends to move. The situation is much like that of a series of pendula tied together so that, when one is set to swinging, the movement is transmitted down the whole line. In a similar way, heat energy introduced on one side of a crystal can appear on the other side shortly thereafter. Metals are special since they also have the ability to carry heat energy via the conduction electrons. Because the electrons move much more rapidly than the atoms, conduction of heat via electrons is generally more efficient than through lattice vibrations. As a consequence it is not surprising that there is a fairly constant ratio between heat conductivity and electric conductivity in metals.

The high reflectivity of metals is a direct consequence of the large number of loosely bound electrons in a metallic structure. An incoming light wave that strikes a surface of a metal causes the loose electrons in the metal to oscillate freely more or less in time with the electric and magnetic pulses that constitute the light wave. Once the electrons are set into back-and-forth oscillation they emit electromagnetic signals much as electrons in bunches sloshing back and forth along a radio antenna send out radio signals. The electromagnetic signals sent out by the moving electrons have the same frequency as the oscillation. Hence, in metals, there is a reemission of the signal that struck the surface of the metal. Crudely, one can think of the free electrons in metals as acting as relay stations for picking up an oscillation signal from an incoming electromagnetic pulse and sending it right back out again.

Like magnesium, most metals have rather simple crystal structures. The most common arrangements are hexagonal close-packing or face-centered cubic. Both of these structures can be regarded as the most efficient ways to pack identical spheres together without wasting much space between spheres. There is, of course, repulsion between the spheres because they all carry net positive charge. What keeps the spheres from flying apart from each other is the negatively charged electron cloud, the total charge of which just neutralizes the positive charges of all the spheres. Because the attraction between positive ions and negative electrons is uniform in all directions of space, there are no preferred directions along which chemical bonds are established.

As we shall see later, the situation is quite different in nonmetals such as sulfur where covalent bonds between the atoms keep the atoms in a more open arrangement in space. In diamond, for example, the four near-neighbor carbon atoms are tetrahedrally arranged around a given carbon, and it is very difficult to move any of the carbon neighbors from their covalently bonded positions.

In metals, on the other hand, it is relatively easy to displace the positive ions from one fixed arrangement to another provided that the average spacing stays the same. It is for this reason that many of the metals are soft, malleable (i.e., can be beaten into sheets), and ductile (i.e., can be drawn out into wire). The group I metals, or alkali metals, are good examples of such behavior. Sodium, for example, can be cut with a knife rather easily, or beaten into a button with a single hammerblow, or extruded through a tiny hole to make wire.

Not all metals, however, show these properties, particularly if there is any great amount of covalent bonding between the cations of the metal structure. The transition metals, to be discussed in the next chapter, have covalent bonds between the metal ions in addition to the regular metal-ion–electron attraction. Such transition metals (for example, iron, tungsten, vanadium) are extraordinarily hard, brittle, and high-melting compared with the simple ideal metals.

15.2 Alkali Metals

The group of elements on the extreme left of the periodic table, group I, consisting of lithium (Z = 3), sodium (Z = 11), potassium (Z = 19), rubidium (Z = 37), cesium (Z = 55), and francium (Z = 87), constitutes the alkali metals. They are called *alkali metals,* after the Arabic word *al-qili,* meaning plant ashes, because ashes of plants are particularly rich in sodium carbonate and potassium carbonate. The term *alkali* is also applied to any substance with marked basic properties.

In their elemental form, the alkali elements are metals *par excellence.* Unfortunately, they are also extremely reactive chemically, so that when a fresh surface is exposed to air there is immediate corrosion by the oxygen and water in the atmosphere to convert the initially shiny surface to dull oxidation products. Were it not for this extreme corrosion, sodium could rival copper and aluminum as a lightweight, efficient electric conductor. Recently a breakthrough has been achieved in the fabrication of sodium wire so that it can be made in the form of polyethylene-encased cables. One such commercial product Nacon is now being used as a high-voltage underground cable, thus illustrating the fact that, once sodium is protected from O_2 and H_2O attack, it can be very effective in power transmission.

All the alkali elements crystallize with a body-centered cubic lattice in which the lattice points are occupied by 1+ ions. Why do the alkali elements prefer to form 1+ ions rather than those of a higher charge? Part of the answer can be seen by noting the properties of the individual alkali atoms as given in Figure 15.2. These properties are for the isolated gas atoms. The column headed "Electronic configuration" indicates the population according to the principal quantum number in the undisturbed neutral atom. As indicated, each of the atoms has one electron in the outermost energy level. The energy required to pull off this valence electron is given in the column headed "First ionization potential." As ionization potentials go, these are relatively small values, indicating that it is fairly easy to pull this one

Figure 15.2 PROPERTIES OF ALKALI ATOMS	Element	Atomic Number	Electronic Configuration	Ionization Potential, eV		M^+ Ionic Radius, nm
				First	Second	
	Lithium	3	$2,2s^1$	5.39	75.6	0.068
	Sodium	11	$2,8,3s^1$	5.14	47.3	0.098
	Potassium	19	$2,8,8,4s^1$	4.34	31.8	0.133
	Rubidium	37	$2,8,18,8,5s^1$	4.18	27.4	0.148
	Cesium	55	$2,8,18,18,8,6s^1$	3.89	23.4	0.167
	Francium	87	$2,8,18,32,18,8,7s^1$			(0.176)

electron off a neutral alkali atom. However, the second ionization potential, the energy required to pull off a second electron, is many times higher than the first ionization potential. This means that, although it is relatively easy to form the M+ ion, it is practically impossible under ordinary conditions to form the M^{2+}. All of this is consistent with the notion that the closed shell of electrons is difficult to break up. The result is that, when alkali atoms come together to form a liquid or solid, M+ ions are formed.

The properties of the alkali-metal atoms shown in Figure 15.2 are illustrative of the general changes expected in going through a group of the periodic table. For example, the radius of the 1+ cation† increases progressively from lithium down. This is expected because of the increasing number of electronic shells populated. Similarly, the ionization potential shows progressive decrease in going down the group. This is consistent with increased size and resulting smaller attraction for the valence electron. Actually, the change in properties in group I is so regular as to give a false sense of confidence about how well periodic-table trends can be predicted. There are traps for the unwary in later groups.

The alkali elements are the most reactive metals known. Practically any oxidizing agent, no matter how weak, can be reduced by them. Quantitatively, their reducing strength (at least for aqueous solutions) is measured by the reduction potential. Figure 15.3 lists the reduction potentials of the alkali metals, along with other properties that characterize the behavior of

Figure 15.3 PROPERTIES OF ALKALI METALS	Element	Reduction Potential, volts	Density, g/cm^3	Melting Point, °C	Boiling Point, °C
	Lithium	−3.05	0.53	186	1336
	Sodium	−2.71	0.97	97.5	880
	Potassium	−2.93	0.86	62.3	760
	Rubidium	−2.93	1.53	38.5	700
	Cesium	−2.92	1.87	28.5	670

these elements. The reduction potential measures the tendency of a substance compared with that of hydrogen to be reduced (i.e., to act as an oxidizing agent). For the alkali metals, the reduction potential is characteristic of the reaction

$$M^+(aq) + e^- \longrightarrow M(s)$$

in which aqueous ions pick up an electron to form solid metal. The very negative values for the alkali metals listed in Figure 15.3 indicate that the

†From X-ray studies of ionic solids it is possible to determine the radius of an ion. There is a problem, however, in that X-ray investigations give only the distance between centers of adjacent atoms. How should this distance be apportioned? The usual procedure is to adopt one ion as a standard and to assume that it has a definite radius in all its compounds. Other radii are then assigned so that the sum of radii equals the observed spacing. A standard may be obtained from a salt like LiI, where Li+ is so small that the spacing can be assumed to be due to large I− ions in contact.

reaction has very little tendency to go in the direction written. In fact it has great tendency to go in the reverse direction. In other words, the alkali metals are excellent reducing agents, having great tendency to lose electrons in chemical reactions. Lithium shows the most negative reduction potential of all common reducing agents, meaning it is the strongest reducing agent of them all. This arises both because of the relative ease of removing an electron (as shown by the small ionization potentials for the alkali elements) and because the small Li^+ ion formed interacts strongly with water molecules of the aqueous solution.

The great reactivity of the alkali metals poses a special problem in their handling. For example; water, although a relatively poor oxidizing agent, has great tendency to attack them. To avoid such problems, alkali metals are usually stored in kerosene or other inert hydrocarbon compounds.

As shown in Figure 15.3, the melting points of the alkali metals are quite low, in which respect they are unlike many other metals, such as iron (which melts at $1535°C$). The explanation for the low melting points lies in the ease of moving the positive ions, the feature that also accounts for malleability and ductility. Unlike solids in which the atoms are covalently bound one to another, the positive ions of a metal are bound together only through their mutual attraction to the negative electron cloud. Thus, their geometric positions are not so rigidly fixed. However, the boiling points are fairly high, showing that it is hard to remove atoms from the metal. This indicates that metallic forces are appreciable even in the liquid state.

Occurrence   The alkali metals occur in nature only as $1+$ ions. Sodium and potassium are most abundant, ranking sixth and seventh, by weight percentage, of all the elements in the earth's crust. Lithium is moderately rare but is found in small amounts in practically all rocks. Rubidium and cesium are rare. Francium is essentially nonexistent, since it has an unstable nucleus and so is radioactive. Trace amounts of it have been prepared by nuclear reactions.

Since most of the compounds of the alkali metals are soluble in water, they are generally found in seawater and in brine wells. However, there are many common rocks which are insoluble complex compounds of the alkali metals combined with Si, O, Al. Presumably as the result of evaporation of ancient seas, there are also large salt deposits which serve as convenient sources of the alkali metals and their compounds.

Sodium ion and potassium ion are among the indispensable constituents of animal and plant tissue. Na^+ is the principal cation of the fluids outside the cells, whereas K^+ is the principal cation inside the cells. Besides filling general physiological roles, such as aiding water retention, these ions have specific functions. For example, Na^+ depresses the activity of muscle enzymes and is required for contraction of all animal muscle. In plants, K^+, but not Na^+, is a primary requirement. As a result, more than 90 percent of the alkali content of ashes is due to potassium. Plants have such a high demand for potassium that, even in soils where the sodium content predominates manyfold, the potassium is taken up preferentially. Since an average crop extracts from the soil about 50 lb of potassium per acre, the necessity of potassium fertilizers is obvious.

Preparation   To prepare alkali elements, it is necessary to reduce the 1+ ion. This can be done chemically or electrolytically. Purely chemical methods would seem impossible, since they require a reducing agent stronger than the alkali metals. However, chemical reduction can be carried out in special cases, as in the reaction of rubidium chloride with calcium at high temperature:

$$Ca(s) + 2RbCl(s) \longrightarrow CaCl_2(s) + 2Rb(g)$$

The reaction occurs in the direction indicated only because the rubidium escapes as a gas out of the reacting mixture, thus preventing the attainment of equilibrium. In the equilibrium state, the concentration of rubidium would be very small.

In practice, the alkali metals are generally prepared by electrolysis of molten alkali compounds. For example, sodium is made commercially in ton quantities by the electrolysis of a fused mixture of $NaCl$ and $CaCl_2$ at about 600°C. ($CaCl_2$ is added to reduce the melting point of the bath.) Sodium metal is formed at the iron or copper cathodes, and chlorine at the carbon anodes. To prevent oxidation of the Na by the chlorine, the electrode compartments are separated by a wire-gauze partition.

Properties and uses   As mentioned, the alkali metals exhibit to a high degree typically metallic properties. Although too expensive and too chemically reactive to be used much for their metallic properties, they do find special application. Besides the underground cables mentioned above, alkali metal as a liquid is used to solve the difficult engineering problem of conducting heat energy from the center of a nuclear reactor to the exterior, where it can be converted into useful work. The expense and difficulty involved in working with alkali metal are compensated for by its excellence as a conductor.

Cesium has the distinction of being the metal from which electrons are ejected most easily by light; such light-induced emission is termed the *photoelectric effect*. For this reason, cesium finds use in the *photocell*, a device for converting a light signal to an electric signal. The basic principle of the photocell is as follows: An evacuated tube contains two electrodes, one of which is coated with cesium metal, cesium oxide, or an alloy of cesium, antimony, and silver. In the absence of light, the tube does not conduct electricity, since there is nothing to carry the charge from one electrode to the other. When struck by light, the cesium-coated electrode emits electrons, which are attracted to the positive electrode, and thus the circuit is completed. Television pickup divices such as the iconoscope and the image orthicon use the photocell principle. Color effects are made possible because the cesium metal has a high response to red light and a low response to blue light, whereas cesium oxide is most sensitive to the blue.

Although all the alkali metals are very good reducing agents, only sodium finds extensive use for this purpose. It is used to make other metals by reducing their chlorides and also in the production of various compounds of carbon. For the latter purpose, sodium is frequently used in the form of its solution in liquid ammonia. It is a remarkable fact that sodium and the other alkali metals dissolve in the waterlike solvent NH_3 to give colored solutions which can be evaporated to give the alkali metal unchanged. In the blue solutions it is assumed that the alkali metal is dissociated into $1+$ ions and electrons. The electrons are associated with NH_3 molecules; therefore, the anions in these solutions can be considered as solvated electrons. More concentrated solutions have a metallic, bronzelike appearance and have very high electric conductivity, indicating that the electrons are extremely mobile. Reducing properties are somewhat diminished in all these solutions compared with the pure alkali metals. The alkali metals form some extraordinary compounds with ammonia. For example, $Li(NH_3)_4$, which is a metallic solid containing lithium atoms tetrahedrally surrounded by NH_3 molecules, has the smallest density (0.5 mg/cm^3) and the lowest melting point ($-185°C$) of any known metallic substance.

Another new and striking use for lithium involves administering the Li^+ ion to human beings suffering from manic-depressive psychoses and other psychiatric sicknesses. This work, still largely experimental, and not without possible dangers, is helping researchers understand the relationship between physiology and human behavior.

Compounds   The alkali metals readily form compounds by reacting with other substances. For example, sodium metal on standing in air becomes covered with sodium peroxide (Na_2O_2). Furthermore, water vigorously attacks any of the alkali metals to liberate hydrogen:

$$2M(s) + 2H_2O \longrightarrow 2M^+ + 2OH^- + H_2(g)$$

Thus, the problem with the alkali metals is not to get them to form compounds but to keep them from doing so.

All the compounds of the alkali metals are ionic, even the hydrides, and all contain the alkali metal as a $1+$ ion. Most of the compounds are quite soluble in water; hence a convenient way to get a desired anion into solution is to use its sodium salt. The alkali-metal ions do not hydrolyze appreciably and do not form complex ions to any appreciable extent. Since the alkali-metal ions are colorless, any color of alkali-metal compounds must be due to the anion.

The hydrides of the alkali metals are white solids prepared by heating alkali metal in hydrogen. The simple oxides, M_2O, are not so easily formed. Of the alkali metals, only lithium reacts directly with oxygen to form Li_2O. When sodium reacts with oxygen, the peroxide Na_2O_2 is formed instead. Potassium, rubidium, and cesium under similar conditions form superoxides of the type MO_2. In order to get the simple oxides, it is necessary to reduce some alkali-metal compound such as the nitrate. For example,

$$2KNO_3(s) + 10K(s) \longrightarrow 6K_2O(s) + N_2(g)$$

The oxides are all basic oxides and react with water to form hydroxides. However, commercially, the hydroxides of the alkali metals are made by electrolysis of aqueous alkali-chloride solutions. For example, as discussed in Section 13.4, sodium hydroxide, or *caustic soda,* as it is often called, is made by electrolysis of aqueous NaCl.

Other important compounds of the alkali metals, such as *washing soda* (Na_2CO_3) and *baking soda* ($NaHCO_3$), are discussed in later chapters in connection with the corresponding anions.

Qualitative analysis   Because the alkali metals do not form many insoluble compounds and because the alkali ions are colorless, it is difficult to detect the presence of these elements by chemical methods. Instead, their presence is usually shown by running flame tests on the sample in question. The simplest way to run a flame test is to shape a piece of fine platinum wire into a loop, dip the loop in HCl solution and heat to remove volatile impurities, and then use the loop to heat the sample in a burner flame. The sodium yellow is extremely intense, so that even traces of it can mask other flame colors. The main reason for cleaning the platinum loop by the HCl treatment is to help expel sodium as NaCl, which at high temperatures is boiled off. (In general, chlorides are more volatile than most other solids.) The potassium flame is colored a delicate violet and can be observed in many cases only through cobalt glass, which filters out interfering colors such as sodium yellow. The flames of K, Rb, and Cs are so similar that definite identification requires examination of the line spectrum with a spectroscope.

15.3 Alkaline-Earth Metals

Group II of the periodic table contains the elements beryllium ($Z = 4$), magnesium ($Z = 12$), calcium ($Z = 20$), strontium ($Z = 38$), barium ($Z = 56$), and radium ($Z = 88$). They are called the *alkaline-earth metals,* because the old alchemists referred to any nonmetallic substance insoluble in water and unchanged by fire as an "earth" and because the "earths" of this group, e.g., lime (CaO) and magnesia (MgO), give decidedly alkaline reactions. Probably the most characteristic features of the group II elements are their good metallic properties, their strength as reducing agents, and their formation of compounds in which they show oxidation state 2+. Many of these compounds are of low solubility. Compared with group I elements, the alkaline-earth elements are both less metallic and poorer reducing agents.

Electronic configurations and some related properties of the alkaline-earth-metal atoms in the isolated gaseous state are shown in Figure 15.4. It should also be noted that radium atoms are radioactive and undergo sponta-

Figure 15.4
PROPERTIES OF ALKALINE-EARTH ATOMS

Element	Atomic Number	Electronic Configuration	Ionization Potential, eV			M^{2+} Ionic Radius, nm
			First	Second	Third	
Beryllium	4	$2,2s^2$	9.32	18.2	153.9	0.035
Magnesium	12	$2,8,3s^2$	7.64	15.0	80.1	0.066
Calcium	20	$2,8,8,4s^2$	6.11	11.9	51.2	0.099
Strontium	38	$2,8,18,8,5s^2$	5.69	11.0	(43)	0.112
Barium	56	$2,8,18,18,8,6s^2$	5.21	10.0	(36)	0.134
Radium	88	$2,8,18,32,18,8,7s^2$	5.28	10.1		0.143

neous nuclear disintegration. In each of the alkaline-earth-metal atoms there are two electrons in the outermost energy level. With the exception of beryllium, the next lower principal quantum level contains eight electrons. The chief difference in going down the group is the stepwise inclusion of sets of 8, 18, and 32 electrons. As expected, there is a corresponding increase in size from Be to Ra. This is illustrated in the last column, which gives experimentally determined values of the M^{2+} cation radius in crystals. The values need to be considered as approximate, since ionic sizes depend on environment, but they do show the expected trend as more levels are populated.

As expected, with increasing size we find a decreasing ionization potential in going down the group. The first ionization potential is the energy required to pull one electron from the neutral, isolated atom. The larger atoms hold their outer electrons less tightly than do small atoms, hence the decreasing ionization potential from Be to Ba. The anomalous increase for radium has not yet been explained. The second ionization potential measures the energy required to pull one electron from the 1+ ion to form a 2+ ion. Because the electron is pulled off a positively charged ion rather than a neutral atom, the second ionization potential of an atom is always greater than the first. The third ionization potential indicates the energy to pull one electron off a 2+ ion to form a 3+ ion and, for a given element, is greater than the second ionization potential.

Inspection of the ionization potentials in Figure 15.4 shows that removal of the third electron from alkaline-earth elements requires very high energies. Such high energies are usually not available in chemical reactions, and therefore 3+ ions of the alkaline-earth elements are not encountered except in some hot stars. In practice, only the 2+ ions of these elements are observed. These ions are relatively easily formed and interact strongly with water in aqueous solutions and with negative ions in solid compounds. For these reasons 1+ ions of the alkaline-earth elements are not encountered, and it is the 2+ ions which are formed under all usual conditions.

Figure 15.5 shows some of the properties of the group II elements in the solid state, as distinguished from the properties of the isolated atoms.

Figure 15.5 PROPERTIES OF ALKALINE-EARTH METALS	Element	Reduction Potential, volts	Density, g/cm^3	Melting Point, °C	Boiling Point, °C
	Beryllium	−1.85	1.86	1280	1500(?)
	Magnesium	−2.37	1.74	650	1100
	Calcium	−2.87	1.55	810	1300(?)
	Strontium	−2.89	2.6	800	1300(?)
	Barium	−2.90	3.6	850	1500(?)
	Radium	−2.92	5(?)	960(?)	1100(?)

In the solid state the elements have typically metallic properties: high luster and good conductivity. They are harder than the group I elements but still can be cut with a hard steel knife. The fairly high melting points are in line with this greater hardness. The boiling points (many of which have not been accurately determined) are higher than those of the alkali metals and suggest that the forces of attraction between the electron cloud and M^{2+} ions are greater than those between the electron cloud and M^+ ions.

The reduction potentials, shown in the second column, are relatively large and negative. They correspond to the reaction

$$M^{2+}(aq) + 2e^- \longrightarrow M(s)$$

and indicate that for aqueous solutions these elements are good reducing agents (reverse reaction). For example, all the alkaline-earth metals have the ability to react with water to release hydrogen by the reaction

$$M(s) + 2H_2O \longrightarrow M^{2+} + H_2(g) + 2OH^-$$

Occurrence In nature, the alkaline-earth elements are found only in compounds as 2+ ions. As discussed in Section 14.12, 2+ ions combine with 2− ions to form compounds less soluble than those of 1+ ions. Consequently, many alkaline-earth compounds are insoluble and, unlike alkali-metal compounds, are found as insoluble deposits in the earth's crust. Most important of these deposits are the silicates, carbonates, sulfates, and phosphates.

Beryllium on a mass basis makes up only 0.0006 percent of the earth's crust. It is very widespread, but only in trace amounts. The only important beryllium mineral found in any quantity is a silicate, beryl, or $Be_3Al_2Si_6O_{18}$. Enormous single crystals of beryl weighing many tons have been found. The gem stone emerald is beryl, colored deep green by trace amounts of chromium.

Magnesium is the eighth most abundant element in the earth's crust, making up about 2 percent of its mass. It is widely distributed, principally as the silicate minerals such as asbestos ($CaMg_3Si_4O_{12}$) and the carbonate, oxide, and chloride. Magnesite ($MgCO_3$) and dolomite ($MgCO_3 \cdot CaCO_3$) are the principal sources of magnesium in addition to seawater and deep salt wells.

Calcium is the most abundant of the group I and group II elements on a mass basis (3.6 percent of the earth's crust) but is outnumbered 6 to 5 on an atom basis by sodium. The principal occurrences of calcium are as the silicates, carbonate, sulfate, phosphate, and fluoride. Calcium carbonate ($CaCO_3$), as the mineral calcite, the most abundant of all nonsilicate minerals, appears in such diverse rocks as limestone, marble, and chalk. Most of these appear to be derived from the skeletons of marine animals which have been laid down on seabeds and consolidated. The mineral gypsum ($CaSO_4 \cdot 2H_2O$) is also very common. It apparently owes its origin in many cases to limestone beds which have been acted on by sulfuric acid produced from the oxidation of sulfide minerals. Phosphate rock is essentially $Ca_3(PO_4)_2$, an important ingredient of bones, teeth, and seashells.

Strontium is relatively rare and ranks twentieth in order of abundance by weight; barium, which makes up 0.05 percent of the earth's crust, is about 2.5 times as abundant. The principal mineral of strontium is strontianite ($SrCO_3$); of barium, barite ($BaSO_4$).

Radium is very rare, but its presence is easily detected by its radioactivity. Because its nucleus spontaneously disintegrates, all the radium presently found is due to the nuclear breakdown of heavier elements, particularly uranium. For this reason, uranium ores such as pitchblende (impure U_3O_8) are principal sources of radium. It has been estimated that the average abundance of radium in the earth's crust is less than 1 part per million million. This makes a uranium mineral which contains $\frac{1}{4}$ g of radium per ton of ore a relatively rich source of radium.

Preparation   Since the alkaline-earth elements occur only as the 2+ ions, preparation of the metals requires a reduction process. Reduction can be accomplished by electrolysis of the molten halides or hydroxides or by chemical reduction with appropriate reducing agents. Beryllium, for example, is made by heating beryllium fluoride (BeF_2) with Mg and also by electrolyzing a mixture of beryllium chloride ($BeCl_2$) and NaCl.

The extraction of magnesium from seawater accounts for the bulk of United States production. In the process, the magnesium ion in seawater (about 0.13 percent) is precipitated as $Mg(OH)_2$ by the addition of lime (CaO). The hydroxide is filtered off and converted to $MgCl_2$ by reaction with HCl. The dried $MgCl_2$ is mixed with other salts to lower the melting point and then electrolyzed at about 700°C to give metal of 99.9 percent purity.

Magnesium can also be prepared by a chemical reduction process in which magnesium oxide, obtained by heating dolomite, is reduced at high temperatures by iron and silicon. Since the reaction is carried out above 1100°C, the boiling point of magnesium, the process produces gaseous magnesium, which escapes from the reaction mixture to condense as a very high purity product.

Properties and uses   All the alkaline-earth metals are good conductors of heat and electricity, but only magnesium finds any considerable use. Surprisingly, this use is based on the structural qualities of magnesium rather than on its electrical properties. Lightest of all the commercially important structural metals, magnesium has relatively low structural strength but this can be increased by alloying it with other elements. The principal elements added are aluminum, zinc, and manganese. The aluminum helps to increase the tensile strength; the zinc improves the working properties (machining); and the manganese reduces corrosion. The use of magnesium alloys is steadily increasing because of modern emphasis on weight reduction in such things as aircraft, railroad equipment, and household goods.

Too rare and costly for most large-scale uses, beryllium is important as a trace addition for hardening other metals such as copper. In the finely powdered form, beryllium (and its compounds) must be handled carefully, since it is extremely toxic.

Calcium, strontium, and barium are more reactive than beryllium and magnesium. The situation is complicated further by the fact that, when exposed to air, they form oxides which flake off to expose fresh surface. The great affinity for oxygen makes these elements useful as deoxidizers in steel production and as "getters" in the production of low-cost electron tubes. Most radio tubes, for example, have a thin deposit of barium metal on the inner wall of the glass or metal envelope. The purpose is to pick up any gases such as oxygen in the tube.

Finely divided magnesium burns rather vigorously to emit very intense light particularly at the higher energies. For this reason, magnesium is used as one of the important light sources for photography. Flashbulbs contain wire or foil of magnesium (or aluminum) packed in an oxygen atmosphere. When the bulb is fired, an electric current heats the metal and initiates the oxidation reaction.

The flame spectra of strontium salts are characteristically red, and those of barium are yellowish green. Strontium and barium salts are frequently used for color effect in flares and fireworks.

Compounds   At ordinary temperatures, the alkaline-earth elements form compounds only in the 2+ oxidation state. With the exception of beryllium, all such compounds are essentially ionic. The alkaline-earth ions are colorless and, except for Be^{2+}, do not hydrolyze appreciably in aqueous solution. Beryllium salts hydrolyze to give acid solutions. Unlike the compounds of group I, many group II compounds are not soluble in water.

1   **Hydrides**   When heated in hydrogen gas, Ca, Sr, and Ba form hydrides. These are white powders which react with H_2O to liberate H_2. Calcium hydride, CaH_2, is used as a convenient, portable hydrogen supply.

$$CaH_2(s) + 2H_2O \longrightarrow Ca^{2+} + 2OH^- + 2H_2(g)$$

2   **Oxides**   The oxides of these elements are characteristically very high melting (*refractory*). They can be made by heating the metals in oxygen or thermally decomposing the carbonates or hydroxides. For example, lime (CaO) is made from limestone ($CaCO_3$) by the reaction

$$CaCO_3(s) \longrightarrow CaO(s) + CO_2(g)$$

Except for beryllium oxide (BeO), which is amphoteric, the oxides of group II are basic. Both lime and magnesia (MgO) are used as linings in furnaces, sometimes specifically to counteract acidic impurities, as in steel production.

3   **Hydroxides**   The hydroxides of group II are made by adding water to the oxides in a process called *slaking*. For example, the slaking of lime produces calcium hydroxide [$Ca(OH)_2$], sometimes called *slaked lime*. The reaction

$$CaO(s) + H_2O \longrightarrow Ca(OH)_2(s) \qquad \Delta H = -67 \text{ kilojoules}$$

is accompanied by a threefold expansion in volume, sometimes to the consternation of building contractors whose lime supplies accidentally get wet. Lime is an important constituent of cement and is also used as an important industrial base, since it is cheaper than NaOH. The hydroxides of the alkaline-earth elements are only slightly soluble in water; however, the solubility increases with increasing ionic size.

4   **Sulfates**   The sulfates of group II range from the very soluble beryllium sulfate to the practically insoluble radium sulfate. This decreasing order is opposite to that observed for the hydroxides. To account for the change of trend, two factors need to be considered: As discussed in

Section 14.12, solubility depends on lattice energy and on hydration energy. For the alkaline-earth sulfates, the lattice energies are all about the same, apparently because the sulfate ion is so large (about 0.3 nm radius) that changing the size of the much smaller cation makes little difference. The difference in solubility must therefore be due to differences in hydration energy. From Be^{2+} to Ba^{2+}, size increases, hydration energy decreases, and the sulfates become less soluble. For the alkaline-earth hydroxides, the lattice energies are not the same but decrease with increasing cation size. Apparently for these hydroxides this is a larger effect than the change in hydration energy. Thus, the hydroxides increase in solubility down the group.

Magnesium sulfate is well known as the heptahydrate, $MgSO_4 \cdot 7H_2O$, or epsom salt. In medicine it is useful as a purgative, apparently because magnesium ions in the alimentary canal favor passage of water from other body fluids into the bowel to dilute the salt.

Calcium sulfate has already been mentioned as the mineral gypsum ($CaSO_4 \cdot 2H_2O$). When gypsum is partially dehydrated,

$$CaSO_4 \cdot 2H_2O(s) \rightleftharpoons CaSO_4 \cdot \tfrac{1}{2}H_2O(s) + \tfrac{3}{2}H_2O(g)$$

it forms plaster of paris, sometimes written $2CaSO_4 \cdot H_2O$.† The use of plaster of paris in making casts and molds arises from the reversibility of the above reaction. On water uptake, plaster of paris sets to gypsum, with an expansion of volume, which makes possible remarkably faithful reproductions.

Barium sulfate and its insolubility have been repeatedly mentioned. Although Ba^{2+}, like most heavy metals, is poisonous, the solubility of $BaSO_4$ is so low that $BaSO_4$ can safely be ingested into the stomach. The use of $BaSO_4$ in taking X-ray pictures of the digestive tract depends on the great scattering of X rays by the Ba^{2+} ion. The scattering of X rays by atoms is proportional to the electron density of the atom. Ba^{2+} contains 54 electrons in a relatively small volume and, hence, scatters X rays more efficiently than ions of lighter elements. Actually $BaSO_4$ is more important as a white pigment.

5   **Chlorides and fluorides**   Beryllium chloride and fluoride ($BeCl_2$ and BeF_2) are unusual in that they do not conduct electricity in the molten state. For this reason, they are usually considered to be molecular rather than ionic salts. The chlorides and fluorides of the other group II elements are all typical ionic solids. Calcium fluoride (CaF_2), occurring in nature as the mineral fluorspar, is quite insoluble in water. The chloride,

---

†We started our discussion in Chapter 1 by noting Lavoisier's interest in the fermentation reaction. For the record, we should note that Lavoisier's detailed chemical research was the more prosaic conversion of gypsum to plaster of paris.

CaCl$_2$, is very soluble in water and, in fact, has such great affinity for water that it is used as a dehydrating agent.

6   Carbonates   All the carbonates of group II are quite insoluble and therefore are found as solid compounds in nature. Calcium carbonate (CaCO$_3$) is found as the most common nonsilicate rock, limestone. The existence of large natural beds of CaCO$_3$ poses a special problem for water supplies, since CaCO$_3$, though essentially insoluble in water, is soluble in water containing carbon dioxide. Since our atmosphere contains an average of 0.04% CO$_2$ at all times, essentially all groundwaters are solutions of CO$_2$ in H$_2$O. These groundwaters dissolve limestone by the reaction

$$CaCO_3(s) + CO_2 + H_2O \rightleftharpoons Ca^{2+} + 2HCO_3^-$$

which produces a weathering action on limestone deposits and results in contamination of most groundwaters with calcium ion and bicarbonate ion, HCO$_3^-$. The dissolving action of CO$_2$-containing water explains the many caves found in limestone regions. These caves abound in weird formations produced partly by the dissolving action and partly by reprecipitation of CaCO$_3$. The optimum conditions for CaCO$_3$ deposition are slow seepage of groundwater, steady evaporation, and no disturbing air currents. In limestone caves, these conditions are ideally met. Groundwater containing Ca^{2+} and HCO$_3^-$ may seep through a fissure in the roof and hang as a drop from the ceiling. As the water evaporates along with the carbon dioxide, the above reaction reverses to deposit a bit of limestone. Later, another drop of groundwater seeps on to the limestone speck, and the process repeats. In time a long shaft reaching down from the roof may be built up in the form of a limestone stalactite. Occasionally drops of groundwater may drip off the stalactite to the cave floor, where they evaporate to form a spire, or stalagmite, of CaCO$_3$. The whole process of dissolving and redeposition of limestone is very slow and may take hundreds of years.

Qualitative analysis   As a group, the alkaline-earth cations (excluding beryllium) can be distinguished from other common cations by taking advantage of the fact that, like group I elements, they form soluble sulfides but, unlike group I elements, they form insoluble carbonates.

Given a solution containing alkaline-earth cations, the barium can be precipitated as yellow BaCrO$_4$ by addition of K$_2$CrO$_4$ in the presence of an acetic acid buffer. From the residual solution (containing Sr^{2+}, Ca^{2+}, Mg^{2+}), light yellow SrCrO$_4$ can be precipitated by subsequent addition of NH$_3$ and alcohol. The BaCrO$_4$ precipitates in the first step and SrCrO$_4$ in the second because BaCrO$_4$ (K$_{sp}$ = 8.5 × 10^{-11}) is less soluble than SrCrO$_4$ (K$_{sp}$ = 3.6 × 10^{-5}). The point of using an acetic acid buffer (Section 12.8) is to keep the H^+ concentration around 10^{-5}, where the chromate concentration, governed by the equilibrium

$$2CrO_4^{2-} + 2H^+ \rightleftharpoons Cr_2O_7^{2-} + H_2O$$

is too low to precipitate Sr^{2+} but high enough to precipitate Ba^{2+}. Subsequent addition of NH_3 reduces the H^+ concentration, thereby increasing the CrO_4^{2-} concentration sufficiently to precipitate $SrCrO_4$, especially in the presence of alcohol which lowers its solubility.

Calcium ion can be separated from magnesium ion by addition of ammonium oxalate to form white, insoluble calcium oxalate (CaC_2O_4). (The K_{sp} of CaC_2O_4 is 1.3×10^{-9}, compared with 8.6×10^{-5} for MgC_2O_4.) Finally, the presence of Mg^{2+} can be shown by adding more NH_3 and Na_2HPO_4, which precipitates white magnesium ammonium phosphate, $MgNH_4PO_4$.

15.4 Hard Water and Ion Exchange

Because limestone is so widespread, most groundwater contains small but appreciable concentrations of calcium ion. The presence of this Ca^{2+} is objectionable because of the formation of insoluble precipitates when such water is boiled or when soap is added. Water that behaves in this way is called "hard" water. It represents an industrial and household problem of the first magnitude.

Hardness in water is always due to the presence of calcium, magnesium, or ferrous (Fe^{2+}) ion. The hardness may be of two types: (1) *temporary, or carbonate, hardness,* in which HCO_3^- ions are present in the water in addition to the aforementioned metal ions; (2) *permanent, or noncarbonate, hardness,* in which the dipositive ions but no HCO_3^- ions are in the water. In either case, the hardness manifests itself by a reaction with soap (but not with detergents) to produce a scum. Soap is a sodium salt of a complicated hydrocarbon acid. The usual soap is sodium stearate ($NaC_{18}H_{35}O_2$) and consists of Na^+ ions and negative stearate ions. When stearate ions are added to water containing Ca^{2+}, insoluble calcium stearate forms:

$$Ca^{2+} + 2C_{18}H_{35}O_2^- \longrightarrow Ca(C_{18}H_{35}O_2)_2(s)$$

This insoluble calcium stearate is the familiar scum or bathtub ring.

Hardness in water is also objectionable because boiling a solution containing Ca^{2+} and HCO_3^- results in the deposition of $CaCO_3$, as in cave formation. In industrial boilers the deposition of $CaCO_3$ is an economic headache, since, like most salts, $CaCO_3$ is a poor heat conductor. Fuel efficiency is drastically cut, and boilers have been put completely out of action by local overheating due to boiler scale.

The major question then is how to soften hard water efficiently and economically. The most direct way to soften water (as is done in many households) is simply to add huge quantities of soap. Eventually, enough stearate ion can be added to precipitate all the objectionable Ca^{2+} as scum, leaving the excess soap to carry on the cleansing action.

Another way to soften water (this works only for temporary hardness) is to boil the water. The reaction

$$Ca^{2+} + 2HCO_3^- \rightleftharpoons CaCO_3(s) + H_2O + CO_2(g)$$

is reversible, but the forward reaction can be made dominant by boiling off the CO_2. Boiling is not practical for large-scale softening.

The third way to soften water is to precipitate the Ca^{2+} out of solution. This can be done by adding washing soda (Na_2CO_3). The added carbonate ion, CO_3^{2-}, reacts with Ca^{2+} to give insoluble $CaCO_3$. If bicarbonate ion is present, the water may be softened by adding a base such as ammonia. The base neutralizes HCO_3^- to produce CO_3^{2-}, which then precipitates the Ca^{2+}. On a large scale, temporary hardness is removed by adding limewater. The added OH^- neutralizes HCO_3^- and precipitates $CaCO_3$ by the process

$$Ca^{2+} + HCO_3^- + OH^- \longrightarrow CaCO_3(s) + H_2O$$

It might seem odd that limewater, which itself contains Ca^{2+}, can be added to hard water to remove Ca^{2+}. Yet it should be noted that when $Ca(OH)_2$ is added, there are 2 moles of OH^- per mole of Ca^{2+}. Two moles of OH^- neutralize 2 moles of HCO_3^- and liberate 2 moles of CO_3^{2-}, thus precipitating 2 moles of Ca^{2+}—one that was added and one that was originally in the hard water.

A fourth method to soften water is to tie up the Ca^{2+} so that it becomes harmless. One way to do this is to form a complex containing Ca^{2+}. Certain phosphates, such as $(NaPO_3)_n$, sodium polyphosphate, presumably form such complexes in which the Ca^{2+} is trapped by the phosphate.

The fifth and most ingenious method of softening water is to replace the offending calcium ion by a harmless one such as Na^+. This is done by the process called *ion exchange*. An *ion exchanger* is a special type of giant

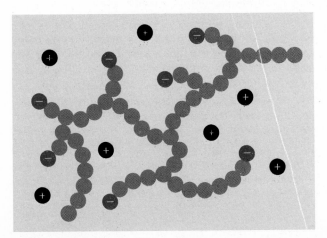

Figure 15.6
Ion exchanger.

molecule, which is pictured in part in Figure 15.6. The colored circles represent groups, usually CH_2, which are covalently bound to each other in the form of a three-dimensional, cross-linked network. The negatively marked circles represent covalently bound groups which carry an excess of negative charge. The molecule is thus a negatively charged network with a very porous structure. The pores are filled with water molecules (not shown) and enough positive ions to give an electrically neutral structure. The identity of the positive ions is not very important, since their only function is to preserve electric neutrality. Consequently, one type of cation such as Ca^{2+} can take the place of another type such as Na^+ without much change in structure. It is this kind of ion exchange which is used in water softening. Hard water containing Ca^{2+} is placed in contact with an ion exchanger whose mobile ion is Na^+. Exchange occurs which can be represented by the equilibrium

$$Ca^{2+} + 2Na_{\ominus}^+ \rightleftharpoons 2Na^+ + {}_{\ominus}Ca_{\ominus}^{2+}$$

where the negative circles represent a negative site on the exchanger. The equilibrium constant for this reaction is usually of the order of 10 or less, and therefore, in order to remove all the Ca^{2+}, it is necessary to run the hard water through a large amount of ion exchanger. This is most conveniently done by pouring the water through a tube, a foot or more high, filled with ion exchanger. Once the exchanger has given up its supply of Na^+, it cannot soften water further. However, it can be regenerated by exposure to concentrated solutions of NaCl, which reverses the above reaction.

The ion exchangers originally used for softening water were naturally occurring silicate minerals called *zeolites*. The giant network of a zeolite is negatively charged and is composed of covalently bound silicon, oxygen, and aluminum atoms. Mobile Na^+ ions in the pores can be readily exchanged for Ca^{2+} ions. Zeolites are very closely related in structure to the clays, which also show ion exchange. Here, ion exchange is important for plant nutrition, since many plants receive nourishment from the soil in this fashion.

Just before World War II, chemists were able to synthesize ion exchangers superior to the zeolites. The most common synthetic exchanger consists of a giant hydrocarbon framework having a negative charge due to covalently bound SO_3^- groups. It has also been possible to prepare ion exchangers in which the giant network is positively charged, the charge being due to covalently bound groups of the type $N(CH_3)_3^+$. Such positively charged networks can function as *anion* exchangers; i.e., they have mobile negative ions which can be displaced by other anions.

Combination of synthetic anion exchangers with cation exchangers has made possible the removal of all ions from a salt solution. If a salt solution containing M^+ and A^- is first run through a cation exchanger whose exchangeable ions are H^+, the salt solution is completely converted to a solution of an acid containing H^+ and A^-. If the acid solution is run through an anion exchanger whose exchangeable ions are OH^-, the anions A^- in the solution are replaced by OH^-. Since in the original solution the number of negative charges is exactly equal to the number of positive charges, equal amounts of H^+ and OH^- are produced in the solution. Neutralization occurs, and pure water results. Water thus "deionized" contains fewer ions than the most carefully distilled water.

15.5   Aluminum

Aluminum is the second element of group III of the periodic table, the other elements of which are boron, gallium, indium, and thallium. The element boron is not a metal and is discussed in Chapter 17; the other elements are metals but are rarely encountered.

Occurrence   Aluminum is the most abundant metal and, in fact, is the third most abundant element (8 percent by weight) of the earth's crust, but it is of secondary importance to iron, partly because of the difficulties in its preparation. It occurs primarily as complex aluminum silicates, such as feldspar ($KAlSi_3O_8$), the constituent of most common rocks, from which it is economically unfeasible to separate pure aluminum. Further, unless the product Al is completely free of iron and silicon, its properties are practically useless. Fortunately, there are natural deposits of oxide in the form of bauxite ($Al_2O_3 \cdot xH_2O$) from which pure Al can be obtained by electrolytic reduction. However, before electrolysis is carried out, it is necessary to remove iron and silicon impurities from the ore.

Preparation   Purification of bauxite is accomplished by the Bayer process, which makes use of the amphoterism of aluminum. The crude oxide is treated with hot NaOH solution, in which the aluminum oxide dissolves because of the formation of aluminate ion $[Al(OH)_4^-]$. Silicon oxide also dissolves (to form silicate ions), but ferric oxide stays undissolved, since Fe_2O_3, unlike Al_2O_3, is not amphoteric. The solution is filtered to remove Fe_2O_3 and cooled. On agitation with air and addition of crystalline aluminum hydroxide as a seed, aluminum hydroxide precipitates, leaving the silicate in solution.

The production of metallic aluminum from the above purified bauxite is carried out by the Hall-Héroult process. Bauxite, dissolved in a molten mixture of fluorides, such as cryolite† (Na_3AlF_6), calcium fluoride, and sodium fluoride, is electrolyzed at about 1000°C in cells like that represented

---

†The mineral cryolite occurs in nature almost exclusively as an enormous geologic dike in Greenland. In appearance, the mineral looks like glacial ice. Since it can be melted in a candle flame, it was thought by the Eskimos to be a special kind of ice. The name cryolite comes from the Greek *krios* (frost) and *lithos* (stone).

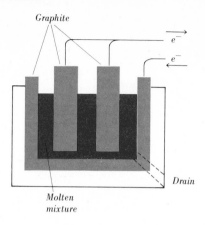

Graphite

e^-

e^-

Drain

Molten
mixture

Figure 15.7

Electrolytic preparation of aluminum.

schematically in Figure 15.7. The anode consists of graphite (carbon) rods dipping into the molten mixture; the cathode is a graphite lining supported by an iron box. The electrode reactions are very complicated and only imperfectly understood. At the cathode, oxyfluoaluminum complex ions are reduced to liquid aluminum (melting point 659.7°C). At the anode, a series of products is formed including oxygen, fluorine, and various carbon compounds of these elements. The carbon anodes gradually corrode away and must be replaced periodically. Continual addition of bauxite and recurrent draining off of the liquid aluminum allow uninterrupted operation. Because the weight of 1 equiv (weight which picks up the Avogadro number of electrons) of aluminum is so low, only 9 g, electric-power consumption is high. Consequently, the process is economically feasible only near cheap sources of electric current.

Properties   Pure aluminum is quite soft and weak but, when alloyed with other metals, becomes quite strong. Because it is so light (density 2.7 g/cm³), aluminum finds extensive use as a structural material. Although chemically active, it resists corrosion because of a self-protective oxide coat. It is also a good conductor of heat and electricity and so is used in cooking utensils and electrical equipment.

Although not so active as group I and II metals, aluminum is an excellent reducing agent, as shown by the reduction potential

$$Al^{3+} + 3e^- \longrightarrow Al(s) \qquad -1.66 \text{ volts}$$

Just as in the case of the alkaline-earth elements, it is the hydration of the ion which enables the reaction to proceed (from right to left). It has been estimated that over 4000 kilojoules of heat is evolved when 1 mole of Al^{3+} ions is hydrated. The main reasons for this great hydration energy are the high charge of Al^{3+} and its small size (0.052-nm radius).

The high oxidation potential indicates that aluminum should reduce water, but the reaction is too slow to detect, probably because of the oxide coat. However, the oxide (being amphoteric) is soluble in acid and in base, and consequently Al liberates hydrogen from both acid and basic solutions. The net reactions may be written

$$2Al(s) + 6H^+ \longrightarrow 2Al^{3+} + 3H_2(g)$$
$$2Al(s) + 2OH^- + 6H_2O \longrightarrow 2Al(OH)_4^- + 3H_2(g)$$

The first of these equations seems to imply that aluminum dissolves in any acid. However, this is not the case. It is true that Al dissolves readily in hydrochloric acid, but in nitric acid no visible reaction occurs. The inertness is attributed to an oxide coat. A coating of Al_2O_3 should be quite stable because of the great strength of the Al—O bond.

Further indication of the great affinity of aluminum for oxygen comes from the high heat of formation of Al_2O_3. When aluminum burns in air to form solid Al_2O_3,

$$2Al(s) + \tfrac{3}{2}O_2(g) \longrightarrow Al_2O_3(s) \qquad \Delta H = -1670 \text{ kilojoules}$$

a large amount of heat is evolved, which can be used effectively in the reduction of less stable oxides. For example, since it requires 197 kcal to decompose 1 mole of Fe_2O_3 into the elements, Al can reduce Fe_2O_3 with energy left over. The overall reaction can be considered to be the sum of two separate reactions,

$$
\begin{array}{lll}
2Al(s) + \tfrac{3}{2}\,O_2(g) \longrightarrow Al_2O_3(s) & \Delta H = & -1670 \text{ kilojoules} \\
\underline{\hphantom{2Al(s)\,}Fe_2O_3(s) \longrightarrow 2Fe(s) + \tfrac{3}{2}\,O_2(g)} & \underline{\Delta H = } & \underline{+820 \text{ kilojoules}} \\
2Al(s) + Fe_2O_3(s) \longrightarrow 2Fe(s) + Al_2O_3(s) & \Delta H = & -850 \text{ kilojoules}
\end{array}
$$

Actually, when the reaction is carried out, the heat evolved is sufficient to produce Fe and Al_2O_3 in the molten state. The production of molten iron by this reaction, frequently called the *thermite reaction,* has been used for welding operations. Because of the high temperature that results (estimated at 3000°C), it has also been used in incendiary bombs.

Often, in the preparation of pure metals from their oxides, the common reducing agents hydrogen and carbon are unsuitable because of the formation of hydrides and carbides. In such cases, aluminum is sometimes used for the reduction, as, for example, in the preparation of manganese and chromium from their oxides. The reduction of oxides with Al is called the Goldschmidt reaction and owes its success to the great stability of Al_2O_3.

Aqueous solutions of most aluminum salts are acid because of hydrolysis of Al^{3+} (Section 12.9). The ion is certainly hydrated, and its formula is probably $Al(H_2O)_6^{3+}$. When base is progressively added, a white, gelatinous precipitate is formed. This precipitate, variously formulated as $Al(OH)_3$ or $Al_2O_3 \cdot xH_2O$, is readily soluble in acid or excess base, but only if freshly precipitated. On standing, it progressively becomes more difficult to dissolve. The explanation suggested for this "aging" is that oxygen bridges are formed between neighboring aluminum atoms. In basic solutions, aluminum forms aluminate ion, $Al(OH)_4^-$.

Compounds   Like other 3+ ions, aluminum ion may be crystallized (usually by slow evaporation of water) from aqueous solutions containing sulfate and singly charged cations to give *alums*. These alums are double salts having the general formula $MM'(SO_4)_2 \cdot 12H_2O$, where M is a single charged cation, such as K^+, Na^+, or NH_4^+, and M' is a triply charged cation, such as Al^{3+}, Fe^{3+}, or Cr^{3+}. Ordinary alum is $KAl(SO_4)_2 \cdot 12H_2O$. Of the 12 hydrate water molecules, 6 are bound directly to the aluminum to give a distinct $Al(H_2O)_6^{3+}$ ion. The other six water molecules are symmetrically placed about the K^+ ion, but there is no distinct $K(H_2O)_6^+$ ion. The crystals of alum are usually large octahedra and have great chemical purity. Because of this purity, $KAl(SO_4)_2 \cdot 12H_2O$ is useful in the dyeing industry, where the alum serves as a source of Al^{3+} uncontaminated by Fe^{3+}. The Al^{3+} is precipitated on cloth as aluminum hydroxide, which acts as a binding agent (mordant) for dyes. The absence of Fe^{3+} is imperative for producing clear colors.

When aluminum hydroxide is heated to high temperature, it loses water and eventually forms Al_2O_3, sometimes called *alumina*. This is a very inert material of high melting point (about 2000°C), which finds use in making containers for high-temperature reactions. Ordinarily, alumina is white, but it can be colored by the addition of such oxides as Cr_2O_3 or Fe_3O_4. Synthetic rubies, for example, can be made by mixing Al_2O_3 and Cr_2O_3 powders and dropping them through the flame of an oxyhydrogen torch. Because of the great hardness of Al_2O_3, such synthetic jewels are used as bearing points in watches and other precision instruments.

Qualitative analysis   Like the alkali and alkaline-earth elements, aluminum cannot be precipitated as the sulfide from aqueous solution. In the usual schemes of analysis, aluminum precipitates as the hydroxide when NH_3 is added to the solution from which H_2S has removed the acid-insoluble sulfides. Aluminum can be separated from other cations which precipitate as sulfides and hydroxides at this point by taking advantage of the fact that of these cations only Al^{3+}, Cr^{3+}, and Zn^{2+} are amphoteric. Zinc can be differentiated from aluminum, either by using the fact that ZnS but not $Al(OH)_3$ precipitates when $(NH_4)_2S$ is added in the presence of an SO_4^{2-}–HSO_4^- buffer or by using the fact that $Zn(OH)_2$ but not $Al(OH)_3$ is soluble in excess ammonia. Chromium can be differentiated from aluminum by oxidizing the chromium in basic solution with H_2O_2 to CrO_4^{2-}, which can be precipitated as yellow, insoluble $PbCrO_4$, or $BaCrO_4$. A possible confirmatory test for aluminum is the formation of a red precipitate from $Al(OH)_3$ and the dye aluminon.

15.6 Tin and Lead

These two elements are the bottom two in group IV, which contains carbon (C), silicon (Si), germanium (Ge), tin (Sn), and lead (Pb). Group IV is a good example of how metallic character increases in going down a group. Carbon, in the form of diamond, is a good insulator; silicon and germanium, when very pure, are also insulators, though they can be made conducting by the presence of impurities as mentioned in Section 7.6. Tin and lead are metals, but one form of tin can be obtained that is nonmetallic. The reason for the increase in metallic character in going down the group is the progressively weaker binding of the outer electrons. Furthermore, the atomic size is increasing, which not only explains why the electron binding is weaker but also suggests that the tendency to form covalent bonds decreases. Small atoms such as carbon generally tend to form covalent bonds and concentrate electron density along preferred bond directions in space; large atoms such as lead tend to spread the electron density more uniformly. Small atoms have few near neighbors in the solid state (4 in the case of carbon in the form of diamond); large atoms have relatively many near neighbors (12 in the case of lead).

Tin   Tin occurs principally as the mineral cassiterite, SnO_2, from which the element is produced by reduction with carbon. The ordinary form of tin, stable at room temperature, is called white tin and is metallic. There is another form, called grey tin, stable below 13°C, which is much less dense and is nonmetallic. In the folklore of Europe there is frequent reference to "tin disease," a mysterious cancerous growth that occasionally appeared on tin organ pipes in cathedrals. Although the superstitious peasants were wont to attribute the pest to the work of Satan, and savants were equally sure it was due to microorganisms, the low ambient temperature so characteristic of these cathedrals was probably just bringing about a structural change to a more stable thermodynamic state of the element. Metallic tin is rather inert and resists corrosion well, partly because of an invisible, tightly adhering, oxide coat. Because of its inertness, tin is widely used as a protective plating on steel, as, for example, in making "tin cans." The steel is coated either by being dipped in molten tin or by being made the cathode in an electrolytic bath which contains dissolved tin salts. Tin-plated steel, although less expensive than zinc-plated steel (called "galvanized iron"), has the disadvantage of rusting very rapidly when the tin coat is punctured.

Two series of tin compounds, stannous (2+) and stannic (4+), are known. The 2+ state is formed when metallic tin is dissolved in acid solution; however, the rate of reaction is rather slow. In solution, the Sn^{2+} ion is colorless and hydrolyzes according to the reaction

$$Sn^{2+} + H_2O \rightleftharpoons SnOH^+ + H^+$$

Gradual addition of base to solutions of stannous salts precipitates a white solid usually described as stannous hydroxide [$Sn(OH)_2$]. Further addition of base dissolves the precipitate to form stannite ion, which is written both as $Sn(OH)_3^-$ and $HSnO_2^-$. Stannite ion is a powerful reducing agent. Furthermore, on standing, solutions of stannite disproportionate to give the oxidation states 0 and 4+,

$$2Sn(OH)_3^- \longrightarrow Sn(s) + Sn(OH)_6^{2-}$$

In acid solution, stannous ion is frequently complexed with anions. For example, in chloride solutions the whole series $SnCl^+$, $SnCl_2$, $SnCl_3^-$, and $SnCl_4^{2-}$ has been identified. Solutions of stannous chloride are frequently used as convenient, mild reducing agents. The reducing species in these solutions is usually represented as Sn^{2+}.

In the stannic state, tin is often represented as the simple Sn^{4+} ion. However, because of its high charge, Sn^{4+} probably does not exist as such in aqueous solution but is extensively hydrolyzed, even in quite acid solutions. When base is added to stannic solutions, a white precipitate forms, which may be $Sn(OH)_4$ or, more probably, a hydrated oxide, $SnO_2 \cdot xH_2O$. The precipitate is soluble in excess base to give stannate ion, usually written $Sn(OH)_6^{2-}$ or SnO_3^{2-}.

Both stannous and stannic sulfides are insoluble in water and can be precipitated by H_2S in acid solution. Stannic sulfide (SnS_2) is a yellow solid which is soluble in high concentrations of sulfide ion. The reaction can be written

$$SnS_2(s) + S^{2-} \longrightarrow SnS_3^{2-}$$

The complex ion, SnS_3^{2-}, is called the *thiostannate ion* and is analogous to stannate ion, SnO_3^{2-}. The dissolving of SnS_2 in excess S^{2-} can be used to distinguish it from another yellow sulfide, CdS. Owing to the stability of the thiostannate ion, brown-black insoluble stannous sulfide, SnS, can be oxidized by the relatively poor oxidizing agent S_2^{2-}, polysulfide ion.

$$SnS(s) + S_2^{2-} \longrightarrow SnS_3^{2-}$$

When solutions of thiostannate are acidified, SnS_2 is precipitated.

Lead   Lead occurs as the mineral galena (PbS), from which the element is produced in several different ways. In one of these, the sulfide ore is roasted in air until it is completely converted to the oxide, which is then reduced with carbon in a small blast furnace.

$$2PbS(s) + 3O_2(g) \longrightarrow 2PbO(s) + 2SO_2(g)$$
$$2PbO(s) + C(s) \longrightarrow 2Pb(l) + CO_2(g)$$

In an alternative process, the sulfide ore is only partially oxidized by air, the product containing a mixture of PbO, PbS, and $PbSO_4$. This mixture is then smelted (i.e., fused) in the absence of air, with the result that the PbS reduces PbO and $PbSO_4$ to lead.

$$PbS(s) + 2PbO(s) \longrightarrow 3Pb(l) + SO_2(g)$$
$$PbS(s) + PbSO_4(s) \longrightarrow 2Pb(l) + 2SO_2(g)$$

The crude lead may contain impurities such as antimony, copper, and silver. If lead of high purity is required, it can be refined by an electrolytic process. Pure lead is a soft, low-melting metal which, when freshly cut, has a silvery luster that rapidly dulls on exposure to air. The tarnishing is due to the formation of a surface coat of oxides and carbonates. Primary uses of lead are in the manufacture of lead storage batteries (Section 13.7), alloys such as type metal and solder, white lead paint (hydrated lead hydroxycarbonate), and as "antiknock" additive to gasoline (tetraethyl lead).

Practically all the common lead compounds correspond to lead in the 2+ state. This state is called *plumbous,* from the Latin name for the element, *plumbum.* When the halide concentration of plumbous solutions is increased, insoluble plumbous halides form. In excess halide ion, the precipitates redissolve, presumably because of the formation of complex ions of the type $PbCl_3{}^-$ and $PbBr_4{}^{2-}$. Unlike two other common insoluble chlorides, AgCl and Hg_2Cl_2, $PbCl_2$ can be dissolved by raising the temperature of its saturated solution.

Plumbous ion hydrolyzes somewhat less than stannous ion. When base is added, white $Pb(OH)_2$ is precipitated. Being amphoteric, it dissolves in excess base to form plumbite ion [$Pb(OH)_3{}^-$ or $HPbO_2{}^-$]. Unlike stannite ion, plumbite is stable in solution. With most 2− anions, Pb^{2+} forms insoluble salts, for example, $PbSO_4$, $PbCO_3$, PbS, $PbCrO_4$, $PbHPO_4$. Lead sulfide is the least soluble of these, and the others convert to it in the presence of sulfide ion.

The principal compound of lead in the 4+, or plumbic, state is PbO_2, lead dioxide. This compound, used in large amounts for the cathode of lead storage batteries, can be made by oxidation of plumbite ion with hypochlorite ion in basic solution. The reaction can be written

$$Pb(OH)_3^- + ClO^- \longrightarrow Cl^- + PbO_2(s) + OH^- + H_2O$$

With acid solutions, PbO_2 is a potent oxidizing agent,

$$PbO_2(s) + 4H^+ + 2e^- \longrightarrow Pb^{2+} + 2H_2O \qquad + 1.46 \text{ volts}$$

which is made even more potent in the presence of concentrated acid and anions which precipitate Pb^{2+}. In very concentrated solutions of base, PbO_2 dissolves to form plumbates, such as PbO_4^{4-}, PbO_3^{2-}, $Pb(OH)_6^{2-}$. Red lead (Pb_3O_4), much used as an undercoat for painting structural steel, can be considered to be plumbous plumbate (Pb_2PbO_4). Its use in preventing corrosion depends on the fact that, as a strong oxidizing agent, it renders iron passive.

Like most heavy metals, lead and its compounds are poisonous. Fairly large doses are required for toxicity, but the danger is amplified because the lead tends to accumulate in the body (central nervous system). The toxicity may be due to the fact that lead and other heavy metals are powerful inhibitors of enzyme reactions. Because of lead toxicity, use of lead compounds in paints and in gasoline additives is being discouraged. In the latter application, lead is volatilized in automobile engines and exhausted to the atmosphere where dangerous levels may build up in areas of heavy automobile traffic.

Analysis of an unknown solution for possible presence of lead or tin depends on the fact that lead and tin precipitate as sulfides in 0.3 N acid solution, although much of the lead may precipitate as white $PbCl_2$ along with $AgCl$ and Hg_2Cl_2 when HCl is added to the original unknown. $PbCl_2$ can be separated from the other two chlorides by leaching with hot water. Addition of K_2CrO_4 and acetic acid to the leach solution gives the confirmatory yellow precipitate, $PbCrO_4$.

Lead sulfide (black) can be separated from tin sulfide, either SnS (brown-black) or SnS_2 (yellow), by treatment with ammonium polysulfide, which converts SnS and SnS_2 to SnS_3^{2-} but leaves the PbS undissolved. PbS can be dissolved with hot HNO_3 and reprecipitated as a white sulfate with H_2SO_4. To confirm, the $PbSO_4$ is dissolved in ammonium acetate and precipitated as $PbCrO_4$.

If, to a solution containing SnS_3^{2-}, HCl is added in excess, the tin stays in solution, probably as a chloride complex. Evaporation (to drive off H_2S) in the presence of an iron wire (to reduce the tin) followed by $HgCl_2$ addition confirms tin if a precipitate of white Hg_2Cl_2 or black Hg is observed.

QUESTIONS

15.1 *Metallic Properties and Structure* Compare and contrast the metal sodium with the nonmetal diamond with respect to each of the following: (a) structure of the solid, (b) observed properties. (c) Explain how your answer to (a) accounts for your answer to (b).

15.2 *Alkali Metals* In going down the alkali-metal group atoms: (a) How do the ionization potentials and ionic radii change? (b) How do the melting and boiling points change? (c) Give reasons for all four trends.

15.3 *Alkaline-Earth Metals* State which member of the alkaline-earth elements best fits each of the following: (a) most abundant, (b) highest ionization potential, (c) extracted from seawater, (d) salts hydrolyze to give acid solutions, (e) oxide is amphoteric, (f) radioactive, (g) used in some photocells.

15.4 *Hard Water* Consider a sample of hard water that contains the following ions: K^+, Ca^{2+}, HCO_3^-, SO_4^{2-}. (a) Which ion, or ions, is responsible for the water being classed as hard and what does that mean? (b) What might be observed if this water is heated to boiling? (c) Is the boiled water hard? Explain.

15.5 *Ion Exchange* Water that is softened by household ion exchangers may be unsuitable for watering household plants because it may contain too high a level of Na^+. (a) How does the Na^+ get there? (b) How could the water be softened, using ion exchange, and not add any positive ions?

15.6 *Aluminum* Write net equations for each of the following changes: (a) Bauxite is dissolved in hot NaOH solution. (b) A dissolved aluminum salt gives a slightly acid solution. (c) Al reduces Cr_2O_3 in the Goldschmidt process. (d) Ordinary alum dissolves in water. (e) Aluminum hydroxide is heated to high temperature. (f) Aluminum is precipitated in qualitative analysis by adding NH_3 to solution.

15.7 *Tin* Write net equations and show clearly the substance oxidized and the substance reduced in each of the following changes: (a) Tin dissolves in strong acid. (b) Stannite ion disproportionates. (c) Stannous sulfide is dissolved by polysulfide, S_2^{2-}.

15.8 *Lead* In one process for preparing lead, galena (PbS) is partially oxidized to PbO and $PbSO_4$. Suppose you were preparing lead this way and in one batch you obtain an equal number of moles of PbS, PbO, and $PbSO_4$. What, and how much, should you add prior to smelting?

15.9* *Electrolysis* Suppose you manufacture magnesium and aluminum by electrolysis. Your only significant cost is the electricity, and you recover no by-products. (a) Which costs you more to make, Al or Mg? (b) How many times as much per pound?

15.10 *Stalactites* What role does CO_2 in the atmosphere play in stalactite formation?

15.11 *Reduction Potentials* Ignoring radium, which ones of the alkaline-earth elements could reduce from solution which of the alkali elements?

15.12 *Metallic Properties* Account for each of the following: (a) The conductivity of aluminum increases as it is purified. (b) The conductivity of aluminum increases as it is cooled. (c) Aluminum is softer than a transition metal. (d) Aluminum foil eventually loses its metallic appearance.

15.13* *Solubilities* K_{sp} for $BaCrO_4$ is 8.5×10^{-11} and that for $SrCrO_4$ is 3.6×10^{-5}. (a) Which is more soluble, $BaCrO_4$ or $SrCrO_4$? (b) How many times as soluble in pure water? (c) How many times as soluble in 0.10 M Na_2CrO_4?

15.14 *Gas Calculation* You want to fill balloons with hydrogen, despite its danger. If each balloon holds 5 liters at 1 atm and 27°C, how many balloons can you fill from the hydrogen generated by 1.00 kg of CaH_2?

15.15 *Reduction Potentials* Given

$$Na^+ + e^- \longrightarrow Na(s) \qquad -2.71 \text{ volts}$$
$$Rb^+ + e^- \longrightarrow Rb(s) \qquad -2.93 \text{ volts}$$
$$Ba^{2+} + 2e^- \longrightarrow Ba(s) \qquad -2.90 \text{ volts}$$

From this list, pick out (a) the best reducing agent, (b) the best oxidizing agent, and (c) which of these species can reduce the Cs^+ ion.

15.16 *Stoichiometry* What is the volume of H_2 at STP produced by the reaction of 0.230 g of sodium and water?

15.17* *Weight Percent* The density of 2.30 M CsCl solution is 1.29 g/ml. What is the percent of CsCl by weight in the solution?

15.18 *Solubility* Account for the difference in solubility trend of sulfates and hydroxides in going down the alkaline-earth metal group.

15.19 *Solubility* What is the effect of each of the following on solubility: (a) low lattice energy, (b) high hydration energy? Under what conditions might a solid have a relatively high hydration energy without having a correspondingly high lattice energy?

15.20 *Analysis* Given that a solution contains one of the following ions: Pb^{2+}, Sn^{4+}, Ag^+. Describe how you could identify an unknown ion.

15.21 *Thermodynamics* For the elements Be, Mg, Ca, and Sr, the corresponding heats of fusion (kilojoules/mole) are 9.79, 9.03, 9.33, and 9.16 respectively. Using the data in Figure 15.5, calculate for each of the elements the ratio of its heat of fusion to its melting point (in K). Based on your finding, comment on the significance of the ratio.

15.22 *Solubility Product* If the solubility product of $Mg(OH)_2$ is equal to 8.9×10^{-12}, what will be the pH of a saturated solution of $Mg(OH)_2(s)$ in water?

15.23 *Chemical Reactions* Write a balanced equation for the following reactions: (a) hydrogen prepared by dissolving magnesium in water, (b) hydrogen prepared by dissolving aluminum in a solution of strong base, (c) reacting $LiNO_3(s)$ with solid lithium, (d) heating potassium metal in hydrogen gas.

15.24 *Rubidium* To prepare alkali elements it is necessary to reduce the ion. Give an equation for the preparation of Rb from RbCl (a) electrochemically, and (b) chemically.

15.25* *Electrolysis* If 0.50 liter of 0.750 M NaCl is electrolyzed for 10.0 min at 0.600 ampere, what should be the pH of the residual solution?

15.26 *Hydration Energy* 4000 kilojoules of heat are evolved when 1 mole of Al^{3+} is hydrated. How much heat will be evolved when the $Al^{3+}(g)$ contained in 15.0 g of $Al(NO_3)_3$ is hydrated?

15.27* *Stoichiometry* A bottle contains 8.50 g of $KMnO_4$ per liter. How many liters will be needed to oxidize 10.0 g of Fe^{2+} in acid solution?

15.28 *Plaster of Paris* How many kilograms of gypsum ($CaSO_4 \cdot 2H_2O$) are needed to make a kilogram of plaster of paris ($CaSO_4 \cdot \frac{1}{2} H_2O$)?

15.29* *Heat of Reaction* Using the data in this chapter, tell how many kilojoules would be liberated in each of the following: (a) slaking 1.00 kg of lime, (b) reaction of 1.00 kg of Al in the thermite reaction. (c) Recalling that it takes 4.184 joules to raise the temperature of one gram of water by one degree, how many liters of water can be heated by 20° in (a) and in (b)?

15.30 *Salts* Consider the following salts: $RbCl$, $PbSO_4$, $CaCl_2$, $AlCl_3$, $SrCO_3$, $PbCl_2$, BaS. (a) Which are insoluble in water at 25°C? (b) Of those which dissolve, which hydrolyzes to give an acid solution? (Write net equation.) (c) Of those which dissolve, which hydrolyzes to give a basic solution? (Write net equation.)

15.31* *Magnesium* As noted, the magnesium-ion content of seawater is 0.13 percent. (a) Assuming a density of 1.0 g/ml for seawater, how many liters of seawater are needed to produce 1.00 kg of Mg? (b) What would be the minimum weight of lime needed in the process? (c) If HCl were introduced as a gas, what is the minimum gas volume at 80.0 atm and 25°C needed in the process? (d) If the final electrolysis occurred at an average current of 5.00 amperes, how long would it take?

15.32 *Solubility* For $Mg(OH)_2$, $K_{sp} = 8.9 \times 10^{-12}$. Calculate the number of moles of $Mg(OH)_2$ which will dissolve in each of the following: (a) 1 liter of water, (b) 200 ml of water, (c) 1 liter of 0.100 M $MgCl_2$, (d) 1 liter of 0.100 M $Ba(OH)_2$, (e) 200 ml of 0.020 M $MgCl_2$ buffered at pH = 9.00.

15.33 *Trends* Compare the elements sodium, magnesium, and aluminum with respect to (a) electronic structure of ion, (b) radius of ion, (c) reduction potential, (d) hydrolysis of salts, (e) solubility of hydroxide in water, acid, and base.

15.34 *Occurrence in Nature* Of the elements discussed in this chapter: (a) Which ones are considered quite poisonous? (b) Which ones occur in nature as water-soluble compounds? (c) Which ones of those in (b) are reasonably abundant? (d) Comparing all three of your answers, what do you make of this (is it just good luck)?

The Transition Elements

In the periodic table of elements there are 10 subgroups containing 61 elements that intervene between main groups II and III. They are called transition elements because they involve an electronic expansion in the next-to-outermost shell before the steady buildup of the outermost electron shell is resumed. All 61 of these elements are metals, due to the presence of one or two electrons in the outermost shell, but their special chemical properties are largely due to incomplete inner shells. In this chapter we consider first the general properties that characterize many of the transition elements and then take up some of the representative elements in detail.

16.1 Electronic Configurations

In the normal buildup of the atoms of the periodic table, electrons are added one at a time while the nuclear charge is stepped up in increments of $1+$. The special position of the transition elements in this atomic buildup comes about because there is a relative shifting of an atom's electronic energy levels as the nuclear charge and the number of electrons in the atom increase. Specifically, for the first row of transition elements from scandium ($Z = 21$) to zinc ($Z = 30$), there is an interchange in the relative position of the $3d$ and the $4s$ energy levels as Z is increased. Whereas the $3d$ level is higher in energy than the $4s$ for $Z = 21$, by the time we get to $Z = 30$, the $3d$ level has dropped to a position considerably lower than the 4s. The reasons for the crossover of the levels are connected with the fact that s electrons penetrate deeply into the core of an atom whereas d electrons are mainly confined to the outer reaches. One way of thinking of the problem is to picture a sudden shrinkage in the average radius of the 3d subshell when Z passes 21 and electrons then begin filling to the 3d subshell.

Figure 16.1 shows the electronic configurations of the first-row transition elements. As can be seen, the number of electrons in the 3d subshell increases steadily but the number in the 4s subshell stays at two. There are

Figure 16.1 ELECTRON CONFIGURATIONS OF FIRST-ROW TRANSITION ELEMENTS	Element	Symbol	Z	Electron Configuration
	Scandium	Sc	21	$1s^2 2s^2 2p^6 3s^2 3p^6 3d^1 4s^2$
	Titanium	Ti	22	——(18)——$3d^2 4s^2$
	Vanadium	V	23	——(18)——$3d^3 4s^2$
	Chromium	Cr	24	——(18)——$3d^5 4s^1$
	Manganese	Mn	25	——(18)——$3d^5 4s^2$
	Iron	Fe	26	——(18)——$3d^6 4s^2$
	Cobalt	Co	27	——(18)——$3d^7 4s^2$
	Nickel	Ni	28	——(18)——$3d^8 4s^2$
	Copper	Cu	29	——(18)——$3d^{10} 4s^1$
	Zinc	Zn	30	——(18)——$3d^{10} 4s^2$

anomalies at Cr ($Z = 24$), which has configuration $3d^5 4s^1$ instead of $3d^4 4s^2$, and at Cu ($Z = 29$), which has configuration $3d^{10} 4s^1$ instead of $3d^9 4s^2$. A steady progression would have predicted that x in $3d^x 4s^2$ should smoothly increase from 1 to 10. The apparent anomalies have been ascribed to special stability associated with half-filled and filled subshells of electrons.

When electrons are removed from an atom, as in the process of ionization, one is tempted to conclude that the last electron to go on in the atom buildup is the first one to come off in the ionization. Since the 3d electrons are added after the 4s, it would appear that on ionization they should be removed before the 4s. However, the prediction is unwarranted because the two processes are different in a major way. In the buildup of the periodic table, the number of electrons is increased at the same time that the nuclear charge is increased. On the other hand, in the ionization process the number of electrons is decreased while the nuclear charge stays constant. The problem is actually a complicated one since interelectron repulsion is an important factor. The experimental fact is that 4s electrons are removed before 3d electrons in the ionization of first-row transition-element atoms.

For the second-row transition elements, yttrium ($Z = 39$) through cadmium ($Z = 48$), the electronic expansion involves the 4d and 5s subshells. For the third-row transition elements, lanthanum ($Z = 57$) through mercury

($Z = 80$), a new problem arises. Not only are the 5d and 6s subshells involved in the expansion but also the 4f subshell, even deeper in the atom, is being filled to 14 electrons. The 14 elements just following lanthanum (from cerium with $Z = 58$ through lutetium with $Z = 71$) belong to this sequence and are called the *lanthanides*. A similar problem involving the 5f expansion occurs with elements 90 through 103 in the last row of the periodic table, giving rise to the *actinide* elements.

16.2   General Properties

Because the electron expansion occurs mainly in a shell other than the outermost, the transition elements show considerable similarity within a horizontal sequence. In fact, in some cases the horizontal similarity through a period is greater than the vertical resemblance down a subgroup.

The most characteristic property of the transition elements is that they are all metals. This is not surprising, since the outermost shell contains so few electrons. However, unlike the metals discussed in the preceding chapter, the transition metals are likely to be hard, brittle, and fairly high-melting. The difference is due partly to the extremely small size of the atoms and partly to the existence of some covalent bonding between the ions. There are exceptions to this general hardness, as in the case of mercury ($Z = 80$), which is a liquid at room temperature and is about as soft as a metal can be.

Another characteristic property of the transition elements is that, in forming compounds, they exhibit many oxidation states, presumably because some of or all the 3d electrons can also be used along with 4s electrons in chemical bonding. Manganese ($Z = 25$), for example, has oxidation numbers of $2+$, $3+$, $4+$, $6+$, and $7+$, corresponding to use of none, one, two, four, and five of the 3d electrons, respectively. In each row of transition elements there is a peaking of maximum state near the middle of the row, after which the maximum state gets smaller in magnitude toward the end of the row. Thus, for the first transition row the maximum state increases regularly ($3+$ for Sc, $4+$ for Ti, $5+$ for V, $6+$ for Cr, $7+$ for Mn) after which there is a falloff ($6+$ for Fe, $4+$ for Co, $3+$ for Ni, $2+$ for Cu, and $2+$ for Zn). The falloff after the peak in the center of the row can be related to the progressive lowering of the 3d subshell energy relative to the 4s, and hence to decreasing availability of d electrons for binding.

Many of the compounds of the transition elements are *paramagnetic*, i.e., are weakly attracted by a magnetic field. Such paramagnetism suggests the existence of unpaired electrons in the compounds. Unpaired electrons are possible, since in forming compounds a transition element may not utilize all its d electrons. Those remaining may exist unpaired in individual d orbitals.

Another characteristic feature of the transition-element compounds is that most of them are colored, both as solid salts and in solution. Frequently in solution the color varies depending on what other ions are in the solution. For example, Fe^{3+} ion in the solid salt $Fe_2(SO_4)_3$ is almost colorless. Yet, when placed in aqueous solution, it becomes very pale yellow, and when a trace amount of KSCN (potassium thiocyanate) is added it turns to a deep blood-red. Color is a subjective property dependent largely on the beholder, and so names of colors do not mean much. However, instrumental analysis of the light transmitted by a solution confirms that different frequencies are absorbed by transition-metal ions in different environments. For example, if "white" light (which can be considered to be composed of a mixture of red, orange, yellow, green, blue, and violet) is passed through a solution containing Fe^{3+} surrounded by six H_2O molecules, it comes out almost white because none of the visible frequencies is absorbed, but when it is passed through a solution containing Fe^{3+} surrounded by six cyanide ions it comes out yellow because some of the blue light has been absorbed. As we will see in the next section, light absorption may result from a partially filled subshell.

A final characteristic of the transition elements is their pronounced tendency to form a large number of tightly bound complex ions. An example is Fe^{3+} surrounded by six cyanide ions in the complex ion $Fe(CN)_6^{3-}$, frequently called ferricyanide ion but more systematically called hexacyanoferrate(III) ion. The idea of a central metal atom with groups, neutral molecules or negative ions, attached to it was suggested first by Alfred Werner in 1893 as an explanation of the existence of a rather mysterious series of compounds such as the following: $CrCl_3 \cdot 6NH_3$, $CrCl_3 \cdot 5NH_3$, $CrCl_3 \cdot 4NH_3$, and $CrCl_3 \cdot 3NH_3$. Why should the compound $CrCl_3$ bind to itself respectively 6, 5, 4, or 3 moles of NH_3? The chemical reactivity of such compounds, sometimes called *coordination compounds*, gives a strong clue to what these structures are. When dissolved in aqueous solution and treated with silver nitrate, 1 mole of $CrCl_3 \cdot 6NH_3$ precipitates 3 moles of insoluble silver chloride, 1 mole of $CrCl_3 \cdot 5NH_3$ precipitates only 2 moles of AgCl, 1 mole of $CrCl_3 \cdot 4NH_3$ precipitates but 1 mole of AgCl, and 1 mole of $CrCl_3 \cdot 3NH_3$ precipitates no AgCl at all. Obviously the three chlorine atoms in $CrCl_3 \cdot 3NH_3$ are bound in an inaccessible way whereas the three chlorine atoms in $CrCl_3 \cdot 6NH_3$ are not locked in the same way.

To explain such coordination compounds, Werner suggested that metal atoms such as chromium could bind to themselves clusters of ions and molecules in such a way that the number of near neighbors was constant, which meant six for octahedral complexes. In the case of $CrCl_3 \cdot 6NH_3$, the actual structure is more accurately represented as $[Cr(NH_3)_6]^{3+}3Cl^-$, indicating that six ammonia molecules are bound octahedrally to and around a central chromium atom, the whole carrying a net charge of 3+, and that three chloride ions are elsewhere in the solid structure but attracted to the tripositive complex ion by normal electrostatic attraction between opposite charges. In $CrCl_3 \cdot 5NH_3$, the constituent species are believed to be $[Cr(NH_3)_5Cl]^{2+}2Cl^-$, indicating there are two kinds of chlorine atoms in the compound, one kind attached directly to the chromium atom in the grouping of six neighbors around the central atom of an octahedral complex and the other kind as free chloride ions. Presumably, only the two free chloride ions are precipitable by addition of Ag^+, the third chlorine being bound in a different way.

Figure 16.2 shows how the different compounds of $CrCl_3$ and NH_3 are composed. Figure 16.3 shows the structures of some of the octahedral groupings. For $Cr(NH_3)_6{}^{3+}$, there is but one way to arrange the six ammonia

Figure 16.2 OCTAHEDRAL CHROMIUM COMPLEXES	Overall Formula	Constituent Species
	$CrCl_3 \cdot 6NH_3$	$Cr(NH_3)_6{}^{3+}3Cl^-$
	$CrCl_3 \cdot 5NH_3$	$Cr(NH_3)_5Cl^{2+}2Cl^-$
	$CrCl_3 \cdot 4NH_3$	$Cr(NH_3)_4Cl_2{}^+Cl^-$
	$CrCl_3 \cdot 3NH_3$	$Cr(NH_3)_3Cl_3$

molecules at the corners of an octahedron centered on a chromium atom. Similarly, for $Cr(NH_3)_5Cl^{2+}$ there is but one way to substitute a chlorine atom for one of the six ammonia molecules. All the six corners of an octahedron are equivalent, and so it makes no difference where we substitute the Cl for the NH_3. (Unfortunately, when one draws an octahedron as we have done here, one gives the impression that the up and down positions are different from the other four disposed around the square. This is purely an accident of drawing since in a true octahedron all six vertices are absolutely equivalent to each other.) For the complex $Cr(NH_3)_4Cl_2{}^+$ there are two really

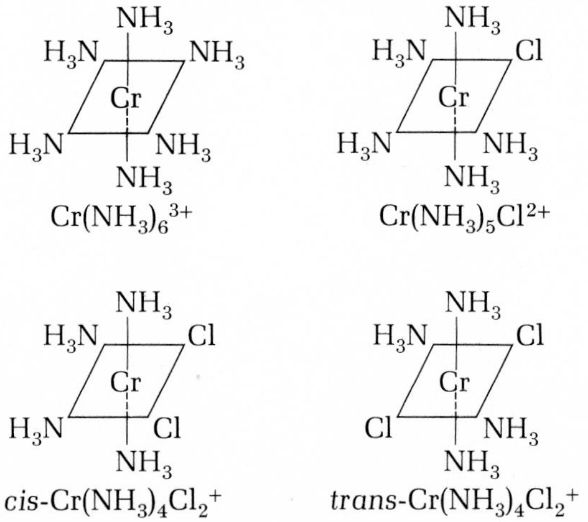

$Cr(NH_3)_6{}^{3+}$

$Cr(NH_3)_5Cl^{2+}$

cis-$Cr(NH_3)_4Cl_2{}^+$

trans-$Cr(NH_3)_4Cl_2{}^+$

Figure 16.3
Coordination complexes of chromium with ammonia and chlorine.

different ways of disposing the six neighbors around the chromium. In one case, called the *trans* complex, the two chlorine atoms are on diagonally opposite corners of the octahedron; in the other, called the *cis* complex, the two chlorine atoms are on adjacent corners of the octahedron.† The *trans* complex and the *cis* complex are two distinct chemical species with different chemical properties. They correspond to different spatial arrangements of the same atoms and are referred to as *isomers* (from the Greek meaning equal parts). As a matter of historical interest, it was the very existence of isomers that gave Werner the major clue to unlocking the secret of the structure of coordination complexes.

16.3 Ligand-Field Theory

There are thousands of different groups, including ions such as Cl^-, Br^-, I^-, and OH^- and neutral molecules such as NH_3 and H_2O, which can be attached to transition-metal atoms to give coordination complexes. The general term for these attaching groups is *ligand,* coming from the Latin word *ligare* meaning "to bind." The six ligands attached to a central metal atom in an octahedral complex can be thought of as setting up a force field on the central atom, and the influence of this force field on the electronic distribution of the central metal atom is the subject of *ligand-field theory.* When properly applied, ligand-field theory gives an excellent explanation of the general transition-element properties described above. We shall examine some of the basic ideas but only in the simplest version.

To understand the effect of ligands on metal atom M to which the ligands are attached, it is necessary to review some features of electron distribution. For *s* electrons, as discussed in Section 3.4, the electron probability distribution is identical along all directions of space outward from the nucleus of M; for *p* electrons, the high-probability regions are along the x, y, or z axes, depending on whether we consider p_x, p_y, or p_z orbitals (recall Figure 3.14). It is the *d* electrons that concern us the most here, since it is their spatial distribution that is characteristically involved in the chemistry of the transition elements. Figure 16.4 shows the five possible orbitals that make up the 3*d* subshell. The five orbitals, labeled d_{z^2}, $d_{x^2-y^2}$, d_{zx}, d_{yz}, and d_{xy}, are shown divided into two groups: d_{z^2}, $d_{x^2-y^2}$ and d_{zx}, d_{yz}, d_{xy}. The two groups differ in that the first group of orbitals has high electron probability *along the* x, y, z axes whereas the second group has high electron probability *between* the x, y, z axes. To distinguish the two groups of orbitals, the d_{z^2}, $d_{x^2-y^2}$ are referred to as *e* orbitals and the d_{zx}, d_{yz}, d_{xy} are referred to as *t* orbitals.

Imagine that an atom containing *d* electrons is placed at the origin of an x, y, z coordinate system and six identical ligands are gradually moved in toward the origin along the x, y, z axes. Figure 16.5 shows the relative placing of the central atom M and the six ligands L. Each of the ligands

†Students of Latin may recall from Julius Caesar's *Commentaries on the Gallic War* that *transalpine* means "across the Alps" and *cisalpine* means "on this side of the Alps."

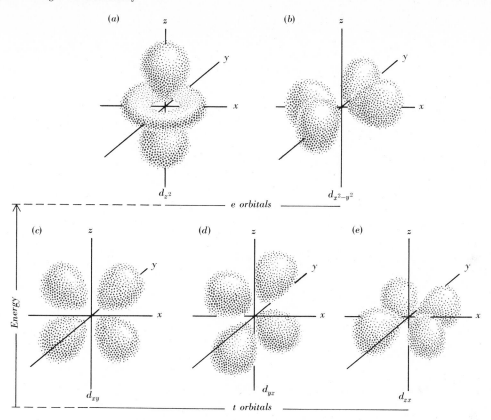

Figure 16.4
Energy separation of 3d orbitals by octahedral ligands.

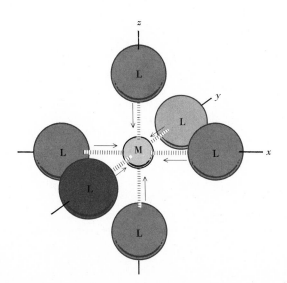

Figure 16.5
Approach of ligands L to a metal atom M to form an octahedral complex.

has its own charge cloud and, as the ligands approach M, there is a repulsive force between the incoming charge cloud and the electrons in the orbitals of M. The repulsion is greater for those orbitals of M that lie along the x, y, z axes than for those orbitals concentrated between the axes. In other words, the incoming ligands exert a greater repulsive force on the M electrons in e-type orbitals (that is, d_{z^2} and $d_{x^2-y^2}$) than on those in t-type orbitals (that is, d_{zx}, d_{yz}, d_{xy}). The net effect is to raise the energy of an electron in an e orbital higher than that of an electron in a t orbital. Whereas all five of the d orbitals had exactly the same energy before the ligands were brought up, the effect of bringing up the ligands is to split the group of five d orbitals into two sets, one of higher energy and one of lower. Clearly it is energetically more favorable other things being equal to place d electrons in the lower-energy t set than in the higher-energy e set.

The splitting of the d orbitals into two sets differing in energy is responsible for the light absorption (or color) shown by transition-element compounds. To understand this process, consider the simplest case, i.e., a transition-element ion with just a single electron in the d orbitals. Such a case is Ti^{3+}. In aqueous solution, Ti^{3+} is surrounded by an octahedron of six H_2O molecules. The single d electron is accommodated in one of the t orbitals since this causes less repulsion between it and the electrons of the bound H_2O ligands. As white light passes through the solution, light energy can raise the electron's energy sufficiently for it to move to one of the d orbitals of the e set. (It now is of higher energy since it is repelled by ligand electrons.) The amount of energy required is a definite amount, and exactly this much energy is removed from the white light. For $Ti(H_2O)_6^{3+}$, the energy required is that of green light (490-nm wavelength). Consequently, the light leaving the solution is missing its green component (which is in the middle of the spectrum of colors) and appears purple (red and violet are at the two ends of the spectrum). Incidentally, the green light is converted to heat as the d electron returns from an e to a t orbital.

Two points should be noted concerning the light-absorption process for transition-metal ions. The first is the fact that for Ti^{3+} the energy difference between t and e orbitals depends on the nature of the ligands. If they concentrate lots of electrons along the bonding directions, the repulsion of an electron in one of the e orbitals will be great and the energy of light required for the transition is great. This means the color observed will change, and we should not be surprised that different Ti^{3+} complexes are of different colors. The other point to note is that if the transition-metal ion differs from Ti^{3+} in having more than one d electron, the situation is complicated by repulsions between the d electrons. However, the basic principle remains: The light absorption results when electrons move from lower-energy to higher-energy d orbitals.

The splitting of the energies of d orbitals also contributes to the stability of the many complex ions formed by transition ions. This comes about because d electrons remaining with the transition ion can be accommodated in the t set of orbitals. As a consequence they do not screen their share of nuclear charge from the ligands. To illustrate, in $Ti(H_2O)_6^{3+}$, the H_2O molecules are attracted not just by a $3+$ ion but instead by something approaching a $4+$ charge since the one d electron is out of the way of the bonding directions. As a consequence, the bonds can be considered either as strong ion-dipole interactions or as covalent bonds (where electron pairs from the O of the H_2O ligand is shared with Ti).

Finally, we should note that ligand-field theory is important in accounting quantitatively for the paramagnetism displayed by transition-metal complexes. Here the problem is that often the measured paramagnetism is less than would be expected from the total number of d electrons. The explanation lies in the fact that in some cases the d electrons do not spread over all five d orbitals but instead pair up in the t set. An example is $Fe(CN)_6^{3-}$, which has five d electrons. Unlike $Fe(H_2O)_6^{3+}$, where all five of the d electrons are unpaired, in $Fe(CN)_6^{3-}$ the five reside in the three t orbitals and so only one can be unpaired and contribute to the paramagnetism. The difference is due to the greater energy separation between e and t which occurs because the CN^- ligands concentrate more electrons along the bond directions than do H_2O ligands. As a consequence an electron in an e orbital of $Fe(CN)_6^{3-}$ is repelled more strongly than one in an e orbital of $Fe(H_2O)_6^{3+}$.

16.4 Lanthanides and Actinides

The scandium subgroup contains the elements scandium ($Z = 21$), yttrium ($Z = 39$), lanthanum through lutetium ($Z = 57$ through 71), and actinium through lawrencium ($Z = 89$ through 103). The 14 elements following lanthanum ($Z = 58$ through 71) are called the *lanthanides*, or *rare-earth elements*. The elements following actinium are called the *actinides*. As usually displayed in the periodic table, the lanthanides all occupy the same position in the sixth period, below yttrium. The actinides occupy a corresponding position in the seventh period. Figure 16.6 indicates the electron configura-

Figure 16.6 ELECTRON CONFIGURATIONS OF SCANDIUM-SUBGROUP ELEMENTS	Element	Z	Electron Population
	Scandium	21	2,8,9,2
	Yttrium	39	2,8,18,9,2
	Lanthanum	57	2,8,18,18,9,2
	↓	↓	↓
	Lutetium	71	2,8,18,32,9,2
	Actinium	89	2,8,18,32,18,9,2
	↓	↓	↓
	Lawrencium	103	2,8,18,32,32,9,2(?)

tions characteristic of the scandium-subgroup elements.

The lanthanide elements correspond to belated filling of the 4f subshell. Since in these elements the 4f subshell is third outermost, changes in its electronic population are well screened from neighboring atoms by the second-outermost and outermost shells. Consequently, all the lanthanides have properties remarkably alike and are difficult to separate from each other. Similar belated filling of the 5f subshell occurs in the actinide series.

The scandium-subgroup elements including the lanthanides and actinides are all typically metallic, with high luster and good conductivity. They are all quite reactive. Many of their compounds such as hydroxides, carbonates, and phosphates are of low solubility. There is some slight tendency for the 3+ ions to hydrolyze in aqueous solution to give slightly acid solution. All these elements are quite rare in nature.

The lanthanides include cerium (58), praseodymium (59), neodymium (60), promethium (61), samarium (62), europium (63), gadolinium (64), terbium (65), dysprosium (66), holmium (67), erbium (68), thulium (69), ytterbium (70), and lutetium (71). Except for promethium, which has an unstable nucleus and is radioactive, they always occur together in combined form. The gradual filling of the 4f subshell in the lanthanides is accompanied by a slight shrinkage in atomic radius, since the nuclear charge increases throughout the sequence. This decrease in size, called the *lanthanide contraction*, leads to subtle differences in the properties of the elements. These slight differences have been exploited in the development of a tedious ion-exchange procedure (Section 15.4) which allows separation of the elements.

Actinium and the actinides are all radioactive in that they have unstable nuclei. The actinides include thorium (90), protactinium (91), uranium (92), neptunium (93), plutonium (94), americium (95), curium (96), berkelium (97), californium (98), einsteinium (99), fermium (100), mendelevium (101), nobelium (102), and lawrencium (103). Experiments to determine the electronic configurations of the actinides are not conclusive, and the assignment of electrons to various energy levels as given in Figure 3.10 is in doubt.

Most members of the actinides are almost nonexistent in nature, and for many years it was thought that there were only 92 elements. However, in 1940 it was found that when $^{238}_{92}U$ was bombarded with neutrons, the nucleus formed, $^{239}_{92}U$, underwent radioactive decay to form successively $^{239}_{93}Np$ and $^{239}_{94}Pu$. In addition to neptunium and plutonium, higher elements have been produced by similar nuclear bombardments. In some cases, instead of neutrons, nuclei of the lighter elements such as helium or carbon are used to bombard nuclei. The actinide elements are not so similar chemically as the lanthanide elements, and so the products of nuclear bombardment can be separated by chemical means. The isolation of plutonium is important, because the element is used as a source of nuclear energy.

16.5   Chromium and Manganese

Chromium   Chromium is one of the less abundant metals (0.037 percent of the earth's crust), but still is approximately 50 times as abundant as Mo and W. Its principal mineral is chromite ($FeCr_2O_4$), some of which is reduced directly by heating with carbon in order to provide ferrochromium (solid solution of Cr in Fe) for addition to alloy steels. Low-chrome steels (up to 1% Cr) are quite hard and strong; high-chrome steels (up to 30% Cr), or stainless steels, are very resistant to corrosion. Most of the remaining chromite is converted to sodium chromate (Na_2CrO_4) by heating it with Na_2CO_3 in air:

$$8Na_2CO_3(s) + 4FeCr_2O_4(s) + 7O_2(g) \longrightarrow$$
$$2Fe_2O_3(s) + 8Na_2CrO_4(s) + 8CO_2(g)$$

The sodium chromate is leached out with acid to form $Na_2Cr_2O_7$, an important oxidizing agent.

Chromium metal is very hard and, although quite reactive in the powdered form, in the massive form is quite resistant to corrosion. Furthermore, it takes a high polish, which lasts because of formation of an invisible, self-protective oxide coat. Consequently, chromium finds much use as a plating material, both for its decorative effect (0.00005 cm thick) and for its protective effect (0.0075 cm thick). Plating is usually accomplished by electrolyzing the object in a bath made by dissolving $Na_2Cr_2O_7$ and H_2SO_4 in water. Since plating will not occur unless the sulfate is present, the sulfate must be involved in some intermediate formed during the electrolysis.

The compounds of chromium are all colored, a fact which suggested the name *chromium,* from the Greek word for color, *chrōma.* The characteristic oxidation states are 2+, 3+, and 6+, represented in acid solution by Cr^{2+} (chromous), Cr^{3+} (chromic), and $Cr_2O_7{}^{2-}$ (dichromate) and in basic media by $Cr(OH)_2$, $CrO_2{}^-$ (chromite), and $CrO_4{}^{2-}$ (chromate).

The chromous ion, Cr^{2+}, is a beautiful blue ion obtained by reducing either Cr^{3+} or $Cr_2O_7{}^{2-}$ with zinc metal. However, it is rapidly oxidized in aqueous solution by air. The reduction potential for

$$Cr^{3+} + e^- \rightleftharpoons Cr^{2+}$$

is −0.41 volt, which means that Cr^{2+} should also be oxidized by H^+, but the latter reaction is very slow. When base is added to solutions of chromous salts, chromous hydroxide precipitates. On exposure to air, $Cr(OH)_2$ is oxidized by O_2 to give $Cr(OH)_3$ (also written $Cr_2O_3 \cdot xH_2O$).

Many chromic salts, such as chromic nitrate $[Cr(NO_3)_3]$ and chromic perchlorate $[Cr(ClO_4)_3]$, dissolve in water to give violet solutions, in which the violet color is due to the hydrated chromic ion, $Cr(H_2O)_6{}^{3+}$. If high concentrations of chloride ion are added, some of the hydrate water is replaced, and the solution slowly turns green because of formation of a chloro complex. Solutions of chromic salts can be kept indefinitely, exposed to the air, without oxidation or reduction. In general, they are slightly acid because of hydrolysis of the chromic ion. This reaction can be written in either of the following ways:

$$Cr^{3+} + H_2O \rightleftharpoons CrOH^{2+} + H^+$$
$$Cr(H_2O)_6{}^{3+} \rightleftharpoons Cr(H_2O)_5OH^{2+} + H^+$$

When base is gradually added to chromic solutions, a green slimy precipitate, which is either $Cr(OH)_3 \cdot xH_2O$ or $Cr_2O_3 \cdot xH_2O$, first forms but then disappears as excess OH^- is added. A deep green color characteristic of chromite ion, written as $CrO_2{}^-$ or $Cr(OH)_4{}^-$, is produced. The precipitation and redissolving associated with this amphoteric behavior can be described as follows:

$$Cr^{3+} + 3OH^- \longrightarrow Cr(OH)_3(s)$$
$$Cr(OH)_3(s) + OH^- \longrightarrow CrO_2{}^- + 2H_2O$$

The green species in the final solution is certainly more complicated than $CrO_2{}^-$ and probably contains more than one Cr atom per ion. When filtered off and heated, the insoluble hydroxide loses water to form Cr_2O_3, chromic oxide or chromium sesquioxide. This is an inert green powder much used as the pigment chrome green.

Chromic ion forms a great number of complex ions. In all of them the chromium atom is surrounded by six other atoms arranged at the corners of an octahedron. A typical octahedral complex, $CrF_6{}^{3-}$, is shown in Figure 16.7. All six octahedral positions are equivalent. Other octahedral complexes

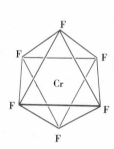

Figure 16.7
Octahedral chromium complex.

of chromic ion are $Cr(NH_3)_6^{3+}$, $Cr(H_2O)_6^{3+}$, $Cr(H_2O)_5Cl^{2+}$, $Cr(NH_3)_4Cl_2^+$, etc. It is characteristic of all these complexes that they form and dissociate very slowly. In potassium chrome alum [$KCr(SO_4)_2 \cdot 12H_2O$], the $Cr(H_2O)_6^{3+}$ complex occurs as a unit in occupying some of the crystal lattice sites.

In the 6+ oxidation state, chromium is known principally as the chromates and dichromates. The chromate ion, CrO_4^{2-}, can be made quite easily by oxidizing chromite ion, CrO_2^-, in basic solution with a moderately good oxidizing agent such as hydrogen peroxide. The reaction is

$$2CrO_2^- + 3HO_2^- \longrightarrow 2CrO_4^{2-} + H_2O + OH^-$$

where the peroxide is written as HO_2^-, because H_2O_2 is an acid and does not exist in basic solution. The chromate ion is yellow and has a tetrahedral structure with four oxygen atoms bound to a central chromium atom.

When solutions of chromate salts are acidified, the yellow color is replaced by a characteristic orange, the result of formation of $Cr_2O_7^{2-}$, dichromate ion,

$$2CrO_4^{2-} + 2H^+ \rightleftharpoons Cr_2O_7^{2-} + H_2O$$

The change is reversed by adding base. The structure of the dichromate ion is shown in Figure 16.8 and consists of two tetrahedra sharing an oxygen atom. Each of the two chromium atoms at the centers of the tetrahedra is bound to four oxygen atoms at the corners. The dichromate ion is a very

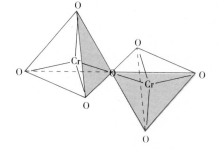

Figure 16.8
Dichromate ion.

good oxidizing agent, especially in acid solution. It will, for example, oxidize hydrogen peroxide to form O_2:

$$Cr_2O_7^{2-} + 3H_2O_2 + 8H^+ \longrightarrow 3O_2(g) + 2Cr^{3+} + 7H_2O$$

It might seem strange that in basic solution hydrogen peroxide oxidizes chromium, whereas in acid solution chromium oxidizes hydrogen peroxide. The reason for this is that, in going from the 3+ to the 6+ state of chromium (Cr^{3+} to $Cr_2O_7^{2-}$), oxygen atoms have to be added, whereas, in going from the 6+ to the 3+ state ($Cr_2O_7^{2-}$ to Cr^{3+}), oxygen is removed. In acid solution,

H$^+$ helps in removing oxygen by forming water; in basic solution, the scarcity of H$^+$ allows addition of oxygen. In the general case of preparing oxy compounds, the change to compounds of higher oxidation state is usually most easily done in basic solution; to go to compounds of lower oxidation state, it is best to work in acid solution.

When solutions of dichromate ion are made very acid, especially in the presence of a dehydrating agent such as concentrated H$_2$SO$_4$, the uncharged species, CrO$_3$, is formed. This deep red solid is chromium trioxide, or, as it is sometimes called, *chromic anhydride*. It is a very powerful oxidizing agent and is used extensively in preparing organic compounds. Solutions of CrO$_3$ in concentrated H$_2$SO$_4$ are used as "cleaning solution" for glass equipment in laboratories. The cleaning action is due to the oxidation of grease.

Manganese   Manganese is not a very common element (abundance, 0.08 percent), but in the earth's crust it is as abundant as carbon and more so than sulfur. The most important minerals are the oxides: MnO$_2$ or pyrolusite, Mn$_2$O$_3$, and Mn$_3$O$_4$. Probably the best way to prepare the metal is by reduction of the oxides with powdered aluminum. However, since most metallic manganese goes into steel production, alloys of manganese are used instead. Two such alloys are ferromanganese (about 80% Mn in Fe) and spiegeleisen (about 20 to 30% Mn and about 5% C in Fe); they are made by reducing mixed oxides of iron and manganese in a blast furnace with carbon or carbon monoxide acting as reducing agent. When added to steel, Mn has two functions. In low amounts, it acts as a scavenger by combining with O and S in the molten iron to form easily removable substances. In high amounts (up to 14 percent), it is alloyed into steel, where it imparts special hardness and toughness such as is needed for resistance to battering abrasion.

In its chemical compounds, manganese shows oxidation states of 2+, 3+, 4+, 6+, and 7+. Most of these compounds are colored and paramagnetic. In the 2+ state, manganese exists as the manganous ion, Mn^{2+}. This is one of the few pink ions in chemistry, and many manganous salts, such as manganous sulfate (MnSO$_4$) and manganous chloride (MnCl$_2$), have pink coloration. Unlike Cr^{2+}, Mn^{2+} is a very poor reducing agent, and neutral or acid solutions of manganous salts can be kept indefinitely exposed to oxygen or other oxidizing agents. When base is added to Mn^{2+}, a white precipitate of Mn(OH)$_2$ is formed. This solid, unlike manganous salts, is promptly oxidized by air to the 3+ state.

In the 3+ state, manganese can exist as the manganic ion, Mn^{3+}, but only in solids and complex ions. Unlike Cr^{3+}, Mn^{3+} is a very powerful oxidizing agent and can even oxidize H$_2$O to liberate oxygen. The reduction potential of

$$Mn^{3+} + e^- \rightleftharpoons Mn^{2+}$$

is positive enough ($+1.51$ volts) that manganic ion can oxidize *itself* to the $4+$ state,

$$2Mn^{3+} + 2H_2O \longrightarrow Mn^{2+} + MnO_2(s) + 4H^+$$

This disproportion can be prevented and the $3+$ state stabilized by (1) complexing the manganese, e.g., with cyanide (CN^-) to give $Mn(CN)_6^{3-}$, or by (2) forming an insoluble salt, such as $MnPO_4$, manganic phosphate, or the hydroxide, written as $Mn(OH)_3$ or $MnOOH$ or even Mn_2O_3.

In the $4+$ state the principal compound of manganese is manganese dioxide (MnO_2). It is not a simple compound, because no matter how careful the preparation, the product always contains fewer than two oxygen atoms per manganese atom. As mentioned in Section 13.7, MnO_2 is the oxidizing agent in the dry cell.

When MnO_2 is heated with basic substances in air, it is oxidized from its original black color to a deep green, the result of conversion to manganate ion, MnO_4^{2-}. Although stable in alkaline solution, this ion (which represents Mn in the $6+$ state) disproportionates when the solution is acidified.

$$3MnO_4^{2-} + 4H^+ \longrightarrow MnO_2(s) + 2MnO_4^- + 2H_2O$$

The MnO_4^- ion is the permanganate ion and shows manganese in its highest oxidation state. It is a very good oxidizing agent, especially in acid solution, where it is usually reduced all the way to manganous ion. Solutions of permanganate salts are frequently used in analytical chemistry to determine amounts of reducing agents by titration. Titration is simplified by using the disappearance of the deep violet color of MnO_4^- as the end-point indicator. The usual procedure is to make the solutions acid, so as to get complete reduction to Mn^{2+}. With neutral or alkaline solutions, MnO_2 is formed; in very basic solutions, MnO_4^{2-}. When $KMnO_4$ is treated with concentrated H_2SO_4, a violently explosive oil, Mn_2O_7, manganese heptoxide, is formed.

The chemistry of manganese illustrates well how the characteristics of compounds change in going from low oxidation state to high. In the low oxidation state, manganese exists as a cation which forms basic oxides and hydroxides. In the higher oxidation states, it exists as anions derived from acidic oxides.

For qualitative analysis, chromium and manganese present in an original unknown mixture as chromate or dichromate and permanganate will be reduced by H_2S in acid solution to Cr^{3+} and Mn^{2+}, forming finely divided sulfur in the process. When the solution, still containing sulfide, is made basic, $Cr(OH)_3$ and MnS precipitate. Treatment with acid and an oxidizing agent serves to remove the sulfide, and subsequent addition of excess NaOH precipitates manganese as a hydroxide and converts the chromium to soluble chromite ion. The appearance of a green color in the solution at this point is a strong indication of the presence of chromium. It can be confirmed by treating the green, basic solution with H_2O_2 to oxidize chromite to yellow chromate, which can be precipitated as yellow $BaCrO_4$ by the addition of barium ion in an acetic acid buffer. The hydroxide precipitate suspected of containing manganese, on treatment with acid and a very strong oxidizing agent such as sodium bismuthate ($NaBiO_3$) or sodium periodate ($NaIO_4$), produces the characteristic violet color of MnO_4^- if manganese is present.

16.6 Iron-Group Elements

The first five transition-element subgroups, headed, respectively, by scandium, titanium, vanadium, chromium, and manganese, show considerable chemical similarity in each vertical column. With the next three subgroups the chemical resemblance along the horizontal sequence is more pronounced than the chemical resemblance down the subgroup. For this reason, the next three subgroups are generally considered in terms of horizontal triads. In the top transition period, the elements iron ($Z = 26$), cobalt ($Z = 27$), and nickel ($Z = 28$) make up the *iron elements;* in the middle period, ruthenium ($Z = 44$), rhodium ($Z = 45$), and palladium ($Z = 46$) are the *light platinum elements;* in the bottom period, osmium ($Z = 76$), iridium ($Z = 77$), and platinum ($Z = 78$) are the *heavy platinum elements.* All these elements are metals with high melting points and high densities. The elements of the first triad are moderately reactive; those of the platinum triads are fairly inert. In the original Mendeleev periodic table all these elements were lumped into what was called *group VIII.* The platinum elements (together with gold and silver) are sometimes referred to as the *noble metals.*

In progressing from left to right in the first transition period, there is a progressive rise in the maximum oxidation state from $3+$ for Sc to $7+$ for Mn. The surprising thing about the following elements is that their maximum oxidation states are less than those of the preceding elements. Furthermore, compounds containing these maximum oxidation states are such strong oxidizing agents that they are not usually encountered. Specifically, iron shows a maximum oxidation state of $6+$, but only the $2+$ and $3+$ states are common; cobalt shows only $2+$ and $3+$; nickel, usually only $2+$. The reason for this return to lower oxidation states is associated with the fact that, at manganese, the $3d$ sublevel has become half-filled. With increasing nuclear charge, it gets progressively harder to break into this half-filled subshell; therefore, effectively, five of the $3d$ electrons are not available for forming compounds. In addition to 18 electrons in the core, iron has electron configuration $3d^64s^2$, cobalt has $3d^74s^2$, and nickel has $3d^84s^2$. Removal of the two $4s$ electrons is relatively easy in each case; hence a $2+$ state is formed. Additional removal of a $3d$ electron would give a $3+$ state. In the case of iron, this happens easily because a half-filled $3d$ level is left; in cobalt and nickel, it does not happen so readily. For cobalt, the 3+ state must be stabilized (e.g., by formation of a complex ion); for nickel, the 3+ state is very rare, and compounds of 3+ nickel are powerful oxidizing agents.

The properties of the iron elements are very similar, as shown in Figure 16.9. The melting points and boiling points are uniformly high; the energies required to pull an electron off the gas atom are nearly the same for the

Figure 16.9 **ELEMENTS** **OF THE** **IRON GROUP**	**Property**	**Fe**	**Co**	**Ni**
	Melting point, °C	1535	1490	1450
	Boiling point, °C	2700	2900	2700
	Ionization potential, eV	7.90	7.86	7.63
	Reduction potential (for M^{2+}), volts	−0.44	−0.28	−0.25
	Density, g/cm^3	7.9	8.7	8.9

three elements; and the reduction potentials are all moderately more negative than that of hydrogen. In addition to the properties listed, these elements are alike in that all are *ferromagnetic;* i.e., they are strongly attracted into a magnetic field and show permanent magnetization when removed from such a field. That these elements are magnetic is not surprising. The electronic configurations would lead us to expect that there would be unpaired electrons in the 2+ ions such as might exist in the metal lattice. However, it is surprising that the magnetization is so large and so persistent. The explanation is that in these metals there are *domains* of magnetization, regions of a million or so ions, all of which cooperatively direct their individual magnetic effects the same way. In an unmagnetized piece of metal, these domains point randomly in all directions in such a way that, in sum, the magnetic effect cancels. When placed in a magnetic field, the domains are turned so that all point in the same direction, giving rise to a large magnetic effect. If the metal is now removed from the field, it remains permanently magnetized unless the domain orientation is disorganized, as by heating or pounding. Of all the elements, only iron, cobalt, and nickel show this kind of magnetism at room temperature. Apparently they are the only ones that satisfy the conditions necessary for domain formation. These conditions are that the ions contain unpaired electrons and that the distance between ions be just exactly right in order that the interaction for lining up all the ions to form a domain may be effective. Manganese metal has most of the properties needed to be ferromagnetic, but the ions of the metal are too close; addition of copper to manganese increases this average spacing, and the resulting alloy is ferromagnetic.

In their compounds, the iron elements behave like typical transition elements. Many of the compounds are colored and paramagnetic, and frequently they contain complex ions.

16.7 Iron

The element iron has an industrial importance which exceeds that of any other metallic element. It is very abundant, ranking fourth in the earth's crust (after O, Si, and Al); it is very common, being an essential constituent of several hundred minerals; it is easy to make by simply heating some of its minerals with carbon; it has many desirable properties, especially when impure. For all these reasons, iron has become such a distinctive feature of civilization that it gives its name to one of the ages in archaeological chronology.

About 5 percent by weight of the earth's crust is iron. Some of this iron is meteoric in origin and occurs in the uncombined, metallic state. However, most of it is combined with oxygen, silicon, or sulfur. The important source minerals are hematite (Fe_2O_3), limonite ($Fe_2O_3 \cdot H_2O$), magnetite (Fe_3O_4), and siderite ($FeCO_3$), usually contaminated with complex iron silicates from which these minerals are produced by weathering. Iron sulfides, such as iron pyrites (FeS_2) or fool's gold, are also quite abundant, but they cannot be used as sources of iron because sulfur, an objectionable impurity in the final product, is hard to remove.

Besides the earth's crust, there is a possibility that the center of the earth may be iron. Indirect evidence based on the study of earthquake waves and tidal action indicates that the core of the earth is liquid and has a density corresponding to that of liquid iron at high pressure.

Iron is practically never produced in a pure state, since it is difficult to make and is too expensive for most purposes. Furthermore, impure iron (steel) has desirable properties, especially when the specific impurity is carbon in carefully controlled amounts. The industrial production of impure iron is carried out on a massive scale in the well-known blast furnace, in which complicated high-temperature reactions occur involving iron ore, limestone, and carbon. As shown in Figure 16.10, the blast furnace is designed for continuous operation. Iron ore, limestone, and coke are added at the top; preheated air or oxygen is blown in at the bottom. As the molten iron forms, it trickles down to a pit at the bottom, from which it is periodically drawn off. All told, it takes about 12 h for material to pass through the furnace. The actual chemical processes which occur in such a furnace are still obscure. It is generally agreed, however, that the active reducing agent is not carbon but carbon monoxide. As the charge settles through the furnace, the coke is oxidized by the incoming oxygen by the reaction

$$2C(s) + O_2(g) \longrightarrow 2CO(g)$$

thus forming the reducing agent CO and liberating large amounts of heat. As the carbon monoxide moves up the furnace, it encounters oxides of iron in various stages of reduction, depending on the temperature of the particular zone. At the top of the furnace, where the temperature is lowest (250°C),

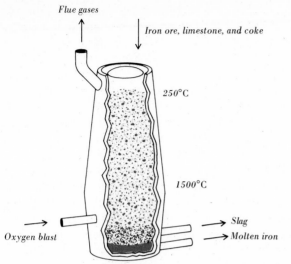

Figure 16.10
Blast furnace.

the iron ore (mostly Fe_2O_3) is reduced to Fe_3O_4 by the reaction

$$3Fe_2O_3(s) + CO(g) \longrightarrow 2Fe_3O_4(s) + CO_2(g)$$

As the Fe_3O_4 settles, it becomes reduced further to FeO,

$$Fe_3O_4(s) + CO(g) \longrightarrow 3FeO(s) + CO_2(g)$$

Finally, toward the bottom of the furnace, FeO is eventually reduced to iron,

$$FeO(s) + CO(g) \longrightarrow Fe(s) + CO_2(g)$$

Since the temperature at the lowest part of the furnace is above the melting point of the impure iron, the solid melts and drips down into the hearth at the very bottom. The net equation for the reduction of Fe_2O_3 can be written as the sum of the last three equations.

$$Fe_2O_3(s) + 3CO(g) \longrightarrow 2Fe(l) + 3CO_2(g)$$

In addition to the foregoing reactions, there occurs the combination of CO_2 with hot carbon,

$$C(s) + CO_2(g) \longrightarrow 2CO(g)$$

and the thermal decomposition of limestone by the reaction

$$CaCO_3(s) \longrightarrow CaO(s) + CO_2(g)$$

Both these reactions are helpful: the former raises the concentration of the reducing agent CO, and the latter facilitates the removal of silica-containing contaminants present in the original ore. Lime (CaO), being a basic oxide, reacts with the acidic oxide SiO_2 to form calcium silicate ($CaSiO_3$). In the form of a lavalike *slag,* calcium silicate collects at the bottom of the furnace, where it floats on the molten iron and protects it from oxidation by incoming oxygen.

Four times a day the liquid iron and molten slag are drawn off through tapholes in the bottom of the furnace. About 1000 tons of impure iron can be produced per day from one furnace. For each ton of iron, there is also produced approximately half a ton of slag. Since slag is essentially calcium aluminum silicate, some of it is put to good use in making cement.

The crude product of the blast furnace (called *pig iron*) contains about 4% C, 2% Si, a trace of sulfur, up to 1% of phosphorus and manganese, and the rest iron. Sulfur is probably the worst impurity (making steel break when worked) and must be avoided, since it is hard to remove in refining operations. The refining operations are technical and of great variety. Only a few of the basic ideas will be discussed here.

When pig iron is remelted with scrap iron and cast into molds, it forms *cast iron.* This can be either gray or white, depending on the rate of cooling. When cooled slowly (as in sand molds, where heat loss is slow), the carbon separates out almost completely in the form of tiny flakes of graphite, giving gray cast iron that is relatively soft and tough. When cooled rapidly (as in water-cooled molds), the carbon does not have a chance to separate out but remains combined in the form of the compound iron carbide (Fe_3C). Such white cast iron is extremely hard and brittle.

Most pig iron is refined into steel by burning out the impurities to leave small controlled amounts of carbon. In the *open-hearth* process (which accounts for most of the United States production), carbon is removed by oxidation with air and iron oxide, the latter being added as hematite and rusted scrap iron. The process is usually carried out on a shallow hearth arranged so that a hot-air blast can play over the surface. In the *basic* open-hearth process, limestone is added to provide CaO for converting oxidation products, such as acidic P_2O_5, into slag. Since it takes about 8 h to refine a batch of steel by this process, there is ample time for continuous testing to maintain quality control. The *Bessemer process* is much more rapid (10 to 15 min) but gives a less uniform product. In this process, molten pig iron taken directly from the blast furnace is poured into an enormous pot, and a blast of air is swept through the liquid mix to give a spectacular tongue of fire as most of the carbon and silicon are burned off. Frequently, the Bessemer and open-hearth processes are combined to take advantage of the good points of each. In this duplex process, a preliminary rapid blowing in the Bessemer converter gets rid of most of the C and Si, with a following slower burn-off in an open-hearth furnace to get rid of P. Elements such as Cr, V, or Mn can be added to produce steels with desired properties. In order to prevent the formation of blowholes (as in Swiss cheese) when the molten steel is poured into ingots, it is necessary for the finished steel to contain some Mn. The function of this Mn is to combine with the oxygen and keep it from bubbling out as the steel solidifies.

The properties of iron in the form of steel are very much dependent on the percentage composition of the impurities present, on the heat treatment of the specimen, and even on the working to which the sample has been subjected. For these reasons, the following comments about iron properties do not necessarily apply to every given sample of iron. Compared with most metals, iron is a fairly good reducing agent, but it is not so good as the preceding transition elements. With nonoxidizing acids, it reacts to liberate H_2 by the reaction

$$Fe(s) + 2H^+ \longrightarrow Fe^{2+} + H_2(g)$$

It also has the ability to replace less active metals in their solutions. For example, a bar of iron placed in a solution of $CuSO_4$ immediately is covered with a reddish deposit of copper formed by the reaction

$$Fe(s) + Cu^{2+} \longrightarrow Fe^{2+} + Cu(s)$$

In concentrated nitric acid, iron, like many other metals (Cr, Mo, Co, Ni, etc.), becomes *passive;* i.e., it loses the ability to react with H^+ and Cu^{2+} as above and appears to be inert. When scratched or subjected to shock, reactivity is restored. It may be that passivity is due to the formation of a thin surface coating of oxide which slows down the rates of oxidation below the limits of detectability. When the film is broken, reactivity is restored. Passivity is important in some methods of preventing corrosion of iron.

The two common oxidation states of iron are 2+ (ferrous) and 3+ (ferric). Under vigorous oxidizing conditions it is also possible to get compounds such as $BaFeO_4$, barium ferrate, but in general the 6+ state is rare. Compounds in which the oxidation state is fractional, as $\frac{8}{3}+$ in Fe_3O_4, can be thought of as mixtures of two oxidation states. Thus, it is frequently convenient to consider Fe_3O_4 as a ferrosoferric oxide which can be written $FeO \cdot Fe_2O_3$.

In the 2+ state, iron exists essentially as ferrous ion, Fe^{2+}. This is a pale green, almost colorless ion which, except in acid solutions, is rather hard to keep, since it is easily oxidized to the 3+ state by oxygen in the air. However, since the rate of oxidation by O_2 is inversely proportional to H_3O^+ concentration, acid solutions of ferrous salts can be kept for long periods. When base is added to ferrous solutions, a nearly white precipitate of ferrous hydroxide [$Fe(OH)_2$] is formed. On exposure to air, $Fe(OH)_2$ turns brown, owing to oxidation to hydrated ferric oxide ($Fe_2O_3 \cdot xH_2O$). For convenience, the latter is often written as $Fe(OH)_3$, ferric hydroxide, and the oxidation can be written

$$4Fe(OH)_2(s) + O_2(g) + 2H_2O \longrightarrow 4Fe(OH)_3(s)$$

However, pure ferric hydroxide has never been prepared.

In the 3+ state, iron exists as the colorless ferric ion, Fe^{3+}. Since solutions of ferric salts are acid, appreciable hydrolysis must take place. This can be written as

$$Fe^{3+} + H_2O \rightleftharpoons FeOH^{2+} + H^+$$

Apparently, the yellow-brown color characteristic of ferric solutions is mainly due to $FeOH^{2+}$. By addition of an acid such as HNO_3, the color can be made to disappear. On addition of base, a slimy, red-brown, gelatinous precipitate forms, which may be written as $Fe(OH)_3$. This can be dehydrated to form yellow or red Fe_2O_3.

In both the 2+ and 3+ states, iron shows a great tendency to form complex ions. For example, ferric ion combines with thiocyanate ion, NCS^-, to form $FeNCS^{2+}$, which has such a deep red color that it can be detected at concentrations of 10^{-5} M. The formation of this complex is the basis of one of the most sensitive qualitative tests for the presence of Fe^{3+}.

For qualitative analysis, all three iron-group elements are generally precipitated as black sulfides insoluble in basic solutions. (If ferric ion were present, it would be reduced by H_2S in acid solution to ferrous ion.) FeS can be separated from CoS and NiS because it dissolves fairly quickly in an Na_2SO_4–$NaHSO_4$ buffer, whereas CoS and NiS are slow to dissolve. Separation of iron from cobalt and nickel can also be achieved by making use of the fact that Fe^{2+} plus an excess of NH_3 forms in air insoluble ferric hydroxide, while Co^{2+} and Ni^{2+} form soluble ammonia complexes. The presence of iron can be confirmed by adding thiocyanate after oxidation of Fe^{2+} to Fe^{3+} with H_2O_2, if necessary. The deep red color of $FeNCS^{2+}$ shows that iron is present. To distinguish cobalt from nickel, the sulfides CoS and NiS can be dissolved in acid solution, boiled with bromine water to destroy H_2S, and the solution treated with potassium nitrite. The appearance of insoluble yellow potassium cobaltinitrite [$K_3Co(NO_2)_6$] shows the presence of cobalt. Nickel can be identified by adding a special reagent, dimethylglyoxime, which from basic solution precipitates the reddish-orange, voluminous solid, nickel dimethylglyoxime.

16.8   Corrosion of Iron

"Corrosion" is a general term applied to the process in which uncombined metals change over to compounds. In the special case of iron the corrosion process is called *rusting*. Economically, rusting is a serious problem, and it has been estimated that one man in seven involved in the production of iron works simply to replace the iron lost by rusting. Still, despite considerable study, corrosion is a mysterious process, and its chemistry is not well understood.

Rust appears to be hydrated ferric oxide with a chemical composition corresponding approximately to $2Fe_2O_3 \cdot 3H_2O$, that is, 3 moles of water per 2 moles of ferric oxide. However, since the water content is not always the same, it is preferable to write $Fe_2O_3 \cdot xH_2O$. Because iron will not rust in dry air or in water that is completely free of oxygen, it would seem that both O_2 and H_2O are required for rust formation. Furthermore, it is observed that rusting is speeded up by the presence of acids, strains in the metal, contact with less active metals, and the presence of rust itself (autocatalysis).

In order to account for the observed facts, the following steps have been proposed as the mechanism by which rusting occurs:

$$Fe(s) \longrightarrow Fe^{2+} + 2e^- \tag{1}$$
$$e^- + H^+ \longrightarrow H \tag{2}$$
$$4H + O_2(g) \longrightarrow 2H_2O \tag{3}$$
$$4Fe^{2+} + O_2(g) + (4 + 2x)H_2O \longrightarrow 2(Fe_2O_3 \cdot xH_2O)(s) + 8H^+ \tag{4}$$

In step (1) ferrous ions are produced by loss of electrons from neutral Fe atoms. However, this process cannot go very far unless there is some way to get rid of the electrons which accumulate on the residual iron. One way to do this is by step (2), in which H^+ ions, either from the water or from acid substances in the water, pick up the electrons to form neutral H atoms. (Normally, we would expect these H atoms to pair up to form H_2 molecules; however, H_2 gas is usually not observed in rust formation and so the H must be used up in another way.) Since iron is a good catalyst for hydrogenation reactions, step (3) now occurs to use up the H atoms. In the meantime, the ferrous ion reacts with oxygen gas by step (4) to form the rust and restore H^+ required for step (2). The net reaction, obtained by adding these steps, is

$$4Fe(s) + 3O_2(g) + 2xH_2O \longrightarrow 2(Fe_2O_3 \cdot xH_2O)(s)$$

Since H^+ accelerates step (2) and is replenished in step (4), it is a true catalyst for the reaction and explains the observation that acids speed up the rate of rust formation. (A remarkable example of this is observed when iron pipes are so located as to be in contact with cinders. Such pipes corrode much more rapidly than they normally would, apparently because weathering of sulfur compounds in the cinders forms sulfuric acid.)

The above mechanism also accounts for many other observations and, in particular, for the process often called electrolytic corrosion. For example, when iron pipes are connected to copper pipes, the iron corrodes much faster than normally. The explanation lies in step (1). Residual electrons accumulating from the dissolution of Fe flow from the iron to the copper, where their energy is lower. This removes the excess negative charge from the iron and allows more Fe^{2+} to leave the metal. A complicating feature which also accelerates the reaction is that H atoms, which now form on the negative copper surface instead of the iron, detach themselves more readily from copper than from iron, thus accelerating step (3).

One of the strongest supports for the stepwise rusting mechanism comes from the observation that the most serious pitting of a rusting iron bar occurs in that part of the bar where the oxygen supply is restricted. The reason for this is that, where the oxygen supply is unrestricted, step (4) promptly occurs to deposit rust before the Fe^{2+} formed by step (1) can move away. This, of course, makes it more difficult for more Fe to dissolve, and the reaction is self-stopping. However, if the oxygen supply is restricted, Fe^{2+} may have a chance to diffuse away from encountering enough oxygen to form rust. This means that the rust may deposit some distance away from the point where pitting occurs. Common examples of this are observed at the edges of overlapping plates or around rivet heads. In the latter case, the rivet shank, although protected from air, is eaten away, but the rust forms where the rivet head overlaps the plate. Apparently, moisture that seeps in allows Fe^{2+} to diffuse out to the surface, where it can react with O_2.

Another slightly different example is found in the well-known waterline deposition of rust on partially immersed steel posts. When a new post is placed in water, pitting usually starts where there are strains in the metal, but the rust forms near the waterline, where oxygen is plentiful. This makes the situation go from bad to worse, since the waterline rust now acts as a screen to keep O_2 from reaching the iron. Self-protection is no longer possible, and severe pitting can now occur where the O_2 supply is restricted.

Although there are still many unanswered questions about rusting, it is clear what must be done to prevent it. The most direct approach is to shut off the reactants O_2 and H_2O. This can be done by smearing grease over the iron to be protected, painting it either with an ordinary paint or better with an oxidizing paint so as to make the iron passive, or plating the iron with some other metal. All these methods are used to some extent. Painting or greasing is probably the cheapest, but it must be done thoroughly; otherwise, rusting may only be accelerated by partial exclusion of oxygen. Plating with another metal is more common when appearance is a factor. Chrome plating, for example, is usually chosen because of its dressy look. Zinc plating, or galvanizing, is actually more permanent. Tin plating looks good and is extremely cheap, but it is not reliable.

The relative merits of metals used for plating depend on the activity of the metal relative to iron and the ability of the metal to form a self-protective coat. Zinc, for example, is a self-protecting metal which reacts with O_2 and CO_2 in the air to form an adherent coating which prevents further corrosion. Furthermore, it has a more negative reduction potential than iron, so that, if a hole is punched in a zinc plating so that both Zn and Fe are exposed to oxidation, it is the Zn that is preferentially oxidized.

The Zn compound forms a plug to seal the hole. Tin also forms a self-protective coat, but tin has a less negative reduction potential than iron, and iron is preferentially oxidized when a tin coating is punctured.

One of the most elegant ways to protect iron from corrosion is by "cathodic protection." In this method, iron is charged to a voltage negative compared with its surroundings. This forces the iron to act as a cathode instead of as the anode required for oxidation and effectively stops corrosion by reversing step (1) of the above rusting mechanism. Actually, zinc plating is a method of cathodic protection, since zinc has a more negative reduction potential than iron and forces electrons onto the iron. In practice, for pipelines and standpipes, cathodic protection is obtained by driving stakes of zinc or magnesium, for example, into the ground and connecting them to the object to be protected. In saltwater, where rusting is unusually severe, the steel plates of ships have been protected by strapping blocks of magnesium to the hulls. These preferentially corrode (since they are acting as anodes) but can easily be replaced, while the iron is essentially untouched.

Cathodic protection also explains why tin plating (as in the ordinary "tin can") is so unlasting. So long as the tin coating in unpunctured, there is no corrosion, since tin is a rather inert metal and can be exposed indefinitely to the atmosphere. Once the coating is punctured (and this happens very easily, because it is very thin), there is real trouble, and the iron is worse off than if the tin plating were not there. The reason for this is that iron, being more active than tin, acts as the anode in setting up cathodic protection for the tin. This, of course, accelerates the dissolving of iron and the formation of rust; hence the rust spreads very rapidly.

16.9 Copper, Silver, and Gold

The elements of the copper subgroup, copper, silver, and gold, have been known since antiquity, because, unlike most of the preceding elements discussed, they are sometimes found in nature in the uncombined, or native, state. Originally decorative in function, they soon were adapted to use in coins because of their relative scarcity and resistance to corrosion. Originally, only silver and gold were used as coins, but then someone discovered the happy coincidence that copper could be added, not only to make the coins cost less but also to increase their life in circulation because of increased hardness. Since then, copper, silver, and gold have been called the *coinage metals*, even though their principal uses are now quite different.

Some of the important properties of these elements are shown in Figure 16.11. They are all typically metallic with rather high melting points and rather high boiling points. The low oxidation potentials indicate that they are not very reactive. According to the electronic configurations, there is, in the ground state of these atoms, one s electron in the outermost energy

Figure 16.11 ELEMENTS OF THE COPPER SUBGROUP						Ionization Potential, eV	Reduction Potential (for M$^+$), volts
Symbol	Z	Electronic Configuration	Melting Point, °C	Boiling Point, °C			
Cu	29	2,8,18,4s^1	1083	2300		7.72	+0.52
Ag	47	2,8,18,18,5s^1	961	1950		7.57	+0.799
Au	79	2,8,18,32,18,6s^1	1063	2600		9.22	+1.7

level. When this electron is removed, the 1+ ion results. This is all that we expect, since the second-outermost shell is filled and presumably is hard to break into. In this respect, these elements resemble the alkali metals and consequently are sometimes classified as a group IB. However, the d electrons in the second-outermost shell are close in energy to the outermost electrons and can be removed with little additional energy, especially if there is some way to stabilize the resulting 2+ or 3+ ions. Apparently, this is exactly what happens. Copper forms 1+ and 2+ compounds; silver forms 1+ and 2+ (although the 2+ state is uncommon); and gold forms 1+ and 3+ compounds. However, even with such a variable oxidation state, the chemistry of these elements is simpler than that of the preceding transition elements.

Copper   Considering its usefulness and familiarity, it is surprising that copper is such a small fraction (0.0001 percent) of the earth's crust. Fortunately, its deposits are concentrated and easily worked. Besides native copper, which is 99.9 percent pure, the element occurs in two principal classes of minerals: sulfide ores and oxide ores. The principal sulfide ores are chalcocite (Cu_2S), chalcopyrite, or copper pyrites ($CuFeS_2$), and covellite (CuS); the principal oxide ores are cuprite (Cu_2O), malachite [$CuCO_3 \cdot Cu(OH)_2$], and tenorite (CuO). About 80 percent of present copper production is from the sulfide ores. In order to make the metal, the minerals are first concentrated by flotation, roasted in air, and then smelted. The roasting and smelting process, represented, for example, by the simplified overall equation

$$2CuFeS_2(s) + 5O_2(g) \longrightarrow 2Cu(s) + 2FeO(s) + 4SO_2(g)$$

produces tremendous quantities of sulfur dioxide, which is generally converted on the spot into sulfuric acid.

The copper product is about 97 to 99 percent pure and must be refined (purified) for most uses. This can be done best in a $CuSO_4$ electrolysis cell, like that sketched in Figure 16.12. In the electrolysis cell, the impure copper is made the anode, and the pure copper the cathode. By careful control of

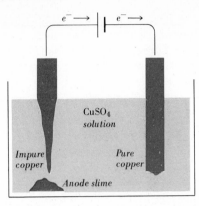

Figure 16.12
Electrorefining of copper.

the electrolysis voltage, the copper can be transferred from the anode to the cathode. The principle of operation can be seen from the following example, in which we consider the purification of a typical bar of copper containing iron and silver as impurities: The iron represents an impurity that is more easily oxidized than copper; the silver, an impurity that is less easily oxidized than copper. By keeping the cell voltage at an appropriate value, only the Fe and Cu are oxidized and go into the solution as ions. The more difficultly oxidized Ag simply drops off to the bottom of the cell as the anode dissolves away. At the cathode, where reduction must occur, the high concentration of Cu^{2+} and the fact that Cu^{2+} is more readily reduced than Fe^{2+} lead to deposition of pure copper. The Fe^{2+} remains in solution, and the solid silver stays at the bottom of the cell. Some common impurities in crude copper are iron, nickel, arsenic, antimony, bismuth (all of which, like iron, are oxidized and remain oxidized) and silver, gold, and traces of platinum metals (all of which, like silver, are not oxidized and collect at the bottom of the cell). The residue at the bottom of the cell beneath the anode is called the *anode slime*. With efficient operation, the recovery of noble metals from the anode slime can pay for the whole refinery operation, leaving the copper as profit.

Metallic copper is malleable, ductile, and, when pure, a very good conductor of heat and electricity. Except for silver, pure copper has the lowest electric resistance of any metal and is used extensively in wires and switches that carry current. It is a poorer reducing agent than hydrogen and does not dissolve in acids unless they contain oxidizing anions. When exposed to the air, it slowly tarnishes with the formation of a green hydroxycarbonate, but this adheres to the metal and protects it from further corrosion. Copper roofs and bronze statues generally turn green because of this effect.

World production of copper is of the order of 5 million tons/year. Most of this goes into the electrical industry, with the remainder used to make alloys. There are over a thousand of these alloys ranging from simple *brass* (copper plus zinc) and *bronze* (copper plus tin) to the more complex and specialized, such as Monel metal (copper, nickel, iron, and manganese).

The compounds of copper correspond to oxidation states of 1+ (*cuprous*) and 2+ (*cupric*). The 1+ state is easily oxidized and is stable only in very insoluble compounds or in complex ions. The 2+ state is the one commonly observed in most copper compounds. The simple cuprous ion, Cu^+, cannot exist in aqueous solution, since it oxidizes and reduces itself by the reaction

$$2Cu^+ \longrightarrow Cu^{2+} + Cu(s)$$

but it can be stabilized in cuprous oxide (Cu_2O), a reddish, insoluble solid. The reddish color observed on metallic copper heated in air is apparently due to a surface coating of Cu_2O. In the classic test for reducing sugars (e.g., glucose, which, unlike sucrose, acts as a mild reducing agent), Cu_2O is formed as a red precipitate when the reducing sugar is heated with an alkaline solution of a cupric salt.

The simple cupric ion, Cu^{2+}, is colorless, and many anhydrous cupric salts are white. However, hydrated cupric salts and their aqueous solutions are blue, owing to the presence of hydrated cupric ion. The water molecules of hydration surround the Cu^{2+} ion, which has a d^9 electron configuration. Six of the nine electrons can be put in the lower t set of d orbitals (Figure 16.4) and three in the upper e set. If it takes yellow-red light to excite one of the t electrons to the e vacancy, the resulting color will appear to be blue. In general, aqueous solutions of cupric salts are acidic because of hydrolysis,

$$Cu^{2+} + H_2O \rightleftharpoons CuOH^+ + H^+$$

but the hydrolysis is not very extensive ($K_h = 1 \times 10^{-6}$). When base is added to these solutions, light blue cupric hydroxide [$Cu(OH)_2$] is formed. The hydroxide is slightly soluble in excess base, and so it might be called slightly amphoteric. When treated with aqueous ammonia solution, $Cu(OH)_2$ dissolves to give a deep blue solution, in which the color is usually attributed to a copper-ammonia complex ion, $Cu(NH_3)_4^{2+}$.

$$Cu(OH)_2(s) + 4NH_3 \longrightarrow Cu(NH_3)_4^{2+} + 2OH^-$$

Like many other complexes of cupric ion, $Cu(NH_3)_4^{2+}$ is paramagnetic, owing to an unpaired electron. It has a planar structure and can be destroyed by heat or by addition of acid. Heat is effective because it boils the NH_3 out of the solution,

$$Cu(NH_3)_4^{2+} \longrightarrow Cu^{2+} + 4NH_3(g)$$

and thus favors dissociation of the complex. Addition of acids results in neutralization of the NH_3 and similarly favors breakup of the complex:

$$Cu(NH_3)_4{}^{2+} + 4H^+ \longrightarrow Cu^{2+} + 4NH_4{}^+$$

One of the least soluble of cupric compounds is cupric sulfide (CuS). This is the black precipitate which is easily prepared by bubbling hydrogen sulfide through a solution of cupric salt. Its K_{sp} is very small, so that not even very concentrated H^+ can dissolve appreciable amounts of it. However, hot nitric acid is a good solvent for CuS, not because H^+ reacts with S^{2-} but because hot nitrate ion (especially in acid solution) is a very good oxidizing agent and oxidizes the sulfide ion to elementary sulfur. The net reaction is

$$3CuS(s) + 2NO_3{}^- + 8H^+ \longrightarrow 3Cu^{2+} + 3S(s) + 2NO(g) + 4H_2O$$

Probably the best known cupric compound is copper sulfate pentahydrate $[Cu(H_2O)_4SO_4 \cdot H_2O]$. In this material, each cupric ion is surrounded by a distorted octahedron of oxygen atoms; four of these lie in a square and belong to four H_2O molecules, and the other two belong to neighboring sulfate groups. The odd H_2O molecule, the fifth one, is not directly bound to the cupric ion but forms a bridge between $SO_4{}^{2-}$ and other H_2O groups. The pentahydrate, or *blue vitriol,* as it is sometimes called, is used extensively as a germicide and fungicide, since the cupric ion is toxic to lower organisms. Its application to water supplies for controlling algae depends on this toxicity.

Silver   Silver is a rather rare element (10^{-8} percent of the earth's crust), occurring principally as native silver, argentite (Ag_2S), and horn silver (AgCl). Only about one-fifth of current silver production comes from silver ores; the rest is mainly a by-product of copper and lead production. The main problem in extracting silver from its ores is to get the rather inert silver (or the very insoluble silver compounds) to go into solution. This can be accomplished by blowing air for a week or two through a suspension of the ore in dilute aqueous sodium cyanide (NaCN) solution. With native silver, the reaction can be written

$$4Ag(s) + 8CN^- + 2H_2O + O_2(g) \longrightarrow 4Ag(CN)_2{}^- + 4OH^-$$

Were it not for the presence of cyanide ion, the oxygen would not oxidize the silver to a higher oxidation state. Metallic silver is a rather poor reducing agent, and hence it is difficult to oxidize it to Ag^+. In the presence of cyanide ion, Ag^+ forms a strongly associated complex ion and is thus stabilized.

Similar reasoning applies to the dissolving of argentite (Ag_2S). This sulfide is very insoluble, and air oxidation by itself is not sufficient to get it into solution. However, in the presence of cyanide ion, solution does occur. In fact, the stability of the complex $Ag(CN)_2^-$ is so great that with high concentrations of cyanide ion the reaction

$$Ag_2S(s) + 4CN^- \longrightarrow 2Ag(CN)_2^- + S^{2-}$$

can be made to proceed to a useful extent without invoking air oxidation. To recover the silver from the residual solutions, it is necessary to use a rather strong reducing agent, such as aluminum metal or zinc metal in basic solution. A possible reaction is

$$Zn(s) + 2Ag(CN)_2^- + 3OH^- \longrightarrow 2Ag(s) + 4CN^- + Zn(OH)_3^-$$

where some of the zinc in the final solution is also present as a cyanide complex. $Zn(CN)_4^{2-}$.

Massive silver appears almost white because of its higher luster. It is too soft to be used pure in jewelry and coinage and is usually alloyed with copper for these purposes. Because of expense it cannot be used much for its best property, electric and thermal conductivity, which is second to none. In the colloidal state, silver appears black, because with very small particles, metallic reflection of the type discussed in Section 15.1 cannot occur.

The compounds of silver are essentially all of the 1+ state, although 2+ and 3+ compounds have been prepared under extreme oxidizing conditions. For example, an oxide believed to be AgO is formed when ozone is passed over elementary silver. The compound decomposes to Ag and O_2 and, in general, behaves as a very strong oxidizing agent. In the 1+ state silver forms the ion Ag^+, sometimes called *argentous ion* after the Latin word for silver, *argentum*. It does not hydrolyze appreciably in aqueous solution, is a good oxidizing agent, and forms many complex ions, for example, $Ag(NH_3)_2^+$, $Ag(CN)_2^-$, $AgCl_2^-$, all of which are linear. When base is added to solutions of silver salts, a brown oxide is formed, which shows little sign of being amphoteric,

$$2Ag^+ + 2OH^- \longrightarrow Ag_2O(s) + H_2O$$

However, the oxide does dissolve in an aqueous solution of ammonia because of formation of the colorless silver-ammonia complex ion, $Ag(NH_3)_2^+$:

$$Ag_2O(s) + 4NH_3 + H_2O \longrightarrow 2Ag(NH_3)_2^+ + 2OH^-$$

Solutions containing $Ag(NH_3)_2^+$ are frequently used as sources of silver for silver plating. They have the advantage of providing low concentrations of

Ag^+, so that reduction by mild reducing agents, such as glucose, slowly deposits a compact silver plate. Evaporation of these solutions leaves dangerous solid residues which are violently explosive. Their composition is not known but has been described as silver amide ($AgNH_2$), silver imide (Ag_2NH), and silver nitride (Ag_3N).

Probably the most interesting of all the silver compounds are the silver halides, AgF, $AgCl$, $AgBr$, and AgI. Except for silver fluoride, which is very soluble in water (up to 14.3 moles/1000 g of H_2O), these halides are quite insoluble. The low solubility is rather surprising, because, in general, salts of $1+$ cations and $1-$ anions are usually soluble. In this respect, AgF is normal; it dissolves much like NaF or KF. The abnormal insolubility of the other silver halides is attributed to the fact that their solid lattices are more stable than expected. The principal reason for this is that there are strong van der Waals attractions between Ag^+ ions and the halide ions, and these attractions are in addition to the ordinary ionic attractions.

Except for AgF, the silver halides are sensitive to light. For this reason, they find use in photographic emulsions. The chemistry of the photographic process is not well understood and is complicated, because it apparently involves defect structures (see Section 7.6). However, the basic steps can be described as follows: When photographic film, consisting of a dispersion of silver bromide in gelatin, is exposed to light, grains of silver bromide are activated, depending on the intensity of the incident light. This is not a visible change and, according to one theory, simply involves migration of an electron from a bromide ion to an interstitial silver ion in the silver bromide emulsion. Whatever the cause, the activated grains react faster with mild reducing agents (developers) than do nonactivated grains. When the exposed photographic emulsion is developed, black metallic silver forms by the preferential reduction of exposed silver bromide grains. The result is that black areas appear on the film where the light was strongest. Since $AgBr$ slowly turns black when exposed to light, the whole film would turn black eventually. However, the photographic image can be fixed by washing out unexposed $AgBr$ grains. Although very insoluble, $AgBr$ will dissolve in solutions containing high concentrations of thiosulfate ion, $S_2O_3^{2-}$, by the reaction

$$AgBr(s) + 2S_2O_3^{2-} \longrightarrow Ag(S_2O_3)_2^{3-} + Br^-$$

Thus, the final step involves soaking the film in a fixing bath, the essential component of which is $Na_2S_2O_3$, or "hypo." The result is a fixed negative image of the exposure. By shining light through the negative on another emulsion, developing, and then fixing it, the light and dark areas are inverted to produce a positive image.

In color photography the processes are much more involved. In one process the film is coated with three emulsion layers, each of which is sensitive to one of the primary colors. On exposure and development, images are formed in each of the three layers. By appropriate choice of dyes and other chemicals, these three images can be colored separately so as to reproduce by superposition the original multicolored pattern.

For qualitative analysis, silver ion can be separated from other cations by adding HCl to precipitate white insoluble AgCl. Two other white insoluble chlorides that may be encountered, $PbCl_2$ and Hg_2Cl_2, may be distinguished by washing with hot water to dissolve the first or adding NH_3 to decompose the second, Addition of NH_3 to Hg_2Cl_2 forms a black color due to formation of Hg and $HgNH_2Cl$, but addition of NH_3 to AgCl dissolves it to form $Ag(NH_3)_2{}^+$ and Cl^-. The presence of silver can be confirmed by adding HNO_3 to the filtrate to decompose the complex and reprecipitate AgCl. To test for Cu^{2+}, H_2S is added in acid solution to precipitate black CuS. Black CuS and black HgS can be distinguished from each other in that CuS is soluble in hot nitric acid but HgS is not.

16.10   Zinc and Mercury

The elements of the zinc subgroup are zinc, cadmium, and mercury. They are more active than the elements of the copper subgroup, but their chemistry is somewhat more simple. The zinc-subgroup elements have a characteristic oxidation state of $2+$, except for mercury, which also forms $1+$ compounds. Some of the more important properties are listed in Figure 16.13.

						Ioni-zation Poten-tial, eV	Reduction Potential (for M^{2+}), volts
Figure 16.13 ELEMENTS OF THE ZINC SUBGROUP	Symbol	Z	Electronic Configuration	Melting Point, °C	Boiling Point, °C		
	Zn	30	$2,8,18,4s^2$	419	907	9.39	-0.76
	Cd	48	$2,8,18,18,5s^2$	321	767	8.99	-0.40
	Hg	80	$2,8,18,32,18,6s^2$	-38.9	357	10.43	$+0.85$

As seen from the electronic configurations, each of these elements has two electrons in the outermost energy level. The situation is reminiscent of that found for the alkaline-earth elements (Section 15.3). The low melting points may at first sight be surprising, but they are not entirely unexpected. In progressing from left to right through the transition sequence, the low point in the atomic volume has been passed, and the atoms have begun to get larger from here on. As the atoms get bigger, they are farther apart, and forces of attraction are weaker. Thus, it becomes easier to melt the elements. Probably of greater importance is the fact that the d shells of the second-outermost shells are filled; therefore, there is little chance for the covalent bonding between ions found in other transition elements. In mercury, the interatomic forces are so weak that its melting point is below room temperature.

Zinc   About a hundred times as abundant as copper, zinc occurs principally as the mineral sphalerite (ZnS), also called *zinc blende*. The metal is prepared by roasting the sulfide in air to convert it to oxide and then reducing the oxide with finely divided carbon. The reactions are

$$2ZnS(s) + 3O_2(g) \longrightarrow 2ZnO(s) + 2SO_2(g)$$
$$ZnO(s) + C(s) \longrightarrow Zn(g) + CO(g)$$

Since the second reaction is carried out at about 1200°C, above the boiling point of Zn, the metal forms as a vapor and must be condensed. Very rapid condensation produces the fine powder known as zinc dust.

Massive zinc has fairly good metallic properties except that it is rather brittle, especially at 200°C, where it can be ground into a powder. It is a moderately active metal and can even reduce water to hydrogen, but only when heated. With acids, ordinary Zn gives the well-known evolution of H_2. Strangely enough, this is very rapid when the Zn is impure but almost too slow to be observed when the Zn is very pure. Apparently the impurities (especially arsenic and antimony) speed dissolving by serving as centers from which hydrogen gas can evolve.

In air, zinc tarnishes but slightly, probably because it forms a self-protective coat of oxide or carbonate. Because it withstands corrosion so well itself and because it can give cathodic protection (Section 16.8) to iron, zinc is often used as a coating on iron to keep it from rusting. Iron protected in this way is called *galvanized iron* and can be made by dipping the iron into molten zinc or by plating zinc on it from an electrolytic bath. The other important use of zinc is in alloys such as the brasses, which are essentially copper-zinc alloys.

In all its compounds, zinc shows only a 2+ oxidation state. The zinc ion, Zn^{2+}, is colorless and not paramagnetic. In aqueous solutions it hydrolyzes to give slightly acid solutions. The hydrolysis usually written

$$Zn^{2+} + H_2O \rightleftharpoons ZnOH^+ + H^+$$

does not proceed as far toward the right ($K_h = 3 \times 10^{-10}$) as that of Cu^{2+} ion. Thus, for equal concentrations, solutions of zinc salts are less acid than those of cupric salts. When base is added to solutions of zinc salts, white zinc hydroxide [$Zn(OH)_2$] is precipitated. This hydroxide is amphoteric, and therefore further addition of base dissolves it to give zincate ion, variously formulated as $Zn(OH)_3^-$, ZnO_2^{2-}, $HZnO_2^-$, or, most probably, $Zn(OH)_4^{2-}$.

Whatever the formula of zincate ion, the equilibrium concentration of Zn^{2+} in basic solution is very small. This means (by the principle of Le Chatelier) that the half-reaction

$$Zn(s) \longrightarrow Zn^{2+} + 2e^-$$

has greater tendency to go to the right in basic solution (small concentration of Zn^{2+}) than in acid solution. Consequently, zinc metal is a stronger reducing agent (has more negative reduction potential) for basic solutions than for acid solutions.

Like other transition elements, zinc has a great tendency to form stable complex ions. For example, zinc hydroxide is easily dissolved in aqueous ammonia because of the formation of a zinc-ammonia complex, $Zn(NH_3)_4^{2+}$. The hydroxide can also be dissolved in cyanide solutions because of the formation of a zinc-cyanide complex, $Zn(CN)_4^{2-}$.

When hydrogen sulfide is passed through solutions of zinc salts which are not too acid, white zinc sulfide precipitates. Although the solubility product of ZnS (1×10^{-22}) is rather small, so that ZnS is essentially insoluble in neutral solutions, addition of acid lowers the sulfide-ion concentration sufficiently so that ZnS becomes soluble. The enhanced solubility of ZnS in acid solution gives a method for separating it from other sulfides such as CuS, Ag_2S, and CdS.

Zinc sulfide is used extensively in the white pigment, lithopone, an approximately equimolar mixture of ZnS and $BaSO_4$. ZnS is also used in making fluorescent screens, because impure ZnS acts as a phosphor; i.e., it can convert energy such as that of an electron beam into visible light. The action of these phosphors is very complex and is closely related to the properties of defects in solid-state structures. The simplest view is that an electron beam impinging on impure ZnS uses its energy to kick electrons out of an impurity center. This electron wanders through the crystal until it finds some other center to which it can return by giving off a flash of light. Many television screens are coated with zinc sulfide phosphor.

Mercury   The only common mineral of mercury is cinnabar (HgS), from which the element is produced by roasting in air:

$$HgS(s) + O_2(g) \longrightarrow Hg + SO_2(g)$$

Unlike any other metal,† mercury is a liquid at room temperature, and its symbol emphasizes this, since it comes from the Latin *hydrargyrum*, meaning liquid silver. The liquid is not very volatile (vapor pressure is 2.4×10^{-6} atm

---

† Cesium metal has a melting point of 28.5°C, and gallium metal has a melting point of 29.8°C. Thus, the uniqueness of mercury as a liquid metal disappears on hot days.

at 25°C), but the vapor is very poisonous, and *prolonged exposure even to the liquid should be avoided.*

Liquid mercury has a high metallic luster, but it is not a very good metal, as it has a higher electric resistance than any of the other transition metals. However, for some uses, as in making electric contacts, the mobility of liquid mercury is such a great advantage that its mediocre conductivity can be tolerated. Furthermore, its inertness to air oxidation, its relatively high density, and its uniform expansion with temperature lead to special uses, as in barometers and thermometers.

Liquid mercury dissolves many metals, especially the softer ones such as copper, silver, gold, and the alkali elements. The resulting alloys, which may be solid as well as liquid, are called *amalgams.* Probably their most distinctive property is that the reactivity of the metal dissolved in the mercury is thereby lowered. For example, in sodium amalgam the reactivity of sodium is so low that sodium amalgam can be kept in contact with water with only slow evolution of hydrogen.

In its compounds, mercury shows both 1+ (mercurous) and 2+ (mercuric) oxidation states. In this respect, it is unlike the other members of the zinc subgroup. The mercurous compounds are unusual because they all contain two mercury atoms bound together. In aqueous solutions, the ion is a double ion corresponding to Hg_2^{2+}, in which there is a covalent bond between the two mercury atoms. Experimental evidence for this is the lack of paramagnetism of mercurous compounds. The ion Hg^+ would have one unpaired electron in its 6s orbital and would be paramagnetic, whereas the ion Hg_2^{2+} would have the two electrons paired as a covalent bond and would not be paramagnetic.

Except for the doubling, mercurous ion behaves much like Ag^+; for example, it reacts with chloride ion to precipitate white mercurous chloride (Hg_2Cl_2), also known as *calomel.* When exposed to light, calomel darkens by partial disproportionation into Hg and $HgCl_2$. Unlike Ag^+, mercurous ion does not form an ammonia complex. When aqueous ammonia is added to Hg_2Cl_2, the solid turns black because of formation of finely divided mercury:

$$Hg_2Cl_2(s) + 2NH_3 \longrightarrow HgNH_2Cl(s) + Hg + NH_4^+ + Cl^-$$

The compound $HgNH_2Cl$, mercuric ammonobasic chloride, is white, but its color is obscured by the intense black. This difference in behavior toward NH_3 provides a simple test for distinguishing AgCl from Hg_2Cl_2.

In the 2+ state, mercury is frequently represented as the simple ion Hg^{2+}, although it is usually found in the form of complex ions, insoluble solids, or weak salts. For example, in a solution of mercuric chloride, the concentration of Hg^{2+} is much smaller than the concentration of undissociated $HgCl_2$ molecules.

Although mercuric sulfide as found in nature is red, when H_2S is passed through a mercuric solution, a black precipitate of HgS is obtained. The color difference may be due to differences in crystal structure. The solubility product of black HgS is very low (1.6×10^{-54}). HgS will not dissolve even in boiling nitric acid. Aqua regia, however, which supplies both nitrate for oxidizing the sulfide and chloride for complexing the mercuric, does take it into solution.

For qualitative analysis, mercurous ion can be separated from other cations by adding HCl to precipitate white insoluble Hg_2Cl_2. If NH_3 is added to the chloride, a black color appears, owing to formation of Hg and $HgNH_2Cl$.

If H_2S is added to an acidic solution containing Hg^{2+} and Zn^{2+}, only black HgS is formed. If the solution is then made basic with NH_3, white ZnS is formed. A confirmatory test for ZnS would be to dissolve it in HCl plus HNO_3, evaporate to dryness, and reprecipitate by addition of H_2S in an SO_4^{2-}–HSO_4^- buffer.

The separation of HgS from CuS makes use of the fact that CuS is soluble in boiling HNO_3 whereas HgS is not. HgS can be confirmed by dissolving in aqua regia and reducing with $SnCl_2$ to Hg_2Cl_2 and Hg. Addition of NH_3 to a solution containing Cu^{2+} gives the blue color characteristic of $Cu(NH_3)_4^{2+}$.

QUESTIONS

16.1 *Electronic Configurations* How many d electrons do you expect for each of the following: Ti, Mn^{3+}, Cu^{2+}, Kr, a half-filled set of t orbitals?

16.2 *Transition-Metal Properties* State what differences you would expect for Cr and Ca with regard to each of the following: (a) hardness, (b) variety of oxidation states, (c) magnetism of compounds, (d) formation of complexes, (e) color of compounds.

16.3 *Ligand-Field Theory* Explain how ligand-field theory can account for each of the following: (a) Cu^{2+} compounds are colored, Zn^{2+} compounds are usually colorless, (b) $K_2Fe(CN)_6$ is diamagnetic, (c) Cr^{3+} binds ligands tightly.

16.4 *Lanthanides* Why are the 14 lanthanide elements so similar in chemical properties?

16.5 *Chromium* Write balanced equations for each of the following changes: (a) air oxidation of chromous hydroxide, (b) the dissolving (in water) of potassium chrome alum, (c) hydrolysis of chromic ion, (d) reduction of dichromate by Fe^{2+} in acid solution.

16.6 *Manganese* (a) Give the formula for an ion or compound of Mn in each of its usual oxidation states. (b) For any of the oxidation states mentioned in (a), which can show disproportionation, under what conditions does disproportionation occur, and how can it be prevented? (c) Write balanced equations for the reduction of MnO_4^- to three different oxidation states in acid, neutral, and basic solution using as reducing agents $SO_2 \longrightarrow HSO_4^-$ for acid, $HSO_3^- \longrightarrow SO_4^{2-}$ for neutral, and $SO_3^{2-} \longrightarrow SO_4^{2-}$ for basic solution.

16.7 *Iron-Group Elements* In one characteristic property the three elements of the iron group differ from all other elements. (a) What is this property? (b) How is it thought to arise? (c) Why do only these three elements show this property?

16.8 *Iron* In producing steel what substances (do not forget gases) go into and what substances come out of (a) a blast furnace, (b) a Bessemer converter, (c) an open-hearth furnace?

16.9 *Corrosion* Consider the mechanism of corrosion of iron and explain whether each of the following is a reactant, a catalyst, and/or an intermediate: (a) H^+, (b) H_2O, (c) H, (d) e^-, (e) Fe, (f) Fe^{2+}.

16.10 *Copper* Compare the reduction potentials of Cu, Ag, and Fe. (a) Which of these three metals is easiest and which hardest to oxidize at the anode of an electrolytic cell? (b) Which of the 2+ ions of these three metals is easiest, and which hardest, to reduce from solution at the cathode of an electrolytic cell? (c) What range of voltage would in principle suffice for the electrolytic purification of Cu from Ag and Fe?

16.11* *Silver* K_{sp} for AgCl is 1.7×10^{-10} and for AgBr, 5.0×10^{-13}. (a) How many times as soluble in water is AgCl as AgBr? (b) How many moles of each will dissolve in 0.50 liter of 0.10 M AgF?

16.12 *Zinc* Write net equations demonstrating each of the following: (a) amphoterism of $Zn(OH)_2(s)$, (b) the solubility of $Zn(OH)_2(s)$ in aqueous NH_3, (c) the solubility of ZnS in acid.

16.13 *Mercury* From Appendix 6 find the reduction potentials relating Hg in the 0, 1+, and 2+ oxidation states. (a) Will the 1+ oxidation state disproportionate (oxidize and reduce itself)? (b) What reducing agents should in principle reduce Hg from the 2+ to the 1+ oxidation state but not take 1+ to the metal? (c) Without using reducing agents mentioned in (b), tell how the 1+ oxidation state can be obtained from the 2+ state?

16.14 *Blast Furnace* In the operation of the blast furnace and in terms of substances added to it, there is but a single reducing agent acting on three oxidizing agents. Write balanced equations for each of these three *net* reactions.

16.15* *Manganese Ores* Calculate the percent Mn by weight in each of the ores (a) MnO_2, (b) Mn_2O_3, (c) Mn_3O_4.

16.16 *Trends* Using data from this and earlier chapters, prepare graphs (property vs. atomic number) for the first-row transition elements (Sc through Zn) showing the following properties: (a) atomic radius, (b) first ionization potential, (c) maximum oxidation state.

16.17 *d Orbitals* Consider four ligands of an octahedral complex as lying in the plane of the X and Y axes. (a) Draw this plane locating the ligands and the d_{xy} and $d_{x^2-y^2}$ orbitals. (b) Describe the difference in repulsive energy between the ligands and one electron in each of the two orbitals drawn. (c) By means of appropriate plane drawings show a similar difference between d_{z^2} and each of d_{zx} and d_{yz}.

16.18 *Transition Elements* Pick transition elements (one each) which answer the following descriptions: (a) exhibits 7+ oxidation state; (b) forms a double 1+ ion; (c) forms a pink ion; (d) forms compounds in several oxidation states, conversion between two of which can be accomplished in both directions by H_2O_2; (e) metal dissolves in dilute but not concentrated HNO_3; (f) is used to galvanize iron.

16.19 *Terms* What is meant by (a) triad, (b) transition metal, (c) covellite, and (d) anode slime?

16.20 *Octahedral Complexes* Given a central atom A with two B ligands and four C ligands arranged octahedrally in the complex AB_2C_4, how many isomers of AB_2C_4 are there? Draw them.

16.21* *Oxidation-Reduction* You are given a mixture that contains cuprous chloride and cupric nitrate. You propose to find out how much there is of each by titrating with permanganate ion in acid solution, which oxidizes Cu^+ to Cu^{2+} and gives Mn^{2+}. If 0.295 g of the mixture requires 12.8 ml of 0.150 M $KMnO_4$, what percentage of the original mixture is CuCl?

16.22 *Thermodynamics* The heats of fusion per gram for Cu, Ag, and Au are 205, 105, and 64.0 joules. The melting points for Cu, Ag, and Au are 1083, 961, and 1063°C. (a) Calculate ΔH when 1 mole of each element melts. (b) Calculate ΔS for melting 1 mole of each element.

16.23* *Stoichiometry* How many liters of oxygen at 20.0°C and 1.50 atm will be produced by oxidation of H_2O_2 by 220 ml of 0.170 M $KMnO_4$ in acid solution?

16.24 *Vanadium* The highest oxidation state of vanadium is 5+. Why is it that a simple V^{5-} ion cannot exist in solution? Illustrate your answer by writing an equation showing what would happen if 5+ were placed in water.

16.25 *Oxidation-Reduction* Write the balanced equations for the following reactions: (a) Cr^{3+} reacts with H_2O_2 in basic medium, (b) $Cr_2O_7^{2-}$ reacts with H_2O_2 in acid medium, (c) Mn^{3+} reacts with water.

16.26 *Solubility* Account for the fact that AgF is soluble in water and other silver halides are insoluble in water.

16.27 *Coordination Compounds* In 1893, an explanation for the rather curious series of compounds $CrCl_3 \cdot 6NH_3$, $CrCl_3 \cdot 5NH_3$, $CrCl_3 \cdot 4NH_3$, and $CrCl_3 \cdot 3NH_3$ was postulated based on the following observations. When dissolved in aqueous solution and treated with silver nitrate, 1 mole of $CrCl_3 \cdot 6NH_3$ precipitated 3 moles of AgCl, 1 mole of $CrCl_3 \cdot 5NH_3$ precipitated 2 moles of AgCl, 1 mole of $CrCl_3 \cdot 4NH_3$ precipitated 1 mole of AgCl, and 1 mole of $CrCl_3 \cdot 3NH_3$ did not precipitate AgCl. Using these results, suggest structures for the above mentioned coordination compounds.

16.28 *Oxidation-Reduction* (a) Write a balanced net equation for the following reaction in neutral solution:

$$Mn^{2+} + S_2O_8^{2-} \longrightarrow MnO_4^- + SO_4^{2-}$$

(b) What will be the concentration of MnO_4^- in a solution made by mixing 50.0 ml of 0.032 M Mn^{2+} with 38.0 ml of 0.035 M $S_2O_8^{2-}$ if the reaction proceeds as in (a)?

16.29* *Oxidation-Reduction* The concentration of an unknown solution of $KMnO_4$ can be determined by mixing an unknown solution with H_2O_2 and measuring the oxygen evolved. (a) Balance the following equation for acid solution.

$$MnO_4^- + H_2O_2 \longrightarrow Mn^{2+} + O_2(g)$$

(b) If 30.0 ml of x M $KMnO_4$ solution liberates 0.150 liter of O_2 at 25°C and 750 mmHg, what is the molarity of the $KMnO_4$ solution? (The oxygen was collected over water.)

16.30 *Cell Voltage* How would the cell voltage of an Al–Au$^+$ cell compare with that of a Na–Ag$^+$ cell?

16.31 *Analysis* You are given three beakers. Each beaker contains either nitrate ions, sulfate ions, or chloride ions. Describe the methods you would use to identify the nitrate, sulfate, and chloride ions?

16.32 *Hydrolysis* Calculate the pH of (a) 1 M $Cu(NO_3)_2$, (b) 1 M $Zn(NO_3)_2$.

16.33 *Qualitative Analysis* How could you distinguish between (a) AgCl and Hg_2Cl_2, (b) CuS and HgS, (c) FeS and CaS, (d) $Mn(OH)_2$ and $Cr(OH)_3$?

16.34 *Corrosion* (a) Explain how cathodic protection serves to reduce corrosion of iron. (b) Using all the reduction potentials given in this chapter, which of the metals covered will provide cathodic protection to iron?

16.35 *Explanations* Give possible explanations for each of the following facts: (a) $Co(NH_3)_6^{3+}$ is diamagnetic. (b) Photographic reducing agents are mild reducing agents. (c) Lu^{3+} is smaller than La^{3+}. (d) MnO_2 is a stronger oxidizing agent in acid than in base.

16.36 *Solubility* For Hg_2Cl_2, K_{sp} is 1.1×10^{-18}. How many moles of Hg_2Cl_2 will dissolve in 1 liter of (a) water, (b) 0.10 M $Hg_2(NO_3)_2$, (c) 0.10 M NaCl?

16.37* *Chromite* The ore chromite, $FeCr_2O_4$, is converted to high-chrome steel containing 30% Cr and 70% Fe by weight. (a) From 1000 kg of chromite, how many kilograms of this steel could be made? (b) The remainder of the Cr is converted to $Na_2Cr_2O_7$. How many kilograms of this will result from the residue?

16.38 *Solubility* For PtS, $K_{sp} = 8 \times 10^{-73}$. Ignoring hydrolysis of S^{2-} and all other reactions, if the Atlantic Ocean is saturated with PtS, how many liters, on average, contain one Pt^{2+} ion?

16.39 *Copper Refining* Suppose you are electrolytically refining a series of crude copper samples containing varying ratios of metals more active than Cu (for example, Fe) to metals less active than Cu (for example, Ag). (a) For those high in inactive metals, should you have to use more or less electric current than for a purer Cu sample? (b) For those high in more active metals, should you have to use more or less current? (c) In the latter case, discuss what happens to solution concentrations and at the cathode after prolonged electrolysis.

16.40 *Blast Furnace* Suppose you operated a blast furnace by continually adding equal weights of Fe_2O_3, $CaCO_3$, and C with just sufficient O_2 to convert all the C to CO. If all the Fe_2O_3 is reduced to Fe and all the $CaCO_3$ is decomposed to CaO, what should be the mole ratio of CO_2 to CO in the exhaust gas?

16.41 *Isomers* Sketch the structures of all possible isomers of the neutral molecular species $Cr(NH_3)_3Cl_3$.

Carbon, Silicon, and Boron

17

Grouped together in the periodic table are three important light elements, carbon, silicon, and boron, that are neither clearly metallic nor nonmetallic. They are sometimes referred to as semimetals in recognition of the fact that when impure or when in the right kind of crystalline modification they often appear to be metals. The elements are important because, between them, they account for most of the compounds in living systems, most of the compounds in ordinary rocks, and most of the key materials in the sophisticated devices of solid-state electronics. Two of the most active areas of current scientific research, molecular biology and materials science, are based on these elements or on specialized techniques derived from their study. In this chapter we consider some of the most characteristic features of C, Si, and B chemistry, although the field of carbon chemistry is so huge that several additional chapters will be needed to give an adequate picture of it.

17.1 Carbon

Carbon is the first element of group IV. Its electron configuration is $1s^2 2s^2 2p^2$ corresponding to two electrons in the K shell and four electrons in the L shell. The chemistry therefore is characterized by four valence electrons per atom. Although not very plentiful in the earth's crust (<0.1 percent), carbon

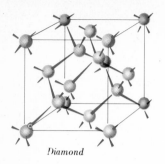

Diamond

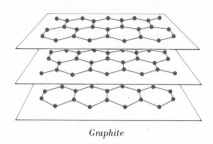

Graphite

Figure 17.1

Allotropic forms of carbon.

is the second most abundant element (oxygen is first) in the human body (17.5 percent). It occurs in all plant and animal tissues, combined with hydrogen and oxygen, and in their geological derivatives, petroleum and coal, where it is combined mostly with hydrogen in the form of hydrocarbons. Combined with oxygen, carbon also occurs in the atmosphere as CO_2 and in carbonate rocks such as limestone. In the free state, carbon occurs to a slight extent as diamond and graphite, the two allotropic forms of the element.

As shown in Figure 17.1, the principal difference between diamond and graphite is that, in the former, each carbon atom has four nearest neighbors whereas, in the latter, each carbon has three. In the diamond lattice the distance between centers of adjacent carbon atoms is 0.154 nm, with each atom bonded to four other atoms at the corners of a tetrahedron. Since each of these carbon atoms in turn is tetrahedrally bonded to four carbon atoms, the result is an interlocked structure extending in three dimensions. The molecule formed is very hard (the hardest naturally occurring substance known) and has a high melting point (3500°C). These properties presumably arise because the positions of the atoms are rigidly determined by the covalent bonds. Furthermore, diamond is a nonconductor of electricity. Since the sharing of four additional electrons fills all the orbitals, it is impossible for another electron to move in on a given carbon atom. In other words, all the pairs of electrons in the diamond structure are localized between specific pairs of C atoms and are not free to migrate through the crystal, because no other C atom can accommodate them. For this reason, diamond is an insulator for electric current.

Diamond is also characterized by a high refractive index; i.e., light rays entering diamond from air are bent strongly away from their original straight-line path. The effect is thought to be primarily due to a slowing down of the light wave by the tightly bound electrons. Because of high refractive index, much of the light falling on a diamond is internally reflected by interior surfaces. The traditional sparkle of gem stones is primarily due to their shapes, which take maximum advantage of this internal reflection. Also, the refraction of different colors of light is not equal; therefore, when held at the proper angle, the diamond reflects only a portion of the spectrum of white light to the eye. This high dispersion effect, which always accompanies high refractive index, explains the brilliant "fire" observed from well-cut diamonds.

In graphite, the structure consists of giant sheetlike molecules which are held to each other, 0.340 nm apart, probably by van der Waals forces. Within the sheets each carbon atom is covalently bound to three neighbors 0.142 nm away, which in turn are also bound to three carbon atoms. Since each carbon has four valence electrons and only three carbons to bond to, there are more than enough electrons to establish single bonds. However, since there can be no preference as to where the last electron should be located (all three neighbors being equivalent), it must be considered as belonging to all three bonds.

The electronic configuration of graphite may be represented as a resonance hybrid of the three contributing structures shown across the top of Figure 17.2. An alternative way to represent the electron distribution is shown at the bottom left of Figure 17.2. Each hexagon has inscribed in it a dashed circle which represents the so-called π electrons. Designating the z direction as the direction perpendicular to the graphite sheet, we can imagine p_z orbitals sticking out above and below the planar net of carbon atoms. As is the case in Figure 4.12 where π-bond formation is illustrated,

Figure 17.2
Various representations of bonding in graphite.

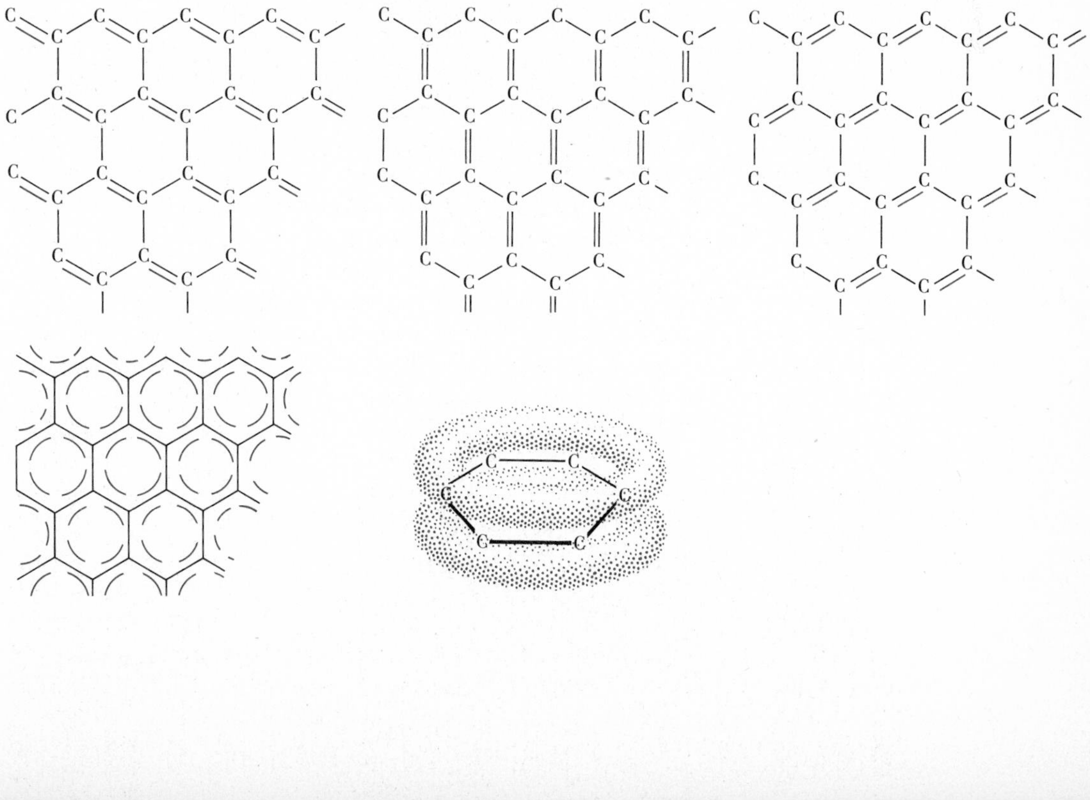

a p_z orbital on one carbon atom can overlap in a side-to-side way with the p_z orbitals on neighboring atoms. Thus the p_z electron of a particular carbon is not localized on that particular carbon atom but is spread out over the whole sheet. A portion of the resulting charge cloud is shown on the lower right of Figure 17.2.

Massive graphite is a soft, gray, high-melting solid with a dull metallic luster and fairly good electric conductivity. The softness is attributed to the weak sheet-to-sheet bonding, which permits adjacent layers to slide over each other. Generally, graphite has good lubricating properties which are sometimes directly ascribed to the ease with which the carbon layers slide over each other. This is not quite exact. Under normal circumstances, the lubricative property of graphite comes about because layers of gas molecules are adsorbed on the graphite sheets. It is generally these gas layers that act as almost frictionless cushions. If the adsorbed gas is pumped off, then graphite loses much of its desirable character as a lubricant. In the vacuum of outer space, for example, graphite becomes fairly sticky and is generally replaced by more suitable lubricating agents. The high melting point of graphite is traceable to the strong covalent binding within the sheets, which makes difficult the disordering necessary for melting. The conductivity and metallic luster stem from the freedom of the π electrons (one per carbon) to move from atom to atom. Because of its high melting point and its electric conductivity, graphite finds extensive use as an industrial electrode material, as, for example, in the electrolytic preparation of aluminum.

Besides massive graphite, there are several porous forms of carbon which apparently are graphite in character. These include coke (made by heating coal in the absence of air), charcoal (made from wood in the same way), and carbon black (soot). They all have tremendous surface areas; for example, a 1-cm cube of charcoal can have a surface area of 500 ft^2, which is equivalent to 2 billion holes drilled through the cube. Since each exposed carbon atom at the surface can use its extra valence electron to bind other atoms, these forms of carbon have strong adsorption properties.

Under normal conditions graphite is the stable form of carbon, and so diamonds ought spontaneously to convert to graphite. However, the rate of conversion from diamond to graphite is too slow to observe. At high pressure, the principle of Le Chatelier predicts that diamond should become stable, since its density (3.51 g/cm^3) exceeds that of graphite (2.25 g/cm^3). By raising the pressure and working at high temperature (to increase the rate), diamonds have been prepared synthetically. Although not generally of gem quality, the synthetic material finds industrial application as an abrasive.

17.2 Compounds of Carbon

At room temperature carbon is rather inert, but at higher temperature it reacts with a variety of other elements. With metals and semimetals, carbon forms solid carbides of complex structure, such as silicon carbide (SiC), iron carbide (Fe_3C), and calcium carbide (CaC_2). Silicon carbide, formed by heating quartz (SiO_2) with graphite, is the industrial abrasive Carborundum. There are at least six different polymorphic forms of solid SiC, one of which has Si and C atoms occupying alternate positions in a diamond lattice. Iron carbide, mentioned in Section 16.7 as the essential constituent of white cast iron, has an extremely complex structure. Calcium carbide, obtained by heating CaO with coke, reacts with water to liberate acetylene,

$$CaC_2(s) + 2H_2O \longrightarrow Ca^{2+} + 2OH^- + C_2H_2(g)$$

and can be used for the preparation of C_2H_2. The formation of acetylene from CaC_2 reflects the fact that the CaC_2 lattice contains Ca^{2+} and C_2^{2-} ions. The arrangement of these ions is the same as that of Na^+ and Cl^- ions in NaCl.

With nonmetals, carbon forms molecular compounds, which vary from simple carbon monoxide to extremely complex hydrocarbons. With the nonmetal sulfur, carbon reacts at high temperature to form carbon disulfide (CS_2). At room temperature, CS_2 is unstable with respect to decomposition to the elements. However, the rate of decomposition is unobservably slow, and liquid CS_2 is a familiar solvent, particularly for such substances as rubber and sulfur. Use of CS_2 as a solvent is hazardous, because it is toxic and highly flammable. When carbon disulfide vapor is heated with chlorine gas, the following reaction occurs:

$$CS_2(g) + 3Cl_2(g) \longrightarrow CCl_4(g) + S_2Cl_2(g)$$

The carbon tetrachloride (CCl_4) thus formed resembles carbon disulfide in being a nonpolar liquid at room temperature, and therefore it is a good solvent for nonpolar solutes. As a cleaning fluid, CCl_4 should be used with great caution, because, although it is not flammable, the liquid can penetrate the skin, and the vapor is quite toxic. In fact, it should never be used in an unventilated space.

With oxygen, carbon forms oxides, the most important of which are carbon monoxide (CO) and carbon dioxide (CO_2). Their electronic formulas are usually represented as follows:

$$:C:::O: \qquad :\overset{\cdot\cdot}{O}::C::\overset{\cdot\cdot}{O}:$$

These oxides are most conveniently prepared by combustion of carbon or hydrocarbons, with CO predominating when the supply of oxygen is limited. As previously indicated (Sections 14.2 and 16.7), carbon monoxide is an important industrial fuel and reducing agent. It is a colorless, odorless gas and is quite poisonous, because it interferes with the normal oxygen-carrying function of the hemoglobin in the red blood cells. Instead of forming a complex compound with oxygen molecules, hemoglobin forms a more stable complex compound with CO (carboxyhemoglobin). The tissue cells are thus starved for lack of oxygen, and death may result. Concentrations of 0.2 percent in air cause unconsciousness in about half an hour and death in about 3 h. Because CO is present in the exhaust gases of automobiles, near-toxic concentrations frequently are approached in congested areas during peak traffic hours. It is a marvel that the human race manages to survive at the same time that it spews more than *five million tons* of carbon monoxide into the atmosphere each year from cars throughout the world.

Unlike CO, CO_2 is not poisonous and, in fact, is necessary for various physiological processes, e.g., the maintenance of the proper pH of blood. Since it is *generated* in respiration and *consumed* in photosynthesis, the concentration in the atmosphere remains fairly constant at about 0.04 percent. The principal sources of commercial CO_2 are (1) the distillery industry, where the fermentation of sugar to alcohol,

$$C_6H_{12}O_6 \xrightarrow[\text{yeast}]{} 2C_2H_5OH + 2CO_2(g)$$

cheaply produces large amounts of by-product CO_2, and (2) the thermal decomposition of limestone to form CO_2 and CaO. The gas is formed conveniently in the laboratory by thermal decomposition of bicarbonates such as $NaHCO_3$ or by the reaction of bicarbonates or carbonates with acid. The gas is rather dense (approximately $1\frac{1}{2}$ times the density of air) and settles in pockets to displace the lighter air. Since it is not combustible itself, it acts as an effective blanket to shut out air in fire fighting. The phase relations of carbon dioxide and the use of CO_2 as a refrigerant have been indicated in Section 9.5.

Compared with most gases, CO_2 is quite soluble in water; at 1 atm pressure and room temperature, the solubility is 0.03 M. (It is twice as soluble in alcohol, where it has the peculiar physiological effect of increasing the rate of passage of alcohol from the stomach to the intestines, where it is taken up by the blood. The heady effect of champagne compared with that of still wine is partly due to this.) The water solutions of CO_2 are acid, with a pH of about 4. Although it has been suggested that this acidity arises primarily from weak carbonic acid (H_2CO_3) formed by the reaction of CO_2 with H_2O, this acid has never been isolated. Recent experiments indicate, in fact, that more than 99 percent of the CO_2 in aqueous solution remains in the form of linear molecules. However, a small amount of CO_2 does react to form H_2CO_3, which can dissociate to H^+ and bicarbonate ion. Thus there are the two simultaneous equilibria

$$CO_2 + H_2O \rightleftharpoons H_2CO_3$$
$$H_2CO_3 \rightleftharpoons H^+ + HCO_3^-$$

which can be combined and written as

$$CO_2 + H_2O \rightleftharpoons H^+ + HCO_3^-$$

The constant for this last equilibrium, loosely called the *first dissociation of carbonic acid,* is 4.2×10^{-7}. The dissociation of bicarbonate ion into H^+ and carbonate ion, CO_3^{2-}, has a constant of 4.8×10^{-11}.

The carbonate and bicarbonate ions are planar ions, containing carbon bonded to three oxygen atoms at the corners of an equilateral triangle. The situation is reminiscent of graphite, with more than enough electrons to form single bonds to all three oxygens; as a result, the electronic distribution is represented as a resonance hybrid. For carbonate ion, the contributing structures are usually written as in Figure 17.3.

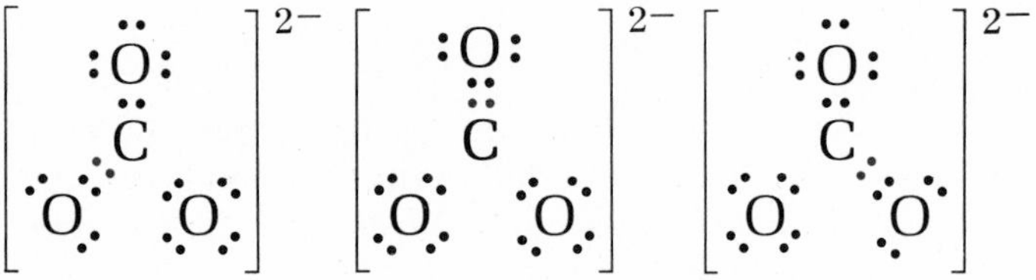

Figure 17.3
Contributing structures of carbonate ion.

Derived from carbonic acid are the two series of salts: bicarbonates, such as $NaHCO_3$, and carbonates, such as Na_2CO_3. The former can be made by neutralizing 1 mole of CO_2 (or H_2CO_3) with 1 mole of NaOH; the latter, by neutralizing 1 mole of CO_2 with 2 moles of NaOH. The net reactions are

$$CO_2 + OH^- \longrightarrow HCO_3^-$$
$$CO_2 + 2OH^- \longrightarrow CO_3^{2-} + H_2O$$

Actually, the compounds are industrially so important that large-scale cheaper methods are used. The most famous is the Solvay process, which uses ammonia to neutralize the acidity of CO_2 and relies on the low solubility of $NaHCO_3$ in cold water for separation. The process is one in which CO_2 (from the thermal decomposition of limestone) and NH_3 (recycled in the process) are dissolved in NaCl solution. NH_3 neutralizes CO_2 by the reaction

$$NH_3 + CO_2 + H_2O \longrightarrow NH_4^+ + HCO_3^-$$

to form HCO_3^-, which precipitates as $NaHCO_3$ if the temperature of the brine is 15°C or lower. On thermal decomposition, $NaHCO_3$ is decomposed to give Na_2CO_3:

$$2NaHCO_3(s) \longrightarrow Na_2CO_3(s) + CO_2(g) + H_2O(g)$$

Sodium carbonate and sodium bicarbonate are industrial chemicals of primary importance. Na_2CO_3, or soda ash, is used, for example, in making glass, where it is used directly, and in making soap, where it is first converted to NaOH, or lye, by addition of $Ca(OH)_2$ and then boiled with animal or vegetable fats. When recrystallized from water, the hydrate $Na_2CO_3 \cdot 10H_2O$, or washing soda, is formed. The mild basic reaction resulting from hydrolysis of carbonate ion,

$$CO_3^{2-} + H_2O \rightleftharpoons HCO_3^- + OH^-$$

is used to supplement soap in laundering. $NaHCO_3$, or baking soda, is a principal component of baking powders, used to replace yeast in baking. Yeast ferments sugars, releasing CO_2 gas, which raises the dough; with baking powder, the CO_2 for leavening is obtained by the action of $NaHCO_3$ with acid substances such as alum.

In addition to the compounds that carbon forms with oxygen, there are numerous compounds in which carbon is bonded to the nonmetal nitrogen. The simplest of these carbon-nitrogen compounds is cyanogen (C_2N_2), made by thermal decomposition of cyanides such as AgCN. At room temperature, cyanogen is a colorless gas with the odor of bitter almonds; it is very poisonous. In many chemical reactions C_2N_2 behaves like the halogens. For example, in basic solution it disproportionates according to the equation

$$C_2N_2(g) + 2OH^- \longrightarrow CN^- + OCN^- + H_2O$$

The cyanide ion, CN^-, resembles chloride ion in that both give insoluble silver salts. Cyanide salts can also be made by the following high-temperature reaction:

$$Na_2CO_3(s) + 4C(s) + N_2(g) \longrightarrow 2NaCN(s) + 3CO(g)$$

Cyanide ion forms many complex ions with transition-metal ions, for example, $Fe(CN)_6^{3-}$. Unlike chloride ion, CN^- combines with H^+ to form a weak acid, HCN, which in solution is called *hydrocyanic acid* (prussic acid). Like cyanogen, HCN is poisonous. However, as we shall see in Section 24.4, HCN is believed to have been a key intermediate in the reactions from which life originated.

The anion OCN^-, formed by the disproportionation of cyanogen, is called the cyanate ion. It exists in many salts, e.g., ammonium cyanate (NH_4OCN). The last compound is of special interest because on heating it is converted to urea, $CO(NH_2)_2$, the principal end product of protein metabolism. As we noted in Chapter 1, the discovery of this reaction by Friedrich

Wöhler in 1828 was a milestone in chemistry. It represented the first time that chemists were able to synthesize in the laboratory a compound previously thought to be produced only in living organisms.

Related to the cyanate ion, OCN^-, is the thiocyanate ion, SCN^-, where the prefix *thio-* indicates that a sulfur atom has replaced an oxygen atom. Salts containing thiocyanate ion can be prepared by fusing cyanides with sulfur. For example, heating NaCN with sulfur produces NaSCN. Like CN^-, SCN^- precipitates Ag^+ and also forms complex ions. One of the most famous is the blood-red $FeNCS^{2+}$ complex used as a detection test for the presence of Fe^{3+}. It might be noted that the product of the reaction

$$Fe^{3+} + SCN^- \longrightarrow FeNCS^{2+}$$

is written with the nitrogen end facing the iron atom. This is to emphasize that in the complex there is a direct Fe-to-N bond.

By far the greatest number of carbon compounds are those formed with hydrogen. There is a fantastic number of them, probably already numbering well over a million and estimated to be increasing at roughly 100,000 new compounds synthesized per year. Those composed exclusively of carbon and hydrogen are called hydrocarbons; those containing additional elements are called hydrocarbon derivatives. Together, hydrocarbons and their derivatives are called *organic compounds,* because at one time it was thought they could be made only by living organisms. The chemistry of organic compounds is taken up in greater detail in Chapters 19 to 24.

The fact that one isotope of carbon, $^{14}_6C$, is radioactive has led to the development of a rather novel method of dating archaeological discoveries. The basic ideas of the method are as follows: Carbon dioxide in the atmosphere contains mostly $^{12}_6C$ and a little $^{13}_6C$, both of which are nonradioactive. In addition, there is a small amount of $^{14}_6C$, which, even though it is constantly decaying, remains rather uniform in abundance, apparently because cosmic rays act on $^{14}_7N$ of the atmosphere to form $^{14}_6C$. Because the rate of decay balances the rate of production, the ratio of $^{14}CO_2$ to $^{12}CO_2$ in the atmosphere does not change with time. Now, it is well known that plants absorb CO_2 from the atmosphere in the process of photosynthesis. So long as the plant is alive and growing, the ratio of $^{14}_6C$ to $^{12}_6C$ atoms in the plant carbohydrates will be the same as that in the atmosphere. However, once the plant has been removed from the life cycle, as, for instance, when a tree is chopped down, the ratio of $^{14}_6C$ to $^{12}_6C$ begins to diminish as the $^{14}_6C$ atoms undergo radioactive decay. The half-life of $^{14}_6C$ is 5570 years; therefore at the end of 5570 years the ratio of $^{14}_6C$ to $^{12}_6C$ becomes half as great as it is in the atmosphere. To determine the age of a wooden relic or, for that matter, of any once-living material, a sample is burned to CO_2 and the ratio of $^{14}_6C$ to $^{12}_6C$ is measured.

17.3 Silicon

Silicon is the second element of group IV, occupying the position just below carbon. Its electron configuration is $1s^2 2s^2 2p^6 3s^2 3p^2$ corresponding to 2,8,4 for successive shell population, compared with $1s^2 2s^2 2p^2$ or 2,4 for carbon. Thus, silicon is like carbon in having four outer electrons it can use for bonding purposes, and we can expect considerable similarity in chemical properties. However, the silicon atom (radius 0.117 nm) is significantly larger than the carbon atom (radius 0.077 nm). The result is that there is a decisive difference in the stability of silicon bonds to oxygen compared with silicon bonds to hydrogen. Like carbon, silicon forms a tetrahedral molecule of the type SiH_4 and a few higher hydrosilicons which contain chains of silicon atoms. However, since Si—O bonds are formed preferentially to Si—H or Si—Si bonds, the chemistry of silicon is primarily concerned with oxygen compounds rather than with hydrosilicons. Furthermore, whereas the smaller carbon atom often forms multiple bonds (i.e., double or triple), silicon invariably forms single bonds. As a result, oxygen-silicon compounds contain Si—O—Si bridges in which oxygen is bonded by single bonds to each of two silicon atoms instead of being bonded by a double bond to one silicon atom. This is quite unlike the case of carbon, where oxygen is frequently found bonded to a single carbon atom as the $>C{=}O$ group.

Silicon is the second most abundant element in the earth's crust (26 percent by weight) and is about as important in the mineral world as carbon is in the organic. As SiO_2, it accounts for quartz, flint, and opal; as complex silicates of aluminum, iron, magnesium, and other metals, it accounts for practically all rocks, clays, and soils.

The preparation of pure silicon is quite difficult. It can be accomplished by the reduction of SiO_2 with Mg or by the reduction of the chloride with Na. Since it is mainly used for addition to steel, it is more usually prepared as ferrosilicon by reduction of mixtures of SiO_2 and iron oxides with coke. The element is a semimetal with a crystal structure like that of diamond. At room temperature it is inert to most reagents but will dissolve in basic solutions to liberate H_2. At elevated temperatures it reacts with many metals such as magnesium to form silicides (such as Mg_2Si).

For the electronics industry, the need is for ultrapure silicon, i.e., silicon that contains less than 0.001 percent foreign elements. As mentioned in Section 7.6, some solids such as silicon and germanium have rather low electric conductivity in the pure state but become better conductors when traces of group III or group V elements are added. The enhanced conductivity increases with the amount of impurity and rapidly increases with increasing temperature. Many important devices such as transistors and solar batteries depend on such electrical properties. To get ultrahigh-purity materials, stringent controls have to be exercised to minimize contamination. A speck of dust, for example, brought into a preparative lab on a shoe can ruin an entire batch of material. Protective devices such as plastic overshoes, gloves, hairnets, flow-through air chambers, and micropore-filtered air con-

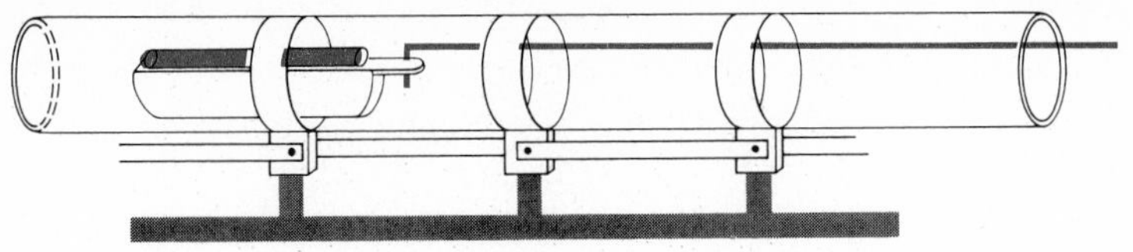

Quartz tube   *Heater loops*

Pull rod

Figure 17.4
Zone-refining technique to purify silicon.

ditioning give environments rivaling most hospital operating rooms.

Still, the major problem in preparing ultrapure materials is not so much contamination but intrinsic impurities in the starting materials. If, for example, elemental silicon is to be made by reducing silicon tetrachloride ($SiCl_4$) with zinc vapor, both the $SiCl_4$ and the Zn starting materials need to be prepurified by the best techniques available, e.g., vacuum distillation, selective adsorption. As a last step in preparing ultrapure material, an ingot of high-purity silicon is subject to a "zone-refining" process. In this process, the high-purity ingot is slowly drawn through a long, inert-gas-filled quartz tube. Heating coils wound in narrow strips around the tube, as shown in Figure 17.4, melt the silicon in narrow zones. Impurities, being more soluble in the liquid, remain in the high-temperature molten zones and get swept to the end of the ingot where they can be cut off and discarded. Zone refining has made it possible to reduce impurity concentrations to one part per billion. Of course, the price goes up as the impurity concentration goes down.

The use of high-purity silicon in solar batteries has been hitherto confined mostly to expensive space vehicles, but the development is an exciting one for the future because solar energy is so abundant and, of course, it is free. There are several types of solar cells but a common one used in spacecraft consists of a thin wafer of ultrapure silicon containing traces of boron (a group III element) giving it a p-type character. One side of the wafer then has phosphorus or arsenic (group V elements) diffused into it, giving it n-type character. When the n-type surface is exposed to the sun, photons of solar energy penetrate to about the depth of the p-n junction and give up their energy in a burst by creating an electron and a "hole" (recall Section 7.6). When either the electron or the hole, depending where the electron-hole pair is created, migrates across the p-n junction a voltage is created. A single cell can generate only about 0.4 volt but with thousands of them placed in series, large power production can be achieved. Unfortunately, the cells are rather expensive, so that even modest power demands are extremely costly.

17.4   Compounds of Silicon

The compounds of silicon are almost all oxy compounds. However, other compounds can be prepared which are unstable with respect to conversion to the oxy compounds. Thus, the hydrosilicons, prepared by reaction of silicides with acid and analogous to the hydrocarbons, are unstable in

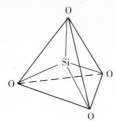

Figure 17.5
Tetrahedral SiO_4 unit.

oxygen with respect to rapid conversion to SiO_2. Silane (SiH_4), for example, is oxidized as follows:

$$SiH_4(g) + 2O_2(g) \longrightarrow SiO_2(s) + 2H_2O$$

Disilane (Si_2H_6), trisilane (Si_3H_8), and compounds up to Si_8H_{18} have been prepared, but they are progressively less stable as the silicon-silicon chain length increases. Derivatives of the silanes, such as silicon tetrachloride ($SiCl_4$), are also known but are unstable with respect to SiO_2.

The silicates (oxy compounds of silicon) have been extensively investigated, and in practically every case the silicon atom is found to be tetrahedrally bonded to four oxygen atoms. As shown in Figure 17.5, four valence electrons from Si and six valence electrons from each O are insufficient to complete the octets of all the atoms. The oxygen atoms may obtain electrons from some other atoms and become negative in the process. This produces the discrete anion, SiO_4^{4-}, found, for example, in the gem-stone mineral zircon ($ZrSiO_4$). Alternatively, the oxygen atoms may complete their octets by sharing electrons with other silicon atoms. Since one, two, three, or four of the oxygen atoms can thus bridge to other silicon atoms, many complex silicates are possible. One bridge oxygen per silicon atom gives $Si_2O_7^{6-}$ analogous to $Cr_2O_7^{2-}$. Two bridge oxygens per silicon lead to formation of extended chains (Figure 17.6) as found in the mineral spodumene [$LiAl(SiO_3)_2$]. The chains are negatively charged anions, because each of the oxygen atoms which is not a bridge atom has picked up an electron to

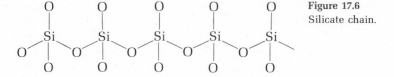

Figure 17.6
Silicate chain.

complete its octet. In the compounds, cations such as Li^+ or Al^{3+} hold the solid together by ionic attractions. With three bridge oxygen atoms per silicon, extended two-dimensional sheets are built up, which can be thought of as sheets of SiO_4 tetrahedra, each sharing three corner atoms with other tetrahedra.

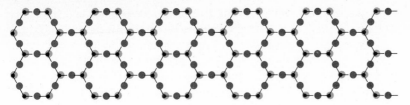

Figure 17.7
Silicate sheet.

A portion of such a sheet is shown in Figure 17.7, where the solid, red circles represent oxygen atoms above the plane of the paper, and the dark circles, silicon atoms in the plane of the paper with oxygen atoms (lighter red) below the plane. The oxygen atoms extending below the plane are negatively charged and are attracted to positive ions, which in turn are attracted to other similar sheet silicate ions. Stacks of sheetlike silicate ions such as these are found in mica and clay minerals. Their compositions are usually further complicated because of partial replacement of silicon atoms by aluminum. A characteristic property of these sheetlike minerals is easy cleavage parallel to the sheets.

In the limit there can be four bridge oxygen atoms per silicon. This leads to the three-dimensional structure found in feldspars [e.g., orthoclase $(KAlSi_3O_8)$], zeolites [e.g., analcite $(NaAlSi_2O_6 \cdot H_2O)$], and silica [e.g., quartz (SiO_2)]. In the feldspars and zeolites, some of the Si (oxidation state $4+$) is replaced by Al (oxidation state $3+$). Consequently, the framework has a net negative charge which must be balanced by cations held in lattice holes. In zeolites, the latticework is more open than in feldspars, and thus the cations can be replaced by ion exchange (Section 15.4). In silica, the framework contains only Si and O atoms and is electrically neutral. If the framework is an ordered one, the silica is crystalline, as in quartz; if it is disordered, as by supercooling molten SiO_2, the silica is noncrystalline. Crystalline silica has a very high melting point, but noncrystalline (or vitreous) silica can be softened at a considerably lower temperature. Thus softened, it can be blown into various forms such as laboratory ware which take advantage of its desirable properties. SiO_2 transmits ultraviolet as well as visible light, has a low thermal coefficient of expansion (only about one-twentieth that of glass or steel so it can be rapidly heated or cooled without fear of breakage due to thermal shock), and is inert to most chemical reagents. However, it is dissolved by solutions of HF to form complex fluosilicate ions, SiF_6^{2-}, and to a limited extent by basic solutions, to form various silicate ions.

Derived from SiO_2 are other silicate systems of practical importance, e.g., glass and cement. Glass is made by fusing SiO_2 (as quartz sand) with basic substances like CaO and Na_2CO_3. Special glasses such as Pyrex contain other acidic oxides (B_2O_3) substituted for some of the SiO_2. Like silica, glass will dissolve in solutions of HF and also is slowly etched by basic solutions.

As a consequence of the latter reaction, it is frequently observed that glass stoppers stick fast in reagent bottles containing basic solutions such as NaOH and Na_2CO_3. Cement, a complex aluminum silicate, is made by sintering limestone and clay at high temperature and grinding the product to a fine powder. When mixed with water and allowed to stand, it sets to a hard, rigid solid by a series of complex reactions. Although these reactions are only imperfectly understood, they seem to involve slow hydration of silicates to form some sort of interlocking structure. The hydration is accompanied by the evolution of considerable heat, which may cause cracking unless provision is made for heat removal.

The high thermal stability of Si—O—Si chains has been exploited in the *silicones,* compounds in which organic residues are bonded to Si atoms in place of negatively charged silicate oxygens. A typical example of a silicone is the chainlike methyl silicone shown in Figure 17.8. Thanks to the methyl groups, this silicone has lubricating properties characteristic of

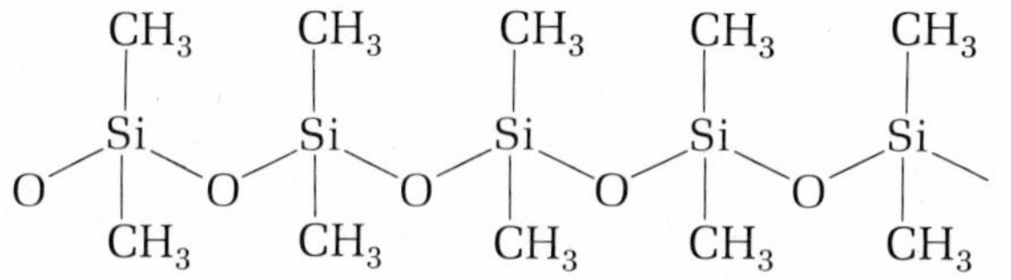

Figure 17.8
Silicone chain.

hydrocarbon oils, but unlike hydrocarbons it is unreactive, even at high temperatures. More complicated silicone polymers are made possible by having oxygen or hydrocarbon bridges between chains. These rubbery materials of which "silly putty" is an example, are used as electric insulators at elevated temperatures.

17.5   Boron

Boron is the first element of group III. It has electron configuration $1s^22s^22p^1$ corresponding to three valence electrons in the outer shell. Normally, a less than half-filled outer shell would make one predict that the particular element is metallic, and this in fact is the case for the element aluminum (Al, $1s^22s^22p^63s^23p^1$) which is just below boron in group III. However, boron is a small atom and, as generally found in the uppermost part of the periodic table, the small size makes for properties quite different from those of other elements of the same group. Both as an element and in its compounds, boron differs markedly from the other members in group III.

In nature, boron is moderately rare (0.0003 percent abundance) and occurs principally as the borates (oxyboron anions), e.g., borax ($Na_2B_4O_7 \cdot 10H_2O$). The element may be produced by reducing the oxide,

B_2O_3, with a metal such as Mg, electrolyzing fused borates, or reducing boron trichloride (BCl_3) with hydrogen at high temperature. Only the last method gives a reasonably pure product.

Massive boron is very hard but brittle. It has a dull metallic luster but in the pure state is a poor conductor of electricity and is not classified as a metal. When its temperature is raised, its conductivity increases. This is unlike metallic behavior; therefore boron and substances like it (silicon and germanium) are *semiconductors*. As mentioned in Section 7.6, the explanation of semiconductivity is that, at room temperature, electrons are bound rather tightly to local centers, but as the temperature is raised, they are freed and are able to wander through the crystal. The higher the temperature, the greater the number of electrons freed; hence the conductivity increases, even though lattice vibrations offer more resistance at the higher temperature. In the case of boron one can imagine that most of the electrons are localized as electron pairs between boron atoms, but as the temperature is raised some of the B—B bonds are broken and the freed electrons can migrate through the structure.

At room temperature, boron is inert to all except the most powerful oxidizing agents, such as fluorine and concentrated nitric acid. However, when fused with alkaline oxidizing mixtures, such as NaOH and $NaNO_3$, it reacts to form borates. Boron also dissolves in molten aluminum, from which there separates on cooling an aluminum boride, AlB_{12}. This same boride is formed when boron oxide is reduced with aluminum and for a long time was considered to be pure boron. In fact, AlB_{12} is still referred to as *crystalline boron*. Other borides such as Mg_3B_2 are known and can be prepared by direct union of the elements.

When magnesium boride reacts with acids, several boron-hydrogen compounds are formed which have puzzled chemists since their discovery. The simplest of these boron hydrides would be BH_3, formed by sharing the three valence electrons of boron with three H atoms. However, this compound is not known as a stable compound. Instead, boron forms a series of hydrides ranging from B_2H_6 (diborane) to $B_{18}H_{22}$. All these compounds are surprising, since there seem to be too few electrons to hold them together. Diborane, for example, has only 12 valence electrons (3 from each B and 1 from each H) for what appears to be 7 bonds (3 bonds in each BH_3 unit and 1 bond between them).

Figure 17.9 shows several of the many interesting structures that have been proposed. Structure (*a*) contains two one-electron bonds; structure (*b*), a no-electron bond; structure (*c*), two hydrogen bridges; and structure (*d*), a protonated double bond, i.e., a double bond between the two boron atoms with two protons embedded in it. Not one of these structures satisfactorily accounts for all the observed properties of diborane. For one thing, the compound behaves as if it were not paramagnetic, which is inconsistent with unpaired electrons as shown in structures (*a*) and (*c*). Furthermore, structure

(a)

(b)

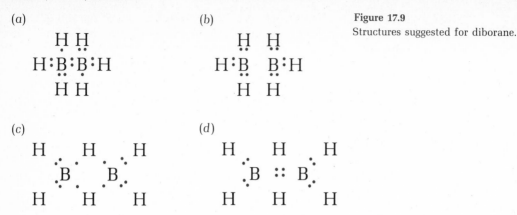

Figure 17.9
Structures suggested for diborane.

(c)

(d)

(b) not only suffers from having a "no-electron bond" but also disagrees with the observation that two of the six hydrogen atoms behave differently from the others. Finally, the experimentally determined boron-boron distance is longer than that predicted from structure (d). The existence of molecules such as diborane emphasizes that the assumption that a pair of electrons is localized in a bond between two atoms is a simplification which is not valid in all cases. In fact, it is becoming increasingly acceptable to describe diborane and other boron hydrides as containing *three-center bonds,* which consist of an electron pair spread out over three atoms. In diborane there are two three-center bonds, each consisting of an electron pair spread over the two boron atoms and one of the middle hydrogen atoms shown in Figure 17.9c or 17.9d.

All the boron hydrides, ranging from gaseous B_2H_6 to solid $B_{18}H_{22}$, inflame in air to give dark-colored products of unknown composition. In the absence of air, they decompose on heating to boron and hydrogen. With water, they react to form hydrogen and boric acid.

The only important oxide of boron is B_2O_3, boric oxide. As already mentioned, it is acidic, dissolving in water to form H_3BO_3, boric acid. Boric acid is an extremely weak acid for which K_I is 6.0×10^{-10}.

The borates, formed either by neutralization of boric acid or reaction of B_2O_3 with basic oxides, are extremely complicated compounds. Although a few, such as $LaBO_3$, contain discrete BO_3^{3-} ions, most contain more complex anions in which boron atoms are joined together by oxygen bridges. As shown in Figure 17.10, the simple BO_3^{3-}, or orthoborate, ion is a planar ion with the three oxygen atoms at the corners of an equilateral triangle. In more complex anions such as the one shown, there are still three oxygen atoms about each boron atom, but some of these are joined to other boron atoms. Other borates are even more complex and may have, in addition to triangular BO_3 units, tetrahedral BO_4 units. This seems to be true for borax,

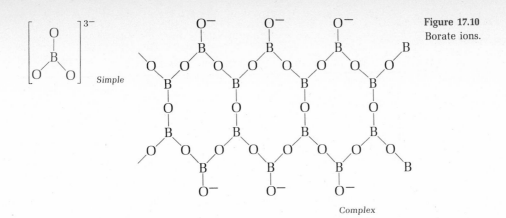

Figure 17.10
Borate ions.

which is the most common of the borates. It is extensively used in water softening, partly because it reacts with Ca^{2+} to form insoluble calcium borate and partly because it hydrolyzes to give an alkaline solution. Because borax dissolves many metal oxides to form easily fusible borates, it is used widely as a flux in soldering operations. By removing oxides such as Cu_2O from the surface of hot brass, the flux allows fresh metal surfaces to fuse together.

The boron halides (BF_3, BCl_3, BBr_3, and BI_3) are also unusual in several respects. Unlike the halides formed by typical metals, these are molecular substances and do not contain ions in the solid state. Also, the boron atom in these molecules has only a sextet of electrons; hence it can accommodate another pair of electrons. This occurs, for example, in the reaction

$$:\ddot{F}: \quad H \qquad :\ddot{F}:H$$
$$:\ddot{F}:B + :N:H \longrightarrow :\ddot{F}:B:N:H$$
$$:\ddot{F}: \quad H \qquad :\ddot{F}:H$$

where the product is sometimes called an *addition compound*.

QUESTIONS

17.1   *Carbon*   Why is it that at low pressure, graphite is the stable form of carbon, but at high pressure, diamond becomes stable?

17.2   *Carbon Compounds*   Assign oxidation numbers to all atoms in the following compounds: CO, CO_2, CaC, $C_6H_{12}O_6$, C_2H_5OH, $NaHCO_3$.

17.3   *Silicon*   Since silicon forms a solid analogous to diamond, why doesn't it form the analog of graphite?

17.4   *Silicon Compounds*   Consider spodumene, $LiAl(SiO_3)_2$. (*a*) What is the oxidation state of Si in spodumene? (*b*) What is the charge on an SiO_3 unit in spodumene? (*c*) Is it a discrete ion? Explain. (*d*) Show that the charge on SiO_3 and the tetrahedral bonding of Si are consistent with the octet rule and the structure exhibited by spodumene.

17.5 *Boron* Write balanced equations for two chemical methods for preparing elemental boron.

17.6 *Boron Compounds* Some boron compounds do not conform to the octet rule. Must $LaBO_3$ be one of these, or can you draw an octet structure for the borate ion in this compound?

17.7 *Silicates* In terms of the elements present tell how the following substances differ: silica, cement, ordinary glass, Pyrex glass.

17.8 *Equations* Write balanced equations for each of the following changes: (a) preparation of silicon carbide, (b) complete titration of sodium carbonate solution with strong acid, (c) heating of ammonium carbonate, (d) dissolving of silica in aqueous HF, (e) dissolving of B_2O_3 in water.

17.9* *Solvay Process* From 1000 kg of NaCl and 1000 kg of $CaCO_3$, how much Na_2CO_3 can you make by the Solvay process (a) if the CO_2 of the last step is sold as a by-product, (b) if the CO_2 is recycled and all converted to Na_2CO_3?

17.10 *Boric Acid* For H_3BO_3, $K_I = 6.0 \times 10^{-10}$. (a) What concentration of boric acid is needed to produce a solution whose pH is 5.00? (b) Will a solution made by dissolving NaH_2BO_3 be acidic or basic? Give the principle equilibrium which accounts for your answer and calculate its K.

17.11 *Silicates* The natural asbestoses belong to a class of minerals called amphiboles which contain the extended ion $(Si_4O_{11}{}^{6-})_n$. Sketch its structure.

17.12 *Soda Ash* (a) When Na_2CO_3 is used in making glass, what happens to the carbon? (b) What happens to it when soda ash is converted to lye, as for making soap? (c) If the conversion is not made, what is the OH^- concentration in 1.0 M solution of Na_2CO_3?

17.13* *Heat of Reaction* The heat evolved when 1.0 g of O_2 reacts with B and Si to form B_2O_3 and SiO_2 is 26 and 27 kilojoules/g, respectively. What is the numerical value of ΔH for the reaction of B with SiO_2 to form B_2O_3 and Si?

17.14 *General* Tell what is meant by, and give an illustration of, each of the following: (a) a three-center bond, (b) a p-n junction, (c) an addition compound, (d) disproportionation.

17.15 *Explanations* Explain why each of the following is true: (a) Graphite is a conductor, diamond is not. (b) Charcoal adsorbs polar molecules more strongly than nonpolar ones. (c) Quartz has a high melting point. (d) Boron conducts better at high temperature than at low temperature.

17.16 *Fermentation* Yeast converts the sugar glucose ($C_6H_{12}O_6$) to alcohol and CO_2 according to the equation $C_6H_{12}O_6 \longrightarrow 2C_2H_5OH + 2CO_2(g)$. How many liters of CO_2 will be produced at 1.00 atm and 35.0°C from 12.0 g of sugar?

17.17* *Composition* What is the weight percent of carbon in the following compounds: (a) CO_2, (b) C_2H_5OH, (c) C_6H_5OH, (d) $CO(NH_2)_2$?

17.18 *Neutralization* (a) Write a balanced, net equation for the reaction of calcium carbide with water. (b) How many milliliters of 0.150 M HCl solution must be added to neutralize 100 ml of water to which 2.60 g of CaC_2 was added?

17.19 *Silicon* How would you go about preparing pure silicon?

17.20 *Carbon Disulfide* When CS_2 is heated with chlorine, the following reaction occurs:

$$CS_2(g) + 3Cl_2(g) \longrightarrow CCl_4(g) + S_2Cl_2(g)$$

If 0.250 mole of CS_2 and 0.600 of Cl_2 are allowed to react in a 10.0-liter tank at 200°C, what is (a) the concentration of CCl_4 and S_2Cl_2 at the end of the reaction? (b) Percent change in the pressure of the system?

17.21 *Bond Strength* Account for the high thermal stability of silicon-oxygen-silicon chains.

17.22 *Reactions* Write balanced equations for the following processes: (a) the reaction of BCl_3 with NH_3, (b) the burning of silane (SiH_4), (c) dissolving CO_2 in water, (d) dissolving cyanogen in strongly basic solution.

17.23* *pH* What is the pH of 120 ml of 1.0 M HCN after it has been "neutralized" with 120 ml of 1.0 M NaOH?

17.24 CO_2 *Cycle* It has been stated that except for limited amounts of nuclear energy all of our energy on earth has come from the sun's energy. Explain.

17.25 *Solid Structures* By use of Figure 17.1 figure out a unit cell for diamond and one for graphite. Tell how many C atoms lie within the unit cell in each case. (Recall that atoms shared between adjacent unit cells must be subdivided.)

17.26 *Bonding* (a) What relationship if any can you see between the "extra pair" of bonding electrons in the carbonate ion and the three-center bond in diborane? (b) Can you draw resonance structures for diborane?

17.27* *Carbonate Equilibria* When distilled water stands in contact with air, CO_2 from the air dissolves to the extent of about 1×10^{-3} M. Using the equilibrium constants of Section 17.2, calculate the concentration of all ionic species (HCO_3^-, CO_3^{2-}, H^+, OH^-) in such distilled water.

17.28 *Ultrapure Silicon* Consider the unit cell for diamond of Figure 17.1 which applies also to Si. In the case of Si, the edge length is 0.542 nm. If there is one impurity atom per billion (10^9) Si atoms, how far apart, on average, are the impurity atoms?

18 Nonmetals

As we have seen the large majority of the elements are metals. Of the remainder, a few have some metallike properties and can be classed as semimetals; examples of these were discussed in the last chapter. Although we have therefore covered most of the elements, those remaining include some of the most common and most important of all. Although as a class they are linked only by their absence of metallic properties, this fact alone tells us something. They lack metallic properties because their outer electrons are bound too tightly to form metallic structures. It is not surprising therefore to find these elements clustered near the upper right-hand corner of the periodic table since the energy of binding outer electrons is found to increase from bottom to top in a group and from left to right in a period.

Two elements, hydrogen and oxygen, discussed in Chapter 14 are non-metals as are nitrogen, phosphorus, sulfur, and the halogens which we discuss in this chapter. From what we already know of the behavior of these elements, we see that their properties are more diverse than the metals. Some are gases, some are oxidizing agents, some form negative ions. All show frequent examples of covalent bonding. Although group properties may be marked, as with the halogens, the nonmetals are chemical individuals to a greater extent than are the metals.

18.1   Nitrogen

Nitrogen is the first element in group V. It is a most important element because it is an essential ingredient of the amino acids that make up proteins in living organisms. Also, it accounts for the bulk of the earth's atmosphere, although the intriguing question of how some bacteria can convert atmospheric nitrogen into components of living matter is still largely unanswered. Nitrogen chemistry is complicated, but it is fundamental to understanding the life process.

Nitrogen is about one-third as abundant as carbon and occurs principally *free* as diatomic N_2 in the atmosphere and *combined* as Chile saltpeter ($NaNO_3$). In plants and animals, nitrogen is found combined in the form of proteins, which average in composition 51% C, 25% O, 16% N, 7% H, 0.4% P, and 0.4% S.

Elementary nitrogen is usually obtained by fractional distillation of liquid air. Since N_2 has a lower boiling point (77 K) than O_2 (90 K), it is more volatile and evaporates preferentially in the first fractions. Very pure N_2 can be made by thermal decomposition of some nitrogen compounds, such as ammonium nitrite (NH_4NO_2):

$$NH_4NO_2(s) \longrightarrow N_2(g) + 2H_2O(g)$$

It is interesting to note that pure nitrogen obtained from decomposition of compounds such as NH_4NO_2 was the key that led to the discovery of the noble gases. Lord Rayleigh, in 1894, was the first to note that nitrogen from the decomposition of compounds was of lower density (1.2505 g/liter at STP) than the residual gas obtained from the atmosphere by removal of O_2, CO_2, and H_2O (1.2572 g/liter at STP). In conjunction with Sir William Ramsay, Rayleigh removed the nitrogen from the air residue by various reactions, such as the combination of nitrogen with hot magnesium to form solid magnesium nitride (Mg_3N_2). After removal of the nitrogen, there was still some gas remaining which, unlike any gas known at that time, seemed completely unreactive.† It was christened argon from the Greek word meaning lazy. Later spectroscopic investigations showed that crude argon, and hence the atmosphere, contains the other noble elements helium, neon, krypton, and xenon. Including the noble gases, the average composition of the earth's atmosphere is as shown in Figure 18.1. In addition to the noble gases listed, there are traces of radon (Rn) in the atmosphere. The concen-

---

†For many years attempts were made to combine the noble gases with other elements, but the results were either negative or inconclusive. Finally, in 1962, the element xenon was combined to form xenon hexafluoroplatinate ($XePtF_6$). Since then, more compounds of xenon and, in addition, of krypton and radon have been prepared. These include XeF_2, XeF_4, XeO_3, $Na_4XeO_6 \cdot 8H_2O$, KrF_2, and others. Some of the noble-gas compounds are surprisingly stable, whereas others are explosively unstable.

Figure 18.1 AVERAGE COMPOSITION OF DRY AIR	Component	Percentage by Volume	Normal Boiling Point, K
	Nitrogen (N_2)	78.03	77.3
	Oxygen (O_2)	20.99	90.2
	Argon (Ar)	0.94	87.4
	Carbon dioxide (CO_2)	0.023–0.050	Sublimes
	Hydrogen (H_2)	0.01	20.4
	Neon (Ne)	0.0015	27.2
	Helium (He)	0.0005	4.2
	Krypton (Kr)	0.00011	121.3
	Xenon (Xe)	0.000009	163.9

tration is very low and variable, because radon is produced by radioactive decay of other elements and is itself unstable to nuclear disintegration. As can be seen from the figure, nitrogen is by far the predominant constituent of the atmosphere.

The N_2 molecule contains a triple bond and may be written

:N:::N:

As noted in Section 4.7, the triple bond consists of a bond (from electron sharing between two p orbitals along the bond axis) plus two π bonds at right angles to each other (from electron sharing, side to side in the other p orbitals). Although very stable with respect to dissociation into single atoms, it is unstable with respect to oxidation by O_2 in the presence of water to nitrate ion, NO_3^-. It is fortunate that this reaction is very slow; otherwise, N_2 and O_2 in the atmosphere would combine with water in the oceans to form solutions of nitric acid. In practice, nitrogen is frequently used when a relatively inert atmosphere is required, as, for example, in incandescent lamp bulbs to retard filament evaporation.

The principal compound of nitrogen is probably ammonia (NH_3). It occurs to a slight extent in the atmosphere, primarily as a product of the putrefaction of nitrogen-containing animal or vegetable matter. Commercially it is important as the most economical pathway for nitrogen "fixation," i.e., the conversion of atmospheric N_2 into useful compounds. In the Haber process, synthetic ammonia is made by passing a nitrogen-hydrogen mixture through a catalyst bed consisting essentially of iron oxides. By using a temperature of about 500°C and a pressure of about 1000 atm, there is approximately 50 percent conversion of N_2 to NH_3:

$$N_2(g) + 3H_2(g) \rightleftharpoons 2NH_3(g) \qquad \Delta H = -92 \text{ kilojoules}$$

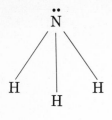

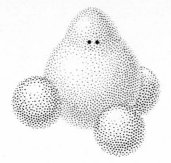

Figure 18.2
Ammonia molecule.

NH$_3$ is a polar molecule, pyramidal in shape, with the three hydrogen atoms occupying the base of the pyramid and an unshared pair of electrons, the apex. The structure, shown in Figure 18.2, leads to a compound, which is easily condensed (condensation temperatures of $-33°$C) to a liquid of great solvent power. In many respects, liquid ammonia is as versatile a solvent as water, and, like water, it can dissolve a great variety of salts. In addition, it has the rather unique property of dissolving alkali metals such as Na to give solutions which contain alkali cations such as Na$^+$, plus electron anions e$^-$ (see Section 15.2).

Ammonia gas is very soluble in water, which is easily explained by the fact that both NH$_3$ and H$_2$O are polar molecules. Not so easy to explain is the basic character of the aqueous solutions formed. At one time it was thought that the NH$_3$ molecules react with H$_2$O to form molecules of the weak base, ammonium hydroxide,

$$\overset{\displaystyle H}{\underset{\displaystyle H}{H:\overset{\cdot\cdot}{\underset{\cdot\cdot}{N}}:H:\overset{\cdot\cdot}{\underset{\cdot\cdot}{O}}:}}$$

which could then dissociate into ammonium ions (NH$_4^+$) and hydroxide ions. However, nuclear magnetic experiments indicate that, in aqueous ammonia solutions, protons jump back and forth so rapidly between nitrogen and oxygen atoms that the distinction between NH$_3$ plus H$_2$O and NH$_4$OH is arbitrary. Thus, the basic nature of aqueous ammonia can be represented by either of the equilibria

$$NH_3 + H_2O \rightleftharpoons NH_4^+ + OH^-$$
$$NH_4OH \rightleftharpoons NH_4^+ + OH^-$$

and K for either is 1.8×10^{-5}. By neutralizing ammonia with acids, ammonium salts can be formed; these contain the tetrahedral NH$_4^+$ ion. They resemble potassium salts, except that they give slightly acid solutions. This can be interpreted as

$$NH_4^+ + H_2O \rightleftharpoons NH_4OH + H^+$$

or

$$NH_4^+ \rightleftharpoons NH_3 + H^+$$

and K for either is 5.5×10^{-10}. Some ammonium salts, such as ammonium nitrate (NH_4NO_3) and ammonium dichromate [$(NH_4)_2Cr_2O_7$], are thermally unstable because they undergo auto-oxidation. As an illustration, NH_4NO_3 sometimes explodes when heated to produce nitrous oxide (N_2O) by the reaction

$$NH_4NO_3(s) \longrightarrow N_2O(g) + 2H_2O(g)$$

Whereas ammonia and ammonium salts represent nitrogen in its lowest oxidation state (3−), the highest oxidation state of nitrogen (5+) appears in the familiar compounds nitric acid (HNO_3) and nitrate salts. Nitric acid is one of the most important industrial acids, and large quantities of it are produced, principally by the catalytic oxidation of ammonia. In this process, called the *Ostwald process,* the following steps are important:

$$4NH_3(g) + 5O_2(g) \xrightarrow{Pt} 4NO(g) + 6H_2O(g)$$

$$2NO(g) + O_2(g) \longrightarrow 2NO_2(g)$$

$$3NO_2(g) + H_2O \longrightarrow 2H^+ + 2NO_3^- + NO(g)$$

In the first step, a mixture of NH_3 and air is passed through a platinum gauze heated at about 800°C. On cooling, the product nitric oxide (NO) is then oxidized to nitrogen dioxide (NO_2), which disproportionates in solution to form nitric acid and NO. By keeping a high concentration of O_2, the remaining NO is converted to NO_2, and the last reaction is driven to the right. To get 100 percent acid, it is necessary to distill off volatile HNO_3.

Pure nitric acid is a colorless liquid, which, on exposure to light, turns brown because of slight decomposition to brown NO_2:

$$4HNO_3 \longrightarrow 4NO_2(g) + O_2(g) + 2H_2O$$

It is a strong acid in that it is 100 percent dissociated in dilute solutions to H^+ and nitrate ion, NO_3^-. Like carbonate ion, nitrate ion is planar. The ion is colorless and forms a great variety of nitrate salts, most of which are quite soluble in aqueous solutions.† Owing to the low complexing ability of nitrate ion, practically all these salts are dissociated in aqueous solution.

In acid solution, nitrate ion is a good oxidizing agent. By proper choice of concentrations and reducing agents it can be reduced to compounds of nitrogen in all the other oxidation states. Possible reduction products are NO_2, HNO_2, NO, N_2O, N_2, NH_3OH^+, $N_2H_5^+$, and NH_4^+. Some metals such as copper and silver, which are too poor as reducing agents to dissolve in HCl, for example, will dissolve in HNO_3. Both acids contain the oxidizing agent H^+, but only the nitric has the additional oxidizing agent NO_3^-. Some metals, such as gold, which are insoluble in HCl and also in HNO_3, are soluble in a mixture of the two acids. This mixture is called *aqua regia* and usually consists of 1 part of concentrated HNO_3 to 3 parts of concentrated HCl. The dissolving power of aqua regia is due to the oxidizing ability of nitrate ion in strong acid plus the complexing ability of chloride ion.

Reduction of NO_3^- usually produces a mixed product. The actual composition of the product depends on the rates of the different reactions. These rates in turn are influenced by the concentration of NO_3^-, the concentration of H^+, the temperature, and the reducing agent used. Thus, for example, in *concentrated* nitric acid, copper reacts with nitric acid to give brown NO_2 gas, but, in *dilute* nitric acid, copper reacts to form colorless NO gas. However, since NO is easily oxidized by air to NO_2, some brown NO_2 fumes may also appear when dilute nitric acid is used.

As can be seen from the above list of reduction products of NO_3^-, compounds of nitrogen are possible in the 4+, 3+, 2+, 1+, 1−, and 2−, as well as 5+ and 3−, states. Some of the more common representative species of these states are discussed below.

1   **The 4+ state**   When concentrated nitric acid is reduced with metals, brown fumes are evolved. The brown gas is NO_2, nitrogen dioxide. Since the molecule contains an odd number of valence electrons (five from the nitrogen and six from each of the oxygens), it should be and is paramagnetic. When brown NO_2 gas is cooled, the color fades, and the paramagnetism diminishes. These observations are interpreted as indicating that two NO_2 molecules pair up (dimerize) to form a single molecule of N_2O_4, nitrogen tetroxide. The equilibrium

$$2NO_2(g) \rightleftharpoons N_2O_4(g) \qquad \Delta H = -61 \text{ kilojoules}$$

is such that at 60°C and 1 atm pressure half the nitrogen is present as NO_2 and half as N_2O_4. As the temperature is raised, decomposition of N_2O_4 is favored. The mixture NO_2–N_2O_4 is poisonous and is a strong

---

† Because of the solubility of the nitrates, it is not usual to find solid nitrates occurring naturally as minerals. The extensive deposits of $NaNO_3$ in Chile occur in a desert region where there is insufficient rainfall to wash them away.

oxidizing agent. As already mentioned in connection with the Ostwald process, NO_2, or more correctly a mixture of NO_2 and N_2O_4, dissolves in water to form HNO_3 and NO.

2 **The 3+ state** The most common representatives of the 3+ oxidation state of nitrogen are the salts called *nitrites*. Nitrites such as $NaNO_2$ can be made by heating sodium nitrate above its melting point:

$$2NaNO_3(l) \longrightarrow 2NaNO_2(l) + O_2(g)$$

They can also be made by chemical reduction of nitrates with substances such as C or Pb. Nitrites are important industrially in the manufacture of dyestuffs. When acid is added to a solution of nitrite, the weak acid HNO_2, nitrous acid ($K_{diss} = 4.5 \times 10^{-4}$), is formed. It is unstable and slowly decomposes by several complex reactions.

3 **The 2+ state** The oxide NO, nitric oxide, is, like NO_2, an odd molecule in that it contains an uneven number of electrons. However, unlike NO_2, NO is colorless and does not dimerize appreciably in the gas phase. In the liquid phase, as shown by a decrease of paramagnetism, some dimerization occurs to form N_2O_2. The existence of simple NO molecules in the gas phase poses a problem in writing an electronic formula. There is magnetic evidence that the odd electron spends half its time with the N and half with the O. This situation is similar to that for O_2. There is a triple bond, as in N_2, but one additional electron, rather than the two extra found in O_2 (Section 4.7).

Nitric oxide can be made in several ways:

$$4NH_3(g) + 5O_2(g) \longrightarrow 4NO(g) + 6H_2O$$
$$3Cu(s) + 8H^+ + 2NO_3^- \longrightarrow 3Cu^{2+} + 2NO(g) + 4H_2O$$
$$N_2(g) + O_2(g) \longrightarrow 2NO(g)$$

The first of these reactions is the catalytic oxidation that is the first step of the Ostwald process for making HNO_3. The second is observed with dilute nitric acid but not with concentrated. The third is extremely endothermic (180 kilojoules) and can occur only when large amounts of energy are added. Apparently, this last reaction occurs when lightning bolts pass through the atmosphere and is one of the paths by which atmospheric nitrogen is made available to plants. In air, NO is rapidly oxidized to brown NO_2:

$$2NO(g) + O_2(g) \longrightarrow 2NO_2(g)$$

Nitric oxide also combines with many transition-metal cations to form complex ions. The most familiar of these complexes is $FeNO^{2+}$, the ferrous nitroso ion, which forms in the brown-ring test for nitrates. When concentrated sulfuric acid is carefully poured into a solution containing ferrous ion and nitrate, a brown layer appears at the junction of the H_2SO_4 and the nitrate-containing solution. The NO for the complex is formed by reduction of NO_3^- by Fe^{2+}.

4 **The 1+ state** When solid ammonium nitrate is gently heated, it melts and undergoes auto-oxidation according to the following equation:

$$NH_4NO_3(l) \longrightarrow N_2O(g) + 2H_2O(g)$$

The compound formed, N_2O, called *nitrous oxide*, or *laughing gas*, has a linear molecule with the oxygen atom at one end. Although rather inert at low temperatures, N_2O decomposes to N_2 and O_2 at higher temperatures. Perhaps because of this decomposition, substances which burn briskly in air actually burn more vigorously in N_2O. Compared with the other oxides of nitrogen, nitrous oxide is considerably less poisonous. However, small doses are mildly intoxicating; large doses produce general anesthesia and in dentistry are frequently used for this purpose. Nitrous oxide has an appreciable solubility in fats, a property which has been exploited in making self-whipping cream. Cream is packaged with N_2O under pressure to increase its solubility. When the pressure is released, the N_2O escapes to form tiny bubbles which produce whipped cream.

5 **The 2− state** In many ways similar to ammonia is the compound hydrazine (N_2H_4). This compound can be made by bubbling chlorine through a solution of ammonia:

$$Cl_2(g) + 4NH_3 \longrightarrow N_2H_4 + 2NH_4^+ + 2Cl^-$$

When pure, N_2H_4 is a colorless liquid at room temperature. Like liquid ammonia, it is a good solvent for many salts and even for the alkali metals. Hydrazine is unstable with respect to disproportionation,

$$2N_2H_4(l) \longrightarrow N_2(g) + 2NH_3(g) + H_2(g)$$

and is violently explosive in the presence of air or other oxidizing agents. It is quite poisonous. Hydrazine has been used as a rocket propellant. For example, the reaction

$$N_2H_4(l) + 2H_2O_2(l) \longrightarrow N_2(g) + 4H_2O(g)$$

which takes place in the presence of Cu^{2+} ion as catalyst, is strongly exothermic and is accompanied by a large increase in volume. The heat liberated expands the gases still further and adds to the thrust.

6 **The 3− state** In addition to ammonia and the ammonium salts, nitrogen forms other compounds in which it is assigned an oxidation state of 3−. These include the nitrides, such as Na_3N, Mg_3N_2, and TiN, many of which can be formed by direct combination of the elements. Some

of these, for example, Na$_3$N and Mg$_3$N$_2$, are quite reactive and combine with water to liberate ammonia. Others, for example, TiN, are very inert and can be used to make containers for high-temperature reactions. The compound nitrogen tri-iodide (NI$_3$) might also be included with the 3— oxidation state of nitrogen, since nitrogen is more electronegative than iodine. At room temperature, NI$_3$ is a solid which is violently explosive and is well known for the fact that even a fly's landing on it can set it off.

The above list of nitrogen compounds is by no means exhaustive, but it serves to indicate the great complexity of nitrogen chemistry. Even more complexity is found in the proteins, the nitrogen compounds which are essential constituents of all living matter. As described in Chapter 22, the proteins are natural high polymers containing the peptide link:

$$
\begin{array}{cc}
\text{H} & \text{O} \\
| & \| \\
\end{array}
$$
$$
\text{—N—C—}
$$

There are a great variety of protein molecules, most of which are of extraordinarily high molecular weight, sometimes as high as a million. The structures of only a few of these many different kinds of protein molecules have been worked out. Furthermore, the synthesis of proteins by organisms remains incompletely understood, although amino acids appear to be involved as intermediates. In nature there is constant interconversion between animal and plant proteins. However, the interconversion is not without loss, because the decay of protein material produces some elementary nitrogen which escapes to the atmosphere. Living organisms, with the exception of some bacteria, are unable to utilize elementary nitrogen for the production of proteins. Thus, in order to maintain life, nitrogen must somehow be restored to a biologically useful form.

The *nitrogen cycle*, which traces the path of nitrogen atoms in nature, is shown in simplified form in Figure 18.3. When plant and animal proteins are broken down, as in digestion and decay, the principal end products are NH$_3$ and N$_2$, which are released to the atmosphere, and various nitrogen-containing ions, which are added to the soil. Ammonia in the atmosphere can be returned to the soil by being dissolved in rain. Elemental nitrogen can be returned by two paths: (1) nitrogen-fixing bacteria which live on the roots of leguminous plants convert N$_2$ to proteins and other nitrogen compounds; (2) lightning discharges initiate the otherwise slow combination of N$_2$ and O$_2$ to form NO, which in turn is oxidized to NO$_2$. The NO$_2$ dissolves

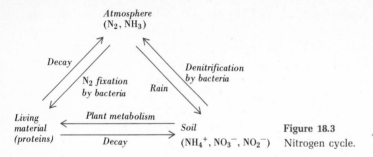

Figure 18.3 Nitrogen cycle.

in rainwater to form nitrates and nitrites, which are washed into the soil. As a final step of the cycle, plants absorb nitrogen compounds from the soil and convert them to plant proteins. Ingested as food, the plant proteins are broken down by animals and reassembled as animal proteins or excreted as waste to the soil. In addition, there are some forms of dentrifying bacteria which convert some of the nitrogen compounds in the soil directly to atmospheric nitrogen.

The nitrogen cycle as outlined above is in precarious balance. Frequently, the balance is locally upset, as, for example, by intensive cultivation and removal of crops. In such cases, it is necessary to replenish the nitrogen by addition of synthetic fertilizers, such as NH_3, NH_4NO_3, or KNO_3.

18.2   Phosphorus

The second element of group V, phosphorus, is considerably more abundant than nitrogen. Its principal natural form is $Ca_3(PO_4)_2$, as found in phosphate rock. Like nitrogen, phosphorus compounds are essential constituents of all animal and vegetable matter. Bones, for example, contain about 60% $Ca_3(PO_4)_2$. Elementary phosphorus can be made by reduction of calcium phosphate with coke in the presence of silica sand. The reaction can be represented by the equation

$$Ca_3(PO_4)_2(s) + 3SiO_2(s) + 5C(s) \longrightarrow 3CaSiO_3 + 5CO(g) + P_2(g)$$

Since the reaction is carried out at high temperature, the phosphorus is formed as a gas, which is condensed to a solid by running the product gases through water. This condensation serves not only to separate the phosphorus from the carbon monoxide but also to protect it from reoxidation by air.

There are several allotropic forms of solid phosphorus, but only the white and red are important. White phosphorus consists of discrete tetrahedral P_4 molecules, as shown on the left in Figure 18.4. The structure of red phosphorus has not yet been completely determined, but there is evi-

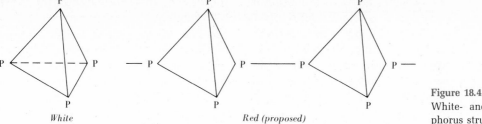

White *Red (proposed)*

Figure 18.4
White- and red-phosphorus structures.

dence that it is polymeric and consists of chains of P_4 tetrahedra linked together, possibly in the manner shown in Figure 18.4. At room temperature, the stable modification of elemental phosphorus is the red form. Because of its highly polymerized structure, it is less volatile, less soluble (especially in nonpolar solvents), and less reactive than white phosphorus. The white form must be handled with care, because it ignites spontaneously in air and is extremely poisonous.

At relatively low temperatures (below $800°C$), phosphorus vapor consists primarily of P_4 molecules. At higher temperatures there is considerable dissociation to give P_2 molecules. Thus, only at very high temperature does elemental phosphorus resemble elemental nitrogen in being diatomic. The favoring of phosphorus molecules that are more complex than those of nitrogen may be attributed to the larger size of the phosphorus atom. In general, large atoms have more difficulty than small atoms do in forming double or triple bonds.

At room temperature, ordinary red phosphorus is not especially reactive, but at higher temperatures it reacts with many other elements to form a variety of compounds. For example, when heated with calcium, it forms solid calcium phosphide (Ca_3P_2). With chlorine, it can form either liquid phosphorus trichloride (PCl_3) or solid phosphorus pentachloride (PCl_5), depending on the relative amount of chlorine present. The three compounds just mentioned illustrate the three most important oxidation states of phosphorus: $3-$, $3+$, and $5+$.

When phosphorus is burned in a limited supply of oxygen, it forms the oxide P_4O_6 (phosphorous oxide). Its structure can be visualized as derived from a P_4 tetrahedron by insertion of an oxygen atom between each pair of phosphorus atoms. P_4O_6 is the anhydride of phosphorous acid, and when cold water is added to it, H_3PO_3 is formed. Phosphorous acid is surprising because, although it contains three hydrogen atoms per molecule, only two can dissociate.

$$H_3PO_3 \rightleftharpoons H^+ + H_2PO_3^- \qquad K_I = 1.6 \times 10^{-2}$$
$$H_2PO_3^- \rightleftharpoons H^+ + HPO_3^{2-} \qquad K_{II} = 7 \times 10^{-7}$$

It has been suggested that the reason for the lack of dissociation of the third H is that it is attached directly to the P instead of to an O.

In the 5+ state, phosphorus exists as several oxy compounds of varying complexity. In contrast with the oxy compounds of nitrogen in the 5+ state, none of these compounds of phosphorus is an especially good oxidizing agent. The least complicated of the phosphorus oxy compounds corresponding to oxidation number 5+ is the oxide, P_4O_{10}, called *phosphoric oxide, phosphorus pentoxide*, or *phosphoric anhydride*. This is the white solid which is usually formed when red phosphorus is burned in an unlimited supply of oxygen or when white phosphorus spontaneously catches fire in air. Although called a pentoxide (because of its simplest formula, P_2O_5), this material both in the vapor and in the most stable solid modification is known to consist of discrete P_4O_{10} molecules. The structure can be visualized as being derived from P_4O_6 by the addition of an oxygen atom extending from each P.

When exposed to moisture, P_4O_{10} turns gummy as it picks up water. The affinity for water is so great that P_4O_{10} is frequently used as an efficient dehydrating agent. With a large amount of water, the acid H_3PO_4, or orthophosphoric acid, is formed. This is a triprotic acid for which the stepwise dissociation is as follows:

$$H_3PO_4 \rightleftharpoons H^+ + H_2PO_4^- \qquad K_I = 7.5 \times 10^{-3}$$
$$H_2PO_4^- \rightleftharpoons H^+ + HPO_4^{2-} \qquad K_{II} = 6.2 \times 10^{-8}$$
$$HPO_4^{2-} \rightleftharpoons H^+ + PO_4^{3-} \qquad K_{III} = 10^{-12}$$

(Like SO_4^{2-} and CrO_4^{2-}, PO_4^{3-} is tetrahedral in structure.) From H_3PO_4, three series of salts are possible, the dihydrogen phosphates, the monohydrogen phosphates, and the normal phosphates. Since $H_2PO_4^-$ in water gives an acid reaction. $Ca(H_2PO_4)_2$ is used with $NaHCO_3$ in some baking powders to produce CO_2. The reaction may be written

$$H_2PO_4^- + HCO_3^- \longrightarrow CO_2(g) + H_2O + HPO_4^{2-}$$

but it does not occur until water is added to the baking powder. Since PO_4^{3-} in water gives a basic reaction, and also since $Ca_3(PO_4)_2$ is rather insoluble, trisodium phosphate is used in water softening.

H_3PO_4 is only one of a series of phosphoric acids that may be formed by the hydration of P_4O_{10}. To distinguish it from other phosphoric acids, H_3PO_4 is called *orthophosphoric acid* and its salts *orthophosphates*. Among the other phosphoric acids are *pyrophosphoric acid* ($H_4P_2O_7$) and *metaphosphoric acid* (HPO_3), both of which can be made by heating H_3PO_4. Pyrophosphates are used for water softening and as complexing agents in electroplating baths. The metaphosphates, with simplest formula MPO_3, exist in a bewildering variety of complex salts. They are all polymeric in structure and can be thought of as being built up of PO_3^- units in such a way that each phosphorus atom remains tetrahedrally associated with four oxygen atoms. The situation in some metaphosphates (illustrated in Figure 18.5) is in some respects comparable with that of the silicate chains (shown in Figure

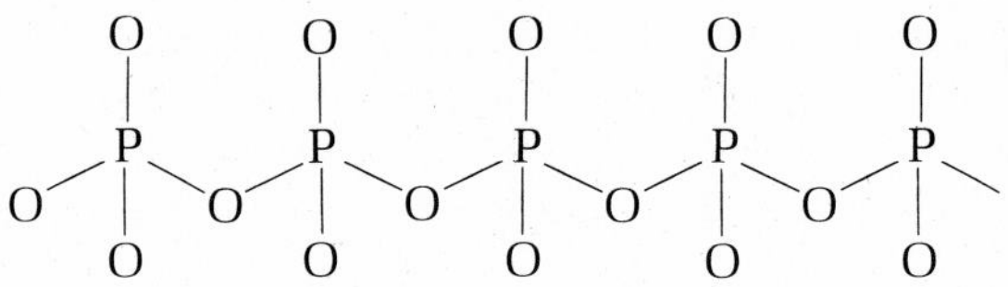

Figure 18.5
Metaphosphate chain.

17.6). In other metaphosphates it is possible to have cyclic polymers such as the structure shown in Figure 18.6 for trimetaphosphate, $Na_3P_3O_9$. When heated above 620°C, $Na_3P_3O_9$ melts to a clear, colorless liquid. If this liquid is cooled suddenly (quenched), it does not crystallize but instead forms a glass (sometimes called *Graham's salt*). The glass is quite soluble in water and in solution can precipitate Ag^+ and Pb^{2+} but not Ca^{2+}. In fact, it seems to form a complex with Ca^{2+}, which makes it impossible to precipitate Ca^{2+} with the usual reagents such as carbonate. Because of the sequestering action on Ca^{2+}, the material has been used extensively in water softening under the trade name Calgon. At one time it was believed that Graham's salt was a hexametaphosphate, $Na_6P_6O_{18}$, but it is a much higher polymer of the type $(NaPO_3)_n$, where n can be as high as 1000.

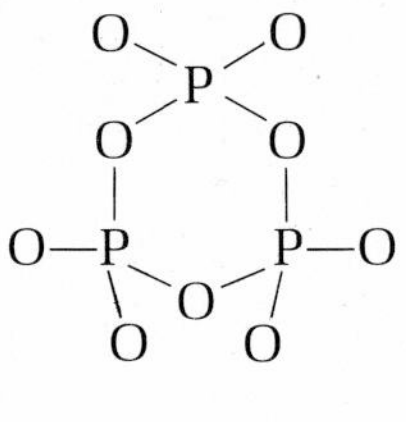

Figure 18.6
Trimetaphosphate.

Like nitrogen, phosphorus is an essential constituent of living cells. It occurs as phosphate groups in complex organic molecules. One of the principal functions of these phosphate groups is to provide a means for storing energy in the cells. For example, when water splits a phosphate group off adenosine triphosphate (ATP) to form adenosine diphosphate (ADP), approximately 42 kilojoules (or 10 kcal, the large calories used in nutrition) of heat is liberated per mole. This energy can be used for the mechanical work of muscle contraction. Further discussions of this interesting subject are found in later chapters.

18.3 Sulfur

Although not very abundant (0.05 percent), sulfur is readily available because of its occurrence in large beds of the free element. These beds, usually located several hundred feet underground, are thought to be due to bacterial decomposition of calcium sulfate. They are exploited by pumping superheated water (at about 170°C) down to the beds to melt the sulfur and blowing the molten sulfur to the surface with compressed air. Since the product is about 99.5 percent pure, it can be used without purification for most commercial purposes. Besides being found as the free element, sulfur occurs naturally in many sulfide and sulfate minerals, such as $CuFeS_2$, Cu_2S, and $CaSO_4 \cdot 2H_2O$.

There are several allotropic modifications of sulfur, the most important being rhombic and monoclinic sulfur, which differ from each other in the symmetry of their crystals. In the rhombic form, which is the stable one at room temperature, sulfur atoms are linked to each other as puckered, eight-membered rings having the configuration shown in Figure 18.7. Above 96°C, monoclinic sulfur is stable; the arrangement of S atoms in it is not known. When heated above the melting point, sulfur goes through a variety of changes. Starting as a mobile, pale yellow liquid, it gradually thickens above 160°C and then becomes less viscous as the boiling point is approached. If the thick liquid, which may be dark red if impurities are present, is poured into water, *amorphous,* or *plastic, sulfur* is produced. X-ray analysis of amorphous sulfur shows that it contains long strings of sulfur atoms. In accordance with this, the change in viscosity with temperature has been attributed to opening of S_8 rings, which then couple up to form less mobile long chains. These in turn are broken into fragments as their kinetic energy is increased.

Although much of the sulfur produced is used directly in insecticides, fertilizers, paper and pulp fillers, and rubber, most of it is converted to

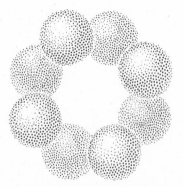

Figure 18.7
S_8 molecule.

industrially important compounds, especially sulfuric acid. Sulfuric acid is produced from sulfur dioxide (SO_2), usually made by burning sulfur in air:

$$S(s) + O_2(g) \longrightarrow SO_2(g)$$

Sulfur dioxide is also a by-product of the preparation of various metals from their sulfur ores. For example, SO_2 is formed in the roasting of the copper ore chalcocite, or Cu_2S:

$$Cu_2S(s) + O_2(g) \longrightarrow 2Cu(s) + SO_2(g)$$

In the *contact process*, which accounts for nearly all the H_2SO_4 production, the SO_2 is oxidized by air in the presence of catalysts such as vanadium pentoxide (V_2O_5) or platinum:

$$2SO_2(g) + O_2(g) \longrightarrow 2SO_3(g)$$

The product, SO_3, or sulfur trioxide, is the anhydride of H_2SO_4, and we would expect the final step in preparing sulfuric acid to be the dissolving of SO_3 in H_2O. However, SO_3 reacts with water to form a fog of H_2SO_4, and the uptake of SO_3 by H_2O is extremely slow. The usual method for circumventing this difficulty is to dissolve the SO_3 in pure H_2SO_4 in a reaction which goes smoothly to produce $H_2S_2O_7$, pyrosulfuric acid. On dilution with water, 100% H_2SO_4 results:

$$SO_3(g) + H_2SO_4 \longrightarrow H_2S_2O_7$$
$$H_2S_2O_7 + H_2O \longrightarrow 2H_2SO_4$$

Pure H_2SO_4 is a liquid at room temperature which freezes at 10°C. In many respects, liquid H_2SO_4 resembles water. For example, it slightly conducts electricity, presumably because, like water, it is dissociated into ions:

$$H_2SO_4 \rightleftharpoons H^+ + HSO_4^-$$

where H^+ is solvated not by H_2O as in H_3O^+, but by H_2SO_4 to give the analog $H_3SO_4^+$. Furthermore, like water it dissolves many substances, even ionic solids. However, H_2SO_4 differs from water in that its extent of dissociation is considerably greater and in that H_2SO_4 may force a proton on any dissolved species. For instance, when acetic acid is placed in pure H_2SO_4, the following reaction occurs:

$$CH_3COOH + H_2SO_4 \longrightarrow CH_3COOH_2^+ + HSO_4^-$$

Pure H_2SO_4 has great affinity for water and forms several compounds, or hydrates, with water, such as $H_2SO_4 \cdot H_2O$ and $H_2SO_4 \cdot 2H_2O$. Ordinary commercially available concentrated sulfuric acid is approximately 93 percent by weight H_2SO_4 and can be thought of as a solution of H_2SO_4 and $H_2SO_4 \cdot H_2O$. The monohydrate may be H_3O^+ and HSO_4^-, and the great heat observed to be liberated when concentrated sulfuric acid is added to water may be due to formation of H_3O^+ and subsequent hydration of it and of HSO_4^-. Frequently, concentrated H_2SO_4 is used as a dehydrating agent, as, for example, in desiccators to keep substances dry. It is also used in the role of a catalyst and a dehydrating agent in reactions for the removal of water. As an example of the latter, H_2SO_4 is used in the manufacture of ethers from alcohols:

$$2C_2H_5OH \xrightarrow{H_2SO_4} C_2H_5OC_2H_5 + H_2O$$

In aqueous solutions, H_2SO_4 is a strong acid, but only for dissociation of one proton. The dissociation constant for the second proton is 1.3×10^{-2}. Because of the dissociation

$$HSO_4^- \longrightarrow H^+ + SO_4^{2-}$$

solutions of HSO_4^-, such as solutions of sodium hydrogen sulfate ($NaHSO_4$), are slightly acid.

Although sulfate is not an especially good oxidizing agent, there is a closely related derivative which is an extremely powerful oxidizing agent. This derivative is produced by electrolytic oxidation of cold, concentrated sulfuric acid and has been assigned the formula $H_2S_2O_8$. The acid, called *peroxydisulfuric acid,* and its salts are peroxy compounds, since they contain an oxygen-oxygen bridge between the two sulfur atoms. In addition to their use as oxidizing agents, the peroxydisulfates are important as intermediates in the preparation of hydrogen peroxide:

$$K_2S_2O_8(s) + 2H_2O + 2H^+ \longrightarrow 2H_2SO_4 + H_2O_2 + 2K^+$$

An interesting derivative of sulfuric acid is found in some of the synthetic detergents used in place of soap. In the preparation of these derivatives, H_2SO_4 reacts with a hydrocarbon as illustrated in the following equation:

The product is called a *sulfonic acid,* and its salts, made by neutralizing the H of the OH group, are called *sulfonates.* In water, sulfonates behave like soap in forming globules for oils to dissolve in but have an advantage over soap in not forming precipitates with calcium ion.

In addition to the oxy compounds that sulfur forms in the $6+$ oxidation state, there are important oxy compounds corresponding to the $4+$ state of sulfur. The simplest of these is the dioxide, SO_2, which is formed either by burning sulfur in air or by reducing sulfates. At room temperature, SO_2 is a gas, but it is quite easily liquefied (boiling point $-10°C$). Sulfur dioxide has a disagreeable, choking odor and is somewhat poisonous. It is especially toxic to lower organisms such as fungi and for this reason is used for sterilizing dried fruit and wine barrels. With water, SO_2 dissolves to give acid solutions, which contain roughly 5% sulfurous acid (H_2SO_3). However, the compound H_2SO_3 has never been isolated pure, and any attempt to concentrate the solution, as by heating, simply expels SO_2. Apparently, H_2SO_3 is a weak diprotic acid, best represented as $SO_2 + H_2O$:

$$SO_2 + H_2O \rightleftharpoons H^+ + HSO_3^- \qquad K_I = 1.3 \times 10^{-2}$$
$$HSO_3^- \rightleftharpoons H^+ + SO_3^{2-} \qquad K_{II} = 5.6 \times 10^{-8}$$

It forms two series of salts, the sulfites, for example, Na_2SO_3, and the hydrogen sulfites, for example, $NaHSO_3$. Addition of concentrated acids to either solid sulfites or solid hydrogen sulfites liberates SO_2 and is a convenient way of making sulfur dioxide in the laboratory. Sulfites, hydrogen sulfites, and sulfurous acid are mild reducing agents and are relatively easily oxidized to sulfates, though sometimes the reaction is quite slow.

When solutions containing sulfite ion are boiled with elemental sulfur, the solid sulfur dissolves in a reaction,

$$S(s) + SO_3^{2-} \rightleftharpoons S_2O_3^{2-}$$

which is easily reversed by addition of acid. The ion formed, $S_2O_3^{2-}$, is called *thiosulfate ion,* where the prefix *thio-* indicates substitution of a sulfur atom for an oxygen atom. Apparently, $S_2O_3^{2-}$ contains two different kinds of sulfur atoms, as found by the following experiment: Solid sulfur containing a radioactive isotope of sulfur was boiled with a solution containing non-radioactive sulfite ions. The thiosulfate ions formed were found to be radioactive, but after acid was added so as to reverse the above reaction, all the radioactivity was recovered as precipitated solid sulfur. The implication is that the same S atom which adds to SO_3^{2-} to form $S_2O_3^{2-}$ is dropped off when acid is added. This can be true only if the added S atom is bound in $S_2O_3^{2-}$ in a way that is unlike the binding of the S atom in SO_3^{2-}. Otherwise, the two S atoms in $S_2O_3^{2-}$ would be identical, and the addition of acid would not preferentially drop off the added radioactive S atom but would have a 50–50 chance of retaining it and its activity in the complex SO_3^{2-}. The structure proposed for $S_2O_3^{2-}$ is, like that of SO_4^{2-}, tetrahedral; it has one sulfur atom at the center of a tetrahedron with the other sulfur atom and the three oxygen atoms at the four corners of the tetrahedron. Thiosulfate is decomposed by acid to give sulfur and HSO_3^-, but the reaction is quite slow, at least so far as formation of visible solid sulfur is concerned. Indeed, when acid is added to thiosulfate solutions, nothing is observed at first. Then a white milkiness develops as colloidal sulfur is produced by gradual agglomeration of S atoms. Thiosulfate ion has the ability to form complex ions with the ions of some metals, especially Ag^+. The silver-thiosulfate complex ion is so stable that thiosulfate solutions can dissolve the insoluble silver halides (see Section 16.9, subsection "Silver," for the use of thiosulfate in the photographic fixing process).

Besides occurring in the positive oxidation states, sulfur forms compounds corresponding to negative oxidation states, especially $2-$. The most familiar of these compounds is probably hydrogen sulfide (H_2S), notorious for its rotten-egg odor. Not so well known is the fact that hydrogen sulfide is as poisonous as hydrogen cyanide and four times as poisonous as carbon monoxide. The presence of H_2S in sewer gas is due to putrefaction of sulfur-containing organic material. The pure compound can be made by bubbling hydrogen gas through molten sulfur. In the laboratory, it is conveniently prepared by interaction of some sulfide such as FeS with acid,

$$FeS(s) + 2H^+ \longrightarrow Fe^{2+} + H_2S(g)$$

or by the warming of a solution of thioacetamide,

Thioacetamide   *Acetamide*

The latter reaction has become extremely popular as a most easily controlled laboratory source of hydrogen sulfide. Like H_2O, H_2S has a bent molecule and is polar; however, it is considerably harder to liquefy (boiling point $-61°C$), presumably because of less hydrogen bonding in the liquid. Gaseous H_2S burns to produce H_2O and either sulfur or sulfur dioxide, depending on the temperature and the oxygen supply. It is a mild reducing agent and can, for example, reduce ferric ion to ferrous ion,

$$2Fe^{3+} + H_2S(g) \longrightarrow 2Fe^{2+} + S(s) + 2H^+$$

During the course of the reaction, the solution becomes milky from the production of colloidal sulfur. In aqueous solution, H_2S is a weak diprotic acid for which the dissociation constants are $K_I = 1.1 \times 10^{-7}$ and $K_{II} = 1 \times 10^{-14}$.

Derived from H_2S are the sulfides, such as Na_2S and HgS. In regard to their solubility in water, the sulfides vary widely, from those which, like Na_2S, are quite soluble in water to those which, like HgS, require drastic treatment to be brought into solution.† Because the sulfides of the group I and group II metals are so soluble, they cannot be precipitated by bubbling H_2S through solutions of their salts. As already discussed in Section 12.10, some sulfides that are insoluble in water can be dissolved simply by raising the H^+ concentration. ZnS, for example, is soluble in 0.3 M H^+, because the H^+ serves to lower the concentration of sulfide ion by combining with it to form H_2S. The net equation can be represented as

$$ZnS(s) + 2H^+ \longrightarrow Zn^{2+} + H_2S$$

CuS, Ag_2S, PbS, and SnS are so insoluble that they cannot be dissolved by H^+ alone. However, hot nitric acid oxidizes sulfide to sulfur and hence lowers the sulfide-ion concentration sufficiently to permit solubility. For CuS, the net reaction can be written

$$3CuS(s) + 8H^+ + 2NO_3^- \longrightarrow 3Cu^{2+} + 3S(s) + 2NO(g) + 4H_2O$$

The least soluble of the ordinary sulfides, mercuric sulfide, is not appreciably soluble in hot HNO_3. In order to "dissolve" it, aqua regia must be used in order that oxidation of the sulfide ion may be accompanied by complexing of the mercuric ion. The net reaction might be written

$$HgS(s) + 2NO_3^- + 4Cl^- + 4H^+ \longrightarrow HgCl_4^{2-} + 2NO_2(g) + S(s) + 2H_2O$$

although the reduction product of nitrate is probably a mixture rather than just NO_2. The differences in solubility behavior of metal sulfides can be used to great advantage in the separation and identification of various elements, as in qualitative analysis.

In addition to sulfides, sulfur forms polysulfides, in which two or more sulfur atoms are bound together in a chain. These polysulfides can be made, for example, by boiling a solution of a soluble sulfide with elemental sulfur. With Na_2S and sulfur, the product is usually described as Na_2S_x and is thought to consist of Na^+ ions and $[S_x]^{2-}$ ions. The polysulfide chains are of varying length and can be considered as being formed by progressive addition of sulfur atoms to sulfide ion. The simplest of the polysulfide chains is the disulfide, S_2^{2-}; it is found in the mineral FeS_2, iron pyrites, or fool's gold. Solid FeS_2 has an NaCl-like structure consisting of an array of alternating Fe^{2+} and S_2^{2-} ions. In acid solution, disulfides (and other polysulfides) break down to form solid sulfur and H_2S.

---

†HgS can be "dissolved" by heating it with aqua regia, but it is questionable that this is a good description of the process. Although the mercury dissolves as $HgCl_4^{2-}$, the sulfur does not dissolve but is oxidized to insoluble elemental sulfur.

In chemical analysis, an unknown solution might contain sulfur as sulfide, sulfate, or sulfite. The sulfide, on addition of acid, generates H_2S gas, which can be detected either by its odor or by its blackening (due to PbS formation) of filter paper wet with a solution of lead acetate. Sulfate, on addition of barium nitrate and acid, produces white insoluble $BaSO_4$. If sulfite is present, it will not precipitate with Ba^{2+} in acid solution; however, if Br_2 is added, the sulfite will be oxidized to sulfate and $BaSO_4$ forms.

18.4   Halogens

Although the chemistry of the group VII elements is somewhat complex, similarities within the group are more pronounced than in any of the other groups except I and II. The elements, fluorine, chlorine, bromine, iodine, and astatine, are collectively called *halogens* (from the Greek *halos*, salt, and *genes*, born), or *salt producers*, because they all have high electronegativity and form negative halide ions such as are found in ionic salts. Except for fluorine, they also show positive oxidation states.

Because of their high electronegativity, the halogens show practically no metallic properties, though solid iodine has a somewhat metallic appearance. Astatine, the heaviest member of the group, may also have some metal properties, but it is a short-lived radioactive element and not enough of it has been prepared to see whether the solid is metallic. Other properties of the group are shown in Figure 18.8. All the atoms have seven electrons in

Figure 18.8 PROPERTIES OF GROUP VII ELEMENTS	Symbol	Z	Electronic Configuration	Melting Point, °C	Boiling Point, °C	Ionization Potential, eV	Reduction Potential, volts (X_2 to X^-)
	F	9	$2,2s^2 2p^5$	-223	-187	17.42	$+2.87$
	Cl	17	$2,8,3s^2 3p^5$	-102	-34.6	13.01	$+1.36$
	Br	35	$2,8,18,4s^2 4p^5$	-7.3	58.78	11.84	$+1.09$
	I	53	$2,8,18,18,5s^2 5p^5$	114	183	10.44	$+0.54$
	At	85	$2,8,18,32,18,6s^2 6p^5$				$+0.2$

the outermost energy level and either gain one electron completely to form a $1-$ ion or share one electron to form a single covalent bond. The former occurs when a halogen atom combines with an atom of low electronegativity to form an ionic bond, as in NaF or KCl; the latter occurs when the halogen atom combines with another atom of similar high electronegativity to form a covalent bond, as in Cl_2. Although the bond between the halogen atoms is fairly strong, the attraction between X_2 molecules is quite weak and due only to van der Waals forces. Going down the group, there is an increasing number of electrons per X_2 molecule, and we would expect van der Waals attraction to increase. Thus, it is not surprising that the boiling points increase in going from F_2 to I_2. At room temperature, fluorine and chlorine are gases; bromine is a liquid; and iodine is a solid.

As indicated by the relatively high values of the ionization potentials, it is fairly difficult to remove an electron from a halogen atom. In fact, it requires more energy to remove an electron from a halogen atom than from any other atom in the same period except for the noble gases (compare values in Figure 3.23). Within the group itself, there is a decrease in the ionization potential; the larger the halogen atom, the less firmly bound are the outermost electrons and the less the energy required to remove an electron from the neutral atom. Of greater significance chemically are the reduction potentials, given in the last column of Figure 18.8. As given, the reduction potentials describe the relative tendency of the half-reaction

$$X_2 + 2e^- \longrightarrow 2X^-$$

to go from left to right in aqueous solution. Since the value $+2.87$ volts for F_2 is the greatest, not only of the halogens but of all common oxidizing agents, F_2 has the greatest tendency of all to pick up electrons and form negative F^- ions. A major reason for this is the small size of F^- (radius 0.133 nm compared to 0.181 nm for Cl^-). The smaller the ion formed, the more strongly it interacts with water molecules of an aqueous solution or with positive ions in the formation of solid products.

Since fluorine is the most electronegative of all the elements, it can show only a negative oxidation state. However, the other halogens also show positive oxidation states in compounds with more electronegative elements. Most of these compounds contain oxygen, which has electronegativity between that of fluorine and chlorine. In oxy compounds, chlorine, bromine, and iodine show a maximum oxidation number of $7+$. In addition, they form compounds in which the halogen atom is assigned oxidation numbers $1+$ and $5+$.

Fluorine   Fluorine is about half as abundant as chlorine and is widely distributed in nature. It occurs principally as the mineral fluorspar (CaF_2), cryolite (Na_3AlF_6), and fluorapatite [$Ca_5F(PO_4)_3$]. Because none of the ordinary chemical oxidizing agents is capable of extracting electrons from fluoride ions, elemental fluorine is prepared only by electrolytic oxidation of molten fluorides, such as KF–HF mixtures. Fluorine is a pale yellow gas at room temperature and is extremely corrosive and reactive. With hydrogen, it forms violently explosive mixtures because of the reaction

$$H_2(g) + F_2(g) \longrightarrow 2HF(g) \qquad \Delta H = -536 \text{ kilojoules}$$

On the skin, it causes severe burns which are quite slow to heal.

Hydrogen fluoride is usually made by the action of sulfuric acid on fluorspar. Because of hydrogen bonding, liquid HF has a higher boiling point (19.5°C) than any of the other hydrogen halides. In aqueous solutions, HF is called hydrofluoric acid and is unique among the hydrogen halides in being a weak, rather than a strong, acid ($K_{diss} = 6.7 \times 10^{-4}$). Also, it can dissolve glass because of the formation of fluosilicate ions, as in the equation

$$SiO_2(s) + 6HF \longrightarrow SiF_6^{2-} + 2H^+ + 2H_2O$$

where glass is represented for simplicity as SiO_2. Other complex ions of fluorine are known, for example, AlF_6^{3-}, in which the small size of the fluoride ion permits relatively large numbers of them to be attached to another atom. Small concentrations of fluoride ion in drinking-water supplies significantly decrease dental decay.

In general, most fluoride salts with 1+ cations are soluble (for example, KF and AgF) and give slightly basic solutions because of the hydrolysis of F^- to HF. With 2+ cations, however, the fluorides are usually insoluble (for example, CaF_2 and PbF_2), but their solubility is somewhat increased in acid solution. The formation of insoluble, inert fluorides as surface coatings is apparently the reason that fluorine can be stored in metal containers such as copper in spite of its great reactivity.

Most amazing of the fluorine compounds are the fluorocarbons. These are materials which can be considered to be derived from the hydrocarbons by substitution of fluorine atoms for hydrogen atoms. Thus, the fluorocarbon corresponding to methane (CH_4) is tetrafluoromethane (CF_4). This compound is typical of the saturated (i.e., containing no double bonds) fluorocarbons in being extremely inert. For example, unlike methane, it can be heated in air without burning. Furthermore, it can be treated with boiling nitric acid, concentrated sulfuric acid, and strong oxidizing agents such as potassium permanganate with no change. Reducing agents such as hydrogen or carbon do not affect it even at temperatures as high as 1000°C. Because of their inertness, the fluorocarbons find application for special uses. For example, $C_{12}F_{26}$ is an ideal insulating liquid for heavy-duty transformers that operate at high temperature. Just as ethylene (C_2H_4) can polymerize to form polyethylene, tetrafluoroethylene (C_2F_4) can polymerize to form polytetrafluoroethylene. The polymerization can be imagined to proceed by the opening up of the double bond to form an unstable intermediate which joins with other molecules to produce a high polymer:

The high polymer is a plastic known commercially as Teflon. Like the other saturated fluorocarbons, it is inert to chemical attack, and it is unaffected even by boiling aqua regia or ozone. Although still rather expensive, fluoro-carbon polymers show considerable promise as structural materials where corrosive conditions are extreme, as in chemical plants. They are also used for coating "greaseless" frying pans and other kitchenware.

Chlorine   Chlorine is the most abundant (0.2 percent) of the halogens and occurs as chloride ion in seawater, salt wells, and salt beds, where it is combined with Na^+, K^+, Mg^{2+}, and Ca^{2+}. On a small scale, the element can be made by chemical oxidation, as with MnO_2,

$$MnO_2(s) + 2Cl^- + 4H^+ \longrightarrow Mn^{2+} + Cl_2(g) + 2H_2O$$

On a commercial scale, chlorine is more economically prepared by electro-lytic oxidation of either aqueous or molten NaCl. The element is a greenish yellow gas (in fact, it gets its name from the Greek *chloros*, green) and has a choking odor. Although not so reactive as fluorine, it is a good oxidizing agent and explodes with hydrogen when mixtures of H_2 and Cl_2 are exposed to ultraviolet light. Most of the commercial chlorine is used as a bleach for paper and wood pulp and for large-scale disinfecting of public water sup-plies. Both these uses depend on its oxidizing action.

The most important compounds of chlorine are those which correspond to the oxidation states $1-$, $1+$, $5+$, and $7+$, although there also are com-pounds of chlorine in the other positive states, $3+$, $4+$, and $6+$. The $1-$ state is familiar as the one assigned to chlorine in HCl and chloride salts. Although HCl can be produced by direct combination of the elements, a more convenient method of preparation is the heating of NaCl with concen-trated H_2SO_4:

$$NaCl(s) + H_2SO_4 \longrightarrow NaHSO_4(s) + HCl(g)$$

Hydrogen chloride gas is very soluble in water, and its solutions are referred to as *hydrochloric acid*. Commercially available concentrated hydrochloric acid is 37% HCl by weight, or 12 M. Unlike HF, HCl is a strong acid and is almost completely dissociated into ions in 1 M solution. Of the common chlorides, silver chloride (AgCl), mercurous chloride (Hg_2Cl_2), and lead chloride ($PbCl_2$) are rather insoluble.

The $1+$ oxidation state of chlorine is represented by hypochlorous acid (HOCl) and its salts, the hypochlorites. Hypochlorous acid is produced to a slight extent when chlorine gas is dissolved in water. Disproportionation of the dissolved chlorine occurs according to the equation

$$Cl_2 + H_2O \rightleftharpoons Cl^- + H^+ + HOCl$$

The yield of products can be greatly increased by tying up the Cl^- and H^+, as by adding silver oxide (Ag^+ to precipitate $AgCl$ and oxide to neutralize H^+). The formula of hypochlorous acid is usually written $HOCl$ instead of $HClO$, to emphasize the fact that the proton is bonded to the oxygen and not directly to the chlorine. The acid is weak, with a dissociation constant of 3.2×10^{-8}, and exists only in aqueous solution. Even in solution it slowly decomposes with evolution of oxygen:

$$2HOCl \longrightarrow 2H^+ + 2Cl^- + O_2(g)$$

$HOCl$ is a powerful oxidizing agent, stronger than MnO_4^-. Hypochlorites, such as $NaClO$, can be made by neutralization of $HOCl$ solutions, but they are produced more economically by the disproportionation of chlorine in basic solution:

$$Cl_2 + 2OH^- \longrightarrow Cl^- + ClO^- + H_2O$$

Commercially, the process is efficiently carried out by electrolyzing cold aqueous $NaCl$ solutions and stirring vigorously. The stirring serves to mix chlorine produced at the anode,

$$2Cl^- \longrightarrow Cl_2 + 2e^-$$

with hydroxide ion produced at the cathode,

$$2e^- + 2H_2O \longrightarrow H_2(g) + 2OH^-$$

so that reaction can occur. Solutions of hypochlorite ion so produced are sold as laundry bleaches, e.g., Clorox.

In aqueous solution, hypochlorite ion is unstable with respect to self-oxidation and, when warmed, disproportionates by the equation

$$3ClO^- \longrightarrow 2Cl^- + ClO_3^-$$

Chlorate ion is pyramidal, the three oxygen atoms forming the base of the pyramid, with the chlorine atom at the apex. The structure and the electronic configuration are shown in Figure 18.9. Probably the most important chlorate

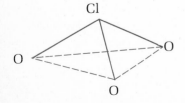

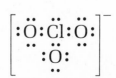

Figure 18.9
Structure of chlorate ion.

salt is $KClO_3$, used as an oxidizing agent in matches, fireworks, and some explosives. Since $KClO_3$ is only moderately soluble in water, it can be precipitated by addition of KCl to chlorate-containing solutions. The chlorate solutions can be produced by electrolyzing hot chloride solutions that are vigorously stirred. Steps in the production can be summarized as follows:

$$2Cl^- + 2H_2O \xrightarrow{\text{electrolyze}} Cl_2 + 2OH^- + H_2(g)$$

$$3Cl_2 + 6OH^- \xrightarrow{\text{stir, heat}} 5Cl^- + ClO_3^- + 3H_2O$$

$$K^+ + ClO_3^- \longrightarrow KClO_3(s)$$

As seen from the equation for the second step, only one-sixth of the chlorine is converted to ClO_3^-, which makes the process seem rather inefficient. However, on continued electrolysis the chloride produced in the second step is reoxidized in the first step.

Unlike hypochlorite ion, chlorate ion is the anion of a strong acid. The parent acid, $HClO_3$, chloric acid, has not been prepared in the pure state, since it is unstable. When attempts are made to concentrate chloric acid solutions, as by evaporation, violent explosions occur. In acid aqueous solutions, chlorate ion, like hypochlorite ion, is a good oxidizing agent, almost the equal of MnO_4^-.

When $KClO_3$ is heated, it can decompose by two reactions,

$$2KClO_3(s) \longrightarrow 2KCl(s) + 3O_2(g)$$
$$4KClO_3(s) \longrightarrow 3KClO_4(s) + KCl(s)$$

the first of which is catalyzed by surfaces, such as powdered glass or MnO_2, from which oxygen can readily escape. In the absence of such catalysts, especially at lower temperatures, the formation of potassium perchlorate ($KClO_4$) is favored. A more efficient method of preparing perchlorates is to use electrolytic oxidation of chlorate solutions. Since $KClO_4$ is only sparingly soluble in water (less than $KClO_3$), it can be made by addition of K^+ to perchlorate solutions. The perchlorate ion has a tetrahedral configuration, as shown in Figure 18.10. The chlorine atom is at the center of the tetrahedron, and the four oxygen atoms are at the corners. In aqueous solutions,

Figure 18.10
Structure of perchlorate ion.

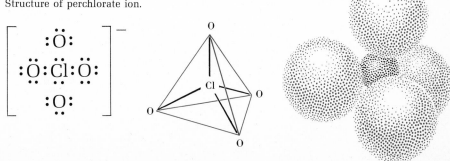

perchlorate ion is potentially a good oxidizing agent, especially in acid solution, but its reactions are so very slow that they are usually not observed. For example, a solution containing ClO_4^- and the very strong reducing agent Cr^{2+} (chromous ion) can be kept for weeks without any appreciable oxidation to Cr^{3+} (chromic ion).

Like chlorate ion, perchlorate ion is an anion of a strong acid. Consequently, in aqueous solution there is practically no association of ClO_4^- with H^+. However, when perchlorate salts are treated with sulfuric acid, pure hydrogen perchlorate ($HClO_4$) may be distilled off under reduced pressure. The anhydrous compound is a liquid at room temperature and is extremely dangerous, because it may explode spontaneously. The danger is especially great in the presence of reducing agents such as organic material (e.g., wood, cloth, etc.). Dilute aqueous solutions of $HClO_4$ are safe and are useful reagents for the chemist. Perchloric acid is probably the strongest of all common acids and in aqueous solution is more completely dissociated than the usual strong acids hydrochloric, sulfuric, and nitric. Why is perchloric a stronger acid than the other oxy acids of chlorine? The dissociation of an oxy acid involves breaking a hydrogen-oxygen bond to form a hydrated hydronium ion and a hydrated anion. The bigger the anion, the less strongly it is hydrated. Consequently, since ClO_4^- is obviously bigger than ClO^-, for example, we might expect $HClO_4$ to be less dissociated than $HOCl$. Since the reverse is true, the bond holding the proton to OCl^- must be stronger than the bond holding the proton to ClO_4^-. That this is reasonable can be seen by noting that oxygen is more electronegative than chlorine; therefore, addition of oxygen atoms to the $HOCl$ molecule pulls electrons away from the H—O bond and tends to weaken it. This picture is supported by observing that the intermediate oxy acid $HOClO$ (chlorous acid) has a dissociation constant, $K_{diss} = 1.1 \times 10^{-2}$, larger than that of $HOCl$, $K_{diss} = 3.2 \times 10^{-8}$. In general, for any series of oxy acids the acid corresponding to the highest oxidation number is the most highly dissociated.

Bromine   Bromine, from the Greek word *bromos* for stink, occurs as bromide ion in seawater, brine wells, and salt beds and is less than a hundredth as abundant as chlorine. The element is usually prepared by chlorine oxidation of bromide solutions, as by sweeping chlorine gas through seawater. Since chlorine is a stronger oxidizing agent than bromine, the reaction

$$Cl_2(g) + 2Br^- \longrightarrow Br_2 + 2Cl^-$$

occurs as indicated. Removal of the bromine from the resulting solution can be accomplished by sweeping the solution with air, because bromine is quite volatile. At room temperature pure bromine is a mobile, but dense, red liquid of pungent odor. It is a dangerous substance, since it attacks the skin to form slow-healing sores.

Although less powerful an oxidizing agent than chlorine, bromine readily reacts with other elements to form bromides. Hydrogen bromide, like HCl, is a strong acid but is more easily oxidized than is HCl. Whereas HCl can be made by heating the sodium salt with H_2SO_4, HBr cannot. The hot H_2SO_4 oxidizes HBr to Br_2, and a nonoxidizing acid, such as H_3PO_4, must be used instead.

In basic solution, bromine disproportionates to give bromide ion and hypobromite ion, BrO^-. The reaction is quickly followed by further disproportionation,

$$3BrO^- \longrightarrow 2Br^- + BrO_3^-$$

to give bromate ion, BrO_3^-. Bromic acid ($HBrO_3$) has never been prepared pure. In aqueous solution, it is a strong acid and a good oxidizing agent. Like Cl and I, Br shows the $7+$ oxidation state, although it is extremely difficult to obtain and eluded chemists until recently. However, use of the fluorocarbons now makes possible utilization of fluorine as an oxidizing agent. This strongest of oxidizing agents is used to oxidize bromate in basic solution to perbromate. Perbromate resembles perchlorate in structure and acid behavior. It is a stronger oxidizing agent (by 0.53 volt) than ClO_4^-, but, like ClO_4^-, is a sluggish oxidizing agent.

One of the important uses of bromine is in making silver bromide for photographic emulsions (Section 16.9, subsection "Silver"). However, the principal use of bromine is in making dibromoethane ($C_2H_4Br_2$) for addition to gasolines which contain tetraethyllead. Tetraethyllead [$(C_2H_5)_4Pb$], added to gasoline as an antiknock agent, decomposes on burning to form lead deposits. The dibromoethane prevents accumulation of lead deposits in the engine, but it acts by getting more compounds into the atmosphere to contribute to the air-pollution problem.

Iodine   Of the halogens, iodine is the only one which occurs naturally in a positive oxidation state. In addition to its occurrence as I^- in seawater and salt wells, it is also found as sodium iodate ($NaIO_3$), small amounts of which are mixed with $NaNO_3$ in Chile saltpeter. The Chilean ore is processed by the reduction of $NaIO_3$ with controlled amounts of $NaHSO_3$. The principal reaction is

$$5HSO_3^- + 2IO_3^- \longrightarrow I_2 + 5SO_4^{2-} + 3H^+ + H_2O$$

In the United States, most of the iodine is produced by chlorine oxidation of I^- from salt wells.

At room temperature, iodine crystallizes as black leaflets with metallic luster. Although, as shown by X-ray analysis, the solid consists of discrete I_2 molecules, its properties are different from those of usual molecular solids.

For example, its electric conductivity, though small, increases with increasing temperature like that of a semiconductor. Furthermore, liquid iodine also has perceptible conductivity, which decreases with increasing temperature like that of a metal. Thus, feeble as they are, metallic properties do appear even in the halogen family.

When heated, solid iodine readily sublimes to give a violet vapor, which consists of I_2 molecules. The violet color is the same as that observed for many iodine solutions, such as those in CCl_4 and in hydrocarbons. However, in water and in alcohol the solutions are brown, presumably because of unusual interactions between I_2 and the solvent. When iodine is brought in contact with starch, a characteristic deep blue color results, which has been attributed to a starch-I_2 complex. The formation of the blue color is the basis for using starch–potassium iodide mixtures as a qualitative test for the presence of oxidizing agents. Oxidizing agents convert I^- to I_2, which with starch forms the colored complex. With very strong oxidizing agents, the color may fade with oxidation of I_2 to a higher oxidation state.

Iodine is only slightly soluble in water (0.001 M), but the solubility is vastly increased by the presence of iodide ion. The color changes from brown to deep red because of the formation of the tri-iodide ion, I_3^-. The tri-iodide ion is also known in solids such as NH_4I_3, where X-ray investigations indicate that the I_3^- ion is linear. No electronic formula conforming to the octet rule can be written for this ion. Apparently, an iodine atom, perhaps because of its large size, can accommodate more than eight electrons in its valence shell. In basic solution, I_2 disproportionates to form iodide ion and hypoiodite ion, IO^-.

$$I_2 + 2OH^- \longrightarrow I^- + IO^- + H_2O$$

Further disproportionation to give iodate ion, IO_3^-, is hastened by heating or by addition of acid. Iodate ion in acid solution is a weaker oxidizing agent than either bromate ion or chlorate ion. Since IO_3^- is a weaker oxidizing agent than ClO_3^-, iodates can be made by oxidizing I_2 with ClO_3^-. Furthermore, iodate salts are not quite so explosive as chlorates or bromates. The greater stability of iodates also is evident in the fact that HIO_3, unlike $HClO_3$ and $HBrO_3$, can be isolated pure (as a white solid); the latter acids detonate when attempts are made to concentrate them. In the 7+ state, the oxy salts of iodine are called *periodates*. They are derivatives of HIO_4 (metaperiodic acid), or H_5IO_6 (paraperiodic acid). In metaperiodates, the iodine is bonded tetrahedrally to four oxygen atoms; in paraperiodates, there are six oxygen atoms bound octahedrally to the iodine atom.

So far as uses are concerned, iodine is less widely used than other halogens. It finds limited use for its antiseptic properties, both as tincture of iodine (solution of I_2 in alcohol) and as iodoform (CHI_3). Since small amounts of iodine are required in the human diet, traces of sodium iodide (1 part per 10^5) are frequently added to table salt.

Analysis   Fluoride differs from Cl^-, Br^-, and I^- in forming an insoluble magnesium salt but a soluble silver salt.

Iodide can be distinguished from Br^- and Cl^- in that its silver salt is insoluble in excess NH_3. Its presence can be confirmed by oxidizing I^- to I_2, which imparts a violet color to CCl_4.

Br^- and Cl^- both form insoluble silver salts, but AgBr is more yellowish than the pure white of AgCl. Furthermore, AgBr dissolves with greater difficulty in excess NH_3 than does AgCl. Finally, when Br^- is oxidized to Br_2 in the presence of CCl_4, the CCl_4 solution is brown, whereas Cl_2 in CCl_4 is yellow.

The oxy anions ClO_3^-, BrO_3^-, and IO_3^- can be detected by reducing them in acid (by adding sulfite or nitrite) and then analyzing for the corresponding halide ions as above.

QUESTIONS

18.1   *Nitrogen*   Although your body relies on nitrogen compounds it cannot utilize N_2 from the air. Describe three distinct routes for the conversion of atmospheric nitrogen to compounds which ultimately form amino acids and proteins in the food you eat.

18.2   *Phosphorus*   Give three examples where the absence of multiple bonding by phosphorus causes a molecule to be more complex than would be true if P, like N, formed multiple bonds.

18.3   *Sulfur*   (a) Write formulas for three compounds of sulfur in three different oxidation states, all of which give acid solutions in water. (b) Write net equations to show the production of acid in each case. (c) By means of K_s show which of the three is relatively most acidic and which is the least.

18.4   *Halogens*   Write balanced equations for each of the following: (a) the dissolving of CaF_2 in acid solution, (b) reaction of chlorate and chloride in acid solution to form chlorine, (c) disproportionation of bromine in base to form bromate as one of the products, (d) oxidation of I_2 by ClO_3^- in acid.

18.5   *Halogens*   Describe and account for the trends in all four properties listed in Figure 18.8.

18.6   *Acids*   Describe, including equations, the commercial processes for making HNO_3 and H_2SO_4. In each case start with the pure element.

18.7   *Structures*   Draw structures for each of the following: P_4O_6, P_4O_{10}, $S_2O_3^{2-}$, NO_3^-, H_5IO_6.

18.8   *Hydrolysis*   Consider solutions of equal concentration of each of the following and arrange them in order of increasing basicity: $NaClO_3$, Na_2S, NaF, Na_3PO_4, NH_4Cl.

18.9* *Cl$_2$ Preparation* If you wanted to prepare in the laboratory 1 liter of Cl$_2$ gas at STP, how many grams of MnO$_2$ would you need?

18.10 *Protein* Using the average protein composition figures of Figure 18.10, calculate how many times there are as many atoms of each element as there are of nitrogen.

18.11 *Haber Process* At 500°C and 1000 atm there is about 50 percent conversion of N$_2$ and H$_2$ to NH$_3$. At 1000°C and 500 atm, would the conversion be more complete or less complete? Explain.

18.12 *Heat of Reaction* Calculate the heat liberated for each of the following: (a) reaction of 1.00 g of fluorine with excess hydrogen, (b) reaction of 1.00 g of hydrogen with excess fluorine, (c) complete dimerization of 1.00 g of NO$_2$.

18.13* *Acidity* Calculate the H$^+$ concentration in an 0.10 M solution of each of the following: (a) HOCl, (b) H$_2$S, (c) NH$_4$Cl.

18.14 *Fertilizer* Commercial fertilizers are given three numbers, for example, 5-10-5. The first number is % (by weight) N, the second is % (by weight) phosphorus (calculated as P$_2$O$_5$), and the third is % potassium (calculated as K$_2$O). Suppose you had an equimolar mixture of the two common fertilizer components KNO$_3$ and Ca(H$_2$PO$_4$)$_2$. What three numbers would you assign?

18.15 *Phosphate* By use of a dot formula for the trimetaphosphate structure pictured in Figure 18.6, find the charge on the ion necessary to complete the octets of all the atoms.

18.16 *Thiosulfate* Suppose that the radiochemical experiment described in Section 18.3 had shown the two S atoms in thiosulfate to be structurally identical. Postulate such a structure for the ion, draw it, and show that it conforms to the octet rule.

18.17* *Hydrazine Rockets* The density of liquid hydrazine is 1.01 g/ml and that of liquid hydrogen peroxide is 1.46 g/ml. Calculate the percentage increase in volume when 1.50 mole of N$_2$H$_4$(l) reacts with 3 moles of H$_2$O$_2$(l) to give products at 550°C and 1.50 atm.

18.18 *Sulfuric Acid* Calculate the concentration of H$^+$, HSO$_4^-$, SO$_4^{2-}$ in a solution labeled 1.2 M H$_2$SO$_4$. Assume the first dissociation is complete and the second has a dissociation constant equal to 1.3×10^{-2}.

18.19* *Sulfuric Acid* Suppose you wanted to make some radioactive sulfuric acid, starting with elemental sulfur containing ^{35}S. (a) Write the chemical equations for the necessary reactions. (b) Calculate what weight of ^{35}S you would need to tag 1.5 percent of the H$_2$SO$_4$ molecules in 90.0 ml of 0.60 M H$_2$SO$_4$.

18.20 *Reactions* Write balanced, net equations for: (*a*) reaction of ammonia with oxygen, (*b*) reduction of nitric acid by Zn to NH_3, (*c*) reaction of copper sulfide with oxygen, (*d*) addition of CuS to solution of nitric acid.

18.21 *Acids* Which of the following is the strongest acid: $HClO_4$, $HClO_3$, $HClO_2$, $HClO$? Explain.

18.22 *Reducing Agents* Arrange the following in order of decreasing reducing strength under conditions where reduction potentials apply: Br^-, I^-, Cl^-, HSO_3^-, HNO_2, H_3PO_3.

18.23* *pH* What is the pH of a solution that is made by dissolving 8.90 g of H_3PO_4 in enough water to make 100 ml of solution?

18.24 *Hydrolysis* Calculate the pH of a solution that is made by dissolving 1.75 g of Na_2HPO_4 in enough water to make 220 ml of solution. (Consider only the first hydrolysis step.)

18.25* *Composition* Calculate the weight percent of phosphorus in each of the following: (*a*) H_3PO_4, (*b*) H_3PO_3, (*c*) Ca_3P_2, (*d*) NaH_2PO_4.

18.26 *Decomposition* How many moles of KCl can be obtained by completely decomposing 2.20 g of $KClO_4$?

18.27* *Heat* How much heat is evolved when 1 g of NO(g) is formed by reaction of N_2 with O_2?

18.28 *Change of State* At 25°C, fluorine and chlorine are gases, bromine is liquid, and iodine is solid. Explain.

18.29 *Analysis* Describe the methods you would use to identify Cl^-, F^-, or I^- if you are given three solutions, each of which contains one of the three ions.

18.30 *Resonance* In order to conform to the octet rule, for which of the following species must resonance be invoked: NO_3^-, P_4, SO_2, ClO^-? Draw dot formulas for all.

18.31 *Lavoisier* One of the few theories of Lavoisier's which proved wrong was his assertion that acid strength increases as oxygen content increases. Where in the chemistry of chlorine could this be disproved? (This case bothered Lavoisier, and he thought better analysis might prove him right. It did not).

18.32 *Science and Society* Detergents contain high phosphate levels to tie up the calcium ion and prevent its tying up the active cleaning agents such as the sulfonates. The phosphate ends up in inland waters where it causes pollution problems. Suggest an alternate solution to the use of phosphates.

18.33* *Chlorate Electrolysis* Suppose you are preparing chlorate electrochemically from aqueous Cl^-. (*a*) How many ampere-hours are needed to prepare 1 kg of $KClO_3$? (*b*) How many liters of H_2 at STP could you get as by-product?

18.34 *Qualitative Analysis* An unknown could contain sulfite, sulfate, chloride, or chlorate. Describe a series of tests which will let you decide which is present and which absent.

18.35 *Equations* Write balanced equations for each of the following: (*a*) preparation of perbromate, (*b*) reaction in acid of paraperiodic acid and hydrazine to form iodine and nitrate, (*c*) the predominate equilibrium reaction that occurs when HPO_4^{2-} is added to water, (*d*) oxidation in basic solution of thiosulfate with chlorate to form sulfate and chloride.

18.36 *Thermodynamics* Consider the exothermic dimerization of NO_2. (*a*) How does the equilibrium constant change as the temperature is raised? (*b*) At high temperature what is the sign of ΔG for the net change? (*c*) What factor causes this? (*d*) Why is the sign of this factor the way it seems to be?

18.37* *Hydrolysis* Calculate the H^+ concentration in the most basic of the following 0.10 M solutions: Na_2HPO_3, NaF, Na_2SO_3, $NaHS$.

18.38 *Nitrogen* Suppose that a totally new microorganism evolved in the sea which could convert, by using sunlight, atmospheric nitrogen in combination with water to ammonia (which dissolved in the sea) and oxygen gas (which replaced nitrogen in the atmosphere). (*a*) What problems would be caused for the continuation of life as we know it? (*b*) What could an international mobilization of scientists do to combat the problems?

Organic Compounds

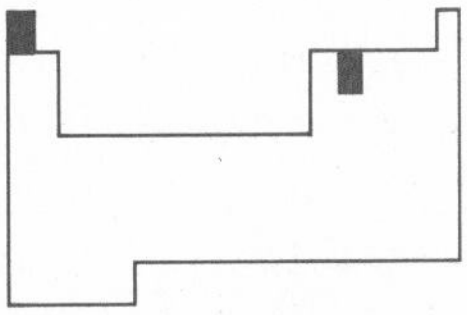

19

Carbon and hydrogen form a vast variety of molecules which are the basis of organic compounds. The close connection between these compounds and organic matter was once interpreted as meaning that organic compounds could be synthesized only by living systems. Since 1828 when Friedrick Wöhler showed that organic compounds can be synthesized in the laboratory, tremendous research effort has been directed toward understanding these compounds and in recent years toward a replication by scientists of natural processes. Through such attempts our knowledge of organic chemistry is probably greater than our knowledge of any other branch of chemistry.

The chemical properties of organic compounds may seem to be special, but no single characteristic is unique to these compounds. However, many of the characteristic properties are shown in greater degree by organic compounds, and the *combination* of necessary chemical properties for life as we know it seems to be uniquely exhibited by the hydrocarbons and their derivatives. In particular, the ability of carbon to form long-chain molecules is greater than that of any other element. Also noteworthy is the fact that organic compounds are slow to undergo chemical change; an unstable atomic arrangement can persist for long periods of time, awaiting the next reaction that will convert it to an even more useful product for some role in a natural process. When reaction occurs, generally most of the molecule remains intact, with reaction occurring only at one site within the molecule.

In this chapter we will discuss first the parent hydrocarbons, of the three types: saturated, unsaturated, and aromatic. Then we will turn to the hydro-carbon derivatives, which can be viewed as hydrocarbon molecules in which one or more hydrogen atoms have been replaced by atoms of other elements. The result of such replacement is to produce a molecule containing *a group of substituent atoms* which bestows characteristic properties on the organic molecule and is called a *functional group*. As a specific example, —O—H is a functional group which when joined to various hydrocarbon residues (i.e., the C and H arrangement that, together with the functional group, comprises the molecule) gives molecules with properties characteristic of alcohols.

Our discussion of hydrocarbon derivatives as defined by their functional groups will concentrate on the general properties bestowed by the functional groups. It is these properties that contribute to the prime role that organic compounds play in living systems. Not only are these compounds important for life but, by working with them, scientists have uncovered a variety of principles and techniques that apply to the production of industrial chemi-cals and have had a marked influence on the quality of life in today's society.

19.1   Saturated Hydrocarbons

The carbon atom has four valence electrons and is expected to form four covalent bonds directed to the corners of a tetrahedron. It matters little whether the bonds are formed to other carbon atoms or to hydrogen atoms, because C and H are very nearly the same in electronegativity. This means that, instead of being restricted to the simplest hydrocarbon CH_4 (methane), a whole series of compounds is possible such as C_2H_6 (ethane), C_3H_8 (propane), and C_4H_{10} (butane). The structural formulas of these are usually written as in Figure 19.1, but of course the molecules are three-dimensional,

Figure 19.1
Some hydrocarbons.

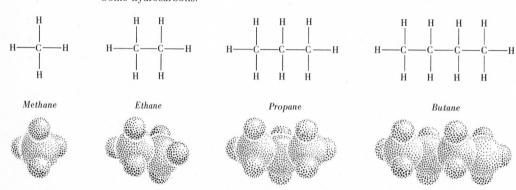

Methane   Ethane   Propane   Butane

as shown in the drawings below the structural formulas. The series of hydrocarbons of this type does not stop with butane but continues almost indefinitely, with each member of the series having the general formula C_nH_{2n+2}. In all the molecules of this series, each carbon atom forms bonds to four other atoms (either hydrogen or carbon). This is the maximum number of atoms to which a carbon atom bonds in its usual compounds; hence, this series of compounds is said to be *saturated,* and the compounds in the series are called *saturated hydrocarbons.* The saturated hydrocarbons are also called *alkanes,* with the ending *-ane* appearing in all their names. Some typical saturated hydrocarbons are listed in Figure 19.2, along with a representative set of physical properties. (Note that after butane these hydrocarbons are named for the number of carbon atoms they contain.) The gradual increase in boiling points makes possible separation of the hydrocarbons.

All the hydrocarbons listed in Figure 19.2 occur in natural gas and petroleum. However, petroleum consists of many more compounds than are

Figure 19.2 SOME STRAIGHT-CHAIN SATURATED HYDROCARBONS, C_nH_{2n+2}	n	Name	Boiling Point, °C	Melting Point, °C
	1	Methane	-161	-182
	2	Ethane	-89	-183
	3	Propane	-42	-188
	4	Butane	-1	-138
	5	Pentane	36	-130
	6	Hexane	69	-95
	7	Heptane	98	-91
	8	Octane	126	-57
	9	Nonane	151	-54
	10	Decane	174	-30

listed in the table.† For example, it includes compounds with 40 or more carbon atoms per molecule. Not only does petroleum contain compounds with formulas not listed in the table, but there may be several different compounds corresponding to the same formula.

---

† The distribution of hydrocarbons in petroleum differs with the age of its formation. Oil shales that are millions of years old, as opposed to those billions of years old, show a preponderance of odd-numbered hydrocarbons over even-numbered ones. The explanation is that, after the appearance of life, petroleum was formed from once-living organisms which, we will see, synthesize organic compounds of only certain compositions.

19.2 Isomers

The formula C_4H_{10} denotes not only the compound *normal butane* (*n-butane*), shown in Figure 19.1 and having physical properties as listed in Figure 19.2, but also the different compound *isobutane*. The difference between these two compounds is shown by their conventional (two-dimensional) structural formulas

Normal butane   *Isobutane*

Although each compound has 4 carbon atoms and 10 hydrogen atoms, the molecular structures are clearly different. Whereas the central carbon atoms in normal butane are each joined to but two other carbon atoms, the central carbon atom in isobutane is joined to three other carbons. As a result of the difference of structure, the two compounds have different properties (isobutane has a boiling point of $-12°C$ and a melting point of $-160°C$, compared with $-1°C$ and $-138°C$, respectively, for normal butane).

Different compounds having the same molecular formula are called *isomers* and to keep track of them the "straight-chain" hydrocarbon is called *normal butane* and its isomer is called *isobutane*. Although it may seem possible that there are other isomers of C_4H_{10}, these are the only ones. Other two-dimensional formulas can be written but they can be shown to be equivalent to one or the other of the above. The problem arises because the two-dimensional formulas do not take into account the three-dimensional nature of these molecules. In a saturated hydrocarbon, each carbon is tetrahedrally surrounded by four atoms, and the molecules can assume various configurations by rotation about individual bonds. The spatial relations can best be seen by the use of molecular models such as those diagramed in Figure 19.3. Of the five atomic arrangements shown, the first four correspond to the same molecule (normal butane) twisted into different shapes; the fifth corresponds to a different molecule (isobutane), and no amount of twisting can convert it to the normal isomer. Isomers always differ in properties but sometimes the differences are so slight that separation is difficult.

For the hydrocarbons having fewer than four carbon atoms there are no isomers. However, for those hydrocarbons with more than four carbon atoms there are increasing numbers of isomers. For example, C_5H_{12} stands for three different isomeric compounds; C_6H_{14} for five; C_7H_{16} for nine. As

Normal butane   *Isobutane*

Figure 19.3
Atomic arrangements of butane molecules. (Small balls represent hydrogen atoms; large balls, carbon atoms.)

Figure 19.4 PROPERTIES OF HEXANE ISOMERS		Boiling Point, °C	Melting Point, °C	Density, g/ml 25°C
Name	**Formula**			
Normal hexane	$\begin{array}{c}\text{H H H H H H}\\ \ \ \ \text{|1 |2 |3 |4 |5 |6}\\ \text{H—C—C—C—C—C—C—H}\\ \ \ \ \text{| | | | | |}\\ \text{H H H H H H}\end{array}$	69	−95	0.655
3-Methylpentane	$\begin{array}{c}\text{H}\\ \text{HCH}\\ \text{H H | H H}\\ \text{|1 |2 |3 |4 |5}\\ \text{H—C—C—C—C—C—H}\\ \text{| | | | |}\\ \text{H H H H H}\end{array}$	63	−118?	0.660
2-Methylpentane	$\begin{array}{c}\text{H}\\ \text{HCH}\\ \text{H | H H H}\\ \text{|1 |2 |3 |4 |5}\\ \text{H—C—C—C—C—C—H}\\ \text{| | | | |}\\ \text{H H H H H}\end{array}$	60	−154	0.649
2,3-Dimethylbutane	$\begin{array}{c}\text{H H}\\ \text{HCH HCH}\\ \text{H | | H}\\ \text{|1 |2 |3 |4}\\ \text{H—C—C———C—C—H}\\ \text{| | | |}\\ \text{H H H H}\end{array}$	58	−129	0.657
2,2-Dimethylbutane	$\begin{array}{c}\text{H}\\ \text{HCH}\\ \text{H | H H}\\ \text{|1 |2 |3 |4}\\ \text{H—C—C—C—C—H}\\ \text{| | | |}\\ \text{H | H H}\\ \text{HCH}\\ \text{H}\end{array}$	50	−100	0.644

the number of carbon atoms increases, the number of isomers increases even faster. There are 75 decanes ($C_{10}H_{22}$), and by the time one reaches $C_{40}H_{82}$ it has been calculated that more than 61 trillion isomers are possible.

It is instructive to consider the differences between the five hexane molecules. These are listed in Figure 19.4. The names given for the isomers reflect structural differences. For the second isomer listed, the name 3-methylpentane implies that there are five carbons in a "straight chain" (pentane) and to the third carbon is attached a methyl group (that is, CH_3 derived from methane). The third isomer differs only in that the methyl group is attached not to the third but to the second carbon from the end. The last two isomers in the table are derived from a "straight chain" of four carbon atoms (butane) with two methyl groups attached either to the second and third carbon atoms or both methyl groups to the second carbon atom. The differences in boiling points are small but the differences are significant. On the other hand, it is seen that the melting points differ more markedly. The reason for this is that molecular shape is extremely important in determining the repeat structure a crystalline solid will adopt, which in turn influences the stability of the solid and hence its melting point. Finally, although there is a difference in density between the isomers it is certainly not very large, as the table shows.

For higher-molecular-weight hydrocarbons (and some other compounds as well) there is, besides the *structural isomerism* just described, another kind of isomerism which can be important. This kind of isomerism is called *optical isomerism* and arises as follows: Consider the hydrocarbon of structural formula

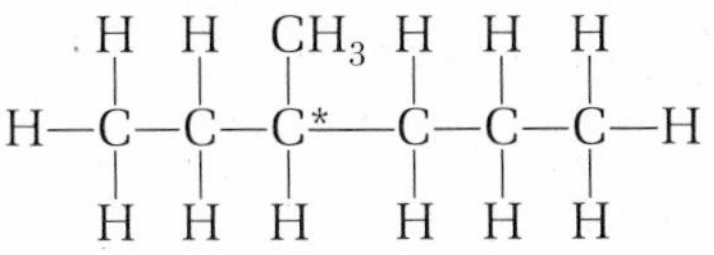

$$\begin{array}{ccccccc} & H & H & CH_3 & H & H & H \\ & | & | & | & | & | & | \\ H- & C- & C- & C^*- & C- & C- & C-H \\ & | & | & | & & | & | & | \\ & H & H & H & & H & H & H \end{array}$$

which can be abbreviated as

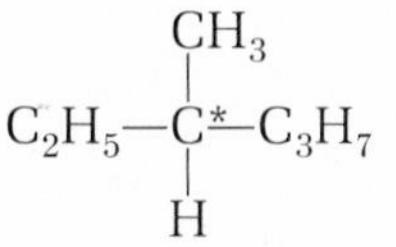

$$\begin{array}{c} CH_3 \\ | \\ C_2H_5-C^*-C_3H_7 \\ | \\ H \end{array}$$

This compound, called 3-methylhexane, is one of the nine structural isomers of heptane, but it in turn consists of two *optical* isomers. To understand this, note that the starred carbon atom is bound to four different groups (C_2H_5, ethyl; CH_3, methyl; C_3H_7, propyl; H, hydrogen). As with any carbon atom in saturated hydrocarbons, the starred carbon is tetrahedral, and its bond arrangement is like that shown in Figure 19.5. The figure shows both optical

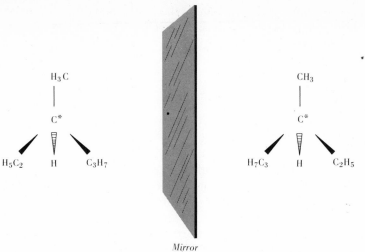

Mirror

Figure 19.5

Optical isomers of 3-methylheptane.

isomers, one of which is the mirror image of the other. It is not possible to convert one of the isomers into the other by any amount of twisting or rotating in space. For example, if the molecule on the left is rotated 180° about the CH_3—C^* bond, the C_3H_7 and C_2H_5 groups change places as in the structure on the right, but the H now projects forward and not backward as in the molecule on the right. The isomer on the left is the mirror image of that on the right in the same way that your left hand is the mirror image of your right. Anyone who has struggled to put on his gloves in the dark knows that no amount of twisting or rotation can convert one hand into its mirror image, the other hand. The two possible configurations are frequently distinguished as D (from *dextro,* right) and L (from *levo,* left).

Optical isomers are so named because they rotate the plane of polarized light differently. This effect on polarized light can be used to show the presence of an optical isomer. However, it is extremely difficult to separate one optical isomer from the other. The reason for this is that, unlike the other isomers that we have been considering, a pair of optical isomers have identical melting and boiling points and identical chemical properties except when reacting with molecules that also show optical isomerism. For example, the solubility of two optical isomers is the same in most solvents but is different in a solvent which is composed of a single optical isomer. In a few cases optical isomers separate themselves by crystallizing into separate kinds of crystals, one kind containing one of the isomers and the other kind containing the other optical isomer. This phenomenon was originally discovered by Louis Pasteur while working on compounds derived from grapes. It preceded his later work on fermentation in which he picked up the challenge of Lavoisier (Chapter 1) to understand fermentation, and in doing so laid the groundwork for modern biochemistry.

Before leaving saturated hydrocarbons, we should note that it is also possible for carbon atoms to form rings, or cycles. The simplest cyclic hydrocarbon is cyclopropane

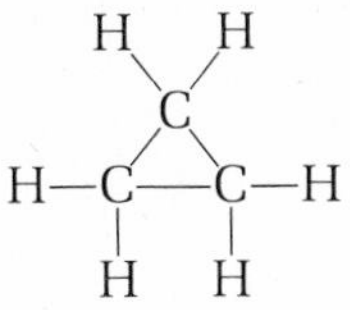

the physical properties of which are listed along with those of some other cyclic hydrocarbons in Figure 19.6. Although the tetrahedral angle preferred by carbon in its saturated hydrocarbons is 109°28′, the C_3 and C_4 cyclic

Figure 19.6 CYCLIC SATURATED HYDROCARBONS	Name	Formula	Boiling Point, °C	Melting Point, °C
	Cyclopropane	C_3H_6	-33	-127
	Cyclobutane	C_4H_8	13	-50
	Cyclopentane	C_5H_{10}	49	-94
	Cyclohexane	C_6H_{12}	81	7
	Cycloheptane	C_7H_{14}	118	-12
	Cyclooctane	C_8H_{16}	149	14

hydrocarbons show a bond angle at carbon within the ring considerably less than this, 60°, and on an average less than 90°, respectively. As a result, such molecules are said to be "strained" and are less stable than other saturated hydrocarbons. Cyclopropane, for example, needs to be handled with care because its mixtures with air (like all hydrocarbon-air mixtures) are explosive.

It is characteristic that all the hydrocarbons will burn in air. A typical reaction is

$$C_7H_{16} + 11O_2 \longrightarrow 7CO_2 + 8H_2O$$

Perhaps the principal use of hydrocarbons is as fuels in supplying our present energy needs. Large deposits of hydrocarbons (natural gas and petroleum) are buried in the earth, and these supplies are tapped by oil and gas wells. Because these deposits are limited, new sources of energy must be found before the hydrocarbons are all burned up and we are deprived not only of fuel but of all the useful organic compounds made from petroleum.

Because the hydrocarbons differ in boiling points they can be separated into a number of fractions for different uses. The lightest ones (containing one or two carbon atoms) constitute natural gas and as such are used without further purification. The next higher hydrocarbons (three and four carbon atoms) are used in bottled gas for supplying cooking and heating needs. The rather wide range of hydrocarbons boiling between 40 and 180°C, consisting of more than a hundred different compounds varying in carbon content from 6 to 10 carbon atoms per molecule, constitute various qualities of gasoline. The best hydrocarbon for use in gasoline for internal-combustion engines contains eight carbon atoms per molecule (octane). Furthermore, if the hydrocarbon chain is branched, it seems that the burning is smoother and hence has better "antiknock" qualities.

It has been found that hydrocarbons can be caused to burn more smoothly in engines if tetraethyllead, $(C_2H_5)_4Pb$, is added to them in quantities of a few milliliters per gallon. It has also been found that the use of tetraethyllead in gasoline is harmful because of the buildup of lead deposits in the combustion chamber. These deposits, however, can be greatly reduced by addition of ethylenedibromide $(C_2H_4Br_2)$. This forms lead halides which serve to remove the lead from the engine but sweep it out into the atmosphere where dangerously high concentrations of lead may accumulate, especially in heavy traffic areas. Therefore, rather than adding tetraethyllead, the less suitable hydrocarbons can be converted to more desirable ones for use as gasoline.

Along with the conversion to more desirable hydrocarbons, the fractional yield of gasoline hydrocarbons from petroleum can be increased. This can be done by breaking down even larger molecules (more than 10 carbon atoms) in a process called *catalytic cracking*. Also, small molecules (fewer than six carbon atoms) can be combined in a process called *catalytic reforming*. For all these hydrocarbon-rearrangement processes, platinum is generally the catalyst, and a major fraction of the world's platinum is used in making catalysts for the petroleum industry.

The fraction of petroleum which boils above 180° is used as kerosene, diesel, and furnace fuel. From this fraction also comes jet fuel, which because it is less volatile is safer to handle than gasoline. The highest boiling fraction (hydrocarbons containing about 18 carbon atoms per molecule and more) is used for lubricants.

19.3   Unsaturated Hydrocarbons

As we saw in Chapter 4, two carbon atoms can share more than two electrons in the formation of multiple, or π (pi), bonds. Molecules which contain carbon-carbon double or triple bonds are called *unsaturated hydrocarbons*. Those containing one or more double bonds are called *alkenes*, or *olefins* (from Latin *oleum,* oil, since vegetable oils contain double bonds). The simplest example of an alkene is ethylene

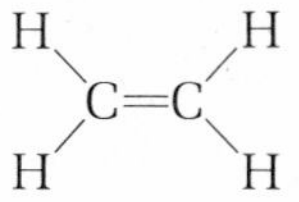

The substitution of a methyl group for one of the hydrogens in ethylene would give *propene.* Substitution for a second hydrogen would give an isomer of *butene.* There are four isomers of butene which are designated as follows:

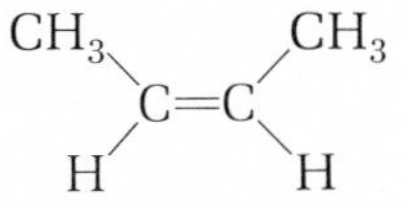

cis-2-Butene

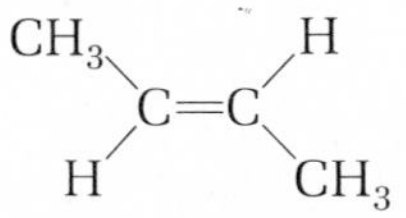

trans-2-Butene

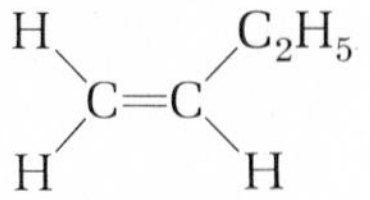

1-Butene

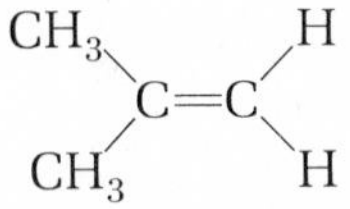

2-Methylpropene

In considering these isomers it is important to note that, whereas rotation about a carbon-to-carbon single bond is relatively easy, rotation about a carbon-carbon double bond is extremely difficult. Two molecules which differ in their configuration about a double bond are not easily convertible from one to the other and so are separate substances. For the butenes shown above, the first two differ only in that the two methyl groups lie either on the same side of the double bond or on opposite sides of the double bond. If rotation were as easy about a double bond as it is about a single bond, these two would be identical. However, for double bonds, rotation is not possible at ordinary temperatures and so the molecules are separate isomers. In naming them, the molecule in which the two methyl groups are on the same side is called *cis* and that in which the methyl groups are on opposite sides is called *trans.* In writing the name of the compound, that of the corresponding saturated hydrocarbon is used except that the ending -ane is changed to -ene. In order to denote the position of the double bond, a number is used. This number is found by counting the carbon-carbon bonds along the chain from the end closest to the double bond. For the *cis* and *trans* butenes, the number is 2. For the third isomer shown, it is the first bond that is double; hence, this isomer is called 1-butene. Finally, the fourth isomer shown consists of a branched chain. To name this isomer, the convention is to pick the longest carbon chain (in this case three) and name the compound accordingly (propene). The position of the branch is shown by a number preceding the name of the substituent hydrocarbon group. The numbering proceeds from the end nearest the double bond. For example,

$$CH_3-CH{=}CH-\overset{\overset{\displaystyle CH_3}{|}}{CH}-CH_3$$

is 4-methyl-2-pentene. The series of hydrocarbons containing a single double bond can be represented by the general formula C_nH_{2n}.

The presence of a double bond in a hydrocarbon introduces several other differences in the structure of the overall molecule. In the first place, the bond angles are affected. Whereas the joining of four groups to a carbon atom places them all at the corners of a tetrahedron, the joining of three groups to a carbon atom, as by formation of one double bond and two single bonds, places the three attached groups at the corners of a triangle, obviously all in the same plane. As an example, ethylene is a planar molecule, in that all six atoms lie in a plane; the bond angles are all 120°. Furthermore, the carbon-carbon distance is shorter for double bonds than for single bonds. In ethane it is 0.154 nm; in ethylene, 0.133 nm. Also, carbon-carbon double bonds are stronger than carbon-carbon single bonds in the sense that more energy is required to break a double bond than to break a single bond. In terms of kilojoules per mole, the energy required to break a carbon-carbon single bond averages about 350 kilojoules whereas that for a double bond averages some 610 kilojoules/mole.

Just as two pairs of electrons can be shared between carbon atoms, three pairs can also be shared to form triple bonds. Compounds containing carbon-carbon triple bonds are called *alkynes*. The simplest such compound is acetylene $HC{\equiv}CH$. In this case all four of the atoms lie in a straight line. The triple bond is shorter still (0.120 nm) than the double bond, and higher energies are required to break it (840 kilojoules/mole). Although the question of rotation about the carbon-carbon triple bond does not arise because only one group can be joined to the carbon at each end of the bond, it should be noted that, as with double bonds, the presence of triple bonds in molecules can create the possibility of isomers differing in the position of the multiple bond. For example, in the triple-bonded molecule C_4H_6, there are two isomers: $HC{\equiv}C-CH_2CH_3$ and $CH_3-C{\equiv}C-CH_3$. The first of these compounds is called 1-butyne (or ethylacetylene); it has a normal boiling point of 8° and a melting point of −126°. The second is called 2-butyne (or dimethylacetylene); it has a boiling point of 27° and a melting point of −32°. It is interesting that within experimental error the heats of combustion of these two compounds are the same, that is, ΔH for the reaction

$$C_4H_6 + 11\tfrac{1}{2}\,O_2 \longrightarrow 4CO_2 + 3H_2O$$

is -2596 kilojoules/mole for either isomer. One can begin to realize the complexities of organic chemistry by noting that a particular hydrocarbon molecule can contain many multiple bonds, some of which may be double and some triple. The number of isomeric possibilities becomes tremendous as the number of carbon atoms increases. Fortunately, it is not necessary to consider in detail all the complexities of the subject in order to obtain a good appreciation for its importance.

19.4   Nomenclature

Before proceeding further, it will be advisable to review the nomenclature used for the hydrocarbons discussed thus far. The systematic nomenclature of organic compounds is based on that for the parent hydrocarbon. Thus, we will summarize the rules which have been agreed upon by IUPAC (International Union of Pure and Applied Chemistry) to untangle most of the problems associated with the naming of complicated compounds. Through the use of such rules, all compounds can be named. Unfortunately, some of the names become more involved than the formulas themselves. In such cases, a "common" or "trivial" name is used instead. Systematic nomenclature has the advantage that only a few rules must be remembered in order to name millions of compounds.

1   The longest hydrocarbon chain in the molecule serves as the basis for the name.

2   If the compound has no double or triple bonds, the ending -*ane* is used; if a double bond is present, -*ene* is used; if two double bonds, -*diene;* if three double bonds, -*triene;* if a triple bond, -*yne.*

3   Carbon atoms of the parent structure are numbered beginning at the end nearest a multiple bond if one is present, or branches (attached hydrocarbon groups) if no multiple bonds are present.

4   Numbers are placed before the name of the parent structure to show the location of multiple bonds, and before each different substituent to show its location.

5   Substituent hydrocarbon groups are named by dropping the suffix -*ane* from the name of the corresponding alkane and adding -*yl.*

6   If more than one substituent is present, they are placed in alphabetical order.

These rules are illustrated by the systematic naming of the following hydro-carbons:

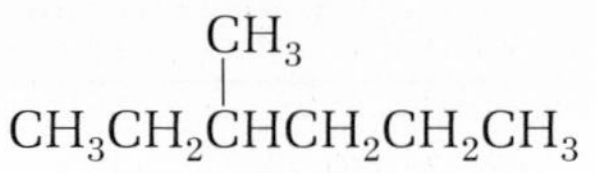

$$\overset{\displaystyle CH_3}{\overset{|}{CH_3CH_2CHCH_2CH_2CH_3}}$$

3-Methylhexane

$$CH_3{=}CHCH{=}CHCH_3$$

1,3-Pentadiene

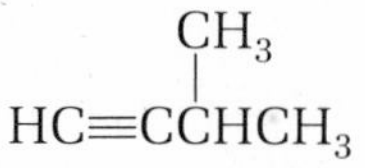

$$\overset{\displaystyle CH_3}{\overset{|}{HC{\equiv}CCHCH_3}}$$

3-Methyl-1-butyne

$$\overset{\displaystyle CH_3 \qquad CH_2CH_3}{\overset{|\qquad\quad\ |}{CH_3CH_2CHCH_2CH(CH_2)_2CH_3}}$$

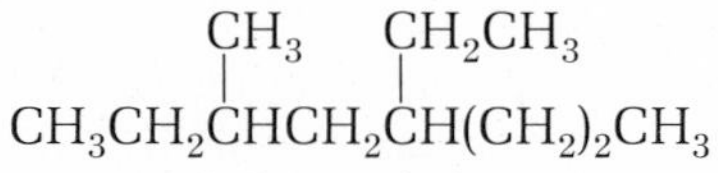

5-Ethyl-3-methyloctane

Note that in each of these cases, the molecule is first "stretched out" to give the longest chain possible. In the first example, the molecule can be pictured as

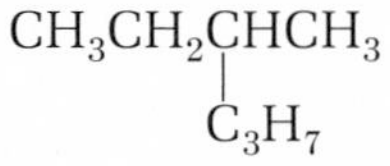

$$\overset{\displaystyle CH_3CH_2CHCH_3}{\overset{|}{\quad\quad\ C_3H_7}}$$

or

$$\overset{\displaystyle CH_3CHCH_2CH_2CH_3}{\overset{|}{\quad\ C_2H_5}}$$

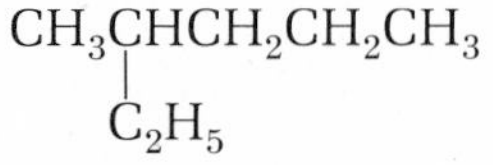

but neither of these gives the longest hydrocarbon chain (6 C's) and so neither is acceptable for naming.

For the second molecule shown, numbering might start from the other end, making it 2,4-pentadiene. This however is wrong since numbering starts at the end closest to a multiple bond, or in the first example where there is none, nearest the branching. If both unsaturation and branching are present, as in the third example, the end nearer the multiple bond is picked to start numbering. Thus 3-methyl-1-butyne is correct, not 2-methyl-3-butyne.

Finally, the last example illustrates that even though the methyl group (as the branch closer to an end of the longest chain) determines the end from which numbering starts, it is placed after "ethyl" in the name since substituents are placed alphabetically.

19.5   Aromatic Hydrocarbons

The combination of multiple bonds and cyclization can produce an important class of compounds called *aromatic hydrocarbons*. They are so named because they are generally characterized by a pleasant aroma. However, the aroma is deceptive in that most of the compounds are extremely poisonous and many are potent carcinogens.

A key to understanding these hydrocarbons is provided by the basic member of the group, benzene, C_6H_6. In the benzene molecule the six carbon atoms are at the corners of a hexagon with the six hydrogen atoms attached to each carbon so that all twelve atoms lie in one plane. The carbon-carbon bonds are all equivalent, and therefore the molecule can be considered as a resonance hybrid (Section 4.5). The contributing structures are usually written as shown in Figure 19.7. It is important to note that the carbon-carbon distances do not alternate between being short and long but are all equal

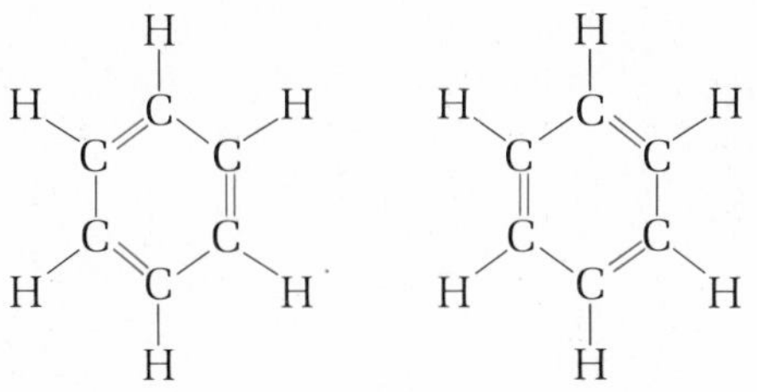

Figure 19.7
Contributing resonance structures of benzene.

to 0.140 nm, intermediate between a single and a double bond. The heat of combustion for benzene has been determined to be 3301 kilojoules/mole, or 146 kilojoules less than would be anticipated for a cyclic compound consisting of alternate single and double bonds. These special properties of benzene have led not only to the original theory of resonance but to a variety of molecular-orbital theories for bonding in these and related compounds (Section 4.7). For simplicity, only one resonance structure is normally drawn for benzene and related compounds, or sometimes the representation of Figure 4.13c is used. In any case, the benzene grouping of atoms is an especially important one which shows up in a number of aromatic compounds.

Hydrocarbon derivatives of benzene consist of molecules in which one or more of the hydrogen atoms have been replaced by hydrocarbon groups. The simplest of these derivatives is toluene ($C_6H_5CH_3$, methylbenzene) in which one of the hydrogens has been replaced by a methyl group. The formulas for several benzene derivatives are illustrated in Figure 19.8. For simplicity only one of the contributing structures of the benzene ring is shown. A further abbreviation is used in the figure in that the C and H atoms of the benzene ring are omitted. These abbreviations are in common usage.

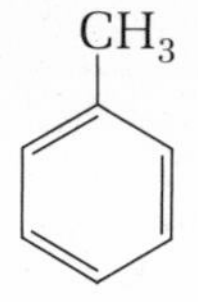

Toluene

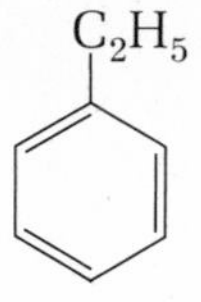

Ethylbenzene

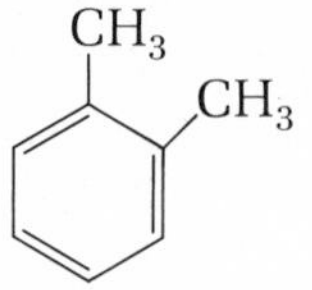

1,2-*Dimethylbenzene*
(*ortho*)

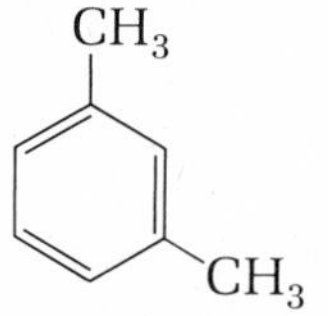

1,3-*Dimethylbenzene*
(*meta*)

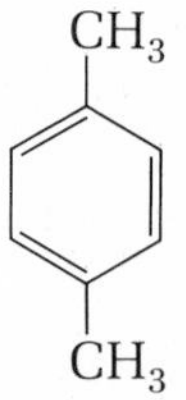

1,4-*Dimethylbenzene*
(*para*)

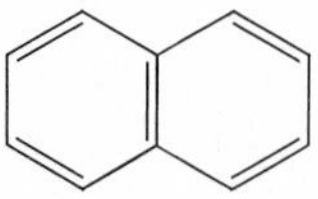

Naphthalene

Figure 19.8
Aromatic hydrocarbons.

It is interesting to note that we have here another example of isomerism in the hydrocarbons. Ethylbenzene and the three dimethylbenzenes are all isomers of molecular formula C_8H_{10}. Although the difference between ethylbenzene and dimethylbenzene is obvious, that between the three dimethylbenzenes lies in the fact that the two methyl groups are, respectively, on adjacent carbon atoms (1,2-dimethylbenzene), on carbon atoms separated by an intervening carbon (1,3-dimethylbenzene), or on carbon atoms which lie directly across the ring from each other (1,4-dimethylbenzene). The numbers designate which C atoms of the ring bear the methyl groups. The isomers are sometimes called *ortho*-, *meta*-, and *para*dimethylbenzene, respectively. That these are indeed isomers and not identical compounds is shown by the fact that their boiling points are different, 144, 139, and 138°C, respectively, as compared with 136°C for ethylbenzene. As in the earlier isomer examples mentioned, there is an even greater difference in the melting points. The melting point of paradimethylbenzene is 13° whereas all the other isomers melt well below zero.

Figure 19.8 also includes the compound naphthalene which consists of two benzene rings fused together along a common edge. This is just one of many examples of compounds which consist of networks of benzene rings. It might be added that in the process of fusing the rings hydrogen atoms are lost, so that as more and more rings are joined together one reaches, in the limit, a substance containing only carbon atoms joined in rings with no hydrogens remaining. This is the substance graphite which consists of sheets of carbon atoms each of which is joined to three other carbons (Figure 17.2).

In naming aromatic compounds, the name of the parent cyclic structure (e.g., benzene, naphthalene) is frequently used. However, it is common practice to call a singly substituted benzene, C_6H_5X, by a name which reflects the presence of the C_6H_5 or *phenyl* group.

19.6   Alcohols and Ethers

As we mentioned at the beginning of this chapter, in addition to the hydrocarbons, organic compounds consist of derivatives of hydrocarbons in which one or more H atoms are replaced by atoms of other elements. Such substitution, to form a functional group attached to a hydrocarbon residue, alters the properties markedly. A good first example is the class of organic compounds called *alcohols*, represented by the shorthand notation ROH. R stands for the hydrocarbon residue that is attached to the OH functional group. We should note that the identity of the hydrocarbon residue R fixes the specific name of the compound and also may modify the class properties. Thus, if the hydrocarbon residue in ROH is the methyl group, CH_3, related to the hydrocarbon methane, the alcohol CH_3OH is called methyl alcohol, or methanol. Figure 19.9 lists a few of the simpler alcohols and describes some of their properties. The names of the last three alcohols in the table

Figure 19.9 ALCOHOLS	Formula	Name	Boiling Point, °C	Melting Point, °C
	CH_3OH	Methyl alcohol	65	−98
	CH_3CH_2OH	Ethyl alcohol	79	−117
	$CH_3CH_2CH_2OH$	n-Propyl alcohol	97	−127
	$CH_3CH_2CH_2CH_2OH$	n-Butyl alcohol	118	−90
	$CH_3CH_2CH_2CH_2CH_2OH$	n-Pentyl alcohol	137	−79

are prefixed by a small letter *n* which stands for *normal*. This means that the alcohol has a structure in which the OH functional group is attached at the end of the chain of carbon atoms. In each case other isomers are possible. For example, there is an isomer of *n*-propyl alcohol called isopropyl alcohol (frequently used as rubbing alcohol) in which the OH functional group is attached not at the end of the string of three carbon atoms but at the middle:

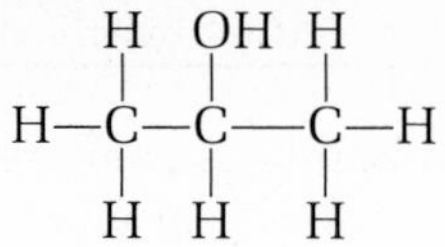

$$
\begin{array}{ccc}
\text{H} & \text{OH} & \text{H} \\
| & | & | \\
\text{H} - \text{C} - \text{C} - \text{C} - \text{H} \\
| & | & | \\
\text{H} & \text{H} & \text{H}
\end{array}
$$

In the IUPAC system of nomenclature discussed in Section 19.4, alcohols are named by replacing the terminal -e of the hydrocarbon name by -ol and prefixing it with a number to indicate the position of the OH group. Thus, n-propyl alcohol is 1-propanol; isopropyl alcohol is 2-propanol. (Organic nomenclature is summarized in Appendix 1.2.)

Alcohols in which the OH group is attached at the end of the chain are called *primary alcohols* and contain the group —CH_2OH. Alcohols in which the OH is attached as in isopropyl alcohol are called *secondary alcohols* and contain the grouping HĊOH. If the hydrocarbon residue contains four or more carbon atoms, a third class of alcohols is possible in which the functional group is joined to a carbon atom that is bound in turn to three other carbons. Such an alcohol is called a *tertiary alcohol;* the simplest of these is tertiary butyl alcohol (or 2-methyl-2-propanol) the structural formula of which can be represented

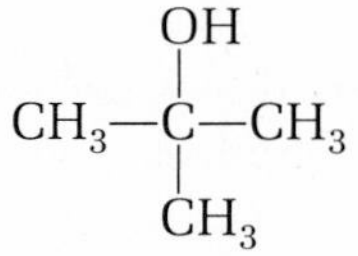

$$
\begin{array}{c}
\text{OH} \\
| \\
\text{CH}_3 - \text{C} - \text{CH}_3 \\
| \\
\text{CH}_3
\end{array}
$$

It is seen from Figure 19.9 that alcohols have a characteristically large liquid range. Compared with a hydrocarbon of similar structure, the alcohol will always have the higher boiling point. In this regard alcohols resemble water, from which they can be considered to be derived through substitution of a hydrocarbon residue for one of the hydrogen atoms. Therefore, it is possible for alcohols, like water, to form hydrogen bonded structures that stabilize the liquid state with respect to the gaseous state. This situation is less complex than in water (Section 14.11) because, as shown in Figure 19.10,

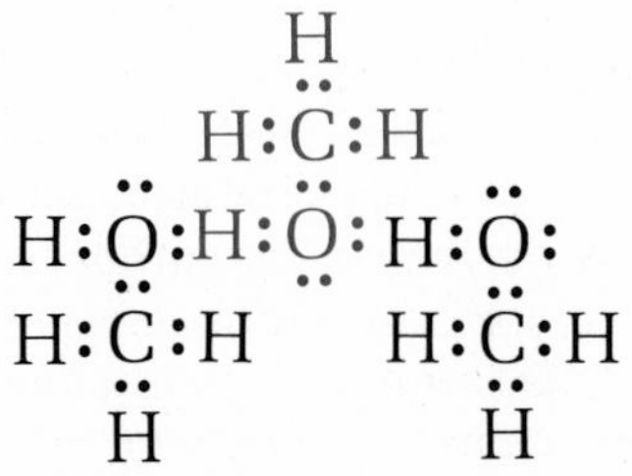

Figure 19.10
Hydrogen bonding in alcohol.

the hydrogen atoms of the methyl groups do not form hydrogen bonds to any appreciable extent.

Because of the possibility of hydrogen bonding, alcohols are among the relatively few nonionic substances that are soluble in water. The first three alcohols listed in the table are completely miscible with water. However, once the hydrocarbon residue becomes larger it exerts greater influence and solubility becomes limited. Thus for pentanol the solubility is but 0.3 mole per liter of water. Higher alcohols are even less soluble.

From the general formula ROH an alcohol might appear to be a base. However, the R—O bond has very little tendency to dissociate in solution. In fact, it has less tendency to dissociate than does the O—H bond. If anything, alcohols are weak acids. However, the acidity of alcohols is even less than that of water. For ethanol the dissociation constant is approximately 7.3×10^{-20}. Weak as it is, it is possible to make salts from alcohols. For example, if sodium is dissolved in an alcohol, hydrogen is liberated:

$$2Na(s) + 2C_2H_5OH \longrightarrow H_2(g) + 2Na^+ + 2C_2H_5O^-$$

Evaporation of the excess alcohol produces a white solid, sodium ethoxide ($NaOC_2H_5$). In water it is completely hydrolyzed to C_2H_5OH:

$$NaOC_2H_5(s) + H_2O \longrightarrow Na^+ + OH^- + C_2H_5OH$$

Just as alcohols can be considered in a formal way to be related to water through replacement of one of the H atoms by a hydrocarbon residue, *ethers,* a second class of organic compounds, can be considered as related to water through replacement of both of the H atoms. Ethers have the general formula R—O—R. In R—O—R the molecule is bent at the oxygen just as it is in H—O—H and R—O—H. Hence ethers, like alcohols and water, must possess a dipole moment. Despite this fact, ethers do not have the waterlike properties characteristic of the lower alcohols. For example, dimethyl ether, CH_3OCH_3, which incidentally is a structural isomer of ethanol, has a boiling point of $-25°C$. The more common diethyl ether, $C_2H_5OC_2H_5$, also has a relatively low boiling point, $35°C$, and neither of them is appreciably soluble in water. The lack of waterlike properties results from the lack of hydrogen bonding and indicates the importance of hydrogen bonding in such properties.

In much of our discussion thus far we have emphasized the relationship between the structure of derivative organic molecules and their parent hydrocarbons. However, we should not draw the false conclusion that this is the method whereby hydrocarbon derivatives are prepared. Methanol, for example, is prepared on a large scale by combining carbon monoxide and hydrogen at high pressures (200 to 300 atm) and high temperatures (300 to 400°C) in the presence of catalysts such as $ZnO-Cr_2O_3$ mixtures:

$$CO(g) + 2H_2(g) \longrightarrow CH_3OH(g)$$

Ethanol, which is sometimes called grain alcohol, results from the fermentation of sugars. In the fermentation process, enzymes, e.g., zymase, which are secreted by yeast cells, act as catalysts for the breakdown of sugar molecules. For the sugar glucose the reaction is

$$C_6H_{12}O_6 \xrightarrow{\text{zymase}} 2CO_2(g) + 2C_2H_5OH$$

The fermentation reaction is carried out on a tremendous scale in the production of alcoholic beverages. The process is a very old one; in fact, the oldest document extant is a Babylonian tablet that gives a recipe for making beer. We have seen (Chapter 1) the importance of the reaction in the development of chemistry.

Beer is essentially a dilute solution of ethyl alcohol and carbohydrates in water; a typical American beer might analyze 4% ethanol, 5% carbohydrates, and 91% H_2O. Beer can be made by fermenting sugar, as from sprouted barley (called malt), and adding sufficient hops (dried flowers of the hop plant) to produce the characteristic flavor. If the sugar comes instead from grape juice, the product is wine. In both cases, the yeast necessary to catalyze the sugar fermentation is eventually killed by the product alcohol so that the percentage content of ethanol is limited. Most wines are about 12 to 14% alcohol by volume; higher alcohol content can be achieved by adding alcohol (to give a "fortified wine" such as sherry) or by distillation. Distillation of wine produces brandy; distillation of fermented grain produces whiskey.†

Numerous organic compounds contain more than one functional group. A simple example is ethylene glycol (1,2-ethanediol, $HOCH_2CH_2OH$), which when drawn is:

$$
\begin{array}{ccc}
\text{H} & & \text{H} \\
| & & | \\
\text{O} & & \text{O} \\
| & & | \\
\text{H}-\text{C}- & \text{C} & -\text{H} \\
| & & | \\
\text{H} & & \text{H}
\end{array}
$$

The molecule contains two alcohol functional groups, and therefore it is not surprising that ethylene glycol is extremely soluble in water. This solubility

---

†Although history records numerous attempts to stamp out the consumption of alcoholic beverages, history also records the importance assigned to alcoholic beverages by various cultures. An illustrative fact from the log of the *Mayflower* is that the Pilgrims landed at Plymouth Rock because "we could not take time for further search . . . our victuals being much spent . . . especially our beer."

allows its extensive use as "permanent antifreeze." The trialcohol glycerol (1,2,3-propanetriol)

$$\begin{array}{ccc} H & H & H \\ | & | & | \\ O & O & O \\ | & | & | \\ H-C-C-C-H \\ | & | & | \\ H & H & H \end{array}$$

plays an important biological role, as we shall see in Chapter 21, in the formation of fats, or lipids. Although glycerol can be prepared from these natural sources, ethylene glycol is made synthetically by some rather interesting reactions. In the first step, ethylene is oxidized in the presence of silver catalyst.

$$H_2C{=}CH_2 + \tfrac{1}{2}O_2 \longrightarrow H_2\overset{\displaystyle O}{\overset{\diagup\,\diagdown}{C}}{-}CH_2$$

The product, ethylene oxide, is an example of an ether in which the oxygen functional group forms part of a cycle. Ethylene oxide is then hydrated to form ethylene glycol

$$H_2\overset{\displaystyle O}{\overset{\diagup\,\diagdown}{C}}{-}CH_2 + H_2O \longrightarrow H_2\overset{\begin{array}{cc} H & H \\ | & | \\ O & O \\ | & | \end{array}}{C}{-}CH_2$$

In this reaction a rather special ether has reacted with water to form an alcohol. It is much more usual, in general, to have the reverse change in which an alcohol is dehydrated to give an ether. In this kind of dehydration reaction two molecules react in the presence of strong acids, which serve both to catalyze the reaction and to remove the product water.

$$C_2H_5OH + HOC_2H_5 \xrightarrow[\text{H}_2\text{SO}_4]{\text{conc.}} C_2H_5{-}O{-}C_2H_5 + H_2O$$

On prolonged standing in air, ethyl ether can become extremely hazardous because it begins to form an explosive peroxide. Consequently before an ether is distilled it should be tested for the presence of peroxides. The presence of peroxides can be detected by formation of a red color (due to $FeNCS^{2+}$) on addition of ferrous sulfate and potassium thiocyanate. The peroxides can be destroyed by addition of reducing agents such as $FeSO_4$. To prevent peroxide formation, ether is often stored with iron wire in it.

Alcohols also can undergo oxidation reactions. Specifically they can be oxidized by oxidizing agents such as sodium dichromate. That the reaction is an oxidation can be seen from the fact that in the process hydrogen atoms are removed from the alcohol. As a specific example, when ethyl alcohol is so oxidized it forms acetaldehyde

Ethyl alcohol   *Acetaldehyde*

for which the balanced equation with aqueous dichromate is

$$3CH_3CH_2OH + Cr_2O_7{}^{2-} + 8H^+ \longrightarrow 3CH_3CHO(g) + 2Cr^{3+} + 7H_2O$$

In carrying out the reaction the mixture is heated, partly because the reaction is a slow one and partly so that the volatile aldehyde product will leave the reaction mixture before it can be oxidized further. (As we shall see, aldehydes themselves are oxidized by the same oxidizing agent sodium dichromate to form carboxylic acids.) A similar oxidation reaction occurs for secondary alcohols; however, the products formed represent a different class of organic molecules. For example, if isopropyl alcohol is oxidized it will form acetone

The product here is called a *ketone* and differs from the product of the oxidation of primary alcohol in that it contains the functional group $-\overset{\displaystyle O}{\overset{\|}{C}}-$ characteristic of ketones instead of the functional group $-\overset{\displaystyle O}{\overset{\|}{C}}H$ characteristic of aldehydes. Tertiary alcohols do not undergo a corresponding oxidation. The reason for this is that oxidation cannot occur at the functional group without breaking the carbon-carbon chain.

Finally, we should note one other compound which appears to be an alcohol but in which the hydrocarbon residue has such profound influence on its properties that it is usually not even considered to be an alcohol. This is the compound phenol

Because the functional group is attached directly to the aromatic ring, the properties of phenol are quite different from those of typical alcohols. Specifically it is much more acidic than an alcohol, having a dissociation constant of 1.3×10^{-10}. In recognition of this fact, phenol is often called carbolic acid. It is used as a general disinfectant, and gram amounts may prove fatal to human beings.

19.7 Aldehydes and Ketones

Aldehydes and ketones, which are obtained by the oxidation of alcohols, contain the C=O, or *carbonyl*, grouping. The carbonyl double bond is a very strong one. It has a bond energy of 750 kilojoules/mole. This is indeed more than twice the bond energy of a carbon-oxygen single bond (370 kilojoules/mole). In contrast, the carbon-carbon double bond has a bond energy of 610 kilojoules/mole, which is less than twice that of the carbon-carbon single bond, 350 kilojoules/mole. Part of the reason for the great strength of the carbonyl bond is the fact that electrons are shared unequally, due to the greater electronegativity of oxygen; the oxygen end is more negative than the carbon end, so that there is additional attraction. This polar character is reflected in the fact that carbonyls are, in general, relatively reactive, despite the strong bond.

The simplest of the carbonyl compounds is formaldehyde, H_2CO, also written HCHO. Like the other lower aldehydes, formaldehyde has a sharp irritating odor. The higher aldehydes, on the other hand, such as $C_8H_{17}CHO$, have flowery odors and so are used in perfumery. Formaldehyde is produced by oxidation of methanol, as by air in the presence of hot copper catalyst. It is a gas at room temperature, having a boiling point of $-21°C$. However, it has a great tendency to polymerize, sometimes with explosive violence, to form a solid:

$$n \; \overset{\displaystyle H}{\underset{\displaystyle H}{\diagdown}}C{=}O \longrightarrow \cdots \overset{\displaystyle H}{\underset{\displaystyle H}{\overset{|}{\underset{|}{C}}}}{-}O{-}\overset{\displaystyle H}{\underset{\displaystyle H}{\overset{|}{\underset{|}{C}}}}{-}O{-}\overset{\displaystyle H}{\underset{\displaystyle H}{\overset{|}{\underset{|}{C}}}}{-}O{-}\overset{\displaystyle H}{\underset{\displaystyle H}{\overset{|}{\underset{|}{C}}}}{-}O \cdots$$

The question might arise as to how the ends of the polymeric chain (which may contain up to 100 units) are terminated. Usually sufficient water is present to add an OH at one end and an H at the other. However, if organic groups are used to cap the ends of the chain, more useful polymers are obtained. Formaldehyde is not unique among the aldehydes in forming polymers.

Formaldehyde is probably best known as its aqueous solution called *formalin.* Because H_2CO is toxic to lower forms of life, formalin finds wide application as a disinfectant and for the preservation of biological specimens. Formaldehyde has been detected in interstellar dust clouds where its presence has been taken to indicate that organic compounds not only can exist in the absence of life but compounds such as formaldehyde may be involved in the emergence of life.

Like other aldehydes, but not ketones, formaldehyde is easily oxidized to a *carboxylic acid* which contains the acid functional group $-\overset{\overset{\displaystyle O}{\|}}{C}-OH$, also written COOH. The oxidation occurs readily in basic solution even with mild oxidizing agents such as silver ion. Reaction with silver ion is frequently employed to determine the presence of aldehyde functional groups in organic molecules. In performing the test, it is necessary to have the silver ion dissolved in a basic solution, as through formation of the complex $Ag(NH_3)_2^+$. On reduction, the silver is deposited as a mirror on the surface of the container. The silver mirror test is thus conveniently employed to distinguish aldehydes. In an alternative test, a reddish deposit of Cu_2O is formed by reduction of a cupric complex by the aldehyde.

Acetone (CH_3COCH_3), also called dimethyl ketone or 2-propanone, is the most important ketone. It can be made by oxidizing or, what amounts to the same thing, by dehydrogenating (removing hydrogen from) isopropyl alcohol in the presence of a copper catalyst. The change can be written

$$H-\overset{\overset{\displaystyle H}{|}}{\underset{\underset{\displaystyle H}{|}}{C}}-\overset{\overset{\displaystyle H}{|}}{\underset{\underset{\displaystyle O}{|}}{C}}-\overset{\overset{\displaystyle H}{|}}{\underset{\underset{\displaystyle H}{|}}{C}}-H \xrightarrow[200-300°C]{\overset{Cu}{catalyst}} H-\overset{\overset{\displaystyle H}{|}}{\underset{\underset{\displaystyle H}{|}}{C}}-\overset{\overset{\displaystyle }{}}{\underset{\underset{\displaystyle O}{\|}}{C}}-\overset{\overset{\displaystyle H}{|}}{\underset{\underset{\displaystyle H}{|}}{C}}-H + H_2$$

Isopropyl   Acetone
alcohol

At room temperature, acetone is a liquid, but it is quite volatile (boiling point 56°C). Like alcohols and aldehydes, it catches fire rather easily. Because it is a good solvent for organic compounds, it is extensively used to dissolve varnishes, plastics, etc. It is characteristic of ketones that they are less easily oxidized than aldehydes and do not give any reaction with mild oxidizing agents such as silver or cupric ion. With a stronger oxidizing agent, cleavage of the carbon-carbon chain occurs.

Both aldehydes and ketones can be reduced to primary and secondary alcohols, respectively. Reducing agents include $LiAlH_4$ (lithium aluminum hydride) or hydrogen on nickel catalyst.

19.8 Acids and Esters

Organic acids contain the carboxyl group COOH. Acetic acid, CH_3COOH, is already familiar to us. It and some of the other simple carboxylic acids are listed in Figure 19.11. Formic acid occurs in ants and gives sting to their bites; acetic acid is in vinegar and in the defense spray of many insects;

Figure 19.11 CARBOXYLIC ACIDS	Formula	Name	Boiling Point, °C	K_{diss}
	HCOOH	Formic acid	101	1.8×10^{-4}
	CH_3COOH	Acetic acid	118	1.8×10^{-5}
	CH_3CH_2COOH	Propionic acid	141	1.3×10^{-5}
	$CH_3CH_2CH_2COOH$	n-Butyric acid	164	1.5×10^{-5}
	$(CH_3)_2CHCOOH$	Isobutyric acid	154	1.4×10^{-5}

propionic acid is formed on heating wood; butyric acid gives rancid butter its terrible smell. All the acids listed are very soluble in water. In fact, all but the last are miscible with water in all proportions. However, as the chain gets longer the solubility drops off, as with the alcohols. Also, like the alcohols, the acids have reasonably high boiling points.

The acid dissociation constants for the carboxylic acids are listed in Figure 19.11. It is seen that except for formic acid the constants are all nearly the same. In general, unless molecules are substituted close to the functional group, the properties of that functional group will not be changed. However, one can see the influence of substitution on acid properties by considering the chlorine-substituted acetic acids. Monochloroacetic acid ($ClCH_2COOH$) has an acid dissociation constant (1.4×10^{-3}) nearly 100 times greater than that of acetic acid. Substitution of a second chlorine atom to give dichloroacetic acid ($Cl_2CHCOOH$) further increases the dissociation constant (3.3×10^{-2}). Finally, substitution of a third chlorine atom to form trichloroacetic acid (Cl_3CCOOH) increases the acid dissociation constant to that of a moderately strong acid (2.0×10^{-1}).

The effect of substituents on properties of a functional group is called the *inductive effect*. It can be understood by noting that the very electronegative chlorine atoms pull electron density toward themselves at the expense of the electron density in the oxygen-hydrogen bond (of the COOH group). Weakening the O—H bond causes a noticeable increase in acid strength. There is a similar effect shown in the series of dicarboxylic acids. Some of these acids are listed in Figure 19.12. The presence of a second carboxylic acid functional group has an inductive effect on the first very similar to chlorine substitution. As the first dissociation constants (K_1) listed in Figure 19.12 show, the effect is greatest when the second carboxylic group

Formula	Name	K_1	K_2
HOOCCOOH	Oxalic acid	5.9×10^{-2}	6.4×10^{-5}
HOOCCH$_2$COOH	Malonic acid	1.5×10^{-3}	2.0×10^{-6}
HOOCCH$_2$CH$_2$COOH	Succinic acid	6.5×10^{-5}	3.3×10^{-6}

Figure 19.12 DICARBOXYLIC ACIDS

is closest to the first. As the two groups are separated progressively by intervening CH$_2$ groups the effect dies out. There is a somewhat similar effect in the second dissociation constant, which is also listed; however, this value is complicated by the difficulty of pulling a proton off an already negative ion.

Salts of carboxylic acids can be formed either by reaction with a base or by direct reaction of the acid with a metal to release hydrogen. The hydrogen evolution is slower than that observed for a strong inorganic acid. Of the salts of organic acids the acetates are commonly encountered in laboratory work. A less obvious example of the salt of an organic acid is soap. These interesting molecules are the salts of long-chain organic acids such as C$_{17}$H$_{35}$COONa (sodium stearate). This substance is remarkable because the stearate ion is practically a hydrocarbon but has a charged group at one end. When placed in water, the ions do not really dissolve, because hydrocarbons are insoluble in polar solvents. Instead, *micelles* are formed in which the hydrocarbon parts of the stearate ions cluster together, as shown in Figure 19.13. The negative charges at the surface of the micelle are dissolved in the water; the hydrocarbon chains in the interior are dissolved in each other. X-ray investigations of soap suspensions show that at low concentrations the micelles are approximately spherical with a diameter of about 5 nm. The cleansing action of soap is thought to stem from the dissolving of grease (essentially hydrocarbon in nature) in these hydrocarbon clusters. Also important in cleansing action is the fact that the large soap micelles modify the structure of water (Section 14.11) by breaking hydrogen

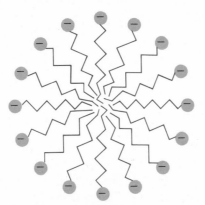

Figure 19.13
Soap micelle consisting of negative carboxyl groups attached to long-chain hydrocarbons.

bonds. The water thus affected is better able to wet foreign particles.

Characteristically organic acids react with alcohols to form compounds called *esters*. Esterification reactions can be described by the general equation

$$R-\overset{\overset{\displaystyle O}{\|}}{C}-O-H + R'-O-H \longrightarrow R-\overset{\overset{\displaystyle O}{\|}}{C}-O-R' + H_2O$$

which shows the splitting out of H_2O and the formation of an ester, RCOOR'. This reaction is catalyzed by the presence of strong acids. The acid catalysis can be understood by postulating a reaction intermediate which is formed in low concentrations from the protonation of the carboxylic acid to form $RC(OH)_2{}^+$. This positive ion attacks the alcohol molecule to remove a proton from the alcohol by combining it with one of the OH groups of the positive ion. Subsequently, the remaining proton is lost to give the ester and regenerate the acid catalyst:

$$R-\overset{\overset{\displaystyle OH}{|}}{\underset{\underset{\displaystyle OH}{|}}{C^+}} + HOR' \longrightarrow H_2O + R-\overset{\overset{\displaystyle +}{}}{\underset{\underset{\displaystyle OH}{|}}{C}}-OR' \longrightarrow R-\overset{}{\underset{\underset{\displaystyle O}{\|}}{C}}-OR' + H^+$$

Most esterification reactions have equilibrium constants $K =$ [RCOOR][H_2O]/[RCOOH][ROH] that are close to unity. In other words, the reaction at equilibrium does not tend to convert acid and alcohol totally to ester. To increase the yield of ester, water can be removed by distillation or by adding a drying agent, or the concentration of one of the reactants, usually the alcohol, can be increased. Since the equilibrium constant is close to unity, the esterification reaction can be readily reversed. The reverse reaction, formation of an acid (usually the salt of the acid) and an alcohol from an ester, is called *saponification* (from the Latin *sapo* for soap). The reason for the name is that natural fats are esters, and soap is made by splitting them to give salts of long-chain organic acids.

As noted previously, carboxylic acids are the end product of the oxidation of primary alcohols. This oxidation can be reversed but only through the use of very strong reducing agents such as lithium aluminum hydride, $LiAlH_4$.

19.9   Amines and Amides

Just as there are organic acids, there are organic bases. The most important of them are compounds called *amines*. The amines can be considered as derivatives of ammonia through replacement of one or more hydrogens of

481

Figure 19.14 AMINES	Formula	Name	Boiling Point, °C	K_b (in H_2O)
	CH_3NH_2	Methylamine	-6	4.4×10^{-4}
	$C_2H_5NH_2$	Ethylamine	17	5.6×10^{-4}
	$(C_2H_5)_2NH$	Diethylamine	56	9.6×10^{-4}
	$(C_2H_5)_3N$	Triethylamine	90	5.7×10^{-4}
	$C_6H_5NH_2$	Aniline	184	3.8×10^{-10}

NH_3 by R groups. Properties of some amines are listed in Figure 19.14. The K_b values listed are for the (base) equilibrium

$$RNH_2 + H_2O \rightleftharpoons RNH_3^+ + OH^-$$

With the exception of aniline (the amine which has a phenyl group attached to nitrogen) all the values are approximately equal and are not very different from the corresponding value for ammonia (1.8×10^{-5}). However, the value for aniline is clearly different, presumably because the functional group NH_2 is attached directly to the aromatic ring. Just as we saw for phenol, the presence of the ring is again critical. In the case of aniline the pair of electrons on the nitrogen atom interacts with the aromatic ring so that the electron pair is less available for accepting a proton. That this is indeed the explanation is shown by the fact that if the aromatic ring is saturated to form the compound $C_6H_{11}NH_2$ the value of K_b rises to 4.4×10^{-4}, which is quite normal.

Another example of modification of base strength by an aromatic ring occurs in cases where the N is a part of the ring. The simplest such example is pyridine

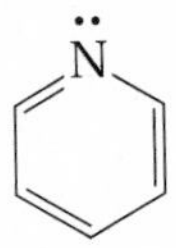

the base strength of which ($K_b = 1.6 \times 10^{-9}$) is also less than that for a more usual amine. Compounds having rings which contain atoms other than C are called *heterocyclic* compounds and those which contain N are especially important biologically. Included in the latter are the nucleic acids (Chapter 23), porphyrins (Chapter 24), and the three amino acids histidine, tryptophane, and proline (Section 22.1). Heterocyclic nitrogen bases are the structural units for the compounds of plant origin called *alkaloids*, which in animals produce a variety of psychophysical reactions. Examples of alkaloids are nicotine, caffeine, codeine, lysergic acid diethylamide ("LSD"), and morphine. The simplest of these, nicotine, will be discussed in the next

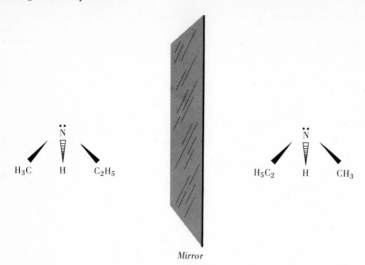

$$H_3C \qquad H \qquad C_2H_5 \qquad\qquad\qquad H_5C_2 \qquad H \qquad CH_3$$

Mirror

Figure 19.15
Mirror images for methylethylamine.

section. It might be noted that the active ingredients of marijuana differ fundamentally from the above compounds in that they are not nitrogen heterocycles.

Also listed in Figure 19.14 are diethylamine, which is a secondary amine (R_2NH), and triethylamine, a tertiary amine (R_3N). For certain secondary and tertiary amines it might be thought that optical isomerism would be observed. In other words, for the compound methylethylamine, as shown in Figure 19.15, the compound is not superimposable on its mirror image. In such a case we would expect two optical isomers. However, they are not observed, because the molecule is not sufficiently rigid to prevent the amine from turning inside out like an umbrella. In doing so, each of the optical isomers is quickly converted into the other.† It seems that this rate of inversion takes place in less than a thousandth of a second, thereby preventing the isolation of optical isomers.

Amines can be prepared by heating a halogenated hydrocarbon, for example RBr, with ammonia in a sealed tube:

$$RBr + 2NH_3 \longrightarrow RNH_2 + NH_4Br$$

More than one of the hydrogens of ammonia can be substituted so that secondary and tertiary amines are also formed. In addition, if a sufficient excess of RBr is used, an interesting compound is formed which has the general formula R_4NBr. It is an ionic substance and contains the ion NR_4^+ which can be considered to be an analog of ammonium ion NH_4^+.

---

†We saw earlier that a familiar analogy of optical isomers are the shapes of left and right gloves. Carrying this analogy further, the umbrella inversion of amines is analogous to turning a left-hand glove inside out to produce thereby a right-hand glove with no preference for the glove being either right-hand or left-hand.

Ammonia will react with a carboxylic acid in a rapid reaction to form the ammonium and carboxylate ions:

$$NH_3 + RCOOH \longrightarrow RCOO^-NH_4^+ \xrightarrow{\text{heat}} R-\overset{\overset{\textstyle O}{\|}}{C}-NH_2 + H_2O$$

If, however, the initial product is heated, loss of water occurs to give an amide ($RCONH_2$). Amides differ from amines in that they are not basic (K_b for an amide is of the order of 10^{-15}), and it is not clear whether the proton is added to the nitrogen or to the oxygen. The lack of basicity of the electron pair of amide nitrogen has been explained by saying that a contributing structure of the amides is one in which the lone pair of electrons is used to form a double bond. This can be represented

$$R-\overset{\overset{\textstyle :\ddot{O}:}{\|}}{\underset{}{C}}-\ddot{N}H_2 \longleftrightarrow R-\overset{\overset{\textstyle :\ddot{O}:^-}{|}}{C}=NH_2^+$$

where the double-headed arrow designates the two contributing resonance structures. Since the actual structure is a combination of these, N does not have full possession of the lone pair of electrons for binding a proton; instead it has a partial positive charge to repel a proton. Consistent with the resonance formulation is the additional fact that compared with ammonia, amides are considerably more acidic (have a greater tendency to lose a proton). The acid dissociation constant of an amide is of the order of 10^{-16} whereas that for ammonia is 10^{-33}.

Just as ammonia will react with acids to form amides, so will primary amines. As we shall see later in discussing proteins, this reaction is an extremely important one in that amide formation is responsible for the

peptide link $\left(-\overset{\overset{\textstyle O}{\|}}{C}-\overset{\overset{\textstyle H}{|}}{N}- \right)$ which holds the proteins together.

As with other functional groups that we have discussed, the NH_2 functional group can occur more than once in a molecule. A simple diamine is ethylenediamine, $H_2NCH_2CH_2NH_2$, which has a K_b of 8.5×10^{-5}. The second constant is 4×10^{-7}, showing that it is always harder to add a second proton to an already positively charged molecule.

Several times we have illustrated the possibility of placing two functional groups of the same kind in the same molecule. It is also possible to place two functional groups that are different in the same molecule. One extremely important example of this occurs in a series of compounds called *amino acids* which, as their names imply, contain both an amine group and an acid group. The simplest of the amino acids is glycine which can be described by either of the following formulas:

$$H_2NCH_2COOH \longrightarrow H_3N^+CH_2COO^-$$

As the arrow indicates, the two functional groups can react with each other so that a proton is transferred from the acidic group to the basic group. In the pH range near 7, glycine exists mostly in the charged form (called the dipolar form) shown above. At higher and lower pH's protons can be removed or added, respectively. In the titration of glycine two dissociation equilibria appear, the first with a dissociation constant 5×10^{-3} and the second with a dissociation constant of 2×10^{-10}. This means that, at very low pH, glycine exists as a positive ion with both ends protonated (that is, $H_3N^+CH_2COOH$) and at very high pH, as a negative ion having lost both protons (that is, $H_2NCH_2COO^-$). In the intermediate pH range the molecule is electrically neutral, although it is correctly described as being a dipolar molecule (that is, $H_3N^+CH_2COO^-$). One end (the nitrogen) carries a positive charge, and the other end (oxygen) carries a negative charge. Frequently various amino acids are characterized by the pH at which the ion shows no net motion in an electric field. This point is called the *isoelectric point*, and for glycine it occurs at a pH of 6.0.

19.10 Natural Products

The term *natural products* is used in organic chemistry to refer to those organic compounds which are synthesized in nature either by plants or animals. The principal aim of natural-products research is to determine the structure of these naturally occurring molecules. The first problem is to isolate the compound of interest so as to obtain it in a pure state. Various techniques are used for the separation, including distillation, chromatography (adsorption on surfaces), and extraction by various solvents. Proof of purity is difficult, but it can be inferred from the reproducibility of chemical analysis of samples purified in different ways. Also, a sharp melting point as contrasted to a wide melting range is taken to indicate isolation of a pure compound.

Once the compound is isolated, it can be chemically analyzed for the constituent elements and its simplest formula determined. Freezing-point lowering of an appropriate solvent (Section 10.4) gives information leading to the molecular formula. At this point the problem really begins. How are the atoms arranged in the molecule? Some functional groups and some atomic arrangements can easily be detected by appropriate instrumental methods, such as infrared absorption. Others show themselves in characteristic chemical reactions. Because organic reactions frequently leave large groups of atoms intact, it is useful to subject the unknown molecule to a series of reactions in the hope that some of these reactions will lead to products of known structure. From the arrangement of atoms in the known products, inferences can be drawn about the structure of the parent molecule. As an example of how this is done, consider the following reactions of the natural product nicotine, which is a nitrogen-containing organic compound found in tobacco:

$$C_{10}H_{14}N_2 \xrightarrow{\text{Fe(CN)}_6{}^{3-}}$$

Nicotine

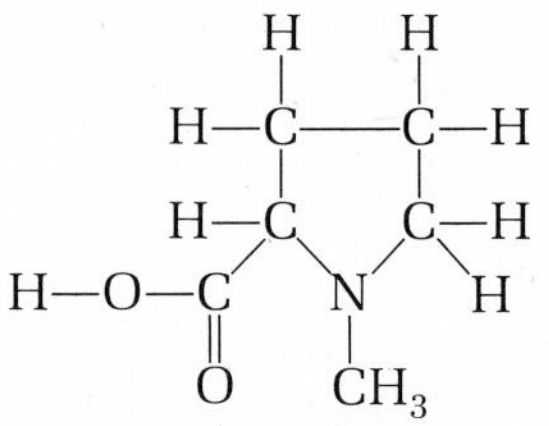

N-Methylpyrrolidine-2-
carboxylic acid

$$C_{10}H_{14}N_2 \xrightarrow{\text{MnO}_4{}^-}$$

Nicotine

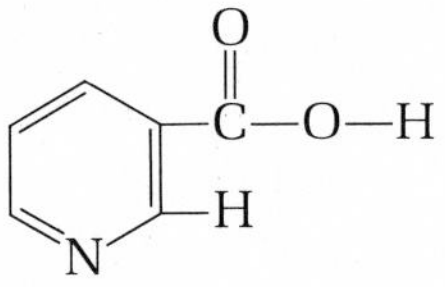

Nicotinic acid

As shown, the two different oxidizing agents [$Fe(CN)_6{}^{3-}$ and $MnO_4{}^-$] each partially convert nicotine to two different products, each of which is a compound of known structure. The products contain different rings of atoms to which COOH groups are attached. Since the COOH group is frequently formed when a group attached to a ring is oxidized, it is reasonable to suppose that nicotine consists of the two rings hooked together and has the structure

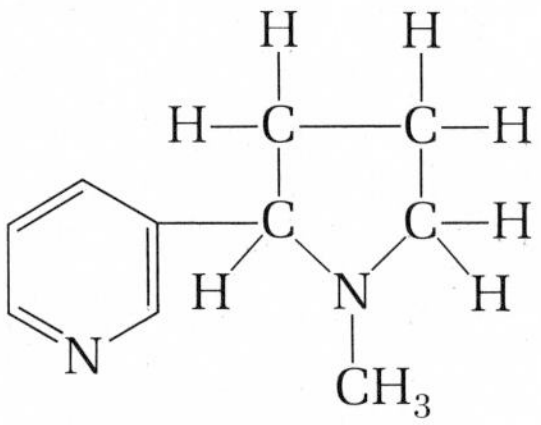

Nicotine

Apparently ferricyanide oxidizes away the left-hand ring, and permanganate the right-hand ring. Ultimate proof of the structure of nicotine and other natural products comes from successful laboratory synthesis starting with compounds of known structure.

QUESTIONS

19.1   *Hydrocarbons*   Indicating the tetrahedral nature of carbon, draw the structure of propane.

19.2   *Isomers*   Draw formulas (like those of Figure 19.4) for all the structural isomers of pentane (C_5H_{12}) and name each.

19.3   *Optical Isomers*   If the four bonds formed by C atoms were in one plane rather than tetrahedral, would a compound such as 3-methylheptane (Figure 19.5) show optical activity? Explain.

19.4   *Unsaturated Hydrocarbons*   Compare and contrast butene with butane in respect to (*a*) molecular formula, (*b*) number of isomers, (*c*) bond angles, (*d*) differences of spatial arrangement of bonding electrons.

19.5   *Acetylene*   Describe the carbon-carbon bond in acetylene by sketching it and contrast the bond with a C—C single bond in terms of its length and the energy needed to break it.

19.6   *Nomenclature*   Draw structural formulas of each of the following: (*a*) 4-methyloctane, (*b*) 3-ethyl-5-methyl decane, (*c*) 1-butyne, (*d*) 3-methyl-2-pentene, (*e*) 4-ethyl-6-methyl-2-octene.

19.7   *Nomenclature*   Give systematic names for each of the following compounds:

$$(a) \quad CH_3CH_2\overset{\overset{\displaystyle CH_3}{|}}{C}HCH_3$$

(*b*) $CH_3CH_2{=}CHCH_2CH_3$

$$(c) \quad CH_3CH_2\underset{\underset{\displaystyle CH_3}{|}}{C}{=}CHCH{=}CH_2$$

$$(d) \quad CH_3\overset{\overset{\displaystyle C_2H_5}{|}}{C}HCH_2\overset{\overset{\displaystyle CH_2C_2H_5}{|}}{C}HCH_2CH_3$$

19.8   *Aromatic Hydrocarbons*   What grouping of atoms is characteristic of simple aromatic compounds? What is special about this grouping?

19.9   *Aromatic Hydrocarbons*   Draw structural formulas for all isomers of trimethylbenzene.

19.10   *Alcohols and Ethers*   Draw all structural isomers of $C_4H_{10}O$ and tell which are ethers and which are primary, secondary, and tertiary alcohols.

19.11 *Fermentation* Using the usual rules for assignment of oxidation numbers, discuss the fermentation reaction as an oxidation-reduction process.

19.12 *Aldehydes and Ketones* Draw formulas for an aldehyde and ketone that are isomers. Show by equations how they may be prepared from starting materials which are also isomers.

19.13 *Acids and Esters* Write balanced equations for each of the following: (a) formation of ethyl acetate, (b) the stepwise neutralization of malonic acid with strong base, (c) the hydrolysis of sodium propionate to give a slightly basic solution.

19.14 *Amines* Calculate the H^+ and OH^- concentrations and pH of each of the following solutions: (a) 0.3 M methylamine, (b) 0.3 M aniline.

19.15 *Amides* (a) What is the functional group of an amide? (b) Why are amides less basic than amines? (c) Would you expect a peptide link to be acid, basic, or essentially neither? Explain.

19.16 *Natural Products* Prior to conducting chemical reactions to determine atomic arrangement in a natural product, what three steps must be performed?

19.17 *Terms* What is meant by each of the following: (a) ketone, (b) hydrocarbon, (c) alcohol, (d) ether, (e) diene, (f) isomer, (g) catalytic cracking?

19.18 *Acids* How would you explain that as you increase the number of chlorine atoms in the acetic acid molecule, the strength of the resulting acid increases?

19.19 *Alcohols* Write the equations of reactions that you would use to synthesize $CH_3C\overset{\overset{\displaystyle OH}{|}}{-}CH\overset{\overset{\displaystyle OH}{|}}{-}CH_3$ from $CH_3CH{=}CHCH_3$.

19.20 *Solubility* Using examples, explain what is meant by "like dissolves like." Are alcohols an exception to this statement?

19.21* *Heat of Reaction* When 1-butyne is burned in oxygen to form CO_2 and water, 54.1 kilojoules of heat per gram of H_2O are given off. How much heat is given off when 75.0 g of 1-butyne are burned completely in oxygen?

19.22 *Ethers* If you were given propanol and asked to make dipropylether, how would you go about doing it?

19.23 *Nomenclature:* Give names for each of the following:

(a) $CH_3CHCH_2CHCH_3$
　　　　$|$　　　$|$
　　　CH_2　CH_3
　　　CH_3

(b) $CH_2{=}CH{-}CH{=}CH{-}CH{=}CH_2$

(c) $CH{\equiv}C{-}CH{-}CH_3$
　　　　　　$|$
　　　　　CH_2
　　　　　$|$
　　　　　CH_3

(d) $CH{\equiv}C{-}CHCH_3$
　　　　　　$|$
　　　　　CH_3

19.24 *Bonds* Differentiate between a double bond and a triple bond in terms of: (a) energy needed to break each, (b) bond length, (c) electron cloud symmetry.

19.25 *Hydrocarbons* Would you expect the heat of combustion of $CH_2{=}CH{-}CH{=}CH{-}CH_3$ to be smaller or greater than the heat of combustion of $CH_3{-}C{\equiv}C{-}CH_2{-}CH_3$? Explain.

19.26 *Nomenclature* Which of the following are incorrect names? Give the correct name in each case. (a) 1-methyl-2-phenylpropane, (b) 4-ethyl-2-pentyne, (c) 2-methyl-5-ethyl-2,4-heptadiene, (d) 5-methyloctane.

19.27 *Terms* What is meant by each of the following: (a) alkaloid, (b) isoelectric point, (c) saponification, (d) inductive effect, (e) formalin?

19.28 *Oxidation of Alcohols* Extensive oxidation of n-butyl alcohol yields an acid while isobutyl alcohol under similar conditions yields a ketone. How do you account for the difference in products?

19.29 *Functional Group* Identify the functional groups in each of the following compounds: (a) phenol, (b) ethylamine, (c) propene, (d) $CH_3CH{=}CHCHO$, (e) acetone.

19.30 *Isomers* Write all isomers of nitroethylbenzene.

19.31 *Amines* (a) How could you convert $C_3H_7NH_2$ to $C_3H_7NH_3Br$? (b) How might you accomplish the reverse conversion?

19.32 *Molecular Formula* Nicotine has the weight percentage composition 74.1% C, 8.64% H, 17.3% N. If 2.43 g of it dissolves in 45.0 g of benzene and lowers its freezing point by 1.63°C, (a) what is the molecular weight of nicotine? (The molal freezing-point lowering constant is 4.90°C.) (b) What is the formula of the compound?

19.33 *Bond Angles* Cyclohexane differs from benzene in several important ways. One of them is the fact that the six C—C—C bond angles in benzene are 120°. In cyclohexane these six angles are essentially the tetrahedral angle of 109°. How is this possible?

19.34 *Isomers* Draw structural formulas for the nine isomers of hexane. Show which ones of these consist of optical isomers.

19.35 *Benzene* As noted in Chapter 4, an important reason for the bonding of atoms is the lowering of energy of electrons when they are allowed to move over larger volumes and not be too constricted. How is this principle involved in accounting for the special stability of benzene.

19.36 *Structural Formulas* Draw structural formulas for each of the following: (a) a diene, (b) a pair of optical isomers, (c) acetone, (d) glycine as it exists in acidic, basic, and neutral solution.

19.37 *Soap* Chemically, what is soap? Is it or is it not soluble in water? Explain.

19.38 *Dicarboxylic Acids* (a) In whole-number pH units compare the pH range over which malonic acid and succinic acid exist as singly-negative ions. (b) Roughly what would you predict for adipic acid, $HOOCCH_2CH_2CH_2COOH$? (c) Why is this trend observed?

19.39 *Naphthalene* Consider the substitution of 2 Cl atoms for any two of the 8 hydrogen atoms in naphthalene (Figure 19.8). How many isomers can thus be formed?

19.40 *Science and Society* Through organic chemistry it has been possible to improve the quality of a number of commonly used items (synthetic fabrics, for example) by substituting synthetic materials for natural products. (a) Where has the energy come from to make the synthetic products and where has it come from to make the natural products? (b) What does your answer to (a) imply for the future? (c) Why do synthetic materials pose greater disposal problems than natural materials?

19.41* *Gasoline and Electric Cars* For octane, the heat of combustion is 5440 kilojoules/mole. (a) How much energy is obtained from burning 1 kg (0.38 gal) of gasoline? (b) Noting that 1 mole of Pb, 1 mole of PbO_2, and 2 moles of H_2SO_4 produce $2 \times 96,500$ coulombs at 2.1 volts, and that coulombs times volts equal joules, calculate the amount of energy which could be obtained from 1 kg of the mixture of $Pb + PbO_2 + H_2SO_4$ appropriate for the lead-storage-battery reaction. (c) An average car's gas tank contains some 40 kg of gasoline. What weight of lead storage battery would be needed to deliver the same amount of energy?

Organic
Reactions

20

All life processes consist of chemical reactions of organic compounds. To understand the life processes means to understand organic reactions and the mechanisms by which they proceed. Biological processes consist of the simultaneous occurrence of many individual chemical reactions. The complexity and sheer number of the reactions makes understanding them a formidable task. However, no individual process is unique; each has features in common with others and with simple organic reactions. Understanding of the complex processes can come from the study of simple individual reactions and the classification of reactions according to relatively few broad types. In this chapter we will approach organic chemistry from the viewpoint of reaction classes as a basis for understanding both synthetic and natural processes.

Any organic reaction involves covalent bonds and the breaking or formation of one or more bonds. In these fundamental processes, electronegativity (Section 4.4) of atoms takes on major significance. For example, when a covalent bond breaks, the shared pair of electrons will tend to remain with the more electronegative atom. Similarly, in forming a new bond an unshared pair will seek out an atom which is electron deficient (owing to its being less electronegative than the atom it is already bound to).

In broad terms, most organic reactions can be classified as

1   *Addition reactions, in which new bonds are formed*

2   *Elimination reactions, in which bonds are broken*

3   *Substitution reactions, in which new bonds are formed and old ones broken*

All these broad classes of organic reactions occur with all classes of organic compounds, i.e., those with various different functional groups. In fact, reaction most often occurs at the functional group. Before starting our discussion of reaction types, it will be well to review the various common functional groups. These are listed, with example compounds, in Figure 20.1. (Nomenclature is summarized in Appendix 1.2.) Note that the common functional groups contain an atom more electronegative than C or H. Hence a charge unbalance exists, and the functional group is the most likely site for reaction to occur.

	Functional Group		Example	
Figure 20.1 **FUNCTIONAL** **GROUPS**	Alcohol	—OH	Ethanol	CH_3CH_2OH
	Ether	—O—	Diethyl ether	$CH_3CH_2OCH_2CH_3$
	Aldehyde	$-\overset{\overset{O}{\|\|}}{C}-H$	Acetaldehyde	$CH_3\overset{\overset{O}{\|\|}}{C}-H$
	Ketone	$-\overset{\overset{O}{\|\|}}{C}-$	Acetone	$CH_3\overset{\overset{O}{\|\|}}{C}CH_3$
	Acid	$-\overset{\overset{O}{\|\|}}{C}-OH$	Propionic acid	$CH_3CH_2\overset{\overset{O}{\|\|}}{C}-OH$
	Ester	$-\overset{\overset{O}{\|\|}}{C}-O-$	Ethyl acetate	$CH_3\overset{\overset{O}{\|\|}}{C}-OCH_2CH_3$
	Amine	$\overset{\|}{N}$	Methyl amine	$CH_3\overset{H}{N}$
	Amide	$-\overset{\overset{O}{\|\|}}{C}-NH_2$	n-Butyl amide	$CH_3CH_2CH_2CH_2\overset{\overset{O}{\|\|}}{C}-NH_2$

20.1 Addition Reactions

In an addition reaction, one reactant is an organic molecule containing a multiple bond. After reaction, the multiple bond is lost and single bonds bind additional atoms. The reacting organic molecule can either contain a functional group, such as the C=O (*carbonyl*) of a ketone, or be an unsaturated hydrocarbon, which might contain, for example, C=C. Typical of the first type is the addition of HCN (hydrogen cyanide) to aldehydes and ketones. To understand how this comes about, it is necessary to realize that in the C=O grouping, oxygen, being more electronegative than carbon, carries a slightly negative charge and C is correspondingly positive (or electron deficient). A reagent having an unshared electron pair (often called a *nucleophile*) can attack the C as follows:

$$1 \quad \overset{\cdot\cdot}{\underset{\cdot\cdot}{O}}{=}\overset{\overset{\displaystyle CH_3}{|}}{\underset{\underset{\displaystyle CH_3}{|}}{C}} \quad :C{\equiv}N: \quad = \quad :\overset{\cdot\cdot}{\underset{\cdot\cdot}{O}}-\overset{\overset{\displaystyle CH_3}{|}}{\underset{\underset{\displaystyle CH_3}{|}}{C}}-C{\equiv}N:$$

$$2 \quad H^+ + :\overset{\cdot\cdot}{\underset{\cdot\cdot}{O}}-\overset{\overset{\displaystyle CH_3}{|}}{\underset{\underset{\displaystyle CH_3}{|}}{C}}-C{\equiv}N: \quad = \quad H-\overset{\cdot\cdot}{\underset{\cdot\cdot}{O}}-\overset{\overset{\displaystyle CH_3}{|}}{\underset{\underset{\displaystyle CH_3}{|}}{C}}-C{\equiv}N:$$

Note that in the initial attack (step 1) the negatively charged CN^- seeks the positive end of the CO group where the C is positive (shown by-δ^+ for partial + charge) because the more electronegative O has the larger share of the bonding electrons (δ^-, partial − charge). After the attack the bonds rearrange with the electronegative O picking up an additional share of electrons as is shown by the negative charge. This negatively charged anion is then attacked by a proton (step 2) to give the final product (called a *cyanohydrin*). This type of attack with addition—which can involve any of a number of aldehydes and ketones, as well as a number of nucleophiles (for example, OH^-, NH_3, HSO_3^-)—gives products of various stabilities. All such products are characterized by the general formula $R_2C(OH)X$, where X stands for the nucleophile.

The above addition to the CO functional group of a ketone is a good example of addition reactions. Note that the functional group, being unsaturated, was important in furnishing a site where addition could occur. It is quite possible for a hydrocarbon to show unsaturation without a functional group and so undergo addition. In fact, for unsaturated hydrocarbons, addition is the usual reaction.

An unsaturated hydrocarbon must contain one or more C-to-C π bonds, as described in Section 4.7 and depicted in Figure 20.2. The electron pair

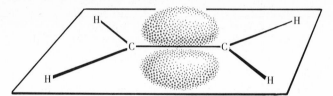

Figure 20.2
π bond in ethylene.

of the π bond is less tightly held between the carbon nuclei than is the other shared pair and so can be used to bind on other atoms. Initiation of such an attack is most probable from atoms which are electron deficient in the sense that they can accommodate additional electrons. To specify this difference from nucleophiles, such reagents which seek additional electrons are called *electrophiles*.† An example of attack on a π bond is afforded by a common test for the existence of such bonds in organic molecules. The test involves mixing an unknown compound with a solution of Br_2 in carbon tetrachloride. Decolorization of the bromine is taken to indicate the presence of a double bond in the compound such that two bromine atoms have added to it:

$$R-CH{=}CH_2 + Br_2 \longrightarrow R-\underset{\underset{\textstyle Br}{|}}{\overset{\overset{\textstyle Br}{|}}{C}}H-CH_2$$

To understand this reaction we need to know that in the presence of water or other nucleophiles, a different product results so that only one Br is added. This is taken to indicate that with Br_2 the reaction seems to occur in the following two steps:

1   $$R-CH{=}CH_2 + Br_2 \longrightarrow R-\overset{+}{C}H-\overset{\overset{\textstyle Br}{|}}{C}H_2 + Br^-$$

2   $$R-\overset{+}{C}H-\overset{\overset{\textstyle Br}{|}}{C}H_2 + Br^- \longrightarrow R-\underset{\underset{\textstyle Br}{|}}{C}H-\overset{\overset{\textstyle Br}{|}}{C}H_2$$

In step 1, the Br_2 molecule attacks the π bond. In the process Br_2 splits to Br^+ and Br^-, with the Br^+ joining to the π bond so that it is now bonded

†Electrophile, like nucleophile, is compounded from the Greek word *philos*, meaning loving. Thus, an electrophile seeks electrons while a nucleophile seeks nuclei. Clearly, they are made for each other.

to one of the C atoms. The resulting product of step 1 is positively charged and is called a *carbonium ion* (a carbon cation). In the second step, Br⁻ (a nucleophile) attacks the positive carbonium ion at the electron-deficient C atom. It is now bound through electron sharing of the nucleophile's electron pair and reaction is complete.

In the presence of water or other nucleophiles, step 2 of the reaction above may not occur. Instead water or the other nucleophile can compete with Br⁻ for the carbonium ion:

$$\text{R}-\overset{+}{\text{C}}\text{H}-\text{CH}_2 + \text{H}_2\text{O} \longrightarrow \text{R}-\underset{\underset{\text{OH}_2^+}{|}}{\text{CH}}-\overset{\overset{\text{Br}}{|}}{\text{CH}_2} \longrightarrow \text{R}-\underset{\underset{\text{OH}}{|}}{\text{CH}}-\overset{\overset{\text{Br}}{|}}{\text{CH}_2} + \text{H}^+$$

In this case, the final product (called a *bromohydrin*) contains two functional groups: one (Br) from the electrophilic attack and one (OH) from the nucleophilic attack. Other nucleophiles, for example, Cl⁻, CN⁻, HSO$_4$⁻, could have been substituted.

Similarly, for addition to unsaturated hydrocarbons, other electrophiles can be used. As an example, let us consider substituting HBr for Br$_2$. In the initial attack on the π bond, the positive (or H) end of HBr will attack:

$$\text{R}-\text{CH}=\text{CH}_2 + \text{HBr} \longrightarrow \text{R}-\overset{+}{\text{C}}\text{H}-\text{CH}_3 + \text{Br}^-$$

This is reasonable both because the positive H seeks the negative electron density and because the pair of shared electrons of HBr end up with the more electronegative Br. As before, the reaction is completed by nucleophilic attack (by Br⁻, H$_2$O, or other nucleophile) on the carbonium ion:

$$\text{R}-\overset{+}{\text{C}}\text{H}-\text{CH}_3 + \text{Br}^- \longrightarrow \text{R}-\underset{\underset{\text{Br}}{|}}{\text{CH}}-\text{CH}_3$$

There is a question about the addition of HBr that does not arise with Br$_2$. This is, which of two possible products will be formed? Is it the one shown above, or is it one with the substituted Br on the end C atom? For a large number of reactions the answer is given by Markovnikov's rule (1870) which states that the H of the added hydrogen halide goes to the C which already had more H. This is as we have shown it, and our mechanism should give a reason for this occurrence. It is that a carbonium ion having the positive charge next to an R group is more stable than one having the positive charge on a C at the end of a chain. Thus, in the initial attack the H⁺ will attach to the C having more H and the positive charge resides on the more substituted C. We should note that there are exceptions to the Markovnikov rule which involve other, less common, mechanisms of addition.

An important example of addition to unsaturated hydrocarbons is their treatment with hydrogen in the presence of suitable metal catalysts such as nickel, platinum, or palladium. In the process the unsaturated hydrocarbon adds sufficient H to become saturated. For example, ethylene is so converted to ethane in this way. As opposed to the ionic mechanism discussed above, hydrogenation seems to be a one-step process with hydrogen (dissolved in the metal catalyst) adding two atoms at a time to the two ends of the double bond.

Finally we should note that aromatic hydrocarbons (e.g., benzene) do not in general undergo addition reactions like unsaturated hydrocarbons. This is the basis of their being considered a separate class of hydrocarbons. Instead, aromatic hydrocarbons undergo substitution reactions and will be discussed later.

20.2 Elimination Reactions

Being the reverse of addition reactions, elimination reactions produce rather than consume unsaturated hydrocarbons. To illustrate, the reaction of HBr with an alkene R—CH=CH$_2$ to form RCHBrCH$_3$ can be reversed by treatment of the alkyl halide with a concentrated solution of KOH in ethanol. The elimination reaction seems to proceed by two mechanisms, with one or the other predominating depending upon the specific reaction conditions. One mechanism is the exact reverse of the addition mechanism discussed above.

$$1 \quad R\overset{\overset{\displaystyle Br}{|}}{-}CHCH_3 \longrightarrow Br^- + \overset{+}{R}CHCH_3$$

$$2 \quad \overset{+}{R}CHCH_3 \longrightarrow RCH=CH_2 + H^+$$

The second is quite similar, but differs in timing sequence. It is called a "concerted" mechanism in that the steps occur simultaneously:

$$R-\overset{\overset{\displaystyle Br}{|}}{\underset{\underset{\displaystyle H}{|}}{C}}-\overset{\overset{\displaystyle H}{|}}{\underset{\underset{\displaystyle H}{|}}{C}}-H \quad {}^-OH \quad = \quad R-\overset{\overset{\displaystyle Br^-}{}}{\underset{\underset{\displaystyle H}{|}}{C}}=\overset{\overset{\displaystyle H}{}}{\underset{\underset{\displaystyle H}{|}}{C}} + H_2O$$

The attack here is by the nucleophile OH$^-$ which extracts a proton, leaving the pair of electrons, which had bound the proton to C, with the carbon atom. That pair then becomes the double bond. Since the C attached to Br cannot accommodate five electron pairs, Br$^-$ leaves, taking with it the fifth electron pair. All this is frequently denoted by arrows which show electron migration during a concerted mechanism:

$$\underset{\substack{| \\ H}}{\overset{\substack{Br \\ |}}{R-C}}-\underset{\substack{| \\ H}}{\overset{\substack{H \\ |}}{C}}-H \quad \overset{OH}{} \longrightarrow \quad \overset{Br^- \quad H_2O}{\underset{\substack{H}}{\overset{R \quad H}{C=C}}}$$

Because the concerted mechanism is initiated by OH$^-$, it predominates over the two-step mechanism in highly basic solution.

Another and specific example of an elimination reaction is the dehydration of ethanol to form ethylene.

$$CH_3CH_2OH \longrightarrow CH_2=CH_2 + H_2O$$

This reaction is catalyzed by concentrated H_2SO_4 which not only acts as a dehydrating agent (for the product H_2O) but furnishes a proton needed in the mechanism:

$$CH_3CH_2OH + H_2SO_4 \longrightarrow CH_3CH_2\overset{+}{O}H_2 + HSO_4^-$$

Here the positive proton attaches to the partially negative O of the —OH. In the charged intermediate, the C—O bond is weakened and subsequently breaks, with release of H_2O.

$$CH_3CH_2\overset{+}{O}H_2 \longrightarrow CH_3\overset{+}{C}H_2 + H_2O$$

Note that H_2O leaves, taking the bonding electrons (O being more electronegative than C) with the C remaining positively charged due to the electron deficiency. The reaction is completed by proton transfer to HSO_4^- which regenerates the catalyst and gives the final product.

$$CH_3-\overset{+}{C}H_2 + HSO_4^- \longrightarrow CH_2=CH_2 + H_2SO_4$$

Pictorially this step can be represented with arrows as follows:

$$\underset{\substack{| \\ H}}{\overset{\substack{H \\ |}}{H-C}}-\underset{\substack{| \\ H}}{\overset{\substack{H \\ +}}{C}} \longrightarrow \underset{\substack{H}}{\overset{H}{C=C}}\overset{H}{\underset{H}{}} + H^+$$

20.3   Substitution Reactions

Of all organic reactions, particularly those responsible for biological reactions, the majority are substitution reactions. A substitution reaction is one in which an atom or group of atoms replaces an atom or group in an organic molecule. Thus, a substitution reaction formally combines addition and elimination. The substitution process may occur by bond breaking preceding formation of a new bond, by bond formation preceding elimination, or by simultaneous bond making and breaking. We will not dwell on these differences, but instead will discuss some of the more common examples of substitution reactions.

The OH functional group of alcohols can be replaced by a halogen atom through reaction of the alcohol in concentrated solution of halogen acid.

$$CH_3CH_2OH + HBr \longrightarrow CH_3CH_2Br + H_2O$$

In the strong acid solution, the alcohol is first protonated:

$$CH_3CH_2OH + H^+ \longrightarrow CH_3CH_2\overset{\overset{\displaystyle H}{|}}{\underset{+}{O}}H$$

In the protonated ethanol molecule, the C atom adjacent to O is somewhat electron deficient. This results from O being more electronegative than C and so it has the major share of the electron pair of the C—O bond. On protonation, O receives a partial positive charge and the electron pair is attracted further toward O. The electron-deficient C is now ripe for attack by a nucleophile, such as Br^-.

$$\longrightarrow BrCH_2CH_3 + H_2O$$

Note that as H_2O leaves, both electrons of the CO bond go with it, thereby neutralizing the oxygen's apparent positive charge. Similarly the sharing of a pair of the Br^- electrons in the new Br—C bond eradicates any net negative charge. In the overall, acid-catalyzed process, an alcohol, ROH, has been converted to an alkyl halide, RX, by substitution of X for OH.

With alkyl halides, such as C_2H_5Br, further substitution reactions can occur. Substitution for Br can be effected by a stronger nucleophile. One example is the so-called Williamson synthesis of ethers. In this case, the stronger nucleophile is $C_2H_5O^-$. It can be generated by reaction of metallic sodium with ethanol.

$$2Na + 2C_2H_5OH \longrightarrow 2Na^+ + 2C_2H_5O^- + H_2(g)$$

Nucleophilic attack of ethoxide ion, $C_2H_5O^-$, on ethyl bromide occurs according to principles we have already seen operate. Note first that in ethyl bromide there is unequal sharing of electrons between C and the more electronegative Br.

$$\begin{array}{c} CH_3 \\ | \\ H_2\overset{}{\underset{\delta+}{C}}\!:\!\!\underset{\delta-}{Br} \end{array}$$

The unequal sharing results in a slight positive charge on C, so that it can serve as the point for attack of the nucleophile, $C_2H_5O^-$:

$$C_2H_5{-}\ddot{\underset{\cdot\cdot}{O}}{:}\overset{CH_3}{\underset{}{\curvearrowright}}\overset{|}{CH_2}(:Br \;=\; C_2H_5{-}O{-}\overset{\overset{CH_3}{|}}{CH_2} + Br$$

Because Br carries away the formerly shared electron pair, it is now a negative Br^- ion, and the new product $C_2H_5OC_2H_5$, diethylether, is electrically neutral.

Esters may also be formed by a substitution reaction of alcohol on an organic acid. It is found that the reaction is catalyzed by acid, if the acid is not present in too high a concentration. If it is too high, protonation of the alcohol occurs, reducing its ability to act as a nucleophile. However, some acid is helpful to protonate the carboxylic acid.

$$\begin{array}{c} O \\ \| \\ R{-}C{-}OH \end{array} + H^+ \;=\; \begin{array}{c} OH \\ | \\ R{-}\underset{+}{C}{-}OH \end{array}$$

The acid is now more readily attacked by the alcohol:

$$\begin{array}{c} OH \\ | \\ R{-}\underset{+}{C}{-}OH \\ \uparrow \\ \ddot{\underset{\cdot\cdot}{O}} \\ R'{-}O{-}H \end{array} \;=\; \begin{array}{c} OH \\ | \\ R{-}C{-}OH \\ | \\ \underset{+}{R'{-}O{-}H} \end{array}$$

The tetrahedral intermediate rearranges by proton transfer:

$$\begin{array}{c} OH \\ | \\ R{-}C(\!{-}\overset{+}{O}H_2 \\ | \\ R'{-}O \end{array} \;=\; \begin{array}{c} OH \\ | \\ R{-}\overset{+}{C} \\ | \\ R'{-}O \end{array} + H_2O$$

Finally it loses a proton to give an ester:

$$
\begin{array}{c}
\text{OH} \\
| \\
\text{R}-\overset{}{\text{C}}^{+} \\
| \\
\text{R}'-\text{O}
\end{array}
\;=\;
\begin{array}{c}
\text{O} \\
\| \\
\text{R}-\text{C}-\text{O}-\text{R}' + \text{H}^+
\end{array}
$$

The reverse of ester formation, ester hydrolysis or saponification, is also a substitution reaction. It is usually carried out in basic solution, where reaction is initiated by nucleophilic attack of OH$^-$ on the positive end of the C—O bond in the ester

$$
\begin{array}{c}
\text{R} \\
| \\
\overset{\delta^-}{\text{O}}\!=\!\overset{\delta^+}{\text{C}} \quad \text{OH}^- \\
| \\
\text{OR}'
\end{array}
\;=\;
\begin{array}{c}
\text{R} \\
| \\
{}^-\text{O}-\text{C}-\text{OH} \\
| \\
\text{OR}'
\end{array}
$$

Electron shifts in the tetrahedral intermediate lead to breaking the C—OR' bond.

$$
\begin{array}{c}
\text{R} \\
| \\
:\!\ddot{\text{O}}\!-\!\text{C}-\text{OH} \\
| \\
\text{OR}'
\end{array}
\;\longrightarrow\;
\begin{array}{c}
\text{R} \\
| \\
\ddot{\text{O}}\!=\!\text{C}-\text{OH} + {}^-\text{OR}'
\end{array}
$$

The reaction is completed by R'O$^-$ picking up a proton from H_2O or from RCO_2H to form the product alcohol, R'OH.

20.4 Aromatic Substitution

Although aromatic compounds such as benzene contain multiple bonds, the stability of their electronic arrangement is usually maintained throughout reaction. Consequently, their predominant reactions are substitution, as, for example, substitution of an atom or group of atoms for one of the H atoms of benzene. Such substitutions are almost invariably electrophilic substitution; i.e., an electron-deficient reagent attacks the benzene ring. This observation is accounted for by noting that the electron-deficient electrophile can utilize the π electrons of the ring as a point of attack. As our first illustration, let us consider a classical case, the Friedel-Crafts reaction between an alkyl halide, such as methyl chloride, CH_3Cl, and benzene.

$$
CH_3Cl +
\begin{array}{c}
\text{H} \\
\text{C} \\
\text{HC} \quad\quad \text{CH} \\
\| \quad\quad\quad | \\
\text{HC} \quad\quad \text{CH} \\
\text{C} \\
\text{H}
\end{array}
\;\longrightarrow\;
\begin{array}{c}
\text{CH}_3 \\
| \\
\text{C} \\
\text{HC} \quad\quad \text{CH} \\
\| \quad\quad\quad | \\
\text{HC} \quad\quad \text{CH} \\
\text{C} \\
\text{H}
\end{array}
+ \; HCl
$$

The Friedel-Crafts reaction is catalyzed by substances such as anhydrous $AlCl_3$ which have strong affinity for Cl^-. It is thought that the function of the catalyst is to generate CH_3^+, a methyl carbonium ion.

$$CH_3Cl + AlCl_3 = CH_3^+ + AlCl_4^-$$

The methyl carbonium ion then attacks benzene, with subsequent loss of a proton.

Other electrophiles also substitute into benzene. For example, Cl_2 or other halogens substitute when catalyzed by $AlCl_3$ or $FeCl_3$. Again it is thought that the catalyst helps generate the electrophile.

1    $Cl_2 + FeCl_3 = Cl^+ + FeCl_4^-$

2    $Cl^+ + C_6H_6 = C_6H_5Cl + H^+$

Another example is furnished by the nitration of benzene through attack by the electrophile NO_2^+, which is apparently generated from HNO_3 in very acid medium:

$$HNO_3 + H^+ = NO_2^+ + H_2O$$

Using arrows, we can designate the reaction course as follows:

It is interesting to ask what happens if the product, nitrobenzene (the solvent which gives shoe polish its characteristic odor), is further nitrated. In the first place, further nitration is made more difficult (slower) by the presence of the first nitro group. The reason is that the nitro group, being quite electronegative, withdraws electron density from the benzene ring, making it less vulnerable to electrophilic attack by other nitro groups. Thus, the nitro substituent is said to "deactivate the benzene ring toward substitution." However, not all the C atoms are deactivated equally. The biggest effect occurs at those C atoms adjacent to the NO_2 and the one opposite it. In terms of the standard nomenclature (Figure 19.8) these are the ortho and para positions. There are a number of theories to explain this fact, one of which is in terms of the contributing resonance structures shown in Figure 20.3. Note that in the bottom three there is positive charge at the ortho and para positions. Consequently, attack by an electrophile is hampered. The fact is that nitration occurs predominantly (93 percent) in the meta position.

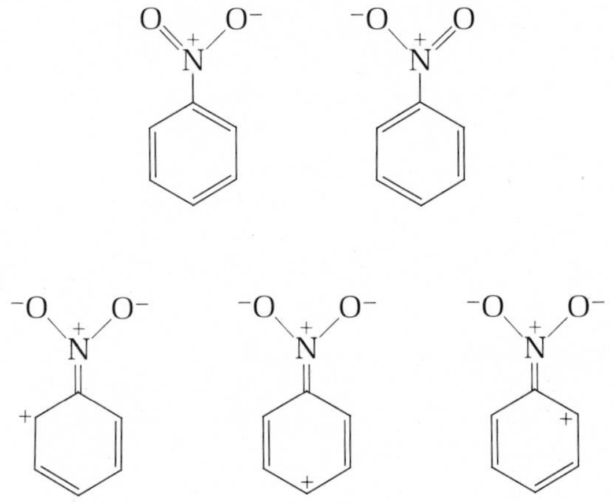

Figure 20.3
Some resonance structures for nitrobenzene.

Other substituents influence differently the reactivity of benzene toward further substitution. For example, toluene (methyl benzene) is more reactive than benzene, with further substitution occurring primarily in the ortho and meta positions. This situation differs from that of nitrobenzene in that the methyl group is thought to contribute electron density to the ring, making the ortho and para positions more reactive by furnishing electrons to bind the added electrophile.

Finally we might note that with a halide-substituted benzene, the reactivity is lowered, but entering groups are directed to the ortho and para positions. These observations show the complexity of aromatic substitution. They are accounted for by saying that the very electronegative halide withdraws electrons from the ring, thereby deactivating all positions. At the same time, there is some tendency toward multiple-bond formation which can stabilize reaction intermediates for substitution in the ortho and para (but

not meta) positions. Examples are

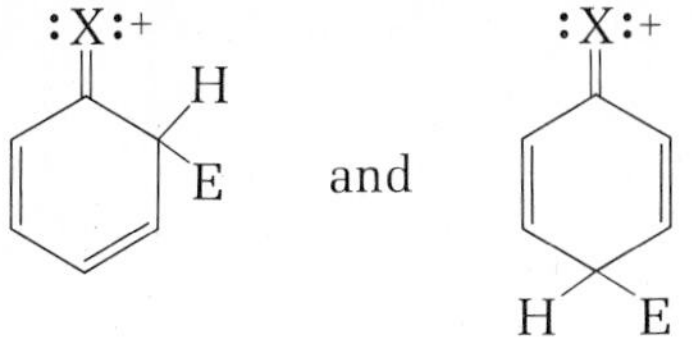

where X is the halogen and E is the electrophile involved in the substitution.

In any case, benzene substituents are frequently classed as being ortho/para directing or meta directing. Examples of ortho/para directors include halides, methyl and other alkyl groups, and OH. Examples of meta directors include NO_2, COOH, and COOR (ester).

20.5   Polymerization

A most important process, both for synthetic formation of plastics and the biological formation of large molecules such as proteins, is polymerization. In this process small molecules link together to form very-high-molecular-weight molecules called *polymers*. The process may occur in a number of ways, some of which are classed as addition reactions (called *addition polymerization*) and some, substitution reactions (called *condensation polymerization*). To illustrate addition polymerization, ethylene polymerizes to form the plastic polyethylene, in a manner which can be visualized as follows:

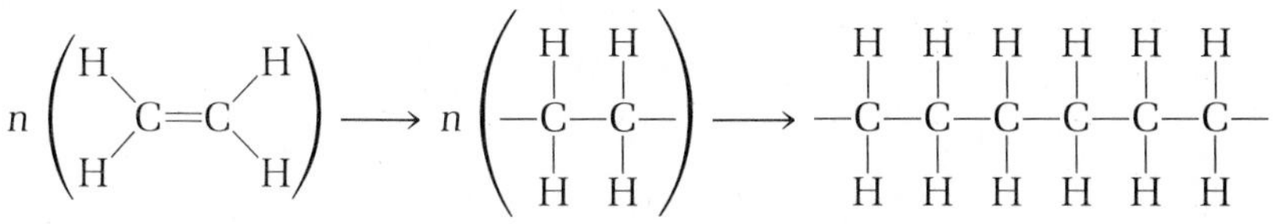

The double bond seems to open up to form an unstable intermediate, which joins with other molecules to produce a high polymer. The term *high polymer* is applied to any large molecule which contains many repeating units.

It should be emphasized that the above equation is not meant to imply a definite mechanism for this reaction. Various possibilities exist for the mechanism of polymerization. One mechanism is the so-called *free-radical mechanism* in which an intermediate would be a free radical such as $R:CH_2\cdot$, where R represents a hydrocarbon chain formed during polymerization and the dot represents an unpaired electron. This so-called free radical can react with a molecule of ethylene as follows

$$R:CH_2\cdot + CH_2 \overset{\frown}{} CH_2 \longrightarrow R:CH_2:CH_2:CH_2\cdot$$

to generate a larger radical which in turn can react further. The end result can be a hydrocarbon chain of very high molecular weight.

Other polymerizations are known to proceed by a mechanism which is called *anionic* (negative-ion mechanism). It can be represented as follows:

$$R\!:\!CH_2\!:\!^- + CH_2\!:\!CH_2 \longrightarrow R\!:\!CH_2\!:\!CH_2\!:\!CH_2\!:\!^-$$

In this reaction the intermediate has a negative charge because it has a pair of electrons in place of the unpaired electron of the free-radical intermediate. This pair of electrons can bind to one end of the ethylene molecule, forcing an electron pair of the double bond to migrate to the opposite end where it can attack an additional ethylene molecule. Eventually the product is again a long-chain hydrocarbon.

Finally it is possible for polymerization to take place by a *cationic* (positive-ion) mechanism:

$$R\!:\!CH_2{}^+ + CH_2\!:\!CH_2 \longrightarrow R\!:\!CH_2\!:\!CH_2\!:\!CH_2{}^+$$

In this case, the positively charged intermediate lacks a pair of electrons and is attacked by an ethylene molecule which uses an electron pair of the double bond to bind itself to the intermediate. The resulting species still lacks a pair of electrons and so reaction continues. For this mechanism, as for the other two, polymerization is eventually terminated by reaction of the intermediate with an appropriate species in trace amounts so as to form a neutral molecule. For the free-radical mechanism polymerization can also be terminated by coupling of two intermediates.

Polyethylene is difficult to polymerize by any of the above mechanisms. Recently, it has been brought about by using transition-element complexes which seem to form complexes both with the growing chain and with a molecule of ethylene to be added to the growing chain.

Derivatives of ethylene in which one of the H atoms has been substituted also undergo polymerization. For example, styrene, in which one H of ethylene is replaced by the phenyl group C_6H_5, polymerizes to form the polymer called polystyrene. The foamed plastic is Styrofoam, an excellent insulator.

In condensation polymerization, the buildup of the polymer occurs not by the addition of a whole molecule to a lengthening chain but by reaction of a lengthening chain with a small molecule, accompanied by elimination of a simpler species such as H_2O. In order that polymerization may occur, two conditions must be met: One end of molecule A must be able to interact with the other end of molecule B so as to split out a small group such as H_2O. Second, both molecules A and B must each contain two functional groups so that, after A and B combine, the free ends can continue to react to extend the polymer.

An example of condensation polymerization is the formation of nylon. Here the reaction is between diaminohexane [$H_2N(CH_2)_6NH_2$] and adipic acid [$HOOC(CH_2)_4COOH$] and involves elimination of water. The reaction can be pictured as follows:

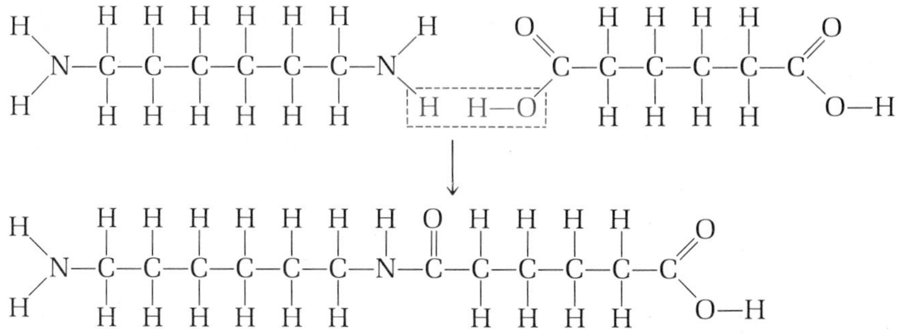

The same reaction occurs at both ends of the molecule to give giant polymers. Therefore, a fundamental difference between condensation polymerization and addition polymerization is that in condensation polymerization the whole batch polymerizes more or less simultaneously. In addition polymerization, individual molecules grow to high molecular weights while others remain unpolymerized. This difference makes condensation polymerization more difficult to control because if the reaction stops too soon, practically all the material is of intermediate molecular weight, too low to be useful. By contrast, if addition polymerization stops too soon, some of the material has grown to useful dimensions with the remainder unreacted. Particularly troublesome in the polymerization of nylon is the fact that the two components, diaminohexane and adipic acid must be mixed in a molar ratio critically near unity. Otherwise the chains will terminate (as the limiting component is used up) short of useful molecular weights.

One way to solve the problem of exact stoichiometry of the components of nylon involves first crystallizing a salt in which the amino groups are protonated ($-NH_3^+$) and the carbonyl groups dissociated ($-COO^-$). Electric neutrality ensures one-to-one stoichiometry between the components. This salt is then heated under reduced pressure (to remove H_2O vapor) and polymerization occurs. Under such conditions it is difficult to study the mechanism of the condensation reaction. However, it is believed to occur in a reaction like that of ester formation in which the nucleophilic N attacks a protonated carboxyl group to form a tetrahedral intermediate.

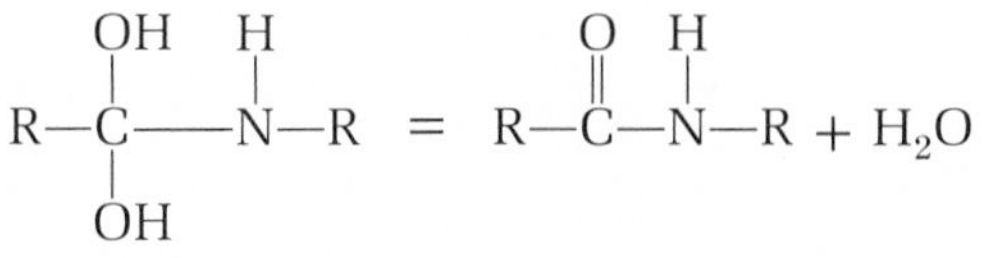

As we will discuss further (Chapter 22), proteins are polymers held together through a similar amide (—CO—NH—) link. Thus, the synthetic fiber nylon bears a chemical resemblance to the natural protein fibers silk and wool. All too often we tend to identify synthetic fibers with cheap products of modern society. However, we should note that the properties of many synthetics are superior to those of natural materials. Even more important is the fact that the supply of natural products is limited and can be replaced only by diverting limited land area from food production and other desirable uses.

Nevertheless, the cheap availability of plastics is creating a serious problem of waste disposal. Unlike iron containers that eventually rust away and paper wrappings that disintegrate in water, plastic containers resist destruction by both the elements and bacteria. Fears have been expressed that in the future a significant fraction of the earth's surface will be completely buried under plastic debris. Still, plastic will burn! The only need is for nonsmogging incinerators. Thus the outlook for plastics is not so grim as for aluminum beer cans and nonreturnable glass bottles, disposal of which represents a real problem.

20.6 Synthesis of Organic Compounds

One of the important aspects of organic chemistry is the synthesis of rather complicated molecules from simpler substances. Synthesis of even a comparatively small molecule such as aspirin (acetylsalicylic acid),

requires a large number of consecutive reactions. To illustrate the general problems involved, we shall consider the steps necessary to make aspirin industrially from simple, familiar chemicals.

As can be seen from its formula, the aspirin molecule contains a benzene ring with two side chains attached. Let us first synthesize benzene. We can do this by heating carbon (as coke) with lime (CaO) to form calcium carbide (CaC_2), treating the calcium carbide with water to form acetylene (C_2H_2), and then heating acetylene at high pressure. Equations for the steps are

$$CaO(s) + 3C(s) \longrightarrow CaC_2(s) + CO(g)$$
$$CaC_2(s) + 2H_2O \longrightarrow C_2H_2(g) + Ca^{2+} + 2OH^-$$
$$3C_2H_2(g) \longrightarrow C_6H_6(g)$$

Once the benzene is obtained, the problem is to introduce the side chains. Since benzene is a rather inert hydrocarbon, somewhat drastic measures are needed to introduce substituents. If it is heated with concentrated sulfuric acid, electrophilic substitution occurs, utilizing the electrophile, SO_3, present in concentrated H_2SO_4.

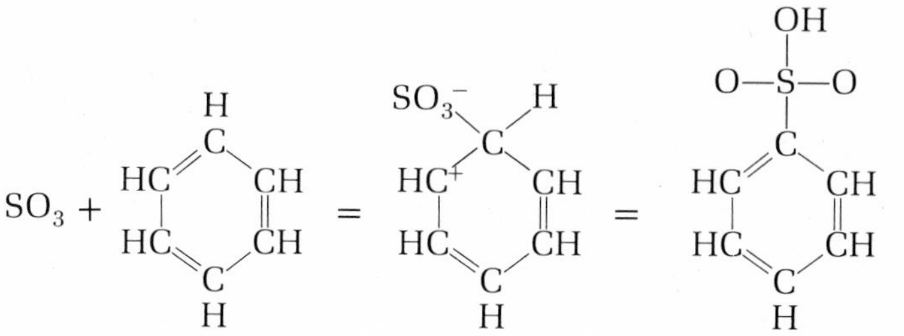

Benzene

Benzenesulfonic
acid

When the product is heated with sodium hydroxide at 350°C, the nucleophile, OH^-, displaces the sulfonate group, leaving a negatively charged oxygen atom.

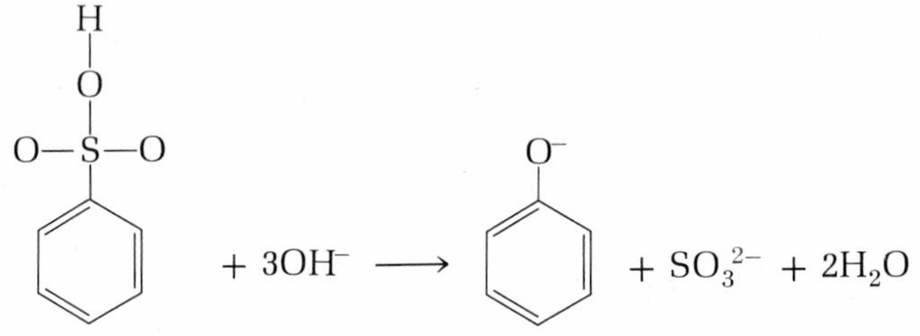

$$+ 3OH^- \longrightarrow \qquad + SO_3^{2-} + 2H_2O$$

Benzenesulfonic
acid

Phenolate
ion

At 150°C the phenolate ion is treated with carbon dioxide under pressure:

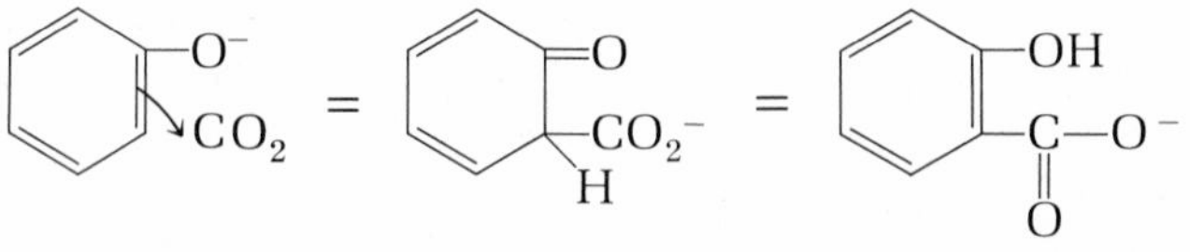

Salicylate ion

Subsequent acidification produces salicylic acid:

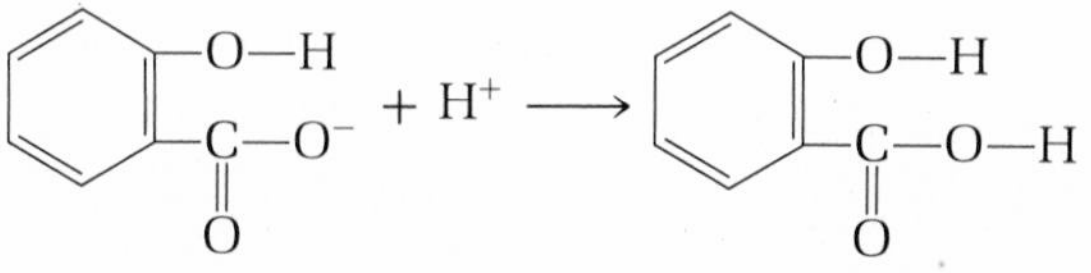

$$+ H^+ \longrightarrow$$

Salicylate
ion

Salicylic
acid

At this stage, the product contains one of the desired side chains (COOH) and is ready for the addition of the other. The reagent necessary for the addition is acetic anhydride, which can be prepared by the reaction of acetic acid with sulfur chloride (S_2Cl_2). Sulfur chloride is formed when chlorine gas is bubbled through molten sulfur,

$$2S + Cl_2(g) \longrightarrow S_2Cl_2(g)$$

Acetic acid results from the air oxidation of acetaldehyde, which in turn comes from the reaction of acetylene with water in the presence of catalysts. Pertinent equations of the reactions leading to acetic anhydride are

$$\underset{Acetylene}{C_2H_2(g)} + H_2O \xrightarrow[\text{catalyst}]{} \underset{Acetaldehyde}{CH_3-\overset{\overset{\displaystyle O}{\|}}{C}-H}$$

$$\underset{Acetaldehyde}{CH_3-\overset{\overset{\displaystyle O}{\|}}{C}-H} + \tfrac{1}{2}O_2 \longrightarrow \underset{Acetic\ acid}{CH_3-\overset{\overset{\displaystyle O}{\|}}{C}-O-H}$$

$$\underset{Acetic\ acid}{8CH_3-\overset{\overset{\displaystyle O}{\|}}{C}-O-H} + 3S_2Cl_2 \longrightarrow \underset{Acetic\ anhydride}{4CH_3-\overset{\overset{\displaystyle O}{\|}}{C}-O-\overset{\overset{\displaystyle O}{\|}}{C}-CH_3} + H_2SO_4 + 6HCl + 5S$$

Now, at long last, we are ready for the final step. Acetic anhydride and salicylic acid are heated together, and aspirin results by the following substitution reaction:

| Salicylic acid | Acetic anhydride | Aspirin | Acetic acid |

QUESTIONS

20.1 *Addition Reactions* (a) What experimental evidence is there that the bromination of ethylene occurs in two steps? (b) Show clearly that the first attack on ethylene in this case is electrophilic and the second is nucleophilic.

20.2 Elimination Reactions Draw electron-dot formulas for ethanol and each of its subsequent intermediates on the way to its formation of ethylene through the elimination reaction in H_2SO_4. Indicate which atom seems to bear net positive charge in the charged intermediates.

20.3 Substitution Reactions Ethers are usually made from alcohols. (a) Although ethanol can be directly converted to diethyl ether through reaction with H_2SO_4 as dehydrating agent, this is not a good synthesis for an unsymmetrical ether such as methyl ethyl ether. Why not? (b) Outline the reaction steps leading to preparation of methyl ethyl ether, starting from appropriate alcohols.

20.4 Aromatic Substitution (a) Why is aromatic substitution usually electrophilic? (b) Tell how you could generate two different electrophiles suitable for aromatic substitution? (c) Why don't aromatic hydrocarbons undergo addition reactions in preference to substitution?

20.5 Polymerization (a) How does a polymer differ from a very-high-molecular-weight compound? (b) How does addition polymerization differ from condensation polymerization? (c) What characteristics are necessary for molecules to undergo each kind of polymerization?

20.6 Synthesis of Organic Compounds Most organic reactions do not result in the complete conversion of starting materials to desired products. For a single reaction, 90 percent conversion is usually considered good. Suppose that a synthesis requires eight successive reactions. If each results in 90 percent conversion to the desired product for use in the next step, what percent of the starting material will end up as desired product after the eight successive reactions?

20.7 Aromatic Substitution (a) How would you go about converting benzene to meta nitrotoluene? (b) How would you alter things if you desired instead the ortho or para isomer?

20.8 Nylon (a) In order to get polymers of nylon having molecular weight of at least 100,000, how many amide links per molecule would have to be formed? (b) For this case, how precisely unity would the stoichiometric ratio of starting components have to be?

20.9 Markovnikov Addition What product would you expect to result from each of the following addition reactions: (a) $HBr + CH_2{=}CHCH_2CH_3$, (b) $HBr + CH_2{=}CHCH_2CH{=}CH_2$, (c) $HBr + CH_3CH{=}C(CH_3)_2$?

20.10 Aspirin Synthesis Consider the series of reactions for the preparation of aspirin given in Section 20.6. (a) In forming phenolate ion, why was benzenesulfonic acid formed first? (b) In converting phenolate to salicylate, what other product might you expect to form? Justify your answer with a resonance structure for the intermediate in this case.

20.11* *Stoichiometry* (a) How many liters at STP of ethylene are needed to make 1.00 kg of polyethylene? (b) How many liters of ethanol (density 0.789 g/ml) are needed to produce this much ethylene by the elimination reaction of Section 20.2?

20.12 *Addition Reaction* In addition to the mechanism presented in Section 20.1 for adding HBr to unsaturated hydrocarbons, the reaction can take place in the presence of catalysts which form $:\ddot{B}r\cdot$ atoms. In this case the Br atom reacts with the double bond to add Br at one end and an unpaired electron at the other. This free radical reacts with HBr to add H to the C bearing the odd electron and produces a $:\ddot{B}r\cdot$ which reacts with another hydrocarbon. (a) Consider the reaction of HBr with $CH_3CH{=}CH_2$ by this mechanism and draw the two possible intermediates and two possible products. (b) It is found that "anti-Markovnikov" addition predominates by this mechanism. Which intermediate is the one formed?

20.13 *Reactions and Structures* Draw structures for the products you would expect to be formed in each of the following reactions: (a) methyl ethyl ketone plus ammonia, (b) acetone and sulfuric acid, (c) 3-methyl-1-butene and HCl.

20.14 *Nucleophile and Electrophile* (a) What is the fundamental distinction between a nucleophile and an electrophile? (b) Give an example of each. (c) Depending on concentration, sulfuric acid contains HSO_4^- or SO_3. One is a nucleophile and one, an electrophile. Which is which? Justify your answer.

20.15 *Terms* State what is meant by, and give an illustration of, the following terms: (a) carbonium ion, (b) nucleophile, (c) electrophile, (d) ketone, (e) aldehyde, (f) Markovnikov's rule.

20.16 *Nucleophiles and Electrophiles* Determine which of the following compounds are nucleophiles and which are electrophiles: (a) ethylene, (b) CN^-, (c) H^+, (d) NH_3, (e) $CH_3\overset{+}{C}HCH_3$, (f) benzene.

20.17 *Electrophilic Substitution* In the reaction of Cl_2 with benzene to produce C_6H_5Cl, $AlCl_3$ acts as catalyst. Account for this fact in terms of a mechanism.

20.18 *Reactions and Structures* Write the structural equations for the following reactions: (a) nitration of nitrobenzene in H_2SO_4, (b) reaction of chlorobenzene with ethylchloride in the presence of $FeCl_3$, (c) reaction of toluene with Br_2 in the presence of $AlBr_3$.

20.19 *Reactivity* Arrange the following compounds in order of increasing reactivity toward the electrophilic reagents: toluene, 1,3-dinitrobenzene, benzoic acid, chlorobenzene, and benzene.

20.20 *Stoichiometry* Calculate the volume of hydrogen gas produced at STP when 5.00 g of sodium metal is added to 250 ml of propanol.

20.21* *pH* The base constant (K_b) for ethylenediamine is 8.5×10^{-5}. What is the pH of (a) 1.0 M; (b) 0.10 M; (c) 0.60 M ethylenediamine solution?

20.22 *Reactions* A student carried out an experiment in which 2-butanol reacted with concentrated HI. The following three compounds were isolated: 2-isobutane, 1-butene, and 2-butene. Explain why there were three products instead of only one.

20.23 *Resonance* Using the resonance structures of benzene, explain why NO_2, COOH, and COOR substituents on the benzene ring are meta directors in electrophilic reactions.

20.24 *Functional Groups* Given that a compound is 55.4% C, 7.7% H, and 36.9% O by weight, and that its molecular weight is less than 200, (a) what is the formula of the compound? (b) What are the possible combinations of functional groups in the molecule?

20.25* *Stoichiometry* Natural rubber is a high molecular weight, linear polyisoprenoid hydrocarbon of the composition $(C_5H_8)_n$ and the general structure

$$\left[CH_2\!-\!\underset{\underset{CH_3}{|}}{C}\!=\!CHCH_2 \right]_n$$

Given that the average molecular weight of a rubber molecule is 300,000, estimate the volume of hydrogen at STP needed to hydrogenate every double bond in 1 mole of rubber.

20.26 *Science and Society* As we run short on petroleum reserves, priorities will have to be set on its use as fuel vs. its use in synthesizing fibers, pharmaceuticals, rubber, food products, etc. With careful attention to natural methods of synthesis and alteration of our way of life, how would you set the priorities? Give all your reasons.

20.27 *Reactions and Structure* Draw structures for the products you would expect to be formed in each of the following reactions: (a) catalytic hydrogenation of 2-methyl-1-butene, (b) reaction of n-butanol and proprionic acid in slightly acid solution, (c) treatment of toluene with methyl chloride in the presence of $AlCl_3$, (d) polymerization of styrene, $C_6H_5CH\!=\!CH_2$.

20.28 *Elimination Reaction* Consider the two mechanisms discussed for the production of $R\!-\!CH\!=\!CH_2$ from $RCHBrCH_3$. (a) Does either mechanism utilize a catalyst? (b) Experimentally, how could you determine whether the reaction was proceeding by which mechanism? (c) Suppose it proceeded by both, how could you alter the relative contributions of the two?

20.29 *Cross-Links* A high polymer consisting of linear molecules usually has the physical properties of a "heavy oil." It can be made more "solidlike" if the linear molecules are linked together in a network. Where this is desired, cross-links are introduced through use of reagents such as divinyl benzene, $CH_2{=}CHC_6H_4CH{=}CH_2$. Show how divinyl benzene could provide cross-links.

20.30 *Stoichiometry* In the synthesis of aspirin how many moles of C and how many moles of Cl_2 are used per mole of aspirin? (*Note:* C is used to make ethylene, which enters the synthesis at two points, and not all of it ends up in the aspirin.)

20.31 *Aromatic Substitution* It can be correctly argued that the explanation given in Section 20.4 for the meta-directing influence of nitro substitution in benzene is incomplete. That explanation concerns only the energy difference of nitrobenzene toward meta and ortho/para substitution prior to reaction. (*a*) What other consideration is important in determining the activation energies for substitution at different positions? (*b*) By means of three resonance structures for each of the three reaction intermediates (ortho, meta, para), show which intermediates are expected to be of higher energy due to the placing of positive charge on adjacent atoms and tell how this will influence the course of substitution reaction.

Fats and Carbohydrates

21

The food we eat is classified as *protein, carbohydrate,* or *fat.* These three classes of organic compounds comprise the material from which living organisms are constructed. As foods they serve both to replenish the structure of the organism and to furnish energy for its function. For the structural aspect, all three classes contain large extended molecules of high molecular weight. For the energy-supplying aspect, the molecules when oxidized furnish relatively high amounts of energy. The energy is greatest for the fats: on complete combustion, fats furnish 38 kilojoules/g; proteins provide 23 kilojoules/g; and carbohydrates, 18 kilojoules/g.†

In animals (including human beings), the structural role is primarily served by proteins, whereas in plants it is the carbohydrates that play these roles. Although fats also have structural functions, particularly in the membranes that separate water-soluble from water-insoluble portions of an organism, the principal role of fats in animals is as an energy reserve. For this purpose, animals convert carbohydrates to fats so that food energy can be stored in this form. There is a second reason for animals to store food as fat: In the oxidation of fat, large amounts of water are formed. This fact is especially important for animals which store fat prior to hibernation. The fat can then serve during hibernation as a source of both energy and water. The camel's hump is another example of water storage through the use of fatty tissue.

† We have consistently used the SI heat unit, the joule. Nutritionists continue to use the large calorie, which is the kilocalorie. To convert kilojoules to large calories, it is merely necessary to divide by 4.184.

We might note that it is the carbohydrates which are the basic energy source in all foods. The reason for saying this is that in photosynthesis plants utilize the sun's energy to form carbohydrates, which then serve directly or indirectly to nourish practically all living organisms. Thus, farmers grow plants to feed their animals who nourish their children who grow more plants. . . .

In our discussion of foods we will start with the fats because their chemistry is simpler than that of carbohydrates or proteins.

21.1 Constitution of Fats

Fats are the most important components of a general class of organic materials called *lipids* which are the *oil-soluble components of living matter.* The lipids include not only fats but also waxes and a few other diverse materials. Fats and waxes are esters of long-chain organic carboxylic acids. Fats may be either semisolid (e.g., lard) or liquid (e.g., olive oil). However, all fats have in common the fact that they are esters of the polyalcohol glycerol (also called glycerin). Waxes, on the other hand, are esters of other alcohols.

A typical fat, stearin, is shown in Figure 21.1. It is a complex ester in that it contains three ester functional groups (COO) per molecule. Stearin can be visualized as being formed by an esterification reaction in which three stearic acid molecules combine with one glycerol molecule as follows:

$$
3C_{17}H_{35}COOH +
\begin{array}{c}
H \\
| \\
H-O-C-H \\
| \\
H-O-C-H \\
| \\
H-O-C-H \\
| \\
H
\end{array}
\longrightarrow
\begin{array}{c}
H \\
| \\
C_{17}H_{35}COO-C-H \\
| \\
C_{17}H_{35}COO-C-H \\
| \\
C_{17}H_{35}COO-C-H \\
| \\
H
\end{array}
+ 3H_2O
$$

| *Stearic acid* | *Glycerol* | *Stearin* |

In the stearin molecule, the hydrocarbon chains ($C_{17}H_{35}$) are probably coiled

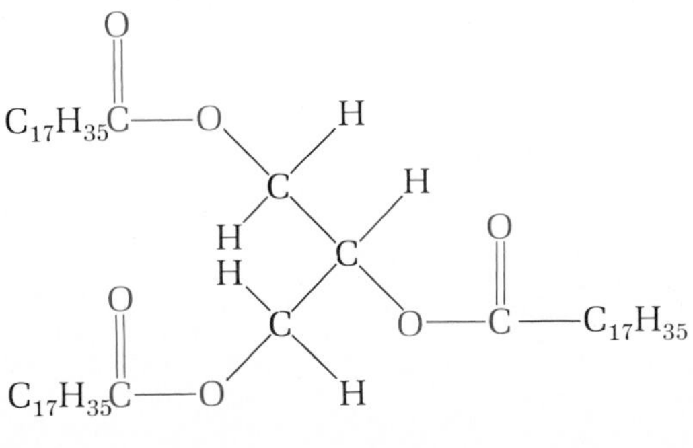

Figure 21.1
Stearin.

up and intertwined in such a way that the whole molecule is rather spherical in shape.

There is no fat that is pure stearin. All the natural fats and oils are mixtures of stearin and other related esters. Depending on the relative percentages of the various component esters, there are beef tallow, lard, olive oil, cottonseed oil, etc. The long-chain acids derived from natural fats are called *fatty acids* and have some striking similarities. In the first place, all consist of a straight-chain hydrocarbon with the carboxylic functional group attached to the terminal carbon atom. With the carbon of the carboxyl functional group, the fatty acids have an *even* number of carbon atoms.† The only known exception to the straight-chain nature and even number of carbon atoms is isovaleric acid, $(CH_3)_2CHCH_2COOH$, which is found in the fat from dolphin and porpoise.

The fatty acids range from the simplest, butyric acid, C_3H_7COOH, from butter fat, to as high as geddic acid, $C_{33}H_{67}COOH$, from Gedda wax. Most of the fatty acids, however, have less than 26 carbon atoms. Stearic acid, which has already been mentioned, is quite common in animal fats. Even more common is palmitic acid, $C_{15}H_{31}COOH$, which composes much of the fat in palm oil and has been reported in all fats.

As will be noticed from the formulas given for the fatty acids, those mentioned thus far contain saturated hydrocarbon residues. There are also found in nature, particularly in vegetable oils, fatty acid esters in which the organic residue is unsaturated, that is, contains one or more multiple bonds and a corresponding deficiency of hydrogen. Of the naturally occurring unsaturated fatty acids, all have 10 or more carbon atoms. Of these the most abundant seem to be oleic acid, $C_{17}H_{33}COOH$, and palmitoleic acid, $C_{15}H_{29}COOH$. Unsaturated fatty acids are by no means uncommon in nature. In fact, oleic acid is the most widely distributed of all the fatty acids; it alone composes 80 percent of the fatty acids in olive oil and is also found in beef tallow and pork fat.

Unsaturation in the fatty acids produces additional interesting regularities among the naturally occurring acids. Oleic acid, which is

$$CH_3(CH_2)_7CH=CH(CH_2)_7COOH$$

can be used to illustrate the regularity. In oleic acid the double bond occurs at exactly the center of the molecule. However, this is not the structural similarity which is carried over to other fatty acids. Instead, it is the fact that the double bond occurs nine carbon atoms from the carboxyl end of the molecule. Because of this structural regularity, which is found in all the

---

†As noted in Section 19.1, petroleum, which was formed after the appearance of life on earth, contains a predominance of *odd*-numbered hydrocarbons. The reason is the loss of carbon dioxide in the formation of petroleum from the *even*-numbered fatty acids of living matter.

other unsaturated fatty acids, there is none in nature that contains fewer than 10 carbon atoms. Another interesting structural consideration is the fact that at the double bond there can exist *cis-trans* isomerism. It turns out that oleic acid is invariably the *cis* isomer:

$$CH_3(CH_2)_7 \diagdown \qquad \diagup (CH_2)_7COOH$$
$$C=C$$
$$\diagup \qquad \diagdown$$
$$H \qquad \qquad H$$

The *trans* isomer does not occur in nature, although it is a known compound, called elaidic acid. If oleic acid is heated to about 200°C it will slowly convert to an equilibrium mixture of *cis* and *trans* isomers, with the *trans* isomer predominating 2:1. Apparently it is not an equilibrium consideration which makes nature form only the *cis* isomer but instead the result of the kinetic mechanism whereby the acid is formed.

In vegetable oils, there commonly occur unsaturated fatty acids that contain more than one double bond per molecule. For example, about one-third of the fatty acid in linseed oil is linoleic acid which has the formula

$$CH_3(CH_2)_4CH=CHCH_2CH=CH(CH_2)_7COOH$$

It is possible to have even more double bonds, as occurs in linolenic acid:

$$CH_3CH_2CH=CHCH_2CH=CHCH_2CH=CH(CH_2)_7COOH$$

The unsaturated fatty acids have a wider liquid range than do the saturated ones. This is illustrated by the fact that oleic acid melts about 55°C lower than stearic acid, for which the melting point is 70°C. There is a similar difference between the esters which make up natural fats and oils. Animal fats, with a higher fraction of saturated fatty acids, tend to be more solid than are vegetable oils.

During the evolution of the art of cooking, a prejudice developed in favor of the solid fat. Thus lard and butter were considered to be superior to the liquid vegetable oils. Because the vegetable oils are considerably more common in nature and hence cheaper, a technology was developed to hydrogenate the unsaturated vegetable oils and make them more like animal fats. For example, cottonseed oil (in which the fatty acid is 33% oleic and 44% linoleic) and corn oil (46% oleic and 42% linoleic) can be hydrogenated by adding hydrogen in the presence of nickel catalyst. Synthetic shortenings such as margarine and Crisco are basically mixtures of vegetable oils, animal fats, and partially hydrogenated vegetable oils. Through controlled hydrogenation the resulting product can have any desired softening temperature.

The prejudice in favor of solid cooking fats rests in part on the fact that unsaturated oils are susceptible to air oxidation to give products having disagreeable odors and tastes. On the other hand, it is believed from statistical evidence that there are certain health advantages in using unsaturated fats for cooking. It is thought that there is a correlation between the intake of saturated fats and the prevalence of arteriosclerotic diseases, which cause three to four times as many deaths as does cancer. These diseases consist of the blocking of critical arteries with a blood clot. The clot lodges where tough tissue has partially obstructed the artery. The prevalence of such blockage seems to be greater in people who regularly eat solid fats.† Although the mechanism for artery blockage is not understood, the dietary correlation is the foundation of the recent trend toward use of polyunsaturated oils in nutrition. Margarine is now produced by using emulsifying agents which make it reasonably solid even at room temperature without resorting to hydrogenation of the vegetable oils.

On a different note we might mention that the unsaturated oils are able to undergo an oxidative polymerization which produces a tough plasticlike material. This is the basis of the use of linseed oil and other so-called "drying oils" in paints. Also, the familiar floor covering, linoleum, is made by binding cork together with thickened linseed oil so that through oxidative polymerization the cork is bound into a tough, lasting material.

21.2   Metabolism of Fats

Fat from food can be transported throughout the body and stored until needed. In times of high carbohydrate intake the body converts carbohydrates to fat; however, the reverse process, conversion of fat to carbohydrate, apparently does not occur. This means that fats will accumulate until they are needed to produce the energy which the body requires to perform its many functions. The fats are thus our long-term energy supply and, as already noted, produce more energy per gram than either carbohydrates or proteins. The reason for this high energy content is the fact that the fats are largely hydrocarbon in composition and hence represent a more reduced state than do the proteins or carbohydrates, which contain higher ratios of oxygen to carbon. It follows then that in the oxidation of food the more reduced fat will produce the greatest amount of energy. Of course, it is the food with the highest fat content which is commonly spoken of as having "the most calories."

The metabolism of foods involves ultimately the burning of the organic matter to carbon dioxide and water. This process can be done in the labora-

---

†The statistical evidence is strong in indicating a direct correlation between arteriosclerosis and percentage of fat in diet, which varies from 17 percent for the South African Bantu to over 40 percent in the United States. However, the correlation is far from perfect and shows some outstanding exceptions. For example, pigeons eat no animal fat at all yet they develop arteriosclerotic lesions just like those found in human beings.

tory only at high temperature. In the body, it is carried out catalytically with a number of enzymes assisting various steps in the process. Each of these steps occurs at low temperature, and because each step involves the chemical reaction of only a few atoms in the molecule, a large number of steps are required. A tremendous amount of research in biochemistry has gone into the elucidation of the various enzymatic steps that occur in metabolism. In the following discussion we simplify the processes somewhat since a thorough discussion would get quite involved if it included all the numerous alternative paths and interlinked processes of delicate control and balance that are required by the living organism. We will discuss only the overall pattern of metabolism in the hope that the general course of the chemistry in the body will be made clear.

The first step in the metabolism of fat involves hydrolysis of the ester to form glycerol and fatty acids. This step is catalyzed by a class of enzymes called *lipases* which can carry out the saponification reaction at body temperature at rates comparable to those produced in the laboratory at high temperatures. After the fat is thus hydrolyzed, the glycerol is converted to glycerol phosphate, which can be utilized along with the carbohydrates in the carbohydrate metabolism that we will discuss later (Section 21.8).

The key part of fat metabolism involves the breakdown of the fatty acid. In addition to various enzymes, an important agent in the metabolism is the rather complex molecule called coenzyme A (or simply CoA) which has the formula $C_{21}H_{35}N_7O_{16}P_3SH$. Although its structure is known, it need not concern us in detail except that we need to note there is a replaceable H attached to the sulfur atom. We will simply abbreviate the entire formula as AH. In the first step of fatty acid metabolism coenzyme A reacts with the fatty acid† according to the equation

$$\underset{\substack{\| \\ O}}{RCH_2CH_2\overset{\displaystyle O}{\overset{\|}{C}}OH} + AH \longrightarrow RCH_2CH_2\overset{\displaystyle O}{\overset{\|}{C}}A + H_2O$$

In this process, coenzyme A has become attached to the fatty acid by replacement of the H and production of water. The coenzyme A is now bonded through its sulfur atom to the carbon of the carbonyl group. The above reaction requires energy and, as we shall see later, energy for such processes is furnished through the participation of adenosine triphosphate, ATP, discussed in Chapter 24.

The second step involves oxidation of the product with a biological oxidizing agent. In order to simplify our discussion, we will denote such

---

† In near-neutral body fluids the fatty acids and other carboxylic acids involved in metabolism are substantially in their ionized form; for example, RCH_2CH_2COOH would be present mainly as the anion $RCH_2CH_2COO^-$. However, for simplicity we follow the usual convention of writing formulas using nondissociated forms.

oxidizing agents by the symbol [O]. It is to be understood that an oxygen atom or molecule does *not* take a direct part in the reaction at this stage; however, ultimately it is oxygen that has produced the biochemical oxidizing agent. The second step of fatty acid oxidation, consequently, can be written

$$RCH_2CH_2\overset{\overset{\displaystyle O}{\|}}{C}A + [O] \longrightarrow RCH{=}CH\overset{\overset{\displaystyle O}{\|}}{C}A + [O]H_2$$

Note that an oxidation has occurred in that two hydrogen atoms have been removed from $RCH_2CH_2\overset{\overset{\displaystyle O}{\|}}{C}A$ to form $RCH{=}CH\overset{\overset{\displaystyle O}{\|}}{C}A$.

The third step involves hydration of the unsaturated product

$$RCH{=}CH\overset{\overset{\displaystyle O}{\|}}{C}A + H_2O \longrightarrow R\overset{\overset{\displaystyle OH}{|}}{C}HCH_2\overset{\overset{\displaystyle O}{\|}}{C}A$$

Subsequently a biochemical oxidation occurs:

$$R{-}\overset{\overset{\displaystyle OH}{|}}{C}H{-}CH_2{-}\overset{\overset{\displaystyle O}{\|}}{C}{-}A + [O] \longrightarrow R{-}\overset{\overset{\displaystyle O}{\|}}{C}{-}CH_2{-}\overset{\overset{\displaystyle O}{\|}}{C}{-}A + [O]H_2$$

At this point a second molecule of coenzyme A reacts with the product $RCOCH_2COA$ to form a molecule similar to the one originally formed from fatty acid plus coenzyme A except that the hydrocarbon chain is two carbon atoms shorter:

$$R{-}\overset{\overset{\displaystyle O}{\|}}{C}{-}CH_2{-}\overset{\overset{\displaystyle O}{\|}}{C}{-}A + AH \longrightarrow R{-}\overset{\overset{\displaystyle O}{\|}}{C}{-}A + CH_3{-}\overset{\overset{\displaystyle O}{\|}}{C}{-}A$$

The other product formed, CH_3COA, is called acetyl CoA since it contains the acetyl group CH_3CO joined to coenzyme A.

The steps of the process repeat with starting material RCOA in place of RCH_2CH_2COA until all the fatty acid chain has been broken down, each time with the removal of two carbon atoms. Since the natural fatty acids consist of an even number of carbon atoms, at the end of the process there is nothing left but acetyl CoA. The acetyl CoA must now be oxidized. This is done in a process, extremely important in all metabolism, called the Krebs citric acid cycle, in honor of Sir Hans Krebs of Oxford who first proposed it in 1937.

21.3 Krebs Citric Acid Cycle

The overall change for the Krebs cycle can be written as

$$CH_3\overset{\overset{\displaystyle O}{\|}}{C}A + 4[O] + 3H_2O \xrightarrow[\text{cycle}]{\text{Krebs}} 2CO_2 + 4[O]H_2 + AH$$

It shows that the acetyl group of acetyl coenzyme A has been oxidized to CO_2 with the production of coenzyme A (AH). In the process the biochemical oxidizing agents [O] are reduced to their reduced form $[O]H_2$, and there is a net consumption of H_2O.

The Krebs cycle consists of a series of consecutive steps in which the product of one reaction becomes the reactant of a subsequent step. The entire process is cyclic (repeats itself) in that the product of the last step, oxalo-acetic acid, feeds the first step. The first step involves the addition of acetyl CoA to oxaloacetic acid to form citric acid. As with practically all biological reactions, this step is catalyzed and controlled by an appropriate enzyme system, in this case called *condensing enzyme*.

$$CH_3-\overset{\overset{\displaystyle O}{\|}}{C}-A + \begin{array}{c} O{=}C-COOH \\ | \\ H_2C-COOH \end{array} + H_2O \longrightarrow \begin{array}{c} H_2C-COOH \\ | \\ HOC-COOH \\ | \\ H_2C-COOH \end{array} + \quad AH$$

| *Acetyl CoA* | *Oxaloacetic acid* | *Citric acid* | *Coenzyme A* |

In this step the acetyl group CH_3CO has been added to the oxaloacetic acid to increase the number of carbon atoms from four to six. This addition can be visualized as resulting from moving a hydrogen from the CH_3CO to doubly bonded O, joining the residual CH_2CO to the adjacent carbon, and adding H_2O to remove coenzyme A.

Next, citric acid is rearranged by the enzyme *aconitase* to give isocitric acid.

$$\begin{array}{c} H_2C-COOH \\ | \\ HOC-COOH \\ | \\ H_2C-COOH \end{array} \longrightarrow \begin{array}{c} H_2C-COOH \\ | \\ HC-COOH \\ | \\ HOC-COOH \\ | \\ H \end{array}$$

| *Citric acid* | *Isocitric acid* |

In step 3, isocitric acid is enzymatically oxidized through removal of two hydrogen atoms:

$$
\begin{array}{c}
\text{H}_2\text{C}-\text{COOH} \\
|\ \ \ \ \\
\text{HC}-\text{COOH} + [\text{O}] \longrightarrow \\
|\ \ \ \ \\
\text{HOC}-\text{COOH} \\
|\ \\
\text{H}
\end{array}
\qquad
\begin{array}{c}
\text{H}_2\text{C}-\text{COOH} \\
|\ \ \ \ \\
\text{HC}-\text{COOH} + [\text{O}]\text{H}_2 \\
|\ \ \ \ \\
\text{O}=\text{C}-\text{COOH}
\end{array}
$$

Isocitric acid *Oxalosuccinic acid*

The product oxalosuccinic acid is then decarboxylated, that is, CO_2 is lost from the middle carbon atom, to form α-ketoglutaric acid:

$$
\begin{array}{c}
\text{H}_2\text{C}-\text{COOH} \\
|\ \ \ \ \\
\text{HC}-\text{COOH} \longrightarrow \text{CO}_2 + \\
|\ \ \ \ \\
\text{O}=\text{C}-\text{COOH}
\end{array}
\qquad
\begin{array}{c}
\text{H}_2\text{C}-\text{COOH} \\
|\ \\
\text{HCH} \\
|\ \ \ \ \\
\text{O}=\text{C}-\text{COOH}
\end{array}
$$

Oxalosuccinic acid *α-Ketoglutaric acid*

In the next step, two things happen: CO_2 is lost and CoA is added with subsequent biochemical oxidation:

$$
\begin{array}{c}
\text{H}_2\text{C}-\text{COOH} \\
|\ \\
\text{HCH} \qquad + [\text{O}] + \text{AH} \longrightarrow \text{CO}_2 + [\text{O}]\text{H}_2 + \\
|\ \ \ \ \\
\text{O}=\text{C}-\text{COOH}
\end{array}
\qquad
\begin{array}{c}
\text{H}_2\text{C}-\text{COOH} \\
|\ \\
\text{HCH} \\
|\ \\
\text{O}=\text{C}-\text{A}
\end{array}
$$

α-Ketoglutaric acid *Succinyl CoA*

Hydrolysis of the resulting succinyl CoA regenerates coenzyme A and produces succinic acid:

$$
\begin{array}{c}
\text{H}_2\text{C}-\text{COOH} \\
|\ \\
\text{HCH} \qquad + \text{H}_2\text{O} \longrightarrow \\
|\ \\
\text{O}=\text{C}-\text{A}
\end{array}
\qquad
\begin{array}{c}
\text{H}_2\text{C}-\text{COOH} \\
|\ \\
\text{H}_2\text{C}-\text{COOH} + \text{AH}
\end{array}
$$

Succinyl CoA *Succinic acid*

Succinic acid is then oxidized to fumaric acid:

$$
\begin{array}{c}
\text{H}_2\text{C}-\text{COOH} \\
|\ \\
\text{H}_2\text{C}-\text{COOH}
\end{array}
+ [\text{O}] \longrightarrow
\begin{array}{c}
\text{HC}-\text{COOH} \\
\|\ \\
\text{HOOC}-\text{CH}
\end{array}
+ [\text{O}]\text{H}_2
$$

Succinic acid *Fumaric acid*

The enzyme *fumarase* now catalyzes the addition of water across the double bond to produce L-malic acid (the L isomer, where L indicates the configuration about the central carbon, of this optically active compound):

$$
\underset{\text{Fumaric acid}}{\begin{array}{c} \text{HC—COOH} \\ \| \\ \text{HOOC—CH} \end{array}} \quad + \text{H}_2\text{O} \longrightarrow \underset{\text{L-Malic acid}}{\begin{array}{c} \text{COOH} \\ | \\ \text{HOCH} \\ | \\ \text{H}_2\text{C—COOH} \end{array}}
$$

Finally L-malic acid is dehydrogenated (i.e., oxidized) to oxaloacetic acid, with which the cycle started.

$$
\underset{\text{L-Malic acid}}{\begin{array}{c} \text{COOH} \\ | \\ \text{HOCH} \\ | \\ \text{H}_2\text{C—COOH} \end{array}} \quad + [\text{O}] \longrightarrow \underset{\substack{\text{Oxaloacetic} \\ \text{acid}}}{\begin{array}{c} \text{COOH} \\ | \\ \text{C=O} \\ | \\ \text{H}_2\text{C—COOH} \end{array}} \quad + [\text{O}]\text{H}_2
$$

If the above steps are added, it will be seen that they do indeed result in the overall equation given earlier for the Krebs cycle. It is common practice in biochemistry to write cyclic processes such as this in the form shown in Figure 21.2. Although they take some getting used to, such diagrams are in common use as pictorial representations of rather complex chemical processes. The convention is that if a reactant is fed into the cycle it is shown entering from outside and a product which leaves the cycle is shown as breaking off from the main stream. Other biochemical processes are cyclic in nature, though none is so basically important as the Krebs cycle.

21.4 Carbohydrates

As we have noted, carbohydrates, although capable of supplying less energy than fats, are a more fundamental food energy source. Carbohydrates represent the first intermediates in the incorporation of carbon and hydrogen into living things through utilization of light energy. This process, called photosynthesis, will be discussed in Chapter 24.

The carbohydrates received their name because the ones that were first analyzed had the empirical formula $C_x(H_2O)_y$, in which carbon appears to be hydrated. Not only is the formula misleading but further studies showed that not all carbohydrates have such a formula. Furthermore, not all organic compounds with such a formula are classed as carbohydrates.

In a plant, carbohydrates form supporting tissues that make up its structure. The structural carbohydrates are called *polysaccharides* and, as we shall see, are composed of a tremendous number of simple monosaccha-

Figure 21.2
Krebs cycle.

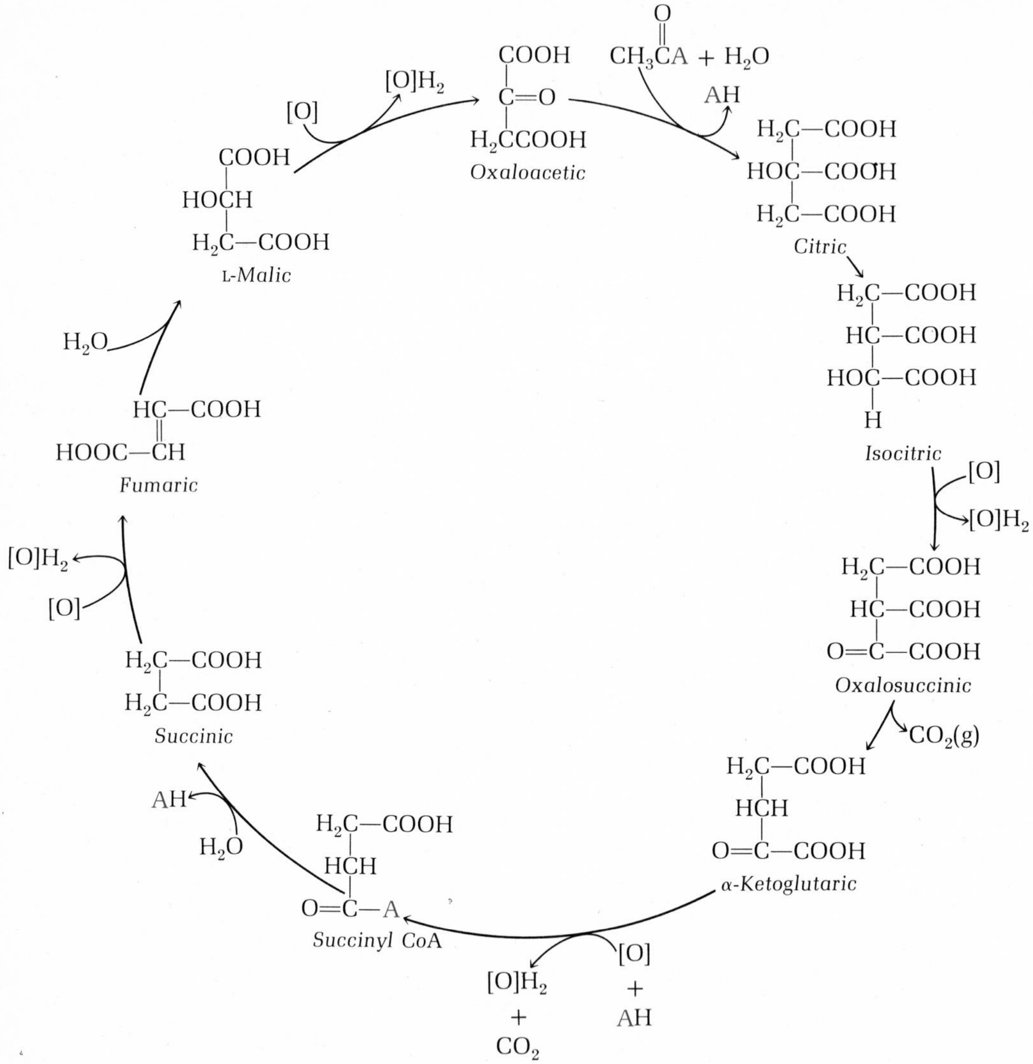

Figure 21.2 Krebs cycle.

ride units. The monosaccharides are *sugars*. (Possibly because they are such excellent sources of quick energy, we have evolved with a "sweet tooth" which has guided us to various natural sources of sugars.) Not all sugars are monosaccharides; some contain more than one saccharide unit and so are called *oligosaccharides* (from the Greek *oligos* meaning few).

21.5 Monosaccharides

In addition to serving as building blocks for carbohydrates, two monosaccharides occur in significant quantities in nature. They are glucose and fructose. Both contain six carbon atoms and have the molecular formula $C_6H_{12}O_6$. Of the large number of isomers of this formula, two in addition to glucose and fructose occur naturally in polysaccharides and oligosaccharides. Interestingly all four of these hexoses are optically active, and in each case the D isomer occurs in nature (as opposed to the L isomer†). The four naturally occurring hexoses are D-glucose, D-galactose, D-mannose, and D-fructose. Their formulas are

D-*Glucose*   D-*Galactose*

D-*Mannose*   D-*Fructose*

(In these diagrams the OH on the extreme right carbon of the ring can point up or down.) It should be noted that all but fructose are in the form of a six-membered ring with oxygen acting as one member of the ring. For fructose the ring is five-membered, and two of the six carbon atoms are attached outside the ring. Each of the formulas shown is asymmetric in the sense that if reflected in a mirror the mirror image would not be identical

†The labels D and L, used to distinguish the two configurations, refer ultimately to the arrangement in the molecule glyceraldehyde. This molecule can be viewed as a tetrahedron with a carbon in the center and the four different groups CHO, OH, H, and CH_2OH at the corners of the tetrahedron. If, when viewed from the outside, the tetrahedron shows H, CHO, and CH_2OH in clockwise arrangement on one of the faces, it is D-glyceraldehyde; if the H, CHO, and CH_2OH are disposed counterclockwise, it is L-glyceraldehyde.

with the original structure. Hence, these sugars are optically active and only one of the two optical isomers is found in nature.

There is a further problem in describing the structure of the monosaccharides, related to our noting above that in the formulas given the OH on the extreme right carbon can point either up or down. The problem can be illustrated by reference to glucose and noting that if it is allowed to stand the ring slowly opens as the following rearrangement occurs:

$$
\begin{array}{ccc}
\text{H}_2\text{COH} & \text{H}_2\text{COH} & \text{H}_2\text{COH}
\end{array}
$$

There are two consequences of the rearrangement. In the first place, when the ring is opened, an aldehyde group is present and consequently glucose, as well as galactose and mannose, gives a positive test for an aldehyde group by reducing Ag(I) or Cu(II). On the other hand, when fructose opens, it forms a ketone and so does not show this test. Consequently, fructose is the only one of the hexoses which is not termed a "reducing sugar." The other complication shown by the above structural equation is that, when the glucose ring recloses, there can result a changed arrangement of groups around the carbon that was the aldehydic carbon. In the process none of the other carbon atoms can change its spatial arrangement or else the ring would not close again. Because change can occur at the aldehydic carbon atom, the arrangement at this position is not permanently fixed. As we shall see, when the monosaccharides are linked into polysaccharides, the arrangement at this position also locks into place.

Glucose is also called *dextrose*. It is the principal sugar in blood, where it occurs at about 0.1 percent. In saline solution glucose is administered intravenously when oral nutrition is not possible. As the most abundant monosaccharide, glucose occurs widely in fruits such as grapes and so is also called *grape sugar*. Fructose also occurs widely in fruits, as well as in honey, but the other hexoses generally occur only as polysaccharides or oligosaccharides.

In addition to the hexoses there are two *pentoses* that occur naturally in complex, high-molecular-weight compounds such as the nucleic acids

(Chapter 23). These pentoses are called D-ribose and D-deoxyribose. Their formulas are

D-*Ribose*   D-*Deoxyribose*

(Again the OH on the extreme right carbon of each ring can be up or down.)

21.6 Oligosaccharides

Oligosaccharides are formed by the linking together of a few (two to ten) monosaccharides. Just as two alcohols can couple together, with the elimination of water to form an ether, so can the "polyalcohol" saccharides. Actually the linkage formed is not called an ether linkage because in oligosaccharides, unlike in ethers, one or both of the carbons attached to the linking O are attached also to another oxygen. This linkage is called a *glycoside* link and is exemplified by two familiar disaccharides. The first of them is sucrose, or ordinary table sugar. It has the structure

D-*Glucose* + D-*Fructose* = *Sucrose*

which is composed of D-glucose with a glycoside linkage to D-fructose.

Important natural sources of sucrose are sugar cane and sugar beets. Although both are about 15% sucrose, the sugar yield per acre from cane is higher than from beets. Because sugar cane grows only in tropical climates, its availability has been limited at various periods in history by political considerations. In fact, the sugar beet industry owes its origin in France to the British blockade during the Napoleonic Wars. Today, about half the world's supply of sucrose is from sugar beets.

Another familiar disaccharide is *lactose,* which makes up about 5 percent of milk. It is a disaccharide of D-galactose and D-glucose:

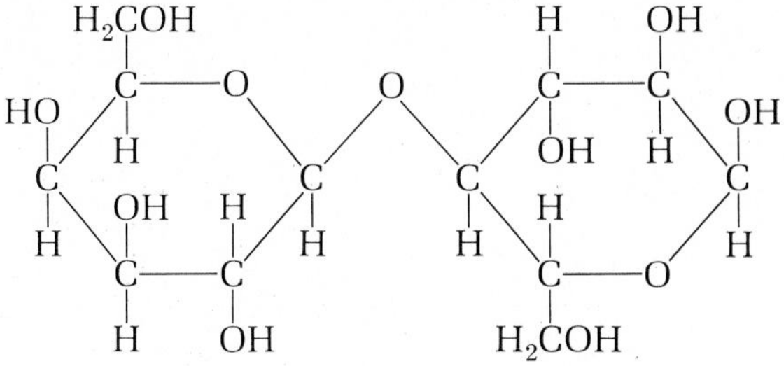

D-*Galactose* + D-*Glucose* = *Lactose*

As we noted earlier, the spatial arrangement of atoms at the carbon next to the ring oxygen is arbitrary for the monosaccharides. This is not the case for the higher saccharides. In the diagrams above, the glycoside link is shown in sucrose as being below the plane of the two rings and in lactose as being above. By convention, the link in sucrose is described as alpha and that in lactose as beta. The relative arrangement of the bridge oxygens with respect to near-neighbor atoms is important because in nature there are enzymes which can cleave an alpha linkage specifically without touching a beta linkage whereas another enzyme exists which is specific for the beta glycoside linkage.

21.7   Polysaccharides

Just as two monosaccharides can link to form a disaccharide, there is nothing to prevent the linkage of additional monosaccharides. For example, one glucose unit can link to a second glucose unit to form the disaccharide called *maltose.* At either end of the maltose unit additional glucoses can link to give large, nearly infinite, chains of glucose units:

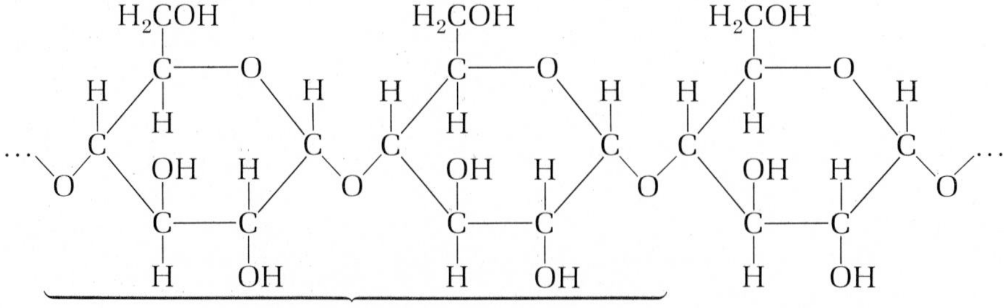

Maltose

The huge linear polymer formed by the end-to-end polymerization of glucose units is one of the forms of starch which represents a reserve energy storage for plants. However, not all starch has this relatively simple structure. Chemical studies on the degradation of starch indicate that there are two types of starch, one, called amylose, consisting of the end-to-end linkage of glucose units, and a second, called amylopectin, which contains both end-to-end linkages and side-chain linkages resulting from joining CH_2OH groups to other glucose units. When this occurs, branches in the chain are formed:

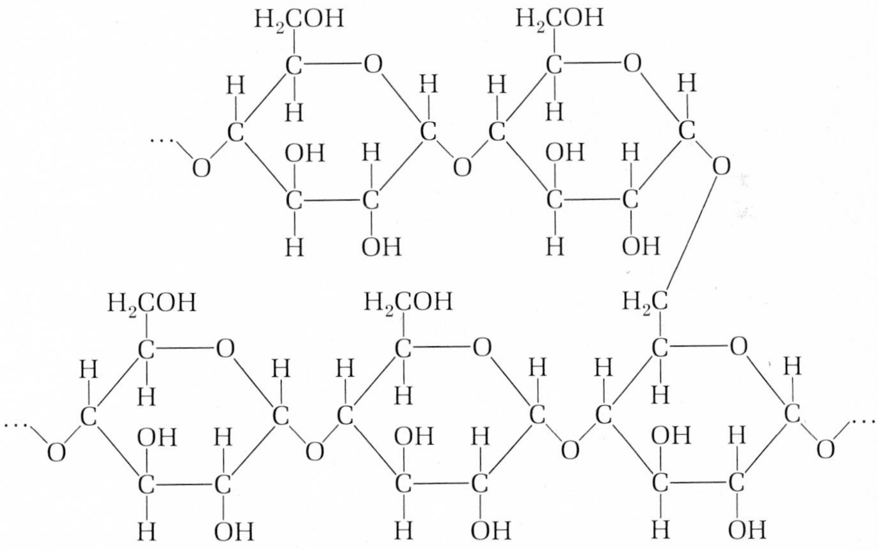

In the branched form of starch one CH_2OH unit out of about 20 or 30 glucose units takes part in the branching. It should be noted that both in the usual end-to-end linkage and in the side-chain linkage the stereochemistry at the glycoside linkage is alpha.

It is estimated that the molecular weight of starch is very high, ranging up to values of about a million. Even higher molecular weights are found in the related polysaccharide, glycogen, which is the form in which carbohydrate is stored in animals. It occurs in muscles and in the liver to be released on metabolic demand. Glycogen, like starch, consists of alpha-linked glucose units. However, it differs from starch in that not only are branched chains present but cross-links are formed which link one polymer chain to another. This kind of network structure is shown in Figure 21.3. Molecular weights of glycogen are estimated to be as high as a hundred million or so. Fortunately, no carbohydrate structures more complicated than glycogen, such as a true three-dimensional network, have yet been encountered.

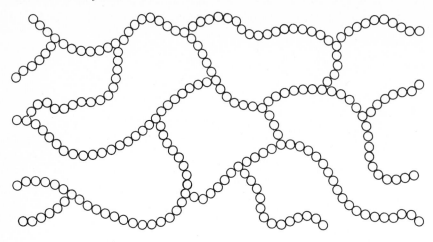

Figure 21.3

Representation of glycogen structure, with glucose unit shown as circle.

In all the polysaccharides discussed thus far, the linkage between monosaccharides has the alpha configuration. However, the beta configuration is also possible. Two glucose units with the beta link

constitute what is called a cellobiose unit. As with the maltose unit, cellobiose can perpetuate itself to give a long-chain polymer. The polymer, which appears to be unbranched, is *cellulose,* the most abundant organic compound in the world. Cotton, for example, is more than 90 percent pure cellulose with polymers having molecular weights on the order of half a million. Because of the different configuration or spatial arrangement at the glycoside linkage, animals are not able to hydrolyze cellulose and use it in their metabolism. However, there are microorganisms which have no such inhibitions in that they possess the appropriate enzyme apparatus for hydrolyzing the beta glycoside linkage of cellulose. Termites, notorious for eating wood, actually owe their ability to digest cellulose to microorganisms which they host in their digestive tracts.

21.8   Carbohydrate Metabolism

Because nearly all our energy supply comes from the sun by way of carbohydrates, the metabolism of carbohydrates can be considered the final step
in introducing solar energy into the life cycle of animals. Basically, this
metabolism involves the enzymatic hydrolysis of polysaccharide to monosaccharide (e.g., starch converted to glucose), followed by the oxidation of
the monosaccharide ultimately to carbon dioxide and water. For each of
these processes there are alternative paths; however, we shall consider only
one. We assume that, as the result of initial hydrolysis, glucose is formed,
and we follow one of the paths for metabolism of glucose.

For the metabolism of glucose, a key reagent is adenosine triphosphate,
$C_{10}H_{16}N_5P_3O_{13}$. As we will discuss later (Section 24.2), adenosine triphosphate (ATP) is an important carrier of chemical energy throughout biological
systems. In this case, however, it is important also in that it actually transfers
a phosphate group to the compound being metabolized. Glucose first reacts
with adenosine triphosphate to form the phosphate ester called glucose-6-
phosphate which is then enzymatically rearranged to fructose-6-phosphate:

Glucose-6-phosphate   *Fructose-6-phosphate*

A second molecule of ATP then reacts to add a second phosphate to the
fructose unit to form fructose-1,6-diphosphate which is subsequently split
by the enzyme *aldolase*:

Fructose-1,6-diphosphate   *Dihydroxy-*   *Glyceraldehyde-3-*
                                    acetone   *phosphate*
                                    phosphate

The two products of the reaction are in equilibrium with each other (i.e., they convert one into the other). It is the glyceraldehyde-3-phosphate which reacts further. As this molecule is removed from the equilibrium, the dihydroxyacetone phosphate rearranges to form glyceraldehyde-3-phosphate so that it can react. The net result is that two molecules of the reactive intermediate glyceraldehyde-3-phosphate result from each of the original molecules of glucose. The next step involves addition of phosphate, this time from inorganic phosphate rather than from ATP, to glyceraldehyde-3-phosphate in a process accompanied by an oxidation:

$$
\begin{array}{c}
HC{=}O \\
| \\
HC{-}OH \\
| \\
HC{-}OPO_3H^- \\
| \\
H
\end{array}
\quad + H_2PO_4^- + [O] \longrightarrow
\begin{array}{c}
O \\
\| \\
C{-}OPO_3H^- \\
| \\
HC{-}OH \\
| \\
HC{-}OPO_3H^- \\
| \\
H
\end{array}
\quad + [O]H_2
$$

Glyceraldehyde-3-
phosphate
1,3-Diphospho-
glycerate

In several additional steps the 1,3-diphosphoglycerate loses phosphate to reform ATP and convert the three-carbon chain to pyruvic acid, $CH_3COCOOH$. Pyruvic acid is then oxidized in the presence of coenzyme A to form acetyl CoA:

$$
\begin{array}{c}
COOH \\
| \\
C{=}O \\
| \\
CH_3
\end{array}
\quad + [O] + AH \longrightarrow
\begin{array}{c}
O \\
\| \\
CH_3{-}C{-}A
\end{array}
\quad + [O]H_2 + CO_2
$$

Pyruvic acid
Acetyl CoA

The acetyl CoA now enters the Krebs cycle (Section 21.3) and is fully oxidized ultimately to carbon dioxide and water. That the overall process results in the liberation of energy is seen from the fact that the original carbohydrate is now completely oxidized. We know that burning of organic compounds releases energy. In the metabolic process, however, part of the released energy goes into a net production of adenosine triphosphate. Overall, *two* molecules of adenosine triphosphate were used to form two molecules of glyceraldehyde-3-phosphate but two molecules of glyceraldehyde-3-phosphate result ultimately in the formation of *four* molecules of adenosine triphosphate; hence, there is for each glucose used a net production of two molecules of adenosine triphosphate. As we shall see in Section 24.2, adenosine triphosphate is the main method for transporting energy within organisms. In carbohydrate metabolism the net formation of adenosine triphosphate represents an important method of storing chemical energy to be drawn on for driving the organism.

On the other hand, the final stages of carbohydrate oxidation are sometimes blocked. This may occur, for example, when heavy exercise has caused a deficiency of oxygen. Under such circumstances, pyruvic acid, $CH_3COCOOH$, is converted to lactic acid, $CH_3CHOHCOOH$. Normally the elevated lactic acid level simply causes the discomfort of muscle fatigue, which quickly dissipates as the liver reconverts the lactic acid to glucose. It has recently been found that excessive buildup of lactic acid in the body can give rise to symptoms of anxiety neurosis (Dalosta's syndrome). It is interesting to speculate that perhaps certain mental disorders arise because relatively simple chemical reactions have gotten out of balance.

QUESTIONS

21.1 *Fats* A sample of fat is separated into several chemically distinct components. In what way are these components the same and in what way do they differ?

21.2 *Oils* (a) How do naturally occurring oils differ from the more solid fats? (b) How can the oils be made to resemble fats? (c) In what sense is this conversion possibly harmful if the product is used as food? (d) Would the converted oil be necessarily any more harmful than a natural fat? Explain.

21.3 *Metabolism of Fats* Why might a metabolic difficulty arise if you ate a synthetic fat composed of a fatty acid having an odd number of C atoms?

21.4 *Krebs Cycle* Consider the diagram of the Krebs cycle shown in Figure 21.2. (a) Where does the overall net reaction start and where does it end? (b) Why doesn't oxaloacetate appear in the equation for the net reaction? (c) How many steps are there in the cycle? (d) Of these steps how many consume reactants as listed in the overall net equation and how many produce products?

21.5 *Carbohydrates* By comparison of the formula for a simple carbohydrate with that for a fat, explain why the heat of combustion per gram is only half as large for the carbohydrate as for the fat.

21.6 *Monosaccharides* (a) Draw the mirror image of D-glucose and show that rotating it in space will not produce D-glucose or any other of the hexoses pictured in Section 21.5. (b) Show that interchange at the one C atom made possible by ring opening does not alter the situation.

21.7 *Oligosaccharides* In the formula given for lactose, show that replacement of the glycoside link with two OH groups gives monosaccharides which have the structures of D-galactose and D-glucose.

21.8 *Polysaccharides* Compare starch, glycogen, and cellulose with respect to each of the following: (a) monosaccharide building unit, (b) nature of glycoside links, (c) cross-links, (d) molecular weight.

21.9 *Carbohydrate Metabolism* One of the steps in carbohydrate metabolism involves the splitting of fructose-1,6-diphosphate. Half the original carbohydrate goes in this step to glyceraldehyde-3-phosphate which feeds the next step. How does the other half of the original carbohydrate get metabolized?

21.10 *Stoichiometry* Given a sample of fatty acid mixture from vegetable oil which has an average molecular weight of 298. If 0.00738 g of H_2 gas is used to completely satuate 1.000 g of the fatty acids, on average, how many $C=C$ double bonds per molecule were there in the oil?

21.11 *Carbohydrate Metabolism* By means of net equations show that the metabolism of glucose to acetyl CoA: (a) requires 2 moles of ATP per mole of glucose, (b) produces 4 moles of ATP per mole of glucose, (c) results in how many moles of acetyl CoA per mole of glucose? What happens to the rest of the C atoms of the glucose?

21.12 *Isomers* How many isomers are there of the same formula as linolenic acid? Sketch them.

21.13 *Heat of Combustion* Which would you expect to have the greater heat of combustion, stearic acid or oleic acid? Give two reasons for your answer.

21.14 *Fructose* Of the naturally occurring hexoses, only fructose fails to reduce Ag(I) or Cu(II). By means of structural formulas for pertinent equilibria, show why fructose is special in this regard.

21.15 *Terms* What is meant by the following terms: (a) lipase, (b) carbohydrate, (c) fatty acid, (d) Krebs cycle, (e) coenzyme, (f) glycoside link?

21.16 *Formulas* Write the formulas for the following compounds: (a) sucrose, (b) linoleic acid, (c) diethyl fumarate, (d) methyl pyruvate.

21.17* *Stoichiometry* How many liters of CO_2 at STP would be produced when 50.0 g of acetyl coenzyme CH_3COA is oxidized?

21.18 *Synthesis* How would you synthesize oxaloacetic acid from α-ketoglutaric acid? What would be your final yield if in each reaction you perform you recover 85 percent of your products?

21.19* *Heat of Combustion* Given the following molar heats of combustion: glucose, 2810 kilojoules; galactose, 2800 kilojoules; fructose, 2820 kilojoules; lactose, 5650 kilojoules; sucrose, 5640 kilojoules. How much heat would be given off by a 1.00-kg mixture that is 30.0% sucrose, 36.5% lactose, 15.2% galactose, 8.00% glucose, and 10.3% fructose by weight?

21.20 *Termites* How are termites able to destroy wood?

21.21 *Metabolism* Which are a better source of quick energy, fatty acids or oligosaccharides? Explain.

21.22 *Mechanism* Propose a mechanism for conversion of D-glucose to D-fructose.

21.23* *Starch* The molecular weight of a certain starch that consists only of glucose units is 1.0×10^6. Estimate the number of glucose *molecules* per mole of starch.

21.24 *Pyruvic Acid* Under normal conditions, pyruvic acid is oxidized to acetyl CoA. This oxidation is sometimes blocked when heavy exercise has caused a lack of oxygen. (*a*) What kind of reaction takes place when oxidation is blocked? (*b*) What is the product of that reaction?

21.25 *Glycoside Link* (*a*) How does a glycoside link differ from an ether link? (*b*) How do an α and β glycoside link differ? (*c*) Show that even though a molecule can be twisted about a C—O bond axis, such twists will not convert an α to a β glycoside link. (*d*) Apart from the fact that lactose has a β and sucrose an α link, what else is different about the glucose unit in the two disaccharides?

21.26 *Fat Metabolism* Noting that the heat of combustion of carbohydrates averages about 47 percent as great per gram as that of fat, estimate what fraction of the heat of combustion of stearin is due to the glycerol.

21.27 *Metabolism* Give a net equation for the complete metabolism of glucose (through the Krebs cycle, too). What relationship is there of this reaction to photosynthesis?

21.28 *Comparison* Discuss fats and carbohydrates with respect to each of the following: (*a*) molecular-weight ranges, (*b*) oxidation number of carbon, (*c*) functional groups, (*d*) solubilities.

21.29* *Glucose* The three equilibrium forms of D-glucose involved in the two simultaneous equilibria drawn in Section 21.5 are called α form (OH down), aldehyde form (ring open), and β form (OH up). At equilibrium in a neutral aqueous solution, the relative amounts of each present are about 33% α, 67% β, and 0.024% aldehyde. (*a*) Calculate K_s for each of the two equilibria as they are written. (*b*) What third equilibrium can you write and what is its K?

21.30 *Metabolic Rate* Mammals and other constant-temperature animals invariably have body temperatures above that of their environments (hence the name *warm-blooded*). (*a*) How are they able to maintain this temperature difference? (*b*) Why are they able to survive a much larger negative temperature difference (environment colder) than positive temperature difference (environment hotter)? (*c*) Why do small animals have higher metabolic rates than large animals?

22 Proteins

Of the classes of biological molecules none is so important for animal life as the proteins. The name itself comes from the Greek *proteios* meaning of the first rank. Proteins are macromolecules which form the structural materials from which animals are constructed. In addition, enzymes, which are biological catalysts that mediate and regulate metabolic processes, are proteins. Even antibodies, which prevent invading viruses and bacteria from attacking the living organism, are themselves proteins.

Some examples of important proteins are keratin, the primary constituent of skin, hair, and fingernails, and collagen, which makes up tendons and developing bones. Both of these proteins are of huge molecular weight; because of their function and structure, they are classified as *fibrous proteins*. Other proteins are called *contractile;* these include myosin and actin, which are structural components of muscle. Of great chemical interest are the metabolic proteins, or enzymes, which are roughly spherical in shape and so are called *globular proteins*. However, the detailed structural properties of the globular protein macromolecule are extremely important in producing the very specific catalysis which enzymes are called upon to perform. Many metabolic processes are regulated by *hormones*, which may be either proteins or the closely related peptides. Hormones regulate metabolism; for example, insulin controls carbohydrate and fat metabolism. All the proteins are polymers consisting of a large number of individual building units called *amino acids*. We will start our discussion of the proteins with the units from which they are built.

22.1 Amino Acids

In 1820 it was discovered that when the protein gelatin was heated with dilute acid the polymer was broken into its individual components. A major component was found to be the molecule glycine, NH_2CH_2COOH, which is the simplest of the amino acids. Structurally it is related to acetic acid CH_3COOH by replacing an H by the amino group NH_2. From 1820 to 1935 various experiments were performed to identify all the amino acids that make up proteins. Although some more have been found in nature, only about 20 occur regularly in proteins. When one considers the staggeringly large number of proteins—each species of animal, bacteria, and virus is composed of its own special proteins—it may seem strange that the number of structural units for all is as small as 20. However, a protein consists of hundreds, millions, or even more amino acid units per molecule, and so it should not be surprising that an enormous number of proteins can be assembled from 20 amino acids.†

All but one of the amino acids can be considered to have the general formula derived from glycine:

$$\underset{\underset{R}{|}}{\overset{\overset{\displaystyle H\quad O}{|\quad \|}}{H_2N-C-C-OH}} \qquad \text{or, better,} \qquad \underset{\underset{R}{|}}{\overset{\overset{\displaystyle H}{|}}{H_3\overset{+}{N}-C-CO_2{}^-}}$$

We should note, as already discussed (Section 19.9), that the amino acids when free do not exist in the form having a free amino group at one end and a carboxylic acid group at the other. Instead there is a transfer of proton from COOH to the more basic NH_2 end of the molecule. Structures for the 20 amino acids are listed in Figure 22.1. In each case the structural factor in common is indicated in blue. The last amino acid listed, proline, is not quite the same general form as the others. In proline the R group links around to the amino group. The other R groups of the amino acids may be acidic (carboxyl group) or basic (amine) and may even contain the additional element sulfur. As we shall see later, sulfur plays an extremely important role in protein structure in serving to form cross-links, which are important in bonding protein chains to each other.

Because the middle carbon atom is bound to four different groups in all amino acids except glycine, an asymmetry results which produces optical

† For a low-molecular-weight protein containing a hundred amino acid residues the number of possibilities for its composition is 20 raised to the 100th power. This is a number which is considerably greater than the number of atoms in the entire universe. It has been stated that the largest physically imaginable number would be the number of neutrons which could be close-packed into the universe. That number is considerably smaller than the number of *possible* amino acid sequences which can be formed from 20 naturally occurring amino acids.

Proteins

$$H_3\overset{+}{N}-\underset{\underset{H}{|}}{\overset{\overset{H}{|}}{C}}-CO_2^-$$

Glycine

$$H_3\overset{+}{N}-\underset{\underset{CH_3}{|}}{\overset{\overset{H}{|}}{C}}-CO_2^-$$

Alanine

$$H_3\overset{+}{N}-\underset{\underset{CH_3CHCH_3}{|}}{\overset{\overset{H}{|}}{C}}-CO_2^-$$

Valine

Figure 22.1
Amino acids.

$$H_3\overset{+}{N}-\underset{\underset{CH_3CHCH_3}{\underset{|}{CH_2}}}{\overset{\overset{H}{|}}{C}}-CO_2^-$$

Leucine

$$H_3\overset{+}{N}-\underset{\underset{CH_3CH_2CHCH_3}{|}}{\overset{\overset{H}{|}}{C}}-CO_2^-$$

Isoleucine

$$H_3\overset{+}{N}-\underset{\underset{CH_2OH}{|}}{\overset{\overset{H}{|}}{C}}-CO_2^-$$

Serine

$$H_3\overset{+}{N}-\underset{\underset{CH_3CHOH}{|}}{\overset{\overset{H}{|}}{C}}-CO_2^-$$

Threonine

$$H_3\overset{+}{N}-\underset{\underset{COOH}{\underset{|}{CH_2}}}{\overset{\overset{H}{|}}{C}}-CO_2^-$$

Aspartic acid

$$H_2\overset{+}{N}-\underset{\underset{COOH}{\underset{|}{(CH_2)_2}}}{\overset{\overset{H}{|}}{C}}-CO_2^-$$

Glutamic acid

$$H_3\overset{+}{N}-\underset{\underset{(CH_2)_4NH_2}{|}}{\overset{\overset{H}{|}}{C}}-CO_2^-$$

Lysine

$$H_3\overset{+}{N}-\underset{\underset{NH}{\underset{\|}{(CH_2)_3NHC-NH_2}}}{\overset{\overset{H}{|}}{C}}-CO_2^-$$

Arginine

$$H_3\overset{+}{N}-\overset{\overset{H}{|}}{C}-CO_2^-$$
$$|$$
$$CH_2$$
$$HC=C$$
$$N\quad NH$$
$$\underset{H}{C}$$

Histidine

$$H_3\overset{+}{N}-\underset{\underset{NH_2}{\underset{CO}{\underset{|}{CH_2}}}}{\overset{\overset{H}{|}}{C}}-CO_2^-$$

Asparagine

$$H_3\overset{+}{N}-\underset{\underset{NH_2}{\underset{CO}{\underset{|}{(CH_2)_2}}}}{\overset{\overset{H}{|}}{C}}-CO_2^-$$

Glutamine

$$H_3\overset{+}{N}-\overset{\overset{H}{|}}{C}-CO_2^-$$
$$|$$
$$CH_2$$

OH

Tyrosine

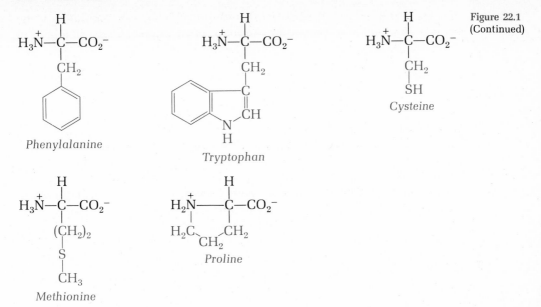

Phenylalanine

Tryptophan

Cysteine

Methionine

Proline

Figure 22.1 (Continued)

activity. This is illustrated in Figure 22.2 in which the two optical isomers of a general amino acid are shown. The two forms are designated as L and D. It is interesting that only the L-amino acids occur naturally in protein structures.

Finally, we should note that, although there are but 20 amino acids that regularly show up in proteins, there are other amino acids which occur free in biological systems. For example, $H_3N^+CH_2CH_2CH_2COO^-$, aminobutyric

Figure 22.2
Optical isomers of a general amino acid.

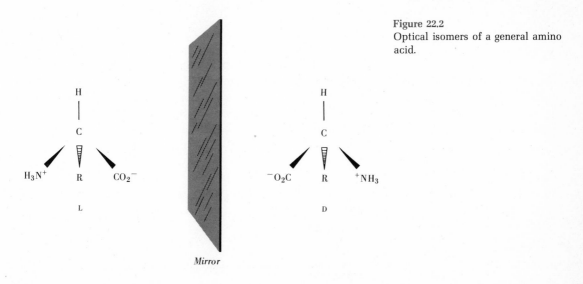

Mirror

acid, is present in brain tissue. Also, ornithine, which is very similar to lysine but contains one less CH_2 in the R group, is an important intermediate in the biosynthesis of urea but is never found in proteins.

22.2 Peptides

Proteins are characterized by the group —NH—CO—, called the *peptide link*. Figure 22.3 shows how the peptide link can be established between two amino acid molecules by splitting out H_2O. Further polymerization is possible, since there is a free NH_2 group (on the far right) and a free acid group (on the far left). The molecule derived from two amino acids, containing a single peptide link, is called a *dipeptide*; from three amino acids, a *tripeptide*, etc. It has become standard practice to call the class of molecules formed from a relatively small number of amino acid residues *peptides* and to reserve the name *protein* for molecules of higher molecular weight, i.e., something greater than 70 amino acid residues.

(a)

Figure 22.3
Formation of a peptide link.

(b)

Peptides have biological importance, particularly as hormones. For example, the pituitary hormone *oxytocin* controls milk secretion in mammals, and *vasopressin* regulates water excretion by the kidneys and influences blood pressure. Both were synthesized by Nobel Prize winner Vincent du Vigneaud. He first identified the sequence of nine amino acids which constitute each of the two hormones and then proceeded to synthesize the hormones. The structure of oxytocin is

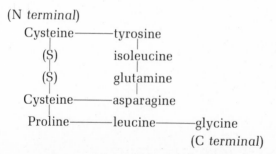

(N *terminal*)
Cysteine———tyrosine
(S)　　　　isoleucine
(S)　　　　glutamine
Cysteine———asparagine
Proline———leucine———glycine
　　　　　　(C *terminal*)

It consists of nine amino acid residues linked in a continuous chain which partly loops back on itself through a disulfide bridge involving the two cysteine residues. The chain of amino acids is linked together by joining the N end of one amino acid with the carboxyl end of the next. At one end of the chain, therefore, there will be a free N and the other end, a free carboxyl. The free nitrogen end is designated as the N terminal and the carboxyl end as the C terminal. In oxytocin, cysteine is the N-terminal amino acid and glycine is the C terminal; however, the C-terminal glycine has been converted to an amide, i.e., the OH of COOH has been replaced by NH_2 to give $CONH_2$. Vasopressin has a very similar structure except that arginine is substituted for leucine and phenylalanine is substituted for the isoleucine.

22.3   Primary Structure of Proteins

The structures just outlined for the two peptides do not tell us anything about the actual shape of the molecule or its configuration in space. They do, however, give extremely important information about the sequence of amino acids that are chemically linked together. In other words, they specify all the covalent chemical bonds in the molecule. Such structures are called *primary structures*. There are additional structural features not obvious from the primary structure which give the molecule its resulting configuration in space.

The first step in determining the primary structure of a protein is to determine its amino acid composition. This can be done by hydrolyzing the protein either enzymatically or by boiling in water with either acid or base serving as a catalyst. The resulting amino acids are then identified and quantitatively assayed. It is found that in proteins there are rarely less than 17 different amino acids. However, the relative amount of the various amino acids will vary considerably. For example, ribonuclease is a protein composed of 124 amino acid residues. All 20 amino acids shown in Figure 22.1 except tryptophan are present. The relative amounts range from 2 leucine residues to 15 serine residues per ribonuclease molecule.

The second step in determining structure is to fix the sequence in which the amino acids are arranged. This is an extremely difficult task. That to date the amino acid sequence is known in more than a thousand proteins is testimony to the fact that proteins are of such great importance. It may also indicate that scientists like to solve puzzles, and amino acid sequence determination represents a good example of puzzle solving in science.

Just as one starts to solve a jigsaw puzzle by working at the edges, so sequence determination starts at the ends of the molecule. At one end of the protein chain there is a free amino group that will undergo reaction with other molecules, for example, 2,4-dinitrofluorobenzene:

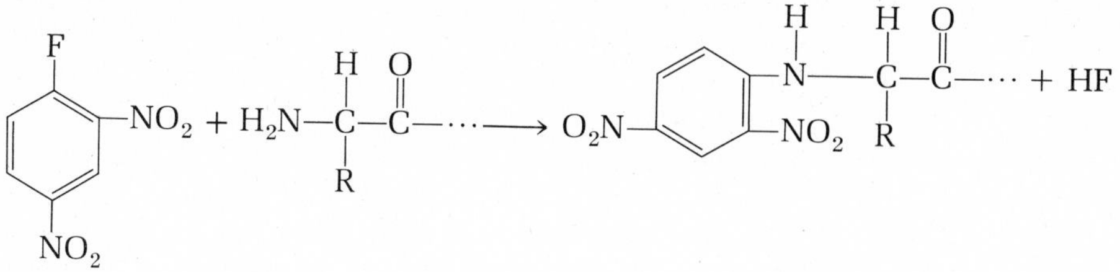

By carrying out this reaction with a protein, we have now managed to tag the end of the protein chain. After hydrolysis of the protein, one looks for the amino acid that is attached to the substituted benzene. Unless this amino acid is one of those which contains an NH_2 in its R side chain (e.g., lysine), it must have been at the end of the molecule. In this way, the N-terminal amino acid can be identified.

Similarly, there are reagents which will react with the carboxyl group at the opposite end of the protein chain. However, another trick is more commonly employed at this end. The enzyme carboxypeptidase has the ability to hydrolyze protein so as to split off mainly just the first amino acid at the C-terminal end of the chain. Therefore, the first amino acid formed in significant concentration after carboxypeptidase hydrolysis will be the residue which lies at the C-terminal end. The process, in principle, can be repeated on the resulting polypeptide, which is now one amino acid residue shorter than before. Unfortunately, after a few of the amino acids have been released, the sample contains various lengths of peptide chain so that hydrolysis does not produce a predominant amino acid.

A different method must be used to complete the sequence determination. This is done by partial hydrolysis of the protein to produce peptides of various lengths. These peptides are then separated from each other and identified as to amino acid content. That this can lead to a sequence determination can be seen by considering the special case of a protein that contains but one threonine residue. From the partial hydrolysis of this protein, suppose the peptide fragments that contain threonine also contain either glycine or serine. We would then know that threonine in the protein is joined on one side to glycine and on the other to serine. If for the tripeptide consisting of these three residues we can show (e.g., by using the previously mentioned dinitrofluorobenzene) that glycine is at the N-terminal end, we must have the sequence "glycine-threonine-serine" in the protein chain. Of course, if we do not have the special case of a single threonine residue per molecule, then we need more information in order to establish the sequence of the chain at each threonine residue. In general, this is the actual case, and a great deal of data is required in order to puzzle out the final sequence.

22.4 Protein Configuration

The sequence of amino acids in the protein molecule fixes the chemical nature of the molecule. However, its biochemical function depends not only on the chemical composition but on the spatial configuration of the entire molecule. An amino acid segment in the protein chain will be disposed in three dimensions in one of a variety of possible orientations. It has been found that the final three-dimensional configuration is fixed not only by the constraints produced by the covalent bonds that hold the amino acid sequence together but also by hydrogen bonding and other similar interactions between various segments of the protein chain, which may be influenced by external factors, such as pH.

A particularly important spatial configuration is the alpha helix, diagramed in Figure 22.4, which seems to be the most stable single configuration of a protein chain. An important feature of the α helix is the fact that it is right-handed; i.e., the amino acid chain forms the pattern of the thread

Figure 22.4
Protein helix.

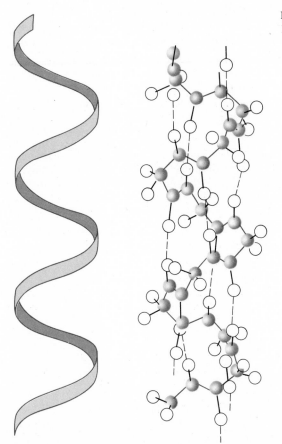

on a right-hand screw. In the alpha helix there are 3.7 amino acid residues per turn of the helix, and each NH group forms a hydrogen bond to the carboxyl group of the third amino acid residue further along the chain. These hydrogen-bonding interactions give the alpha helix its stable configuration. However, in the presence of water, hydrogen bonds can be formed not only within the protein molecule itself but also with the solvent; hence, the configuration of proteins frequently changes in solution.

A further complication is that, even if we can establish the fraction of a protein molecule that is in α-helix form (this can be done by a variety of physical measurements), we still do not have the whole story on its structure. The various helical segments of the protein molecule can hydrogen-bond with side groups; this and other interactions may fold the entire molecule into a more compact form. Generally, the alpha-helix structure is referred to as the protein's *secondary structure,* and the folding of helical segments into additional structural configurations is called its *tertiary structure.*

The tertiary structure of the protein myoglobin was the first one determined. This was done by X-ray diffraction studies by Nobel laureates J. C. Kendrew and M. F. Perutz. Figure 22.5 shows diagrammatically the tertiary structure of the protein myoglobin. Of the 151 amino acids in myoglobin, 118 are components of right-handed alpha helices. There are, in fact, 8 helical segments which vary from 7 to 24 residues in number. These in turn are

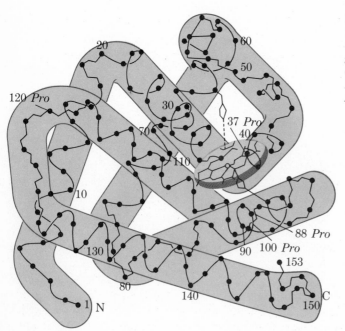

Figure 22.5
Structure of myoglobin. Numbers refer to amino acid subunits along the chain. The gray disk is the iron-containing grouping. (From "The Hemoglobin Molecule" by M. F. Perutz. Copyright © 1964 by Scientific American, Inc. All rights reserved.)

(a)   *(b)*

Figure 22.6

Quaternary structure of hemoglobin (*a*) from above; (*b*) side view. Round disks represent iron-containing groupings that bind oxygen to the molecule. (*From "The Hemoglobin Molecule" by M. F. Perutz. Copyright © 1964 by Scientific American, Inc. All rights reserved.*)

folded into the compact structure shown, which is very roughly describable as a triangular prism some 4.5 by 3.5 by 2.5 nm. Myoglobin is an enzyme which contains one iron atom per molecule.

In addition to tertiary structure, proteins may have additional structure called *quaternary structure*. This arises if two or more protein molecules are bonded together to form a unit. For example, hemoglobin, the oxygen carrier in human blood, consists of four subunits (each similar to myoglobin) bound together in a tetrahedral arrangement as indicated in Figure 22.6.

22.5 Biochemical Function of Proteins

Just as fats and carbohydrates can be burned in the body to produce energy, proteins can also be burned. However, because the body itself is largely composed of proteins it is of greater importance that the proteins in food are rearranged into the many different proteins that constitute the organism. This rearrangement is necessary because our bodies are able to synthesize only about half the amino acids used to make proteins. The other half must be synthesized by plants and microorganisms and fed to us. The latter amino acids are called the *essential amino acids*.

The essential amino acids need not be fed in the form of free amino acids, since our bodies are able to hydrolyze proteins into their constituent parts. This process is done by so-called *proteolytic enzymes*, which degrade the protein chain. One of these, pepsin, was first recognized in 1783, making it the earliest protein known. Pepsin, the protein-cleaving enzyme in the stomach, serves as a good example of the fact that it is proteins themselves which make possible the very fundamental chemical reactions that sustain life. Other enzymes, all of which are proteins, have been repeatedly mentioned as important regulators of metabolic processes.

One of the most active areas of recent research has been the attempt to understand the mechanism of enzyme-catalyzed reactions. Such reactions include all varieties of transformations, polymerization, depolymerization, molecular rearrangement, oxidation-reduction reactions, etc. In many instances the enzyme resembles a simpler nonprotein catalyst in speeding a chemical process. For example, although a number of metal ions are known to catalyze the decomposition of hydrogen peroxide into oxygen and water, the protein enzyme *catalase* is many times more efficient. In general, enzymes are truly amazing in the efficiency of their catalysis.

In addition to catalytic efficiency, enzymes possess other features which make them truly special. One of these is their great specificity. We have already noted that the enzymes which hydrolyze starch are specific for one stereochemical configuration of the glycoside linkage and will not act on the other configuration. This is a general phenomenon of all enzymatic catalysis; i.e., it is specific for a particular molecular configuration. The reason for this specificity is believed to be the fact that in the first step of an enzymatic process the reactant combines with the enzyme to form some sort of a molecular complex

$$E + S \rightleftharpoons ES$$

The reactant molecule is called a substrate and in the above equation is abbreviated S. In the formation of the enzyme-substrate complex (ES) the protein structure of the enzyme provides what is called an *active site* which can bind only to a substrate of the proper configuration. Following the binding of the substrate to the enzyme surface, reaction promptly occurs to convert the substrate to product and leave the enzyme surface free for attachment of another reactant molecule. In many reactions more than one reactant is involved and so of course the enzyme must bind both reactants in close proximity to each other so that the reaction can occur.

Although the mechanism of enzyme activation is still slowly being worked out, it seems that, for a large number of reactions, the reaction itself is accompanied by a change of protein configuration. This change of protein configuration forces reaction to occur. Furthermore, because protein configuration can be controlled by external conditions, the enzymatic process in question can be externally regulated. This is the third characteristic of enzyme reactions: they can be controlled by the biological system so as to coordinate the large number of chemical reactions that must occur simultaneously in order to bring about the various biological processes.

A simple example of the complexities of the biological functions of proteins is afforded by hemoglobin. As Figure 22.6 shows, the molecule consists of four subunits. Each subunit is similar to myoglobin (Figure 22.5) in that it contains an iron atom which can bind an oxygen molecule. The function of this oxygen binding is to transport O_2 from the lungs to various points in the organism where it can be utilized. In connection with the oxygen binding and release, the protein structure changes subtly so as to pump oxygen on and off the iron atom. In connection with this structural change, it is interesting to note that the four subunits of hemoglobin do not behave independently. The structure of one subunit influences the oxygen binding to a second subunit, so that once oxygen is bound to one of the units the binding of the second, third, and fourth is made easier. Such influences of structure on chemical reactivity are fundamental to the control of biological processes. Coupled with the chemist's natural interest in the relationship between molecular structure and chemical reactivity, the potential chemical control of biological processes serves as a stimulus for the intensive research which is under way in this area.

Each of the many steps in all the many reactions occurring in a living organism requires one or more enzymes. Each enzyme in turn is a specific protein which owes its properties to its configuration. The properties are ultimately determined by the amino acid sequence of the protein. A particularly interesting problem, then, is the synthesis of each of the many protein molecules from amino acids. Protein synthesis, one of the truly fundamental steps in the formation of living organisms, is one to which we shall return. First, however, we must discuss, in Chapter 23, an important class of compounds called nucleic acids, which play a vital role in protein synthesis.

QUESTIONS

22.1 *Amino Acids* (*a*) Write the general formula for an amino acid. (*b*) Give an example of an amino acid in which the R group is acid. (*c*) Give an example where R is basic.

22.2 *Optical Activity* (*a*) Why isn't glycine optically active? (*b*) Is aminobutyric acid?

22.3 *Peptides* (*a*) From chemical analysis for the relative amounts of elements present, could you tell whether a given substance was a peptide, an amino acid, or a protein? (*b*) What information would you need to distinguish the three? Tell semiquantitatively how the three differ in this property.

22.4 *Vasopressin* (*a*) Indicate the primary structure of vasopressin. (*b*) Why do you need to indicate which end is which?

22.5 *Primary Protein Structure* (a) To fully characterize a protein, why isn't it sufficient to know its composition in terms of the constituent amino acids? (b) What chemical reactions are useful in determining primary protein structure? (c) If you know the sequence of amino acids, could you draw a protein's configuration? Explain.

22.6 *Protein Configuration* Distinguish clearly between the secondary, tertiary, and quaternary structure of a protein by telling whether each involves interaction (such as hydrogen bonding) between amino acid residues in different molecules, in the same molecule relatively near each other in amino acid primary sequence, or in the same molecule but relatively far apart in primary sequence.

22.7 *Alpha Helix* Protein chains can form the alpha helix no matter what particular sequence of amino acids is present. (a) Why is this? (b) Explain why the extent of alpha-helix formation changes as solvent is changed? (c) Would you expect greater or lesser extent of alpha-helix formation at high vs. low temperature? Explain.

22.8 *Enzyme* In many cases the active site of an enzyme is believed formed by two or more amino acids which are quite far apart in primary sequence. Tell how this is possible in terms of protein conformation.

22.9 *Biochemical Function of Proteins* Tell what is meant by each of the following: (a) proteolytic enzymes, (b) essential amino acids, (c) enzyme specificity, (d) enzyme-substrate complex.

22.10 *Red Cells* A normal human red blood cell contains 265 million hemoglobin molecules of molecular weight 66,000. What weight of hemoglobin is this?

22.11* *Composition* Proteins generally are 16 to 18% N by weight. Which amino acid has the highest percent N by weight? What is its percent N?

22.12 *Peptides* Compare the simplest formulas of oxytocin and vasopressin.

22.13 *Myoglobin* In the myoglobin structure which of the following pairs of amino acid residues is closest together and which pair is farthest separated: (a) 105 and 121, (b) 17 and 118, (c) 41 and 51, (d) 13 and 74?

22.14 *Proline* Despite its structure, which is somewhat different from the other amino acids, proline can take part in both peptide-link and alpha-helix formation. Show that this should be true.

22.15 *Serine* Draw three-dimensional representations of the two optical isomers of serine. Which occurs in nature?

22.16 *Hemoglobin* (a) What is the function of hemoglobin in our blood? (b) How is the function achieved?

22.17 *Amino Acids* What are some chemical reagents that are used for the identification of amino acids in proteins?

22.18 *Proteins* How do fibrous proteins differ from contractile proteins?

22.19* *Composition* What is the percentage composition of elements in a linear unbranched protein that contains one of each of the amino acids listed in Figure 22.1?

22.20 *Optical Activity* Why might it be that only L-amino acids occur naturally?

22.21 *Heat of Combustion* Show that it is reasonable that the heat of combustion of protein is intermediate between that for fats and for carbohydrates.

22.22 *Alpha Helix* By reference to Figure 22.4, draw a portion of an alpha helix showing which atom is N, which O, and where the R groups are.

22.23* *Peptides* Consider all possible peptides, 10 amino acid residues long, that can be made from any combination of the 20 amino acids. (*a*) What range of molecular weights will these peptides exhibit? (*b*) How many distinct peptides are possible?

22.24 *Equilibria* In acid solution glycine exists as $H_3N^+CH_2COOH$. As mentioned in Section 19.9, this species dissociates in two steps with dissociation constants of 5×10^{-3} and 2×10^{-10}. (*a*) Write an equilibrium for the two-step dissociation of protonated glycine and calculate the equilibrium constant. (*b*) Calculate from your answer to (*a*) the isoelectric point for glycine, i.e., the pH at which the concentration of the positively charged form equals the concentration of the negatively charged form.

22.25* *Sequence* Suppose you have a peptide composed of one each of the amino acids glycine, alanine, valine, and leucine. (*a*) How many possible structures are there for the peptide? (*b*) Through reaction with 2,4-dinitrobenzene you learn that glycine is N terminal. Now how many possibilities are there? (*c*) Through use of carboxypeptidase, you learn that alanine is C terminal. Now how many possibilities remain? (*d*) How would you proceed to determine which possibility is correct?

Nucleic Acids 23

A key attribute of a living organism is its ability to reproduce itself. Not only must reproduction link generation to generation, but there must be a regular renewing of structural materials during the growth and constant rebuilding of the living organism. A classic topic of great importance in biology has been the study of genetics, since the genes contain information that serves as the complete plan for the reproduction of the organism and all its functional parts. As we have seen, both the structural and operational parts of animals are largely protein in nature; thus, it is the job of the genes to provide the information that dictates protein synthesis.

During cell division discrete physical structure units called *chromosomes* become visible in the nucleus of the dividing cell. Chromosomes contain the *genes*, or genetic information centers, which set the pattern for the chemical nature of the developing organism. It is known that the chromosomes consist of chemical substances called *nucleic acids*. As we shall see, the nucleic acids both contain the genetic information and are involved as the templates for protein biosynthesis from amino acids.

The nucleic acids themselves are polymers consisting of simpler units. Because there are various functions for the various nucleic acids, the number of constituent units in them varies from as few as 80, in so-called transfer RNA (ribonucleic *acid*), to numbers as high as 10^8 or greater, in chromosomal DNA (deoxyribonucleic *acid*). Viruses are also nucleic acids and have the property of being able to invade a host organism and cause it to reproduce virus rather than host material.

23.1 RNA

Ribonucleic acids are polymers held together by phosphate linkages which bridge substituted sugar molecules. The phosphate links are shown in Figure 23.1. As is seen in the figure, each phosphate linkage carries a negative charge which is neutralized by a counter ion, a metal ion such as sodium or magne-

Figure 23.1
RNA polymer.

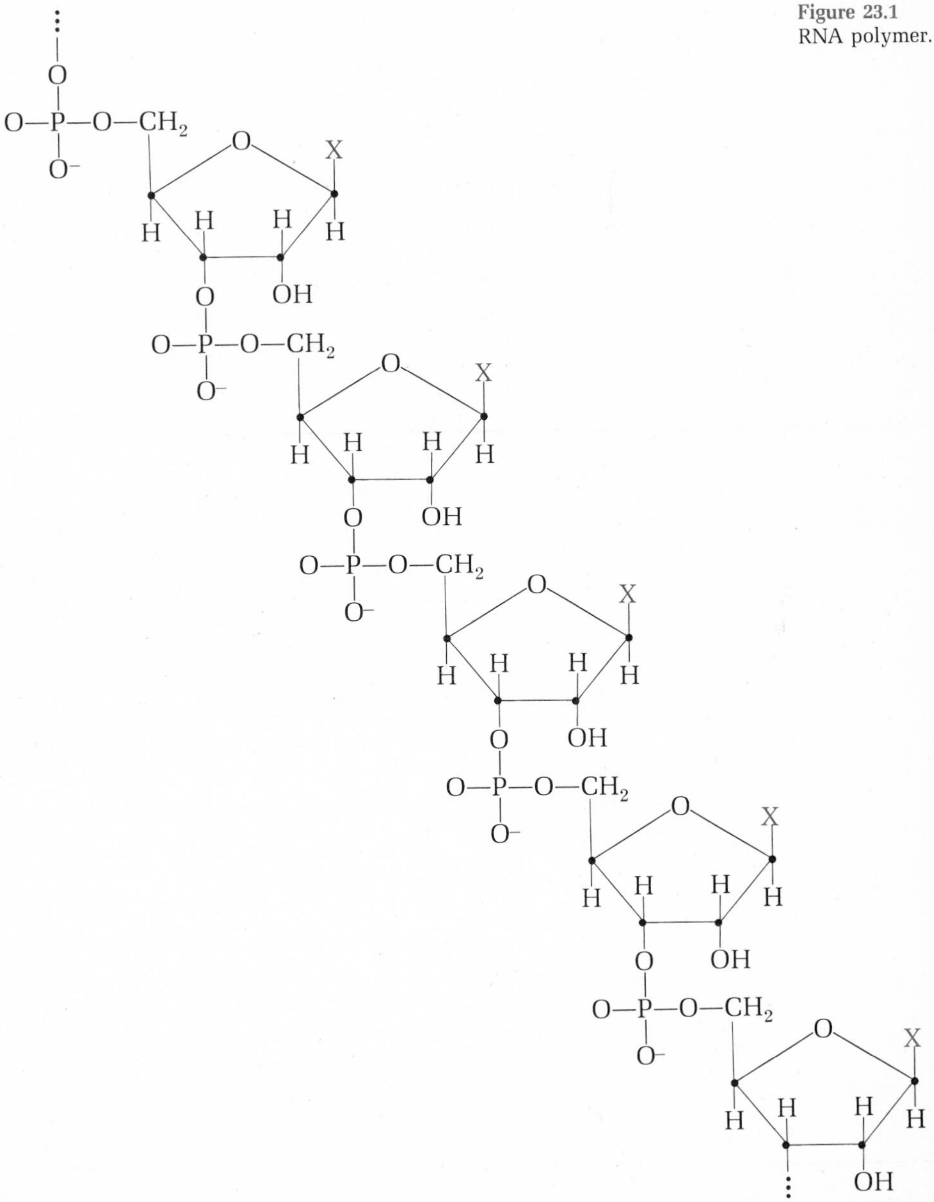

sium or a more complex organic molecule which might be a substituted ammonium ion. The repeating unit phosphate-sugar-X is called a *nucleotide*. As shown in Figure 23.1, all the nucleotides of RNA contain ribose (a pentose) as the sugar unit.

The X groups which are attached to the ribose are either derivatives

of pyrimidine   or of purine

Although neither pyrimidine nor purine occurs as such in nucleic acids, the bulk of the nucleotides of RNA contain only four derivatives. These four are listed in Figure 23.2. It can be seen that adenine and guanine are substituted purines whereas uracil and cytosine are substituted pyrimidines. We will abbreviate the entire nucleotide which contains the adenine residue as A; that containing cytosine as C; guanine, G; uracil, U. It should be noted that each of the residues is a nitrogen base, but in addition to these bases there occur in RNA a few other so-called "strange bases," the function of which is not understood at this time.

By methods similar to those used for determining amino acid sequences in proteins (Section 22.3), the nucleotide sequence of one RNA has been determined. This is the case of a particular RNA obtained from yeast, for which the sequence is as given in Figure 23.3. The task of determining the

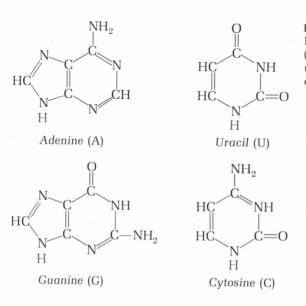

Adenine (A)

Uracil (U)

Guanine (G)

Cytosine (C)

Figure 23.2
RNA nucleotide substituents (attached at X positions in Figure 23.1 by replacement of colored H).

$$G-G-G-C-G-U-(CH_3G)-C-G-C-U-A-G-(H_2U)-C$$
$$U-U-C-C-C-U-C-(CH_3GCH_3)-C-G-C-G-A-(H_2U)-G-G$$
$$I-G-C-(CH_3I)-Z-G-G-G-A-G-A-G-U-C-U-C-C$$
$$U-C-A-G-G-C-C-U-U-A-G-C-Z-T-G-G$$
$$C-G-U-C-C-A-C-C-A$$

Figure 23.3
RNA from yeast. Substituent bases indicated by symbols as defined in Figure 23.2. "Strange substituent bases" are I for inosine, T for thymine, and Z for pseudouridine. Some bases are modified by substitution of methyl groups or addition of hydrogen as indicated in parentheses, G end terminates with phosphate group; A end, with ribose.

sequence for the nucleic acids is even more difficult than that of determining the sequence in amino acids because there are fewer components.† Hence, the job that was performed by Nobel Prize winner R. W. Holley and his co-workers was properly hailed as one of the masterpieces of modern research.

23.2 Structure of DNA

Deoxyribonucleic acid is structurally similar in many respects to ribonucleic acid. One principal difference is in the identity of the sugar unit. As Figure 23.4 shows, the sugar ring of DNA contains no OH groups but only H atoms (compare Figure 23.4 with Figure 23.1). Although the absence of OH is reflected in the name *deoxyribonucleic acid*, there are other structural differences of greater significance. There is a difference in the identity of the attached nitrogen bases. Again four bases predominate with three of them being the same as before (adenine, cytosine, and guanine) but instead of uracil there is now a methyl derivative of uracil called thymine:

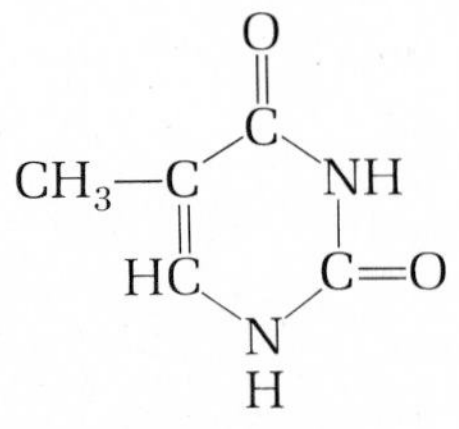

Thymine (T)

†Imagine trying to piece together a jigsaw puzzle. If the puzzle is printed in only four colors, each piece will be more difficult to fit in place than if a larger number of colors are used.

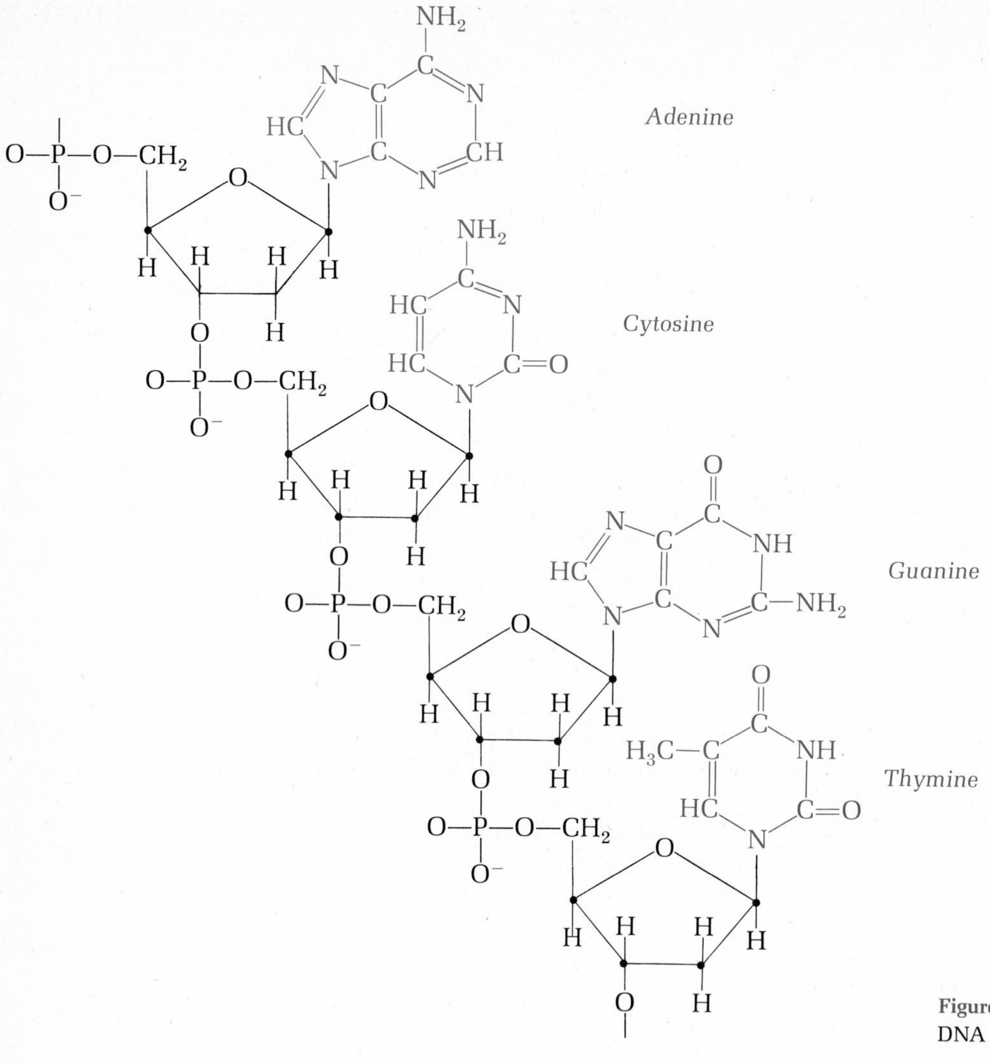

Figure 23.4
DNA polymer.

Another important difference between DNA and RNA is the relationship between the relative numbers of the four nucleotides. Although in RNA the relative numbers may vary without apparent restraint, in DNA there is a definite restraint in that the number of A nucleotides equals the number of T nucleotides and the number of G nucleotides equals the number of C nucleotides. Because A and G are purines whereas C and T are pyrimidines, it follows that there will be one purine for each pyrimidine.

The A = T and G = C conditions reflect the structural nature of the DNA molecule. As has been shown by Francis Crick, James D. Watson, and Maurice Wilkins, it is a double helix, two representations of which are given

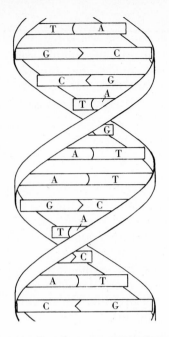

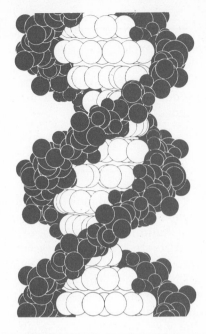

Figure 23.5
DNA double helix composed of two sugar-phosphate chains held together by base pairs as shown.

in Figure 23.5. In the helix the nitrogen bases bridge one sugar-phosphate strand to the other by forming cross-links between the two strands. The cross-links are of constant length because each involves one pyrimidine group bound to one purine group. If there were two purines, the cross-link would be too long and if there were two pyrimidines, it would be too short. The cross-links are thought to be held together by hydrogen bonds which are specific for binding A to T and for binding G to C:

Thus, A is said to be the complement of T, and G the complement of C. It follows therefore that in the final structure there must be one A for each T and one G for each C. It furthermore follows that the arrangement of bases in one of the two chains fixes the arrangement of bases in the other. It might be noted that there are certain viruses which contain DNA that does not have the double-stranded helical arrangement; appropriately it is called single-stranded DNA.

23.3   Biological Role of Nucleic Acids

DNA is the informational molecule of an organism. It contains all the hereditary information that the organism needs. In other words, it must contain within it the building plan for each of the many proteins that constitute the organism. Furthermore, it is necessary that the DNA molecule be able to replicate itself as the organism grows. For this purpose the double-stranded nature of DNA is vital. The two strands are complementary to each other, so that each contains all the information. During cell division, the strands unzip and separately produce two molecules each containing all the information that was coded into the original molecule. During the process no information is lost.†

The information which is contained in the DNA molecule is given by a code written in the sequence of bases along each of its chains. The code ultimately must be used to describe the sequence in which the 20 amino acids are to be used in building the particular protein in question. In order to code for 20 amino acids, clearly the four DNA bases cannot be used one at a time. In other words, there cannot be a one-to-one correspondence between a nucleotide and an amino acid. Furthermore, two nucleotides next to each other on a strand would not do the job either. The number of possible combinations of two adjacent nucleic acids would be 4^2 or 16. (The first nucleotide could be any one of four, making four possibilities. The second could also be any one of four. The total number of possibilities is the product of these two, or 4×4.) Since 16 possibilities are insufficient for the task, it is assumed that the code demands three adjacent nucleotides. For three adjacent nucleotides, there are 64 possible combinations ($4 \times 4 \times 4$), more than sufficient to code for 20 amino acids. Much work has been done in order to decipher the genetic code of DNA. We shall return to this subject later.

In making use of the information contained in the DNA molecule for protein synthesis, it is RNA that actually does the work. Corresponding to

---

†Viruses do not replicate themselves by cell division, and so double-stranded DNA is not involved. Rather, a single strand serves as a template on which the host organism constructs replicas of the virus DNA molecule.

the portion of a DNA molecule that contains the information for a specific protein chain, a complementary RNA chain is constructed. This copy, which is called *messenger RNA*, contains bases complementary to those of a portion of the DNA strand. For example, if the original DNA strand contains the sequence of bases T–G–C the corresponding messenger RNA strand would be –A–C–G–. This sequence can now serve to specify the amino acid that is to be added next in the synthesis of the protein chain. Hence, messenger RNA serves as a template for protein synthesis. In reading the genetic code another type of RNA is important. It is the so-called *transfer RNA*, which recognizes a portion of the messenger RNA chain and brings the indicated amino acid to the messenger RNA template. For example, the transfer RNA designated in Figure 23.3 is specific for the amino acid alanine. It is the job of the transfer RNA to bring alanine to that portion of the messenger RNA template that requires alanine. For each of the 20 amino acids there is at least one transfer RNA which will insert the proper amino acid at the appropriate position. All the transfer RNA's have a common terminal group C–C–A and a common internal group –G–T–Z–C–G–. (Z stands for the "strange base," pseudouracil, which is like uracil except that it is attached to the sugar through a bond to carbon rather than nitrogen.)

It is thought that protein synthesis occurs in a stepwise manner starting from the N-terminal end and proceeding as a transfer RNA adds the next appropriate amino acid to the growing protein chain. However, we should not lose sight of the fact that the order of events is dictated by the original DNA molecule. In order to understand the code contained in it, experiments have been performed with synthetic messenger RNA, in which synthetic nucleic acids are added to a synthesizing system. In the first experiment, the RNA was made up of only uracil nucleotides. It was found that the protein which was synthesized was polyphenylalanine; that is, it contained only phenylalanine amino acid residues. Hence, U–U–U codes for phenylalanine. Similar experiments using nucleotides containing two different residues have verified the expectation that three adjacent bases are needed to code for a particular amino acid.

The triplet of adjacent nucleotide residues on the messenger RNA is called a *codon*. It seems that more than one codon can indicate a particular amino acid. For example, the triplet A–A–A indicates lysine but so does the triplet A–A–G. G–A–A indicates glutamic acid, but so does G–A–G. The reason for this redundancy in the genetic code is not fully understood. There seem to be differences between a codon that codes for an amino acid terminal in the protein chain and that which codes for the same amino acid internal in the chain. Some of this may be connected with the observation that apparently one messenger RNA molecule can dictate consecutive production of more than one protein molecule. This would require punctuation in the genetic code such that one protein molecule could be terminated and another one started in an end-to-end manner on the messenger-RNA template.

One of the many amazing facts concerning the transmittal of information at a molecular level is that the molecules are able to transmit the code with a very few mistakes. In one investigation, it was found that the frequency of error was less than 3 per 100,000, or 99.997 percent accuracy.

QUESTIONS

23.1   *Nucleotide* (*a*) What repeating units and what varying units comprise a nucleotide? (*b*) How do these units differ between RNA and DNA?

23.2   *Phosphate Links* The phosphate links in nucleotides carry a negative charge. (*a*) By means of dot formulas and knowing the number of electrons in neutral P, O, C, and H atoms, show that each phosphate unit must have picked up one extra electron. (*b*) Formally at least where does this electron come from?

23.3*   *RNA* (*a*) What percent of the bases shown in the RNA structure of Figure 23.3 differ from any of those of Figure 23.2? (*b*) Does the presence of these modified and "strange bases" make sequence determination easier or harder? Explain.

23.4   *DNA* Suppose you analyzed a sample of DNA and found that the four bases A, T, G, and C were present in four unequal molar amounts. (*a*) What would you conclude about the structure of this DNA? (*b*) What would you guess was the source of this DNA? (*c*) How does it function biologically, as opposed to DNA in which A = T and G = C?

23.5   *Biological Role of Nucleic Acids* (*a*) Describe the difference in function between DNA, transfer RNA, and messenger RNA. (*b*) How would you expect the molecular weights of the three to differ?

23.6   *Codons* (*a*) Chemically, what is a codon? (*b*) Biologically, what is the function of a codon? (*c*) Mathematically, show that there are more codons than common amino acids.

23.7   *Helix* (*a*) What is the essential physical difference in overall structure of a DNA helix and the alpha helix of a protein? (*b*) What kind of chemical bonding holds each together? Tell how the two are alike in this regard and in what way they differ.

23.8   *DNA* (*a*) What are the basic units that make up a DNA molecule? (*b*) Propose a simplified mechanism for transmission of genetic code from double helix DNA molecule to RNA messenger.

23.9   *Viruses* (*a*) What is the chemical nature of viruses? (*b*) How do viruses replicate themselves?

23.10 *Base Pairing* (*a*) What two structural features cause guanine to pair with only cytosine? (*b*) Illustrate by drawing the other two possibilities (G–A, G–T) and show what is wrong with each. (*c*) Repeat for adenine by drawing the one pair not yet considered.

23.11* *RNA Nucleotides* What is the largest percentage of nitrogen in the RNA nucleotide substituents?

23.12 *Composition* Assuming potassium as the positive ion, calculate the simplest formula and percentage composition of the all uracil RNA nucleotide mentioned in Section 23.3.

23.13 *Structure* Taking into account the tetrahedral nature of C in the sugar ring and of P in PO_4, make a three-dimensional projection of the cytosine-sugar-phosphate portion of the DNA polymer shown in Figure 23.4.

23.14 *Double Helix* The energy of a hydrogen bond is roughly a tenth that of a covalent bond. By reference to Figure 23.5 and the base pairs involved, estimate the total H bond energy in terms of equivalent number of covalent bonds per complete turn of the double helix.

23.15 *Genetic Code* A portion of messenger RNA has the following sequence of bases starting at the beginning end:

GAGUUUGAAAAAAAGUUU. . .

What amino acids (in order) are called for?

Bioenergetics and the Origin of Life

24

In addition to the characterization of the molecules responsible for life processes, there are two related problems which are important. The first relates to the assembly of molecules in the cell, the complex building unit of a living organism. The need for the intricate architecture found in the cell will become increasingly apparent as we explore in this chapter the second problem, i.e., the energy relations within an organism which allow it to function. These energy relations, called *bioenergetics,* not only control the present nutritional economy of life on our planet but also made life possible in the beginning. However, before getting on to these cosmic questions, we must first complete our discussion of the chemical reactions in which food is converted in living organisms into energy, carbon dioxide, and water.

In our discussions in Chapter 21 we focused on the reducing agents (fats and carbohydrates) in the life cycle of animals and merely noted that oxidizing agents were required. In this chapter we discuss those oxidizing agents and the chemical means by which the energy of oxidation-reduction reactions is stored and transported to drive a living organism. A characteristic feature of a biological organism is that the oxidizing agent, oxygen, reacts in one part of the organism while the reducing agent, organic food, reacts in another part. The energy liberated by the indirect reaction is then distributed to all parts of the organism.

24.1 Electron Transport

The first step in the oxidation of foods involves the extraction of two hydrogen atoms from the organic material. For example, in the citric acid cycle outlined in Section 21.3 there were four oxidation steps in each of which two hydrogen atoms were removed during the course of the progressive oxidation of the molecules. It is a hallmark of such biochemical oxidations that many additional reactions are required before the ultimate oxidizing agent, O_2, plays its role. One reason for the many series of steps is that the reactions have to occur at body temperature in a controlled fashion for efficient energy utilization.

Each hydrogen-removal step is controlled by a specific enzyme called a *dehydrogenase*. However, as we know, enzymes are catalysts and not themselves oxidizing agents. The dehydrogenases catalyze the transfer of hydrogen from the molecule being oxidized to the appropriate oxidizing agent. By far the most general biological oxidizing agent for such reactions is the molecule nicotinamide adenine dinucleotide, abbreviated NAD (or, sometimes, NAD^+). The structural formula of NAD is

Because the structure of NAD is so complex, biochemists use a shorthand based on three or four letter abbreviations. To do otherwise would make the subject even more complicated.

As the structure of NAD shows, the molecule contains two ribose rings bound by a diphosphate group. In turn one ribose is joined to a grouping called nicotinamide and the other to an adenine ring system. The reduction of NAD involves only the nicotinamide part of the molecule. We can thus abbreviate this reaction by the equation

$$NAD \qquad + 2H \longrightarrow \qquad NADH \qquad + H^+$$

The addition of two hydrogen atoms furnishes two protons and two electrons. Both electrons and one of the protons are added to the NAD molecule; the other proton is absorbed by one of the many buffers which serve to keep constant pH in biological systems.

The fact that only one of the hydrogen atoms shows up in the reduced form of NAD has caused a disagreement in the nomenclature used by biochemists. One convention ignores the fact and abbreviates the reduced form as $NADH_2$. This nomenclature has the virtue of showing that two equivalents of reduction have occurred. An alternative nomenclature shows the charge on nicotinamide by writing NAD^+, with NADH used for the reduced form. Both conventions seem somewhat cumbersome in that the initials perform functions best handled by entire formulas. Hence we adopt a simpler convention in which the oxidized form is called NAD and the reduced form, NADH.

Although NAD is commonly found in animals, in plants a related molecule performs the same function. It differs from NAD in that a phosphate group is substituted on the ribose ring farthest removed from nicotinamide. To indicate this difference the molecule is abbreviated as NADP.

Following the reduction of NAD to NADH, the next event to occur in the oxidation chain is a transfer of hydrogen from NADH to a closely related molecule called flavin adenine dinucleotide, FAD. This molecule differs from NAD in that the nicotinamide has been replaced by a so-called *flavin* grouping which can be reduced by accepting two hydrogen atoms:

$$FAD \qquad + 2H \rightleftharpoons \qquad FADH$$

In general, this reaction occurs at the surface of a specific protein enzyme with the FAD bound tightly to the enzyme surface. For this reason, the FAD-enzyme entity is called a *flavoprotein,* and the FAD itself is often called a *flavin coenzyme.* We should note that human beings are apparently unable to synthesize the flavin portion of the molecule and so must obtain it in food. A substance which must be so supplied is called a *vitamin.* Riboflavin, which contains the flavin grouping bound to the five carbon atoms of ribose, is called vitamin B_2. The minimum daily requirement of this vitamin amounts to some 2 mg/day. It is found in milk, egg yolk, liver, etc.

To summarize the steps thus far encountered we have

$$NAD + 2H \longrightarrow NADH + H^+$$
$$NADH + H^+ + FAD \longrightarrow NAD + FADH$$

Although this is the general chain of events, it is not the only oxidation mechanism which occurs. Of the four oxidation steps in the citric acid cycle, three start in this manner, but the fourth (the oxidation of succinate to fumarate) bypasses the NAD step; instead, hydrogen atoms are transferred directly to FAD.

Although the oxidation chain is far from complete, these are the steps which involve transfer of H atoms. Subsequent oxidation-reduction reactions involve transfer of electrons. In these subsequent steps, iron atoms are oxidized and reduced between the 2+ and 3+ oxidation states. In other words, FADH is now used to reduce Fe^{3+} to Fe^{2+}. However, the iron is not in the form of the free ion; instead, it is bound in the structure shown in Figure 24.1. The ring system surrounding the iron atom is called a *porphyrin,* a grouping encountered repeatedly in natural systems. Variations in the substituents marked R_1 and R_2 give different molecules which function serially in passing the electron along. These molecules are called *cytochromes* (labeled *a, b, c,* etc.); they differ from each other in the oxidation

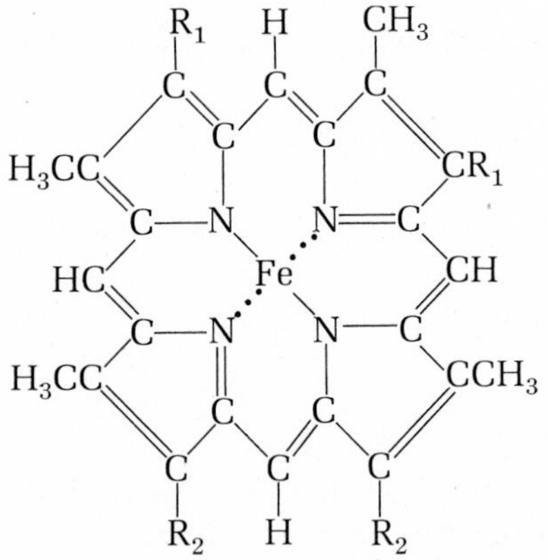

Figure 24.1
Iron porphyrin.

potential of the iron. All the cytochromes include protein bound either directly to the iron atom or at the position of the R groups. Although the exact sequence of reactions of the cytochromes is not yet understood, it must be at least as complicated as the following: First, FADH reduces an iron atom in cytochrome b; reduced cytochrome b then reduces cytochrome c; reduced cytochrome c reduces cytochrome a. Finally, oxygen enters the scene. Yet another iron-containing protein, called an *oxygenase*, acts as a catalyst for the reduction of the oxygen molecule to water by the ferrous ions of reduced cytochrome.

The entire chain of events (no doubt, there are more than we have described here) can be summarized in terms of the oxidation-reduction reactions given in Figure 24.2. We note that in the overall oxidation-reduction process a series of reactions are all occurring simultaneously, so that at one

Figure 24.2 HALF-REACTIONS FOR BIO-CHEMICAL REDUCTION OF O_2	Half-Reactions	Reduction Potential, volts
	$O_2 + 4H^+ + 4e^- \longrightarrow 2H_2O$	0.82
	$Fe^{3+}_{cyt\ a} + e^- \longrightarrow Fe^{2+}_{cyt\ a}$	0.29
	$Fe^{3+}_{cyt\ c} + e^- \longrightarrow Fe^{2+}_{cyt\ c}$	0.26
	$Fe^{3+}_{cyt\ b} + e^- \longrightarrow Fe^{2+}_{cyt\ b}$	0.00
	$FAD + 2e^- \longrightarrow FADH$	-0.185
	$NAD + 2e^- \longrightarrow NADH$	-0.320

end of the chain an organic molecule is to be oxidized by NAD. The oxidized NAD is provided ultimately by reduction of O_2 at the far end of the chain. Thus, O_2 must be the best oxidizing agent in the chain, and as a consequence it has the highest reduction potential and stands at the top of the half-reactions listed in Figure 24.2. These show that O_2 is reduced to H_2O by oxidizing Fe^{2+} cytochrome a to Fe^{3+} cytochrome a. Fe^{3+} cytochrome a oxidizes Fe^{2+} cytochrome c to Fe^{3+} cytochrome c which then oxidizes Fe^{2+} cytochrome b to Fe^{3+} cytochrome b. The shuttling continues as Fe^{3+} cyctochrome b oxidizes FADH to FAD, which oxidizes NADH to NAD. The net result is that NAD is formed and ready to extract H atoms from organic food due to a corresponding reduction of O_2 to H_2O. All the intermediates have shuttled either H atoms, in the case of NAD/NADH and FAD/FADH, or electrons in the case of the cytochromes. In this regard we should point out that although the cytochromes contain the porphyrin structure shown in Figure 24.1, a structure also contained in hemoglobin, we should not confuse cytochromes with hemoglobin. They perform entirely different functions. In hemoglobin, the bound ferrous ion combines with an oxygen molecule to transport it throughout the organism and ultimately release it as an oxygen molecule. The cytochromes, on the other hand, transport electrons to the oxygen molecule.

24.2 Energy Transport

In the overall oxidation of NADH by O_2, a large amount of energy is liberated. In terms of free energy, this amounts to 210 kilojoules/mole of NADH oxidized. This is the principal energy-producing mechanism of living organisms, and the question arises as to how this energy is gathered and transported to other parts of the organism. It is converted into chemical energy through the formation of compounds, the most important of which is adenosine triphosphate (ATP).

ATP

Again we see adenine bound to a ribose which now is bound to a string of three phosphates. In other words, the molecule bears great similarity to half of the NAD molecule. The principal difference is that three phosphates are now bound. It is the binding of the third phosphate in the chain which requires the expenditure of free energy. The reaction can be abbreviated

For the reaction, the free energy used is 33 kilojoules/mole. This is the way in which energy is gathered from the oxidation of food and stored in the formation of ATP. When the energy is needed, the reaction is reversed. Hydrolysis occurs to form adenosine diphosphate (ADP) and 33 kilojoules of usable energy.

The gathering and storing of energy through formation of ATP is called *oxidative phosphorylation*. It accompanies the oxidation of NADH and is the way the 210 kilojoules of energy released by oxidizing NADH is stored. It would seem from the energetics that, if the process were 100 percent efficient, one could form six ATP molecules. ATP requires 33 kilojoules/mole; NADH liberates 210 kilojoules/mole. However, not all this energy is so used, since only three molecules of ATP are actually formed. Even so, with an overall efficiency of about 50 percent the process is as efficient as the very best heat engines that we have been able to construct with far simpler systems.

The three molecules of ATP that are formed during the oxidation of NADH are formed concurrently with three of the oxidation-reduction sequences outlined in Figure 24.2. Examination of the reduction potentials listed indicates that there is a large potential difference between the values for the first two couples listed. Similarly, there is a large potential difference between cytochrome b and c and another between FAD and NAD. Each of these potential differences is big enough to drive the formation of one molecule of ATP, since the 33 kilojoules required for formation of ATP corresponds to just under 0.2 volt. Therefore, sufficient energy is available so that an ATP molecule is formed at each of the three steps noted, with the first of these being the least efficient. That is consistent with the chemistry of O_2 in that oxidations by O_2 are difficult and often require large activation energies. Furthermore, as we shall see, oxygen is a relatively new oxidizing agent in the evolution of living substances, and it may well be that in further evolution this step will become more efficient.

The ATP which is formed along with the burning of food is then available to the organism to drive a variety of other reactions on which it relies. These include muscle contraction; the biosynthesis of proteins, nucleic acids, complex carbohydrates, and lipids; and a variety of other functions. Finally, we should note that, although animals must derive their ATP from the oxidation of organic matter, plants are able to utilize the energy of visible light directly to form ATP and hence are not dependent on the presence of chemical fuels for continued existence.

24.3 Photosynthesis

From energy considerations it is clear that animal life as we know it is absolutely dependent on plant life. The reason for this is that animals gain their nutritional energy through oxidation of plant-produced food. Plants are able to synthesize this food, largely carbohydrates, through utilization of the sun's energy in a process called *photosynthesis*. It has been estimated that, if photosynthesis were to stop completely, all animal life (including human beings) would be dead in 25 years. Thus the process of photosynthesis is of vital interest to us all.

Basically, the photosynthetic process employs the energy of light to perform the following reaction:

$$nCO_2 + nH_2O \longrightarrow (CH_2O)_n + nO_2$$

The $(CH_2O)_n$ indicates a general carbohydrate; plants are able to synthesize a variety ranging from the simple sugar glucose to the polymer cellulose. The chemical process is a highly unfavorable one, requiring 469 kilojoules of energy per mole of CO_2 consumed. Consequently, the photosynthetic process is able to store large amounts of energy in relatively compact masses of chemical fuel. Such chemical fuel produces practically all the energy required by today's civilization. The food we eat and the fuels we burn owe their existence to the photosynthetic process. As the result of extensive research efforts, we know much but not all the details of this fundamental process.

It is known that the sun's light is absorbed by a natural green pigment called chlorophyll. Chlorophyll is quite similar to the porphyrin shown in Figure 24.1 except that the ring system has been slightly modified and the iron atom at the center has been replaced by magnesium. Chlorophyll molecules arranged in stacks serve as "antennas" for receiving the sun's electromagnetic energy and passing it on for ultimate conversion to chemical energy through two separate processes. In the first of these, the oxidizing agent NADP (nicotinamide adenine dinucleotide phosphate, Section 24.1) is photochemically reduced to NADPH. In this process the plants make use of the energy gathered from sunlight. The analogous reduction in animals is that in which NAD is reduced by chemical reducing agents (food) to NADH.

The second photochemical process performed by plants is the photochemical oxidation of water to O_2. This is also an energy-storing process. In the preceding section, we saw that animals derive their energy supply from reaction of oxygen with NADH; plants in two separate processes are able to *produce* the corresponding pair of substances, O_2 and NADPH. It might be noted that plants are not the only photosynthesizing organisms. Certain bacteria and algae are able to perform at least the first of the two processes, i.e., the production of NADPH. The ability to form oxygen evolved later.

Once the NADPH has been formed it is then available to take part in the reduction of carbon dioxide. This is not a photochemical act and can be carried out in the dark so long as excess NADPH is present. Plants are extremely efficient in CO_2 conversion. Some plants in a single day can convert as much CO_2 to carbohydrates as is equivalent to 25 percent of their own weight. The corresponding storage of energy is equally impressive. It is estimated that the world's plants in one year store more than 4×10^{18} (billion billion) kilojoules of energy. This amounts to 1×10^{12} (million million) barrels of petroleum, or about 50,000 barrels per second. (Present world use of petroleum is less than 1000 barrels per second.)

The incorporation of carbon dioxide by plants is the best understood part of the photosynthetic process. This has been made possible by the use of radioactively labeled carbon dioxide, the path of which through the photosynthetic process was skillfully traced by Nobel laureate Melvin Calvin. The key step is a reaction in which a phosphate derivative of a five-carbon carbohydrate (I) reacts with CO_2 and H_2O as follows:

$$
\begin{array}{c}
\text{H} \\
| \\
\text{H}-\text{C}-\text{O}-\text{PO}_3\text{H}^- \\
| \\
\text{C}=\text{O} \\
| \\
\text{H}-\text{C}-\text{OH} \qquad + \text{ CO}_2 + \text{H}_2\text{O} \longrightarrow 2 \\
| \\
\text{H}-\text{C}-\text{OH} \\
| \\
\text{H}-\text{C}-\text{O}-\text{PO}_3\text{H}^- \\
| \\
\text{H}
\end{array}
\qquad
\begin{array}{c}
\text{H} \\
| \\
\text{H}-\text{C}-\text{O}-\text{PO}_3\text{H}^- \\
| \\
\text{H}-\text{O}-\text{C}-\text{H} \\
| \\
\text{C}-\text{O}-\text{H} \\
\| \\
\text{O}
\end{array}
$$

(I)   (II)

The acid functional group COOH in (II) is then reduced in an enzyme-catalyzed reaction to form an aldehyde functional group CHO. This resulting reduction product (III) then rearranges to (IV):

$$
\begin{array}{c}
\text{H} \\
| \\
\text{H}-\text{C}-\text{O}-\text{PO}_3\text{H}^- \\
| \\
\text{H}-\text{O}-\text{C}-\text{H} \\
| \\
\text{C}-\text{H} \\
\| \\
\text{O}
\end{array}
\qquad \longrightarrow \qquad
\begin{array}{c}
\text{H} \\
| \\
\text{H}-\text{C}-\text{O}-\text{PO}_3\text{H}^- \\
| \\
\text{C}=\text{O} \\
| \\
\text{H}-\text{C}-\text{O}-\text{H} \\
| \\
\text{H}
\end{array}
$$

(III)   (IV)

The rearranged product (IV) reacts with an unrearranged molecule (III) to form a six-membered carbohydrate derivative (V):

$$
\begin{array}{c}
\text{H} \\
| \\
\text{H}-\text{C}-\text{O}-\text{PO}_3\text{H}^- \\
| \\
\text{H}-\text{O}-\text{C}-\text{H} \\
| \\
\text{C}-\text{H} \\
\| \\
\text{O}
\end{array}
\; + \;
\begin{array}{c}
\text{H} \\
| \\
\text{H}-\text{C}-\text{O}-\text{PO}_3\text{H}^- \\
| \\
\text{C}=\text{O} \\
| \\
\text{H}-\text{C}-\text{O}-\text{H} \\
| \\
\text{H}
\end{array}
\; \longrightarrow \;
\begin{array}{c}
\text{H} \\
| \\
\text{H}-\text{C}-\text{O}-\text{PO}_3\text{H}^- \\
| \\
\text{C}=\text{O} \\
| \\
\text{H}-\text{O}-\text{C}-\text{H} \\
| \\
\text{H}-\text{C}-\text{O}-\text{H} \\
| \\
\text{H}-\text{C}-\text{O}-\text{H} \\
| \\
\text{H}-\text{C}-\text{O}-\text{PO}_3\text{H}^- \\
| \\
\text{H}
\end{array}
$$

(III)   (IV)   (V)

Reaction of the final product with water leads to glucose. Other reactions between the final product and intermediates in its formation regenerate the five-carbon molecule needed in the first step.

The result of the changes shown above is to introduce CO_2 into the cycle and to lengthen the carbon chain from five carbons to six carbons. It must be emphasized that this is a portion of the overall process and, in fact, the regeneration of (I) constitutes a major portion of the known cycle.

24.4   The Origin of Life

For centuries people have asked: "How did it all begin?" They have been particularly interested in how they came into being and how living organisms began. With our present knowledge of the biochemical nature of organisms, some intelligent guesses can be made and many of these subjected to scientific study. The past few decades have seen ever-increasing interest in such studies and a steady evolution of ideas concerning the origin of life.

Figure 24.3 gives an approximate time scale for the events which preceded the emergence of human beings. The age of the earth is set at about 4.7 billion years based on the distribution of radioactive isotopes in rocks. The next important date (about 2.5 billion years ago) marks the appearance of complex organic compounds. Our knowledge of these comes from analysis of organic matter in rocks, such as oil-bearing shales. Similarly, the analysis of such rocks shows derivatives of chlorophyll present as long ago as 1 billion years. Both these benchmarks greatly precede the appearance of the oldest fossils, which are 600 million years old. We will concentrate our attention on the periods preceding these earliest fossils.

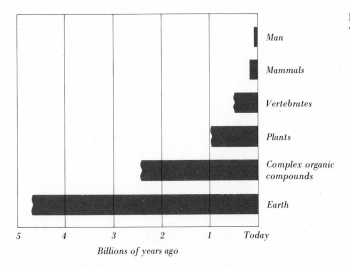

Figure 24.3
Time scale for origin of life on earth.

Most authorities agree that the original atmosphere of the earth was a reducing one; that is, there was no oxygen present and, instead, the gases which the earth was able to hold included some or all of the following: H_2O, CO, CH_4, H_2, NH_3, H_2S, N_2, and possibly CO_2. It is reasoned that such a collection of gases could react to form a variety of organic compounds under the stimulus of an electric discharge (e.g., lightning) or an absorption of ultraviolet energy from the sun. A number of scientists have performed experiments in which some of or all the proposed gases of the original atmosphere are mixed and subjected to electric discharges or ultraviolet light. Such experiments have been found to produce exactly those compounds on which life is based. Perhaps most spectacular is the finding that 17 of the 20 amino acids fundamental to proteins were formed in such experiments. It is possible that the other amino acids were also formed but that the methods of separation and detection failed to show them. The purines and pyrimidines necessary for forming nucleic acids were likewise produced as were a variety of simple sugars.

The key intermediate in forming the carbon-nitrogen compounds is apparently HCN. It has been shown that HCN forms rapidly in the irradiated mixtures and in relatively high concentration. With continuing irradiation the HCN reacts further to form the other compounds, particularly the amino acids. It is reasoned that, as the ages went by, increasingly complex organic compounds formed, in a process which has been labeled "chemical evolution" to produce an organic soup which covered much of the earth's surface. Confirmation of this concept of chemical evolution comes from the fact that not only are simple molecules detected in the spectra emitted by stars and interstellar dust clouds† but meteorites have been found to be rich in complex hydrocarbons. For example, a meteorite which fell at Orgueil, France, in 1864, contained 6 percent organic material with complex compounds having as many as 29 carbon atoms of a molecular-weight distribution similar to that of the hydrocarbons found in butter. In addition to long-chain fatty acids, other meteorites have been found to contain porphyrins, purines, pyrimidines, and amino acids.

The next problem in chemical evolution involves the formation of polymeric molecules (proteins, carbohydrates, lipids, and nucleic acids). Two possible hypotheses have been offered for this condensation. The basic problem is the fact that all the condensations involve removal of water:

For proteins:

$$2\ \overset{\overset{\displaystyle H}{|}}{\underset{\underset{\displaystyle R}{|}}{H_3\overset{+}{N}-C}}-CO_2^- \longrightarrow \overset{\overset{\displaystyle H}{|}}{\underset{\underset{\displaystyle R}{|}}{H_3\overset{+}{N}-C}}-\overset{\overset{\displaystyle O}{\|}}{C}-\overset{\overset{\displaystyle H}{|}}{N}-\overset{\overset{\displaystyle H}{|}}{\underset{\underset{\displaystyle R}{|}}{C}}-CO_2^- + H_2O$$

† In our own solar system spectroscopic studies show CH_4 to be present on Jupiter, Uranus, Neptune, Saturn, Venus, and Mars. The latter two also have C_2H_4 and C_2H_2 in their atmospheres.

For carbohydrates:

$$2 \quad \text{(glucose)} \longrightarrow \text{(disaccharide)} + H_2O$$

For lipids:

$$R-\overset{\displaystyle O}{\overset{\|}{C}}-OH + HOR \longrightarrow R-\overset{\displaystyle O}{\overset{\|}{C}}-O-R + H_2O$$

For nucleic acids:

$$2 \quad \text{(nucleotide)} \longrightarrow$$

$$\text{(dinucleotide)} + H_2O$$

One hypothesis recognizes the fact that we do not know the temperature which prevailed. It has been shown that if the temperature is above that of the boiling point of water these condensation reactions take place spontaneously. On the other hand, if, as many feel more likely, liquid water were present, then reactions that eliminate water would not be favored thermodynamically. However, recent experiments suggest that certain simple carbon-nitrogen compounds might be able to bring about the polymerizations. Hence, the polymerization problem does not seem insurmountable.

Less well understood is the question of organization of the molecules into the appropriate sequences necessary for formation of life. It has been postulated that such organization of the simple molecules could occur either at the surface of inorganic minerals or on droplets of immiscible liquids suspended in the organic soup. Of particular importance is the fact that long times were available for the evolutionary process and all possible arrangements, even quite improbable ones, could form, given sufficient time. If one such arrangement led to an extremely primitive organism that could catalyze the production of others like it, then the most fundamental step in the evolution of life would have been achieved. However, protection and nourishment of such borderline organisms present additional problems. The primitive organism would have to have been protected against ultraviolet light. Just as ultraviolet light can provide the activation energy for the formation of the organism, ultraviolet light can provide activation energy for further reactions which would result in its destruction. The reason that this was such a serious problem is the fact that the earth's atmosphere at that stage did not contain oxygen. At present, oxygen in the upper atmosphere converts to ozone on absorption of ultraviolet light from the sun, thereby shielding present-day organisms from dangerous radiation. Prior to the existence of oxygen in the atmosphere it would have been necessary for an organism to find other means of protection. This could have come about if the organism lived under water where it would have been suitably protected.

The problem of nourishment is important for the following reason. Since the early organisms were formed from the organic soup that came from the earth's original atmosphere, the process of multiplication could not continue for long. The entire supply of nutrients would soon be used up and life would cease. In other words, there was no mechanism for restoring a suitable atmosphere by bringing in energy sources from outside. The ultraviolet light or electric discharges responsible for chemical evolution merely provided activation energy for the process of converting chemical compounds into ones more stable than the original. This would be a one-way street and the system would soon run down.

Both these problems, protection and nourishment, were solved by the emergence of photosynthetic organisms. It is perhaps significant that a molecule easily synthesized from the earliest chemicals is porphyrin, which can serve as the light-gathering pigment for photosynthetic organisms. Probably the original photosynthetic organisms evolved in the sea (even today, most photosynthesis occurs within the top layers of the ocean). It is thought that they were not capable of utilizing water as their reducing agent because there exist today certain primitive photosynthetic bacteria and algae that are not capable of oxidizing water to oxygen. Instead, such organisms use other reducing substances, even hydrogen, for their photosynthetic apparatus.

It would have been necessary only that the original photosynthetic organism be able to photoreduce a compound such as NADP so that it could store energy for chemical reactions. Eventually, of course, such organisms did evolve the ability to oxidize water to oxygen. As the process increased, the earth's atmosphere gradually accumulated oxygen, until levels approaching the present 20 percent were obtained. Even at lower concentrations the oxygen of the atmosphere would have served to reduce ultraviolet radiation so that higher forms of life could develop. Gradually there developed the present biological balance in which plants use carbon dioxide and produce oxygen while animals use oxygen and produce carbon dioxide.

This most basic of economic systems, the balance of carbon dioxide and oxygen utilization by plants and animals, is the economy which must be more intelligently utilized if we are to feed even the present population of the earth. Present-day agricultural methods are not capable of providing adequate diets for a large fraction of the world's population. Yet, the only necessary raw ingredient, solar energy, is present in more than adequate amounts. On average, each square centimeter of the earth's surface receives 1000 kilojoules of light energy per year. If all this light energy could be utilized with the maximum overall efficiency of the photosynthetic process, 25 percent, it would require no more than 2 yd^2 to nourish a human being. Consequently, *in theory* the surface of the earth could sustain a population of about 300 billion, a figure which compares with the present estimated population of 4 billion. Of course, there are numerous problems such as the energy required for the operation of society, including distribution of the foods once they are grown, the fact that people want to live in those areas that are best suited for growing food, and the fact that most people at present do not enjoy the taste of algae grown in the seas. However, all these problems seem small in comparison with death by starvation, a fate which awaits an increasing portion of the world's peoples if the present rate of population expansion continues. Basic science has already uncovered much of the knowledge necessary both to control the world's population growth and to feed increasing numbers of people. The crucial task that remains is to see that this knowledge is intelligently applied to solve our most fundamental problem: our need for food.

QUESTIONS

24.1 *Electron Transport* (a) How can NAD undergo two equivalents of reduction while picking up only one H atom? (b) Using R to designate portions of NAD and FAD which are identical, write a balanced structural equation for the reduction of FAD by NADH.

24.2 *Iron Porphyrin* For the iron porphyrin diagramed in Figure 24.1 to be electrically neutral, what oxidation state must the iron be in? Justify your answer.

24.3 *Energy Transport* Show clearly the role of ATP in transporting energy from one portion of a living organism to another.

24.4 *Photosynthesis* Write a balanced equation for the conversion of intermediate V of Section 24.3 to glucose.

24.5 *Origin of Life* (a) What two critical functions for the evolution of life were served when an oxygen atmosphere was formed? (b) How is it believed that the oxygen atmosphere was formed?

24.6 *Origin of Life* (a) What do the formation of carbohydrates, proteins, and nucleic acids have in common? (b) Why might these reactions have been very improbable?

24.7 *Terms* What is meant by each of the following terms: (a) oxidative phosphorylation, (b) bioenergetics, (c) photosynthesis, (d) dehydrogenase, (e) oxygenase?

24.8 *Origin of Life* What is the reason for believing that the first organisms lived in water?

24.9 *Porphyrin* The function of porphyrin in the human body differs from its function in photosynthetic organisms. What are these two functions?

24.10 *Photosynthesis* Describe the two photochemical processes employed by plants to convert energy from the sun into chemical energy.

24.11 *Origin of Life* According to the hypotheses of chemical evolution, what single substance is formed during the formation of the polymeric molecules: proteins, lipids, carbohydrates, and nucleic acids?

24.12 *Energy* During chemical evolution, what were the initial and final sources of energy for the formation of complex molecules?

24.13 *NAD* Write equations for the sequence of reactions needed to form the NAD molecule.

24.14 *Chlorophyll* How is chlorophyll related to porphyrin?

24.15 *Electron Transport* (a) By reference to Figures 21.2 and 24.2 trace the flow of electrons from isocitric acid to O_2. (b) Write an overall balanced reaction. (c) Compare the concentrations of NAD, FAD, and the three oxidized cytochromes before and after the reaction is complete.

24.16 *Reduction Potentials* The potentials listed in Figure 24.2 include that for the FAD/FADH couple when they are free. When bound to protein, this potential rises to near zero. (a) Will that affect the chain of events? Explain. (b) What does this say about the efficiency of ATP production?

24.17* *Photosynthesis* Assuming an overall 25 percent efficiency for photosynthesis, how many liters of O_2 at STP can be formed by photosynthesis from the sun's energy (1000 kilojoules) which fall on 1 cm^2 of earth each year?

24.18 *Photosynthesis* Since CO_2 conversion to carbohydrate requires energy, how is it possible for plants to carry out this conversion (to a limited extent) in the dark?

24.19 *Chemical Evolution* Write balanced equations which could account for the ultimate conversion of atmospheric CH_4, N_2, and H_2O to (a) glycine, (b) glucose, (c) stearic acid.

24.20 *Biochemical Similarities* For each of the following molecules give at least three examples of important biological molecules which contain it or a close derivative: (a) adenine, (b) ribose, (c) porphyrin.

24.21* *Bioenergetics* For one complete revolution of the Krebs cycle (Figure 21.2) involving consumption of 1 mole of acetyl coenzyme A, there is produced (a) moles of NADH, (b) kilojoules, (c) moles of ATP, (d) energy stored as ATP.

24.22 *Bioenergetics* Calculate the efficiency of metabolizing glucose (in terms of fraction of total energy available which is stored as ATP) from the following facts. Heat of combustion of glucose is 469 kilojoules/mole of CO_2 formed. Metabolism of glucose produces two molecules each of CO_2, ATP, NADH and acetyl coenzyme A. Metabolism of 1 mole of acetyl coenzyme A in the Krebs cycle produces 4 moles of NADH.

24.23 *Science and the Future* After the accumulated reserves of fossil and nuclear fuel are exhausted, we will have to use current solar energy for both food and external energy sources. Based on the figures given in this chapter and (stated) assumptions concerning energy requirements for a reasonable way of life, estimate the maximum population which our planet could sustain.

Appendix 1　Chemical Nomenclature

1.1　Inorganic

The names of the elements are listed inside the back cover of this book. Some of the elements have Latin names (argentum for silver, aurum for gold, cuprum for copper, ferrum for iron, plumbum for lead, and stannum for tin) which appear in the names of compounds of these elements.

Compounds composed of but two elements have names derived directly from the elements. Usually the more electropositive element is named first, and the other element is given an *-ide* ending. Thus, we have sodium chloride (NaCl), calcium oxide (CaO), and aluminum nitride (AlN). If more than one atom of an element is involved, prefixes such as *di-* (for 2), *tri-* (3), *tetra-* (4), *penta-* (5), and *sesqui-* ($1\frac{1}{2}$) are used. For example, AlF_3 is aluminum trifluoride, Na_3P is trisodium phosphide, and N_2O_4 is dinitrogen tetroxide. When the same two elements form more than one compound, the compounds can be distinguished as in the following examples:

	$FeCl_2$	$FeCl_3$
a	Iron dichloride	Iron trichloride
b	Ferrous chloride	Ferric chloride
c	Iron(II) chloride	Iron(III) chloride

In (a) distinction is made through the use of prefixes; in (b) the endings -ous and -ic denote the lower and higher oxidation states, respectively, of iron; in (c), the Stock system, Roman numerals in parentheses, indicate the oxidation states. In a given series of compounds, the suffixes -ous and -ic may not be sufficient for complete designation but may need to be supplemented by one of the other methods of nomenclature. For example, the oxides of nitrogen are usually named as follows:

N_2O	Nitrous oxide
NO	Nitric oxide
N_2O_3	Dinitrogen trioxide, or nitrogen sesquioxide
NO_2	Nitrogen dioxide
N_2O_4	Dinitrogen tetroxide, or nitrogen tetroxide
N_2O_5	Dinitrogen pentoxide, or nitrogen pentoxide

Compounds containing more than two elements are named differently, depending on whether they are bases, acids, or salts. Since most bases contain hydroxide ion (OH^-), they are generally called hydroxides, e.g., sodium hydroxide (NaOH), calcium hydroxide [$Ca(OH)_2$], arsenic trihydroxide [$As(OH)_3$]. The naming of acids and of salts derived from them is more complicated, as can be seen from the following series:

Acid	Sodium Salt
$HClO$, hypochlorous acid	$NaClO$, sodium hypochlorite
$HClO_2$, chlorous acid	$NaClO_2$, sodium chlorite
$HClO_3$, chloric acid	$NaClO_3$, sodium chlorate
$HClO_4$, perchloric acid	$NaClO_4$, sodium perchlorate
H_2SO_3, sulfurous acid	Na_2SO_3, sodium sulfite
H_2SO_4, sulfuric acid	Na_2SO_4, sodium sulfate

When there are only two common oxy acids of a given element, the one corresponding to the lower oxidation state is given the -ous ending and the other the -ic ending. If there are more than two oxy acids, of different oxidation states, the prefixes hypo- and per- may also be used. As indicated in the above example, the prefix hypo- indicates an oxidation state lower than that of an -ous acid, and the prefix per- an oxidation state higher than that of an -ic acid. For salts derived from oxy acids the names are formed by replacing the ending -ous by -ite and -ic by -ate. Salts derived from polyprotic acids (for example, H_3PO_4) are best named so as to indicate the number of hydrogen atoms left unneutralized. For example, NaH_2PO_4 is monosodium dihydrogen phosphate, and Na_2HPO_4 is disodium monohydrogen phosphate. Frequently, the prefix *mono-* is left off. For mono-hydrogen salts of diprotic acids, such as $NaHSO_4$, the presence of hydrogen may also be indicated by the prefix *bi-*. Thus, $NaHSO_4$ is sometimes called sodium bisulfate, though the name sodium hydrogen sulfate is preferred.

Complex cations, such as $Cr(H_2O)_6^{3+}$, are named by giving the number and name of the groups attached to the central atom followed by the name of the central atom, with its oxidation number indicated by Roman numerals in parentheses. Thus, $Cr(H_2O)_6^{3+}$ is hexaaquochromium(III). Complex anions, such as $PtCl_6^{2-}$, are named by giving the number and name of attached groups followed by the name of the element with an *-ate* ending and its oxidation number in parentheses. Thus, $PtCl_6^{2-}$ is hexachloroplatinate(IV). If the attached groups (*ligands*) are not all alike, it is customary to name the ligands in the same order in which they should be written in the formula; i.e., anion ligands generally precede neutral ligands. If more than one kind of anion ligand is present, the order is H^- (*hydrido*), O^{2-} (*oxo*), OH^- (*hydroxo*), other monoatomic anions (in order of increasing electronegativity of the elements, for example, F^-, *fluoro*, last), polyatomic anions (in order of increasing number of atoms), organic anions (in alphabetical order). If more than one kind of neutral ligand is present, the order is H_2O (*aquo*), NH_3 (*ammine*), other inorganic ligands (in order of increasing electronegativity of their central atom; for example, CO, *carbonyl*, precedes NO, *nitrosyl*), organic ligands (in alphabetical order). To indicate the numbers of each kind of ligand, Greek prefixes are used: *mono-* (usually can be omitted), *di-*, *tri-*, *tetra-*, *penta-*, *hexa-*, *hepta-*, *octa-*. Instead of these prefixes, *bis-* (twice), *tris-* (thrice), *tetrakis-* (four times), etc., may be used, especially when the name of the ligand itself contains a numerical designation (e.g., ethylenediamine, frequently abbreviated "en"). Some examples of the application of the above rules follow:

$CrCl_2(H_2O)_4^+$	dichlorotetraaquochromium(III)
$CrCl_4(H_2O)_2^-$	tetrachlorodiaquochromate(III)
$Cr(H_2O)(NH_3)_5^{3+}$	aquopentaamminechromium(III)
$Ga(OH)Cl_3^-$	hydroxotrichlorogallate(III)
cis-$PtBrCl(NO_2)_2^{2-}$	*cis*-bromochlorodinitroplatinate(II)
trans-$Co(OH)Clen_2^+$	*trans*-hydroxochlorobisethylenediaminecobalt(III)
$Mn(CO)_3(C_6H_6)^+$	tricarbonylbenzenemanganese(I)

In the case of complex-ion isomerism, the names *cis* or *trans* may precede the formula or the complex-ion name to indicate the spatial arrangement of the ligands. *Cis* means the ligands occupy adjacent coordination positions; *trans* means opposite positions.

1.2   Organic

The key rules recommended by the International Union of Pure and Applied Chemistry (IUPAC) are summarized as follows:

1   Choose as the parent carbon skeleton the longest sequence of C atoms that contains the principal functional group.

2   Name the parent structure using the name of the alkane that contains the same number of C atoms as the chosen structure. Replace -ane by -ene for double bond or -yne for triple bond. If a functional group is present drop the final e and add suffixes as follows:
   -ol for alcohol (OH)
   -al for aldehyde (CHO)
   -one for ketone (CO)
   -oic acid for acid (COOH)

3   Use prefixes, in alphabetical order, to denote other substituents.

4   Locate substituents and points of unsaturation by numbering the C atoms of the parent skeleton with the following criteria used in decreasing order of priority.

   a   Assign the C atom of the principal functional group the number 1 if it is terminal.
   b   Assign numbers so that the location of the principal functional group is as low as possible if the group is nonterminal.
   c   Assign numbers so that substituents are located by lowest possible numbers. If there are two kinds of substituents, give low-number preference to the first named.

5   If an attached side chain bears substituents, it too must be numbered starting with the C atom which is attached to the parent carbon skeleton. Names of substituents on the side chain and numbers locating them are enclosed in parentheses with the name of the side chain:

$$
\begin{array}{c}
\text{CH}_3 \\
| \\
\text{CH}_3\text{CHCH}_2\text{CH}_3
\end{array}
$$

2-Methylbutane

$$
\begin{array}{c}
\text{CH}_3 \\
| \\
\text{CH}_3\text{CH}_2\text{CHCH}=\text{CH}_2
\end{array}
$$

3-Methyl-1-pentene

$$
\begin{array}{c}
\text{CH}_3 \\
| \\
\text{CH}_3\text{CH}_2\text{CHCHCH}_2\text{OH} \\
| \\
\text{CH}_3
\end{array}
$$

2,3-Dimethyl-1-pentanol

$$
\begin{array}{c}
\quad\quad\quad\quad\text{O} \\
\quad\quad\quad\quad|| \\
\text{CH}_3\text{CH}_2\text{CH}=\text{CHCCH}_3
\end{array}
$$

3-Hexene-2-one

Esters are named by replacing the suffix of the parent acid -*oic acid* by -*oate*:

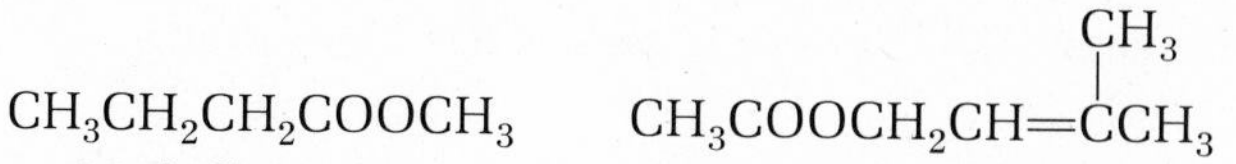

$$CH_3CH_2CH_2COOCH_3$$

Methylbutanoate

$$CH_3COOCH_2CH=\overset{\overset{\displaystyle CH_3}{|}}{C}CH_3$$

3-Methyl-2-butenylethanoate

Cyclic aliphatic hydrocarbons are named by prefixing *cyclo-* to the name of the corresponding open-chain hydrocarbon having the same number of C atoms as the ring:

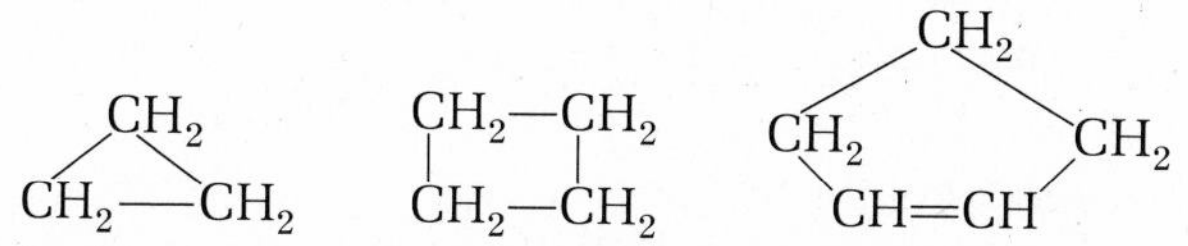

Cyclopropane　　*Cyclobutane*　　*Cyclopentene*

Rings containing atoms other than C (heterocycles) as well as aromatic rings are usually designated by trivial (nonsystematic) names:

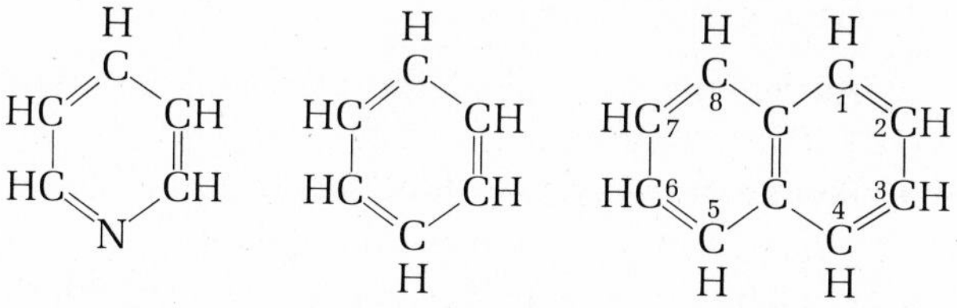

Pyridine　　*Benzene*　　*Naphthalene*

To locate substituents, rings are numbered clockwise around the periphery as shown for naphthalene.

Appendix 2 Mathematical Operations

2.1 Exponential Numbers

Multiplication by a positive power of 10 corresponds to moving the decimal point to the right; multiplication by a negative power of 10 corresponds to moving the decimal point to the left.

1.23×10^4 is 12,300
1.23×10^{-4} is 0.000123

Numbers expressed with powers of 10 cannot be added or subtracted directly unless the powers of 10 are the same.

$$1.23 \times 10^4 + 1.23 \times 10^5 = 1.23 \times 10^4 + 12.3 \times 10^4$$
$$= 13.5 \times 10^4$$
$$1.23 \times 10^{-4} - 1.23 \times 10^{-5} = 1.23 \times 10^{-4} - 0.123 \times 10^{-4}$$
$$= 1.11 \times 10^{-4}$$

When powers of 10 are multiplied, exponents are added; when divided, exponents are subtracted.

$$(1.23 \times 10^4) \times (1.23 \times 10^5) = (1.23 \times 1.23) \times (10^4 \times 10^5)$$
$$= 1.51 \times 10^9$$
$$\frac{1.23 \times 10^{-4}}{1.23 \times 10^{-5}} = \frac{1.23}{1.23} \times \frac{10^{-4}}{10^{-5}} = 1.00 \times 10$$

In taking square roots of powers of 10, the exponent is divided by 2; in taking cube roots, by 3.

Square root of 9×10^4 is 3×10^2
Cube root of 8×10^{-12} is 2×10^{-4}

2.2 Proportionality

If one property A of a substance is related to, or depends on, another property B, then A is said to be *proportional to B*. The simplest kind of proportionality is *direct proportionality*, in which any change in B produces an equal percentage change in A. (An example of a direct proportionality is the relation observed between mass and volume. Since the mass of a substance depends on the volume of it taken, we say that mass is directly proportional to volume.) Mathematically, direct proportionality is indicated as follows:

$A \propto B$
$A = kB$

The symbol $\propto$ is read "is directly proportional to" and is called a *proportionality sign*. In the second expression, the proportionality sign has been replaced by an equals sign and the proportionality constant k. The constant k represents the ratio of A to B and indicates that the ratio A/B is a constant. Thus, if B is doubled, A must also be doubled in order to maintain k constant; if B is halved, A is halved, etc. For the direct proportionality between mass m and volume V, we can write

$m \propto V$
$m = kV$

where the proportionality constant k is the density (m/V) of the substance. Since the density of a substance does not depend on how much mass or volume of that substance we take, we can also write

$$\frac{m_1}{V_1} = \frac{m_2}{V_2} = \frac{m_3}{V_3}$$

where the subscripts 1, 2, and 3 refer to different samples of the same substance.

A more complicated proportionality is the *inverse proportionality* such as that found between volume and pressure of a gas. It is observed that doubling the pressure on a gas halves its volume. Mathematically, this inverse proportionality is expressed in any of the following ways:

$$V \propto \frac{1}{P} \quad \text{or} \quad V \propto P^{-1}$$

$$V = \frac{k}{P} \quad \text{or} \quad PV = k$$

$$P_1V_1 = P_2V_2 = P_3V_3$$

The proportionality constant k has a fixed value as long as only P and V change. Its value depends on the mass of the gas sample and its temperature. In other words, the inverse proportionality between P and V holds only for constant temperature and mass of sample.

In addition to direct and inverse proportionality, other proportionalities are possible. For example, the distance d traveled by a freely falling object is proportional to the square of the time of fall t:

$$d = kt^2$$

For a time twice as long, the object travels four times as far. Another example of a proportionality is the observation that the rate of diffusion R of a gas is inversely proportional to the square root of its molecular mass M. This can be written in several ways:

$$R = \frac{k}{\sqrt{M}}$$

$$R = \frac{k}{M^{1/2}}$$

$$R = kM^{-1/2}$$

2.3 Graphs

Often the best way to represent the dependence between two properties is to plot a graph in which the values of one property are shown along one axis, and values of the other property are shown along an axis perpendicular to the first. As shown in Figure 1, each point represents the result of an observation. Distance along the horizontal axis (abscissa) is the value of B; distance along the vertical axis (ordinate) is the corresponding value of A. By convention, the abscissa represents the property that is independently varied (i.e., chosen at random), and the ordinate represents the dependent property that is observed as a result. For example, in Figure 1, B might represent various values of the pressure placed on a gas, and A the values of the volume observed as a result.

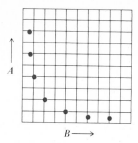

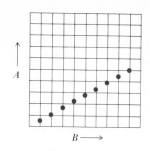

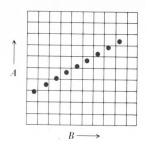

Figure 1
Graph of A vs. B.

Figure 2
Graph of a direct proportionality.

Figure 3
Graph of $A = kB + c$.

The relation shown in Figure 1 is an inverse proportionality, and the points lie on a curve whose equation is $A = k/B$. For a direct proportionality $(A = kB)$ the points would fall on a straight line, as shown in Figure 2. The straight line goes through the origin (the point for which both A and B are zero), and its slope is determined by the numerical value of k. If, as in Figure 3, the straight line does not pass through the origin, its equation is $A = kB + c$. The relation between A and B is still a linear one, but it is not a direct proportionality. The value of c is constant and is equal to the value of A when B is zero. Figure 3 is related to Figure 2 by a simple shift of origin. For gases, the linear relation between volume and Celsius temperature (Figure 6.9) can be converted to a direct proportionality by changing the origin from 0 to $-273°C$.

2.4 Logarithms

The logarithm of a number is the power to which the number 10 must be raised to equal the number. Thus, the log of 10,000 (or 10^4) is 4, and that of 0.01 (or 10^{-2}) is -2. Most numbers are not integral powers of 10, and therefore their logarithms are not immediately obvious. For such numbers, logs may be determined by use of log tables. A simple log table is given in Figure 4.

To get the logarithm of the number 5.7, look for 5 in the first vertical column and then move across to the column headed by 0.7. The log of 5.7 is 0.756. To get the logarithm of any number not lying between 1.0 and 10, first write the number in exponential form (Appendix 2.1), so that it consists of a number between 1.0 and 10 multiplied by a power of 10. Thus, 570 becomes 5.7×10^2. Because the logarithm of a product $a \times b$ is the log of a plus the log of b, we can write

$$\log (5.7 \times 10^2) = \log 5.7 + \log 10^2$$
$$= 0.756 + 2$$
$$= 2.756$$

The principal use of logarithms in this text is in connection with pH, defined as the negative of the logarithm of the hydrogen-ion concentration. For a hydrogen-ion concentration of 0.00036 M, the pH is found as follows:

$$\log 0.00036 = \log (3.6 \times 10^{-4})$$
$$= \log 3.6 + \log 10^{-4} = 0.556 - 4 = -3.444$$
$$pH = 3.444$$

Figure 4 LOGARITHMS	0.0	0.1	0.2	0.3	0.4	0.5	0.6	0.7	0.8	0.9
1	000	041	079	114	146	176	204	230	255	279
2	301	322	342	362	380	398	415	431	447	462
3	477	491	505	519	532	544	556	568	580	591
4	602	613	623	634	644	653	663	672	681	690
5	699	708	716	724	732	740	748	756	763	771
6	778	785	792	799	806	813	820	826	833	839
7	845	851	857	863	869	875	881	887	892	898
8	903	909	914	919	924	929	935	940	945	949
9	954	959	964	969	973	978	982	987	991	996

Sometimes, the reverse procedure is required. For example, if a solution has a pH of 8.50, its hydrogen-ion concentration can be found:

$$pH = 8.50$$
$$\log [H^+] = -8.50 = 0.50 - 9$$
$$[H^+] = 3.2 \times 10^{-9}$$

The number 3.2 is the antilog of 0.50 (the number whose log is 0.50). Antilogs are obtained by using Figure 4 in reverse, i.e., by looking up the logarithm in the body of the table and then finding the number which corresponds to it.

2.5 Quadratic Equations

A quadratic equation is an algebraic equation in which a variable is raised to the second power but no higher and which can be written in the form

$$ax^2 + bx + c = 0$$

The solution of such an equation is

$$x = \frac{-b \pm \sqrt{b^2 - 4ac}}{2a}$$

where the $\pm$ sign indicates that there are two roots. Thus, the equation obtained in Example 1 of Section 12.4

$$\frac{(2n)^2}{(1.000 - n)(1.000 - n)} = 45.9$$

when rewritten gives

$$41.9n^2 - 91.8n + 45.9 = 0$$

for which the roots are

$$n = \frac{-(-91.8) \pm \sqrt{(-91.8)^2 - 4(41.9)(45.9)}}{2(41.9)}$$

$$= 0.772 \text{ or } 1.42$$

The second root ($n = 1.42$) is inadmissible from the nature of the problem (n cannot be greater than 1.00, which represents all the reactants present). The first root ($n = 0.772$) must be the correct one.

2.6 Significant Figures

Numbers which express the result of a measurement such that only the last digit is in doubt are called *significant figures*. The number of significant figures used to express a measured result depends on how precisely the measuring instrument is calibrated. This is illustrated by the example in Figure 5. What can be determined, by measurement, about the length of line AB? The upper ruler gives 3.6 cm; the line is certainly between 3 and 4 cm

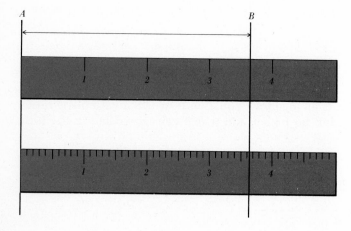

Figure 5
Accuracy of line measurement to illustrate significant figures.

and appears to be about six-tenths of the way from 3 to 4. The lower ruler gives 3.63 cm; the line is certainly between 3.6 and 3.7 cm and appears to be about three-tenths of the way from 3.6 to 3.7. The first measurement (3.6 cm) has two significant figures; the second measurement (3.63 cm) has three significant figures. Because of the difference in the two measuring instruments, there are two ways of expressing the length of the same line. These represent different degrees of precision. The first measurement, 3.6 cm, is precise to 1 part in 36, or approximately 3 percent. The second, 3.63 cm, is precise to 1 part in 363, or about 0.3 percent.

Understanding significant figures is important in calculations involving measured quantities. For example, adding a 0.101-cm line to a 3.63-cm line does not give one line 3.731 cm long, but rather 3.73 cm long. This is true because the measurement 3.63 cm specifies nothing about the third decimal place. The measurement is not precise enough to distinguish between 3.630 cm and 3.633 cm, for instance. When 0.101 cm is added, the total length might be 3.731 cm or 3.734 cm. To keep from giving any misinformation, the total length should be stated as 3.73 cm. Similarly, in calculating the area of a square measured to be 1.2 cm on edge, the result should be specified not as 1.44 cm^2 but as 1.4 cm^2. Since the edge is known only to 1 part in 12, or about 8 percent, the area cannot possibly be calculated to 1 part in 144, or 0.7 percent.

A simple rule that usually holds is that the number of significant figures resulting from multiplication or division of measured quantities is the same as the number of significant figures in the least precisely known quantity. In rounding off answers to get the proper number of significant figures, the usual convention is to increase the last significant digit if the following discarded digit is 5 or greater. For example, to two significant figures the number 0.011863 is 0.012. Zeros used only to show the position of a decimal point are not counted as significant figures.

Appendix 3 Definitions from Physics

3.1 Velocity and Acceleration

When an object changes its position, it is said to undergo a *displacement*. The rate at which displacement changes with time is called the *velocity* and has the dimensions of distance divided by time (e.g., miles per hour, centimeters per second). *Acceleration* is the rate at which velocity changes with time and has the dimensions of velocity divided by time (e.g., centimeters per second per second, or cm/s^2).

3.2 Force and Mass

Force can be thought of as a push or pull on an object which tends to change its motion, to speed it up or slow it down or to cause it to deviate from its path. *Mass* is a quantitative measure of the inertia of an object to having its motion changed. Thus, mass determines how difficult it is to accelerate an object. Quantitatively, force and mass are related by the equation

$F = ma$

where F is the force which produces acceleration a in mass m. If m is in grams and a is in centimeters per second per second, then F is in gram · centimeters per second per second, or dynes. Weight is an expression of force and arises because every object has mass and is being accelerated by gravity. Pounds weight is therefore a measure of force.

3.3 Momentum and Impulse

In dealing with collision problems it is useful to have terms for describing the combined effect of mass and velocity and its change with time. *Mass times velocity, mv,* is called the *momentum* and determines the length of time required to bring a moving body to rest when decelerated by a constant force. Thus, for a particle of momentum mv to be stopped by a constant force F, the required time t is mv/F.

The *impulse* is defined for the case of a constant force as Ft, where t is the time during which the force F acts. Thus, for the stopping of a particle originally of momentum mv by force F in time t, the impulse is just

$$Ft = F\frac{mv}{F} = mv$$

This is true if the particle comes to a complete rest. If, however, the particle bounces back, as it would on collision with a rigid wall, the particle is reflected from the wall with momentum $-mv$ (the minus sign indicating that the velocity is now in the opposite direction). The total impulse, counting the time for deceleration to zero and acceleration to $-mv$, is twice what it was before, or $2mv$.

In considering the pressure exerted by a gas, impulse comes in as follows: The pressure, or force per unit area, is the rate of collision per unit area times the effect of each collision:

$$\text{Pressure} = \frac{\text{force}}{\text{area}} = \frac{\text{number of collisions}}{\text{time} \times \text{area}} \times \text{?}$$

$$\text{?} = \frac{\text{force} \times \text{time}}{\text{number of collisions}} = \text{impulse per collision}$$

3.4 Work and Energy

When a force F operates on (e.g., pushes) an object through a distance d, *work W* is done:

$$W = Fd$$

If force is expressed in dynes (gram · centimeters per second per second) and distance in centimeters, then work has the dimensions dyne · centimeters $(g \cdot cm^2/s^2)$, or ergs. One erg is thus the work done in moving one gram through one centimeter so as to increase its velocity by one centimeter per second all in one second. (For reference, 1 erg is approximately the work a fly does in one pushup.) To scale the unit up, 10^7 ergs is called a joule.

Energy is the ability to do work, and the dimensions of energy are the same as those of work. *Kinetic energy* is the energy a body possesses because of its motion and mass. It is equal to one-half the mass times the square of its velocity. *Potential energy* is the energy a body possesses because of its position or arrangement with respect to other bodies. For example, the potential energy of a rock depends on its distance from the earth. As the rock falls toward the earth, its kinetic energy increases and its potential energy decreases.

3.5 Electric Charge and Electric Field

Electric charge is a property assigned to objects to account for certain observed attractions or repulsions which cannot be explained in terms of gravitational attraction between masses. Electric charge can be of two types, *positive* and *negative*. Objects which have the same type of electric charge repel each other; objects with opposite charges attract each other. A *unit of charge* can be defined as the quantity of electric charge which at a distance of one centimeter from another identical charge produces a repulsive force of one dyne in a vacuum. This unit of charge is called the *electrostatic unit* (esu). An electron has a negative charge of 4.80×10^{-10} esu. The practical unit of electric charge is the *coulomb*, which is almost exactly 3×10^9 esu. In coulombs, the charge of an electron is 1.60×10^{-19} coulomb.

An *electric field* is said to exist at a point if a force of electrical origin is exerted on any charged body placed at that point. The *intensity* of an electric field is defined as the magnitude of the electric force exerted on a unit charge. Any electrically charged body placed in an electric field moves unless otherwise constrained. The direction of a field is usually defined as the direction in which a positive charge would move.

3.6 Voltage and Capacity

An electric capacitor is a device for storing electric charge. In its simplest form a capacitor consists of two parallel, electrically conducting plates separated by some distance. The capacitor can be charged by making one plate positive and the other plate negative. To transfer a unit positive charge from the negative plate to the positive plate, work must be done against the electric field which exists between the charged plates. Therefore, the potential energy of the unit charge is increased in the process. In other words, there is a change in potential energy in going from one plate to the other. This difference in potential energy for a unit charge moved from one plate to the other is called the *potential difference,* or the *voltage,* of the capacitor. Voltage, or potential difference, is not restricted to capacitors but may exist between any two points as long as work must be done in transferring an electric charge from one point to the other. The potential difference between two points is said to be one volt if it requires one joule to move one coulomb of charge from one point to the other. To move an electron through a potential difference of one volt requires an amount of energy, called the *electron volt,* equal to 1.6×10^{-19} joule.

Capacity is the term used to describe quantitatively the amount of charge that can be stored on a capacitor. It is equal to the amount of charge that can be stored on the plates when the voltage difference between the plates is one volt. In general, the amount of charge a capacitor can hold is directly proportional to the voltage; the capacity is simply the proportionality constant.

$$Q = CV$$

If Q, the charge, is 1 coulomb and if V, the voltage, is 1 volt, then C, the capacity, is 1 farad. The capacity of a capacitor depends on its design (area of the plates, distance between them) and on the nature of the material between the plates. For a parallel-plate capacitor, the capacity is given approximately by the following equation:

$$C = \frac{KA}{4\pi d}$$

where A is the area of the plates, d is the distance between the plates, and K is the dielectric constant of the material between the plates. For a vacuum, the dielectric constant K is exactly equal to 1; for all other substances, K is greater than 1. Some typical dielectric constants are 1.00059 for air at STP, 1.00026 for hydrogen gas at STP, 1.0046 for HCl gas at STP, 80 for liquid water at 20°C, 28.4 for ethyl alcohol at 0°C, 2 for petroleum, 4 for solid sulfur.

3.7   Electric Current

A collection of moving charges is called an *electric current.* The unit of current is the *ampere,* which corresponds to a flow of one coulomb of charge past a point in one second. Since current specifies the rate at which charge is transferred, the product of current multiplied by time gives the total amount of charge transferred:

$$Q = It$$

If the current I is in amperes (coulombs per second) and the time t is in seconds, the charge Q is in coulombs.

The current that a wire carries is directly proportional to the voltage difference between the ends of the wire. The proportionality constant is called the *conductance* of the wire and is equal to the reciprocal of the *resistance* of the wire.

$$I = \frac{1}{R} V \quad \text{or} \quad V = IR$$

If V is the potential difference in volts and I is the current in amperes, R is the resistance in ohms.

There are two important kinds of current, direct and alternating. Direct current (dc) implies that the charge is constantly moving in the same direction along the wire. Alternating current (ac) implies that the current reverses its direction at regular intervals of time. The usual house current is 60-cycle alternating current; i.e., it goes through 60 back-and-forth oscillations per second.

3.8   Coulomb's Law

The interaction force between electric charges can be written in terms of Coulomb's law:

$$F = \frac{q_1 q_2}{Kr^2}$$

where F is the force, q_1 and q_2 are the two charges, r is the distance between them, and K is the dielectric constant for the medium between the two charges. In a vacuum, K is unity and so disappears from the above equation. If q_1 and q_2 are of the same sign (e.g., both are positive), then F is positive. This means there is a repulsion between the charges. If q_1 and q_2 are of different signs (e.g., one is positive and the other is negative), F is negative, and there is attraction between the charges.

If q_1 and q_2 are expressed in electrostatic units (esu) and r is given in centimeters, then F will be in dynes.

Appendix 4 Conversion Factors and SI Units

Fundamental Constants	
Avogadro number, N	6.0225×10^{23} molecules/mole
Boltzmann constant, k	1.3805×10^{-16} erg/deg
Electron charge, e	1.6021×10^{-19} coulomb
Electron mass, m	9.1091×10^{-28} g
Faraday constant, $\mathcal{F}$	9.6487×10^{4} coulombs/equiv
Gas constant, R	8.2057×10^{-2} liter $\cdot$ atm/mole $\cdot$ deg
Planck constant, h	6.6256×10^{-27} erg $\cdot$ s
Speed of light, c	2.9979×10^{10} cm/s

Conversion Factors
1 kilometer (km) = 1000 meters = 0.62137 mile
1 meter (m) = 100 centimeters = 39.370 inches (in.)
1 centimeter (cm) = 10 millimeters (mm) = 0.39370 inch
1 kilogram (kg) = 1000 grams = 2.2046 pounds (lb)
1 gram (g) = 1000 milligrams (mg) = 0.035274 ounce (oz)
1 liter (l) = 1000 milliliters (ml) = 1.0567 quarts (qt)
1 atomic mass unit (amu) = 1.6604×10^{-24} g
1 angstrom (Å) = 1×10^{-8} cm = 1×10^{-10} m = 1×10^{-1} nm
1 electron volt (eV) = 1.6021×10^{-19} joule
1 calorie (cal) = 4.1840×10^{7} ergs = 4.1840 joules
1 liter $\cdot$ atmosphere (liter $\cdot$ atm) = 24.217 cal = 101.32 joules
1 kilojoule = 239.0 cal = 0.2390 kcal

Most of the units used in this text are SI units, as recommended by the International Committee on Weights and Measures. The International System of Units (usually designated SI, after Système International) is constructed from the seven base units that follow.

Name of Unit	Quantity Measured	Symbol for Unit
Meter	Length	m
Kilogram	Mass	kg
Second	Time	s
Ampere	Electric current	A
Kelvin	Thermodynamic temperature	K
Candela	Light intensity	cd
Mole	Amount of substance	mol

Decimal fractions or multiples of these units are indicated as follows.

Fraction	Prefix	Symbol	Multiple	Prefix	Symbol
10^{-1}	Deci-	d	10^1	Deka-	da
10^{-2}	Centi-	c	10^2	Hecto-	h
10^{-3}	Milli-	m	10^3	Kilo-	k
10^{-6}	Micro-	μ	10^6	Mega-	M
10^{-9}	Nano-	n	10^9	Giga-	G
10^{-12}	Pico-	p	10^{12}	Tera-	T
10^{-15}	Femto-	f			
10^{-18}	Atto-	a			

Derived from the base units are these specially named units.

Name of Unit	Quantity Measured	Symbol for Unit	Definition of Unit
Newton	Force	N	$kg \cdot m/s^2$
Joule	Energy	J	$kg \cdot m^2/s^2$
Watt	Power	W	$J/s = kg \cdot m^2/s^3$
Coulomb	Electric charge	C	$A \cdot s$
Volt	Electric potential difference	V	$J\,A^{-1}\,s^{-1} = kg \cdot m^2\,s^{-3}\,A^{-1}$

Name of Unit	Quantity Measured	Symbol for Unit	Definition of Unit
Ohm	Electric resistance	Ω	$V/A = kg \cdot m^2\,s^{-3}\,A^{-2}$
Farad	Electric capacitance	F	$A \cdot s/V = A^2 \cdot s^4\,kg^{-1}\,m^{-2}$
Hertz	Frequency	Hz	s^{-1} (cycle per second)

Use of the above units is recommended. There are also certain decimal fractions and multiples of SI units, having special names, which do not belong to the International System of Units and whose use is to be *progressively discouraged*. Among them are the following:

Name of Unit	Quantity Measured	Symbol for Unit	Definition of Unit
Angstrom	Length	Å	$10^{-10}\,m = 10^{-8}\,cm$
Dyne	Force	dyn	$10^{-5}\,N$
Bar	Pressure	bar	$10^5\,N/m^2$
Erg	Energy	erg	$10^{-7}\,J$

There are other units that are not fractions or multiples of SI units, nor can they be defined exactly in terms of SI units. The following do not belong to the International System of Units and are recommended to be *abandoned*.

Name of Unit	Quantity Measured	Symbol for Unit	Definition of Unit
Inch	Length	in.	$2.54 \times 10^{-2}\,m$
Pound	Mass	lb	$0.453502\,kg$
Atmosphere*	Pressure	atm	$101,325\,N/m^2$
Torr	Pressure	Torr	$(101,325/760)\,N/m^2$
Millimeter of mercury	Pressure	mmHg	$13.5951 \times 980.665 \times 10^{-2}\,N/m^2$
Calorie	Energy	cal	$4.184\,J$

*Use of this unit is sanctioned for a limited period of time.

Appendix 5 Vapor Pressure of Water

Temperature, °C	Pressure, mmHg	atm	Temperature, °C	Pressure, mmHg	atm
0	4.6	0.0061	23	21.1	0.0278
1	4.9	0.0065	24	22.4	0.0295
2	5.3	0.0070	25	23.8	0.0313
3	5.7	0.0075	26	25.2	0.0332
4	6.1	0.0080	27	26.7	0.0351
5	6.5	0.0086	28	28.3	0.0372
6	7.0	0.0092	29	30.0	0.0395
7	7.5	0.0099	30	31.8	0.0418
8	8.0	0.0105	35	42.2	0.0555
9	8.6	0.0113	40	55.3	0.0728
10	9.2	0.0121	45	71.9	0.0946
11	9.8	0.0129	50	92.5	0.1217
12	10.5	0.0138	55	118.0	0.1553
13	11.2	0.0147	60	149.4	0.1966
14	12.0	0.0158	65	187.5	0.2467
15	12.8	0.0168	70	233.7	0.3075
16	13.6	0.0179	75	289.1	0.3804
17	14.5	0.0191	80	355.1	0.4672
18	15.5	0.0204	85	433.6	0.5705
19	16.5	0.0217	90	525.8	0.6918
20	17.5	0.0230	95	633.9	0.8341
21	18.7	0.0246	100	760.0	1.0000
22	19.8	0.0261	105	906.1	1.1192

Appendix 6 Reduction Potentials

Half-Reactions	E^0, volts
$F_2(g) + 2H^+ + 2e^- \longrightarrow 2HF$	$+3.06$
$F_2(g) + 2e^- \longrightarrow 2F^-$	$+2.87$
$O_3(g) + 2H^+ + 2e^- \longrightarrow O_2(g) + H_2O$	$+2.07$
$Ag^{2+} + e^- \longrightarrow Ag^+$	$+1.98$
$Co^{3+} + e^- \longrightarrow Co^{2+}$	$+1.82$
$H_2O_2 + 2H^+ + 2e^- \longrightarrow 2H_2O$	$+1.77$
$MnO_4^- + 4H^+ + 3e^- \longrightarrow MnO_2(s) + 2H_2O$	$+1.70$
$Au^+ + e^- \longrightarrow Au(s)$	$\sim +1.7$
$HClO_2 + 2H^+ + 2e^- \longrightarrow HClO + H_2O$	$+1.64$
$HClO + H^+ + e^- \longrightarrow \frac{1}{2} Cl_2(g) + H_2O$	$+1.63$
$Ce^{4+} + e^- \longrightarrow Ce^{3+}$	$+1.61$
$H_5IO_6 + H^+ + 2e^- \longrightarrow IO_3^- + 3H_2O$	$+1.6$
$MnO_4^- + 8H^+ + 5e^- \longrightarrow Mn^{2+} + 4H_2O$	$+1.51$
$Mn^{3+} + e^- \longrightarrow Mn^{2+}$	$+1.51$
$BrO_3^- + 6H^+ + 5e^- \longrightarrow \frac{1}{2} Br_2 + 3H_2O$	$+1.50$
$Au^{3+} + 3e^- \longrightarrow Au(s)$	$+1.50$
$Cl_2(g) + 2e^- \longrightarrow 2Cl^-$	$+1.36$
$NH_3OH^+ + 2H^+ + 2e^- \longrightarrow NH_4^+ + H_2O$	$+1.35$
$Cr_2O_7^{2-} + 14H^+ + 6e^- \longrightarrow 2Cr^{3+} + 7H_2O$	$+1.33$
$2HNO_2 + 4H^+ + 4e^- \longrightarrow N_2O(g) + 3H_2O$	$+1.29$

Half-Reactions	E^0, volts
$Tl^{3+} + 2e^- \longrightarrow Tl^+$	$+1.25$
$MnO_2(s) + 4H^+ + 2e^- \longrightarrow Mn^{2+} + 2H_2O$	$+1.23$
$O_2(g) + 4H^+ + 4e^- \longrightarrow 2H_2O$	$+1.23$
$ClO_3^- + 3H^+ + 2e^- \longrightarrow HClO_2 + H_2O$	$+1.21$
$IO_3^- + 6H^+ + 5e^- \longrightarrow \frac{1}{2}I_2 + 3H_2O$	$+1.20$
$ClO_4^- + 2H^+ + 2e^- \longrightarrow ClO_3^- + H_2O$	$+1.19$
$PuO_2^+ + 4H^+ + e^- \longrightarrow Pu^{4+} + 2H_2O$	$+1.15$
$Br_2 + 2e^- \longrightarrow 2Br^-$	$+1.09$
$N_2O_4(g) + 2H^+ + 2e^- \longrightarrow 2HNO_2$	$+1.07$
$Br_2(l) + 2e^- \longrightarrow 2Br^-$	$+1.07$
$PuO_2^{2+} + 4H^+ + 2e^- \longrightarrow Pu^{4+} + 2H_2O$	$+1.04$
$N_2O_4(g) + 4H^+ + 4e^- \longrightarrow 2NO(g) + 2H_2O$	$+1.03$
$V(OH)_4^+ + 2H^+ + e^- \longrightarrow VO^{2+} + 3H_2O$	$+1.00$
$HNO_2 + H^+ + e^- \longrightarrow NO(g) + H_2O$	$+1.00$
$Pu^{4+} + e^- \longrightarrow Pu^{3+}$	$+0.97$
$NO_3^- + 4H^+ + 3e^- \longrightarrow NO(g) + 2H_2O$	$+0.96$
$2Hg^{2+} + 2e^- \longrightarrow Hg_2^{2+}$	$+0.92$
$2NO_3^- + 4H^+ + 2e^- \longrightarrow N_2O_4(g) + 2H_2O$	$+0.80$
$Ag^+ + e^- \longrightarrow Ag(s)$	$+0.80$
$Hg_2^{2+} + 2e^- \longrightarrow 2Hg(l)$	$+0.79$
$Fe^{3+} + e^- \longrightarrow Fe^{2+}$	$+0.77$
$O_2(g) + 2H^+ + 2e^- \longrightarrow H_2O_2$	$+0.68$
$UO_2^+ + 4H^+ + e^- \longrightarrow U^{4+} + 2H_2O$	$+0.62$
$MnO_4^- + e^- \longrightarrow MnO_4^{2-}$	$+0.56$
$H_3AsO_4 + 2H^+ + 2e^- \longrightarrow HAsO_2 + 2H_2O$	$+0.56$
$I_2 + 2e^- \longrightarrow 2I^-$	$+0.54$
$Cu^+ + e^- \longrightarrow Cu(s)$	$+0.52$
$VO^{2+} + 2H^+ + e^- \longrightarrow V^{3+} + H_2O$	$+0.36$
$Fe(CN)_6^{3-} + e^- \longrightarrow Fe(CN)_6^{4-}$	$+0.36$
$Cu^{2+} + 2e^- \longrightarrow Cu(s)$	$+0.34$
$UO_2^{2+} + 4H^+ + 2e^- \longrightarrow U^{4+} + 2H_2O$	$+0.33$
$Cu^{2+} + e^- \longrightarrow Cu^+$	$+0.15$
$Sn^{4+} + 2e^- \longrightarrow Sn^{2+}$	$+0.15$
$S(s) + 2H^+ + 2e^- \longrightarrow H_2S(g)$	$+0.14$

Half-Reactions	E^0, volts
$HSO_4^- + 3H^+ + 2e^- \longrightarrow SO_2 + 2H_2O$	$+0.11$
$P(s) + 3H^+ + 3e^- \longrightarrow PH_3(g)$	$+0.06$
$UO_2^{2+} + e^- \longrightarrow UO_2^+$	$+0.05$
$2H^+ + 2e^- \longrightarrow H_2(g)$	Zero
$Pb^{2+} + 2e^- \longrightarrow Pb(s)$	-0.13
$Sn^{2+} + 2e^- \longrightarrow Sn(s)$	-0.14
$Mo^{3+} + 3e^- \longrightarrow Mo(s)$	~ -0.2
$Ni^{2+} + 2e^- \longrightarrow Ni(s)$	-0.25
$V^{3+} + e^- \longrightarrow V^{2+}$	-0.26
$H_3PO_4 + 2H^+ + 2e^- \longrightarrow H_3PO_3 + H_2O$	-0.28
$Co^{2+} + 2e^- \longrightarrow Co(s)$	-0.28
$Tl^+ + e^- \longrightarrow Tl(s)$	-0.34
$In^{3+} + 3e^- \longrightarrow In(s)$	-0.34
$Cd^{2+} + 2e^- \longrightarrow Cd(s)$	-0.40
$Cr^{3+} + e^- \longrightarrow Cr^{2+}$	-0.41
$Eu^{3+} + e^- \longrightarrow Eu^{2+}$	-0.43
$Fe^{2+} + 2e^- \longrightarrow Fe(s)$	-0.44
$Ga^{3+} + 3e^- \longrightarrow Ga(s)$	-0.53
$U^{4+} + e^- \longrightarrow U^{3+}$	-0.61
$Cr^{3+} + 3e^- \longrightarrow Cr(s)$	-0.74
$Zn^{2+} + 2e^- \longrightarrow Zn(s)$	-0.76
$TiO^{2+} + 2H^+ + 4e^- \longrightarrow Ti(s) + H_2O$	~ -0.9
$V^{2+} + 2e^- \longrightarrow V(s)$	~ -1.2
$Mn^{2+} + 2e^- \longrightarrow Mn(s)$	-1.18
$Zr^{4+} + 4e^- \longrightarrow Zr(s)$	-1.53
$Al^{3+} + 3e^- \longrightarrow Al(s)$	-1.66
$Hf^{4+} + 4e^- \longrightarrow Hf(s)$	-1.70
$U^{3+} + 3e^- \longrightarrow U(s)$	-1.80
$Be^{2+} + 2e^- \longrightarrow Be(s)$	-1.85
$Th^{4+} + 4e^- \longrightarrow Th(s)$	-1.90
$Pu^{3+} + 3e^- \longrightarrow Pu(s)$	-2.07
$Sc^{3+} + 3e^- \longrightarrow Sc(s)$	-2.08
$\frac{1}{2} H_2(g) + e^- \longrightarrow H^-$	-2.25
$Y^{3+} + 3e^- \longrightarrow Y(s)$	-2.37

Half-Reactions	E^0, volts
$Mg^{2+} + 2e^- \longrightarrow Mg(s)$	-2.37
$Ce^{3+} + 3e^- \longrightarrow Ce(s)$	-2.48
$La^{3+} + 3e^- \longrightarrow La(s)$	-2.52
$Na^+ + e^- \longrightarrow Na(s)$	-2.71
$Ca^{2+} + 2e^- \longrightarrow Ca(s)$	-2.87
$Sr^{2+} + 2e^- \longrightarrow Sr(s)$	-2.89
$Ba^{2+} + 2e^- \longrightarrow Ba(s)$	-2.90
$Ra^{2+} + 2e^- \longrightarrow Ra(s)$	-2.92
$Cs^+ + e^- \longrightarrow Cs(s)$	-2.92
$Rb^+ + e^- \longrightarrow Rb(s)$	-2.93
$K^+ + e^- \longrightarrow K(s)$	-2.93
$Li^+ + e^- \longrightarrow Li(s)$	-3.05

Appendix 7 Equilibrium Constants, K_c

$CrOH^{2+}$	5×10^{-11}	$H_2AsO_4^-$	5.6×10^{-8}
$CuOH^+$	1×10^{-8}	$HAsO_4^{2-}$	3×10^{-13}
$ZnOH^+$	4×10^{-5}	H_2O	1.0×10^{-14}
H_3BO_3	6.0×10^{-10}	H_2S	1.1×10^{-7}
$CO_2 + H_2O$	4.2×10^{-7}	HS^-	1×10^{-14}
HCO_3^-	4.8×10^{-11}	H_2SO_3	1.3×10^{-2}
$HC_2H_3O_2$	1.8×10^{-5}	HSO_3^-	5.6×10^{-8}
HCN	4.0×10^{-10}	HSO_4^-	1.3×10^{-2}
$NH_3 + H_2O$	1.8×10^{-5}	H_2Se	1.9×10^{-4}
HNO_2	4.5×10^{-4}	H_2SeO_3	2.7×10^{-3}
H_3PO_3	1.6×10^{-2}	$HSeO_3^-$	2.5×10^{-7}
$H_2PO_3^-$	7×10^{-7}	H_2Te	2.3×10^{-3}
H_3PO_4	7.5×10^{-3}	HF	6.7×10^{-4}
$H_2PO_4^-$	6.2×10^{-8}	$HOCl$	3.2×10^{-8}
HPO_4^{2-}	1×10^{-12}	$HClO_2$	1.1×10^{-2}
H_3AsO_4	2.5×10^{-4}		

SOLUBILITY PRODUCTS			
$Mg(OH)_2$	8.9×10^{-12}	NiS	3×10^{-21}
MgF_2	8×10^{-8}	PtS	8×10^{-73}
MgC_2O_4	8.6×10^{-5}	$Cu(OH)_2$	1.6×10^{-19}
$Ca(OH)_2$	1.3×10^{-6}	CuS	8×10^{-37}
CaF_2	1.7×10^{-10}	AgCl	1.7×10^{-10}
$CaCO_3$	4.7×10^{-9}	AgBr	5.0×10^{-13}
$CaSO_4$	2.4×10^{-5}	AgI	8.5×10^{-17}
CaC_2O_4	1.3×10^{-9}	AgCN	1.6×10^{-14}
$Sr(OH)_2$	3.2×10^{-4}	Ag_2S	5.5×10^{-51}
$SrSO_4$	7.6×10^{-7}	ZnS	1×10^{-22}
$SrCrO_4$	3.6×10^{-5}	CdS	1.0×10^{-28}
$Ba(OH)_2$	5.0×10^{-3}	Hg_2Cl_2	1.1×10^{-18}
$BaSO_4$	1.5×10^{-9}	Hg_2Br_2	1.3×10^{-22}
$BaCrO_4$	8.5×10^{-11}	Hg_2I_2	4.5×10^{-29}
$Cr(OH)_3$	6.7×10^{-31}	HgS	1.6×10^{-54}
$Mn(OH)_2$	2×10^{-13}	$Al(OH)_3$	5×10^{-33}
MnS	7×10^{-16}	SnS	1×10^{-26}
FeS	4×10^{-19}	$Pb(OH)_2$	4.2×10^{-15}
$Fe(OH)_3$	6×10^{-38}	$PbCl_2$	1.6×10^{-5}
CoS	5×10^{-22}	PbS	7×10^{-29}

Appendix 8 Atomic and Ionic Radii

Ac^{3+}	0.118 nm†	Bi^{5+}	0.074 nm	Cr^{6+}	0.052 nm
Ag^0	0.134	Br^0	0.114	Cs^0	0.235
Ag^+	0.126	Br^-	0.196	Cs^+	0.167
Ag^{2+}	0.089	Br^{5+}	0.047	Cu^0	0.117
Al^0	0.125			Cu^+	0.096
Al^{3+}	0.051	C^0	0.077	Cu^{2+}	0.072
Am^{3+}	0.107	C^{4+}	0.016		
Am^{4+}	0.092	Ca^0	0.174	Dy^0	0.160
As^0	0.121	Ca^{2+}	0.099	Dy^{3+}	0.092
As^{3+}	0.058	Cd^0	0.141		
As^{5+}	0.046	Cd^{2+}	0.097	Er^0	0.158
Au^0	0.134	Ce^0	0.165	Er^{3+}	0.089
Au^+	0.137	Ce^{3+}	0.107	Eu^0	0.185
Au^{3+}	~ 0.09	Ce^{4+}	0.094	Eu^{3+}	0.098
		Cl^0	0.099		
B^0	0.081	Cl^-	0.181	F^0	0.064
B^{3+}	0.023	Cl^{5+}	0.034	F^-	0.133
Ba^0	0.198	Cl^{7+}	0.027	Fe^0	0.117
Ba^{2+}	0.134	Co^0	0.116	Fe^{2+}	0.074
Be^0	0.089	Co^{2+}	0.072	Fe^{3+}	0.064
Be^{2+}	0.035	Co^{3+}	0.063		
Bi^0	~ 0.15	Cr^0	0.118	Ga^0	0.125
Bi^{3+}	0.096	Cr^{3+}	0.063	Ga^{3+}	0.062
				Gd^0	0.162

†One nanometer is equal to 10^{-9} meter.

Atomic and Ionic Radii

Gd^{3+}	0.097 nm	Na^+	0.097 nm	Ru^0	0.125 nm
Ge^0	0.122	Nb^0	0.134	Ru^{4+}	0.067
Ge^{2+}	0.073	Nb^{4+}	0.074		
Ge^{4+}	0.053	Nb^{5+}	0.069	S^0	0.104
		Nd^{3+}	0.104	S^{2-}	0.184
Hf^0	0.144	Ni^0	0.115	S^{4+}	0.037
Hf^{4+}	0.078	Ni^{2+}	0.069	S^{6+}	0.030
Hg^0	0.144	Np^{3+}	0.110	Sb^0	0.141
Hg^{2+}	0.110	Np^{4+}	0.095	Sb^{3+}	0.076
Ho^0	0.158			Sb^{5+}	0.062
Ho^{3+}	0.091	O^{2-}	0.140	Sc^0	0.144
		O^0	0.066	Sc^{3+}	0.081
I^0	0.133	Os^0	0.126	Se^0	0.117
I^-	0.220	Os^{4+}	0.069	Se^{2-}	0.198
I^{5+}	0.062			Se^{4+}	0.050
I^{7+}	0.050	P^0	0.110	Se^{6+}	0.042
In^0	0.150	P^{3+}	0.044	Si^0	0.117
In^{3+}	0.081	P^{5+}	0.035	Si^{4+}	0.042
Ir^0	0.127	Pa^{4+}	0.098	Sm^0	0.162
Ir^{4+}	0.068	Pb^0	0.154	Sm^{3+}	0.100
		Pb^{2+}	0.120	Sn^0	0.140
K^0	0.203	Pb^{4+}	0.084	Sn^{2+}	0.093
K^+	0.133	Pd^0	0.128	Sn^{4+}	0.071
		Pd^{2+}	0.080	Sr^0	0.191
La^0	0.169	Pd^{4+}	0.065	Sr^{2+}	0.112
La^{3+}	0.114	Pm^0	0.163		
Li^0	0.123	Pm^{3+}	0.106	Ta^0	0.134
Li^+	0.068	Po^0	0.153	Ta^{5+}	0.068
Lu^0	0.156	Pr^0	0.164	Tb^0	0.161
Lu^{3+}	0.085	Pr^{3+}	0.106	Tb^{3+}	0.093
		Pr^{4+}	0.092	Tb^{4+}	0.081
Mg^0	0.136	Pt^0	0.130	Tc^0	0.127
Mg^{2+}	0.066	Pt^{2+}	0.080	Tc^{7+}	0.056
Mn^0	0.117	Pt^{4+}	0.065	Te^0	0.137
Mn^{2+}	0.080	Pu^{3+}	0.108	Te^{2-}	0.221
Mn^{3+}	0.066	Pu^{4+}	0.093	Te^{4+}	~0.07
Mn^{4+}	0.060			Te^{6+}	0.056
Mn^{7+}	0.046	Ra^{2+}	0.143	Th^0	0.165
Mo^0	0.130	Rb^0	0.216	Th^{4+}	0.102
Mo^{4+}	0.070	Rb^+	0.147	Ti^0	0.132
Mo^{6+}	0.062	Re^0	0.128	Ti^{3+}	0.076
		Re^{4+}	~0.07	Ti^{4+}	0.068
N^0	0.070	Re^{7+}	0.056	Tl^0	0.155
N^{3+}	0.016	Rh^0	0.125	Tl^+	0.147
N^{5+}	0.013	Rh^{3+}	0.068		
Na^0	0.157				

Atomic and Ionic Radii

Tl^{3+}	0.095 nm	V^{3+}	0.074 nm	Zn^0	0.125 nm
Tm^0	0.158	V^{4+}	0.063	Zn^{2+}	0.074
Tm^{3+}	0.087	V^{5+}	0.059	Zr^0	0.145
				Zr^{4+}	0.079
U^0	0.142	W^0	0.130		
U^{4+}	0.097	W^{4+}	0.070		
U^{6+}	0.080	W^{6+}	0.062		
V^0	0.122	Y^0	0.162		
V^{2+}	0.088	Y^{3+}	0.092		

Appendix 9   References

Additional information and background material can be found in the following books, which are listed by the chapters to which they most apply:

1,2 *A Source Book in Chemistry, 1400–1900,* by H. M. Leicester and H. S. Klickstein (McGraw-Hill).
Textbook of Inorganic Chemistry by J. R. Partington (Macmillan).
Nuclear and Radiochemistry by G. Friedlander, J. W. Kennedy, and J. Miller (Wiley).
Nuclei and Radioactivity by G. R. Choppin (Benjamin).

3,4 *Atomic Spectra and Atomic Structure* by G. Herzberg (Dover).
Nature of the Chemical Bond by L. Pauling (Cornell).
Chemical Bonding by A. L. Companion (McGraw-Hill).
Electrons and Chemical Bonding by H. B. Gray (Benjamin).

5 *Chemistry Problems* by M. J. Sienko (Benjamin).

6 *Introduction to Molecular Kinetic Theory* by J. H. Hildebrand (Reinhold).
Kinetic Theory of Gases by W. Kauzmann (Benjamin).

7 *Seven Solid States* by W. J. Moore (Benjamin).
Solid State Chemistry by N. B. Hannay (Prentice-Hall).

8 *Physical Chemistry* by G. Barrow (McGraw-Hill).
Physical Chemistry by G. Castellan (Addison-Wesley).

9 *Elementary Chemical Thermodynamics* by B. H. Mahan (Benjamin).
Elements of Chemical Thermodynamics by L. Nash (Addison-Wesley).

10   *Electrolyte Solutions* by R. A. Robinson and R. H. Stokes (Butterworth).

Metal Ions in Solution by J. P. Hunt (Benjamin).

11   *How Chemical Reactions Occur* by E. L. King (Benjamin).

12   *Chemistry Problems* by M. J. Sienko (Benjamin).

13   *Introduction to Electrochemistry* by E. H. Lyons, Jr. (Heath).

Oxidation Potentials by W. M. Latimer (Prentice-Hall).

14–18   *Inorganic Chemistry* by M. J. Sienko, R. A. Plane, and R. E. Hester (Benjamin).

Advanced Inorganic Chemistry by F. A. Cotton and G. Wilkinson (Wiley).

Inorganic Chemistry by C. S. G. Phillips and R. J. P. Williams (Oxford University Press).

19,20   *Organic Chemistry* by R. T. Morrison and R. N. Boyd (Allyn and Bacon).

Basic Organic Chemistry by L. F. Fieser and M. Fieser (Heath).

Organic Chemistry by D. J. Cramm and G. S. Hammond (McGraw-Hill).

Organic Chemistry: A Contemporary View by P. L. Cook and J. W. Crump (Heath).

Introduction to Organic Chemistry by C. H. DePuy and K. L. Rinehart, Jr. (Wiley).

Basic Principles of Organic Chemistry by J. D. Roberts and M. C. Caserio (Benjamin).

A Guide to Understanding Basic Organic Reactions by R. C. Whitfield (Houghton Mifflin).

21–23   *Basic Biological Chemistry* by H. R. Mahler and E. H. Cordes (Harper & Row).

Modern Topics in Biochemistry by T. P. Bennett and E. Frieden (Macmillan).

Introduction to Modern Biochemistry by P. Karlson (Academic).

Concepts in Biochemistry by F. J. Reithel (McGraw-Hill).

24   *The Origin of Life* by J. Keosian (Reinhold).

Biochemical Predestination by D. H. Kenyon and G. Steinman (McGraw-Hill).

Chemical Evolution by M. Calvin (Oxford University Press).

Appendix 10 Answers to Selected Questions

Chapter 1

1.11 (a) 7530 joules, (b) 1340 joules
1.17 1.34 g/cm^3
1.19 0.701 joule/g · °C
1.21 (b) The kinetic energy is quadrupled.
1.27 (a) 1160 g, (b) 293 g

Chapter 2

2.7 55.9 g
2.9 (a) 39.10 g, (b) 0.3762 mole and 2.26 × 10^{23} atoms
2.15 (a) 6.02 × 10^{25}, (b) 1.50 × 10^{25}, (c) 25
2.17 (a) 8.07 × 10^{22}, (b) 2.53 × 10^{24}, (c) 4.69 × 10^{23}, (d) 9.03 × 10^{24}
2.19 Atomic weight = 63.6, copper
2.21 3.26 × 10^{19} atoms
2.29 10 half-lives, 143 days

Chapter 3

3.13 1740 kilojoules
3.21 (a) 490 kilojoules, (b) 82 kilojoules, (c) 346 kilojoules
3.29 (b) 4.04 cm^3, (c) 13.8 g/cm^3

Chapter 4

4.23 (b) 3.04 times

Chapter 5

5.5 (a) 5.55, (b) 1.66, (c) 0.0625
5.13 (a) 20.4 g, (b) 7.51, (c) 27.2 g
5.17 (a) 41.3 g, (b) 37.5 g
5.19 (a) 15.1 g, (b) 8.0 g, (c) 4.20
5.21 (a) 0.178, (b) 0.0907, (c) 1.909
5.23 C$_7$H$_5$NO$_3$
5.25 (a) 2 equiv/mole, (b) equiv of Fe = 27.9 g, equiv of O = 8 g

Chapter 6

6.5   (*a*) 174°C, (*b*) −124°C
6.9   (*a*) $g/V = MP/RT$, (*b*) ½, ½
6.17 0.0227
6.21 5.9 liters
6.23 8.2 liters
6.25 4
6.27 (*a*) 3.28×10^{23}, (*b*) 3.13×10^{23},
        (*c*) 1.05×10^{24}
6.29 20.8 liters
6.31 (*b*) 0.452 mole of C_5H_8, 0.147
        mole of C_5H_{12}
6.33 2.00 atm
6.35 13.6 g

Chapter 7

7.11 0.406 nm
7.17 0.128 nm
7.21 (*b*) Twice as many
7.23 AB_3C
7.25 (*a*) 3.000 cm, (*b*) 1.064×10^8,
        (*c*) 6.026×10^{23}

Chapter 8

8.17 (*b*) 0.744, (*c*) 0.668
8.19 0.087
8.21 1200 g
8.25 (*a*) 28.2 cm², (*b*) 6.13×10^3 m²,
        (*c*) 5.89×10^4
8.27 6×10^{23}

Chapter 9

9.11 (*a*) 16.7 joules, (*b*) 113 joules,
        (*c*) 151 joules
9.13 4.78 kilojoules
9.15 37.9 kilojoules
9.17 (*a*) 1.18 kilojoules, −0.360
        kilojoule; (*b*) above 274 K

9.19 −14.9 kilojoules
9.21 7.39 kilojoules

Chapter 10

10.5   −0.744°C, 100.21°C
10.11 (*a*) 50.0 ml, (*b*) 800 ml, (*c*) 40.0 ml
10.13 (*a*) 0.400, (*b*) 0.100, (*c*) 0.0200,
         (*d*) 0.0100
10.15 125 ml
10.19 (*a*) −80.0 kilojoules, (*b*) 267
         joules/mole · deg
10.21 16.5% $NaNO_3$, 59.2% K_2SO_4,
         24.3% H_2O
10.23 326 amu
10.25 765 ml
10.27 0.480 N
10.29 (*a*) 47.3 g, (*b*) 75.0 g
10.31 (*a*) 3.42, (*b*) 4.16
10.33 −6.86°C
10.39 0.080 M Na^+, 0.24 M Ca^{2+},
         0.56 M Cl^-
10.41 (*a*) 4.85 M KCl, 6.19 M NaCl
10.45 0.0100 M Cl^-, 0.0225 M Na^+,
         0.0125 M OH^-

Chapter 11

11.11 (*b*) 2.00×10^{-3} s⁻¹
11.19 Second order
11.25 (*a*) 0.10, (*b*) 0.18, (*c*) 0.003, (*d*) 0.20,
         (*e*) 0.025

Chapter 12

12.3   6.75
12.7 1×10^{-13}, 1×10^{-11}
12.11 (*a*) 6.0×10^{-3}, (*b*) 7.2×10^{-5}
12.13 100
12.19 (*a*) 1.55×10^{-6}, (*b*) 1.55×10^{-4}
12.21 (*a*) 1.0, (*b*) 12.0, (*c*) 3.0
12.23 1.2×10^{-3}

12.25 4.0×10^{-10}

12.29   From 0.60 to 2.7%

12.31   (a) 12.4, (b) 0.398, (c) 0.824

12.35   10

12.41   (a) 7.0×10^{-4}, (b) 1.4×10^{-5} mole

12.45   (a) $[C_2H_3O_2^-]$ is 1.8 times $HC_2H_3O_2$

12.47   (a) 1.0 M H^+, 1.0 M HSO_4^{2-};
        (b) 1.3×10^{-2} M SO_4^{2-}

12.49   0.06 M

12.51   (a) 2.5×10^{-10} and 1×10^{-6};
        (b) Zn^{2+}: 5×10^{-6}, 0.005%; Cu^{2+}:
        3×10^{-4}, 0.3%

Chapter 13

13.5   (b) 5 ml, (c) 620 liters

13.7   0.61 g

13.9   4.00 volts

13.15   (a) 0.0104 mole, (b) 0.735 g,
        (c) 0.232 liter

13.17   1090 coulombs

13.27   (a) 185 g H_2 and 6612 g Cl_2, (b) $1.23
        for H_2 and $0.034 for Cl_2

13.33   (a) 1:14:29, (b) 1:15:55,
        (c) 1:16:04, (d) 1:16:06,
        (e) 1:16:15, (f) 1:17:42

Chapter 14

14.1   (a) 1.8×10^{33}, (b) 1.1×10^{57}

14.7   (a) 53.3%, (b) 48.0%, (c) 43.2%,
        (d) 59.3%

14.13   5.55 m

14.15   0.750 liter

14.17 1.4×10^{17}

14.19   (a) 1.93×10^{17}, (b) 4.19×10^{10} kg

14.21   13.3

14.29 5.0×10^4 liters

14.31   5740 s

Chapter 15

15.9   (b) 1.35 times

15.13   (b) 6.5×10^2, (c) 4.2×10^5

15.17   30.0%

15.25   11.87

15.27   0.665 liter

15.29   (a) 1200 kilojoules, (b) 15,700
        kilojoules, (c) 14.3 and 188 liters

15.31   (a) 770 liters, (b) 2300 g, (c) 25.2
        liters, (d) 441 h

Chapter 16

16.11   (a) 18, (b) 8.5×10^{-10} mole AgCl,
        2.5×10^{-12} mole AgBr

16.15   (a) 63.2%, (b) 69.6%, (c) 72.0%

16.21   12.9%

16.23   1.50 liters

16.29   0.0781 M

16.37   (a) 356 kg, (b) 901 kg

Chapter 17

17.9   (a) 530 kg, (b) 1060 kg

17.13   24 kilojoules/mole B

17.17   (a) 27.3%, (b) 52.2%, (c) 76.6%,
        (d) 20.0%

17.23   11.5

17.27 2×10^{-5} M HCO_3^-,
        5×10^{-11} M CO_3^{2-}, 2×10^{-5} M H^+,
        5×10^{-10} M OH^-

Chapter 18

18.9   3.88 g

18.13   (a) 5.7×10^{-5}, (b) 1.0×10^{-4},
        (c) 7.4×10^{-6}

18.17   288,000%

18.19   0.028 g

18.23   1.1

18.25   (a) 31.6%, (b) 37.8%, (c) 34.1%,
        (d) 25.8%

18.27   3.0 kilojoules

18.33   (a) 1310, (b) 547

18.37 7.5×10^{-11}

Chapter 19

19.21   4060 kilojoules
19.41   (a) 4.76×10^4 kilojoules,
        (b) 630 kilojoules, (c) 3.0×10^3 kg
        (3.3 tons!)

Chapter 20

20.11   (a) 800, (b) 2.08
20.21   (a) 11.97, (b) 11.46, (c) 11.86
20.25 1×10^5 liters

Chapter 21

21.17   2.77 liters
21.19 1.62×10^4 kilojoules
21.23 3.3×10^{27}
21.29   (a) 7.3×10^{-4}, 2.8×10^3; (b) α to β,
        $K = 2.0$

Chapter 22

22.11   32.15%
22.19   53.7% C, 6.6% H, 17.0% N, 20.1% O,
        and 2.7% S
22.23   (a) 588 to 1878, (b) 20^{10}
22.25   (a) 24, (b) 6, (c) 2

Chapter 23

23.3   (a) 12.3 percent
23.11   52%

Chapter 24

24.17   12 liters
24.21   (a) 4, (b) 840, (c) 12, (d) 400

Index

Periodic Table of the Elements

Group	I	II					Transition elements						III	IV	V	VI	VII	0
Period 1	H 1																	He 2
2	Li 3	Be 4											B 5	C 6	N 7	O 8	F 9	Ne 10
3	Na 11	Mg 12											Al 13	Si 14	P 15	S 16	Cl 17	Ar 18
4	K 19	Ca 20	Sc 21	Ti 22	V 23	Cr 24	Mn 25	Fe 26	Co 27	Ni 28	Cu 29	Zn 30	Ga 31	Ge 32	As 33	Se 34	Br 35	Kr 36
5	Rb 37	Sr 38	Y 39	Zr 40	Nb 41	Mo 42	Te 43	Ru 44	Rh 45	Pd 46	Ag 47	Cd 48	In 49	Sn 50	Sb 51	Te 52	I 53	Xe 54
6	Cs 55	Ba 56	* 57–71	Hf 72	Ta 73	W 74	Re 75	Os 76	Ir 77	Pt 78	Au 79	Hg 80	Tl 81	Pb 82	Bi 83	Po 84	At 85	Rn 86
7	Fr 87	Ra 88	† 89–103	Ku 104	Ha 105													

*	La 57	Ce 58	Pr 59	Nd 60	Pm 61	Sm 62	Eu 63	Gd 64	Tb 65	Dy 66	Ho 67	Er 68	Tm 69	Yb 70	Lu 71
†	Ac 89	Th 90	Pa 91	U 92	Np 93	Pu 94	Am 95	Cm 96	Bk 97	Cf 98	Es 99	Fm 100	Md 101	No 102	Lr 103